PRAISE FOR THE FIRST EDITION

"Prior to Congress, I spent over 20 years as an educator; serving as a science teacher, school board member, principal, and education researcher. *Einstein Fellows: Best Practices in STEM Education* is an outstanding publication, featuring some of the best STEM educators in the nation, and insights and guidance on what really works to improve learning in and out of the classroom. Broad implementation of these best practices has the potential to improve STEM education both here in the United States and across the globe. I urge teachers, school administrators, my colleagues in Congress, and education leaders across the nation to give it [a] read; you won't be disappointed."
—U.S. Congressman Mike Honda, California's 17th District

"In a community where we struggle to define 'STEM,' this book provides a clear vision with tangible exemplars to help define the role of science, technology, engineering, and mathematics in our nation's education system. *Einstein Fellows* provides research paired with practical programming and resources in a groundbreaking way that speaks to practitioners, administrators, researchers, and policymakers. This book is STEM for the 21st century."
—Sarah Young, K–12 Science Specialist, Utah State Office of Education

"[The contributors to] *Einstein Fellows* represent a group of master STEM educators who are leaders and education innovators in the classroom and beyond. This collection of essays represents decades and decades of experience and expertise boiled down to best practices current STEM educators can use immediately. *Einstein Fellows: Best Practices in STEM Education* is highly recommended for any educator seeking proven practices from some of our nation's best STEM educators."
—DaNel Hogan, STEM Coordinator, Office of Pima County School Superintendent

"The Albert Einstein Distinguished Educator Fellowship provided this group of STEM educators the opportunity to deepen their knowledge of best practices in STEM education research on teaching and learning. Their insights, distinctive because of their experiences both in the classroom and at the federal-level shaping STEM education policy, provide a unique view into what engaging STEM learning can look like. This collection of essays shares relevant instructional practices and strategies that recognized educators have found to be successful in classrooms around the country. This book could not come at a more critical time, considering our country's dire STEM workforce needs."
—Cindy L. Hasselbring, Special Assistant to the State Superintendent, Maryland State Department of Education

"I applaud the [contributors to] *Einstein Fellows*…for putting forward their best thinking about doing STEM. They offer a useful guide to educators who seek STEM clarity in the form of practices that can be readily adopted in their own classrooms or informal learning settings. Through their writing, the fellows give those of us who work in policy and advocacy roles greater insight into the multiple and subtle variations in meaning that can only be understood in context or inflection or when accompanied by gestures as STEM is spoken in schools."
—Tom Peters, Executive Director, South Carolina Coalition for Mathematics and Science

"This collection of essays provides concrete examples to bring STEM alive in classrooms, including informal learning experiences and the integration of engineering design, which is critical to implementing the Next Generation Science Standards. Since 1990, the Albert Einstein Distinguished Educator Fellowship program has brought accomplished STEM educators from across the nation to Washington, D.C., to inform policy and programs with their knowledge. In turn, they learn the intricacies of federal education policies and gain access to myriad educational resources. I encourage future fellows to continue this tradition and blaze new trails."
—Ioannis Miaoulis, President & Director, Museum of Science, Boston;
Founder, National Center for Technological Literacy

"This book captures the diverse, collective wisdom of over a dozen distinguished and experienced science teachers. If you are a STEM educator, you may want to save yourself time and buy two copies right away as you will surely want to give one of your copies to a younger colleague."
—Stephen Pompea, Education & Public Outreach Department Head,
National Optical Astronomy Observatory

"*Einstein Fellows: Best Practices in STEM Education* is a uniquely positioned book because it synthesizes solutions for many pressing issues in STEM education from some of the most influential teachers, and does so in a practical way. The breadth of topics is comprehensive—from girls in STEM, to gaming, to research experiences for teachers, to sustainability, among others—and each chapter delves deep to offer tried and true practices from expert teachers. This book will be useful for policymakers, teacher educators, STEM industry professionals, as well as teachers."
—Erin E. Peters-Burton, Associate Professor of Science Education
and Educational Psychology, George Mason University

"In *Einstein Fellows: Best Practices in STEM Education*, the topics covered are exceptionally important. Each chapter covers subject matters exceptionally well.... Overall, it is a timely arrival of a book that everyone including parents, students, politicians, and practicing professionals must read and understand their roles in improving the society at large and their authentic participation in educating the young minds early and maintain the discipline at later ages is critical to the society. I applaud the authors for putting forward an excellent book focusing on their ideas of best methods for improving STEM education more realistically."
—Dhadesugoor R. Vaman, Texas A&M Regents Professor, Prairie View A&M University, Texas;
Chief Technology Officer, Digital Compression Technology, Virginia

"This informative collection of essays provides overviews of research insights in the field of STEM education coupled with the wisdom of the teachers who apply it in their classrooms. Each essay—whether it focuses on problem-based learning, engaging girls, interdisciplinary learning, research experiences or informal learning—moves from the theoretical to the personal as the teacher authors provide practical examples for everybody engaged in the valuable work of educating the children of America."
—Arthur Eisenkraft, Distinguished Professor of Science Education, Professor of Physics, Director,
Center of Science and Math in Context (COSMIC), University of Massachusetts—Boston

MORE PRAISE FOR
BEST PRACTICES IN STEM EDUCATION

"This STEM publication with chapters prepared by young theoreticians and practitioners of STEM pedagogical knowledge is a seminal work. The diversity of topics for the effective involvement of all youth in the structure of science makes this publication a necessary resource in all schools. This is a very significant contribution to STEM education."
—H. Prentice Baptiste, Regents and Distinguished Professor, New Mexico State University; President, National Association for Multicultural Education

"STEM teachers, this book is for you. You'll catch glimpses of your future self in these stories. They're not about rocket science—they're about deep commitment to students' thinking and learning."
—Colleen Megowan-Romanowicz, Senior Fellow, American Modeling Teachers Association

Best Practices in STEM Education

Critical Pedagogical Perspectives

M. Cathrene Connery and
Greg S. Goodman, *General Editors*
Vol. 27

The Educational Psychology series is part of the Peter Lang Education list.
Every volume is peer reviewed and meets
the highest quality standards for content and production.

PETER LANG
New York • Bern • Berlin
Brussels • Vienna • Oxford • Warsaw

Best Practices in STEM Education

Innovative Approaches from Einstein Fellow Alumni, *Second Edition*

EDITED BY
Tim Spuck,
Leigh Jenkins,
Terrie Rust
& Remy Dou

PETER LANG
New York • Bern • Berlin
Brussels • Vienna • Oxford • Warsaw

Library of Congress Cataloging-in-Publication Data

Names: Spuck, Tim, 1966- editor.
Title: Best practices in STEM education: innovative approaches from Einstein Fellow alumni /
edited by Tim Spuck, Leigh Jenkins, Terrie Rust and Remy Dou.
Description: Second Edition. | New York: PETER LANG, [2018]
Series: Educational Psychology: Critical Pedagogical Perspectives; volume 27 | ISSN 1943-8109
Previous edition: Einstein Fellows. 2014.
Includes index.
Identifiers: LCCN 2018002493 | ISBN 978-1-4331-5416-4 (Paperback: alk. paper)
ISBN 978-1-4331-5098-2 (Ebook pdf) | ISBN 978-1-4331-5361-7 (Epub)
ISBN 978-1-4331-5362-4 (mobi)
Subjects: LCSH: Science—Study and teaching—United States.
Technology—Study and teaching—United States.
Engineering—Study and teaching—United States.
Mathematics—Study and teaching—United States.
Classification: LCC LB1585.3 .E425 2018 | DDC 507.1/073—dc23
LC record available at https://lccn.loc.gov/2018002493
DOI 10.3726/b13307

Bibliographic information published by **Die Deutsche Nationalbibliothek**.
Die Deutsche Nationalbibliothek lists this publication in the "Deutsche
Nationalbibliografie"; detailed bibliographic data are available
on the Internet at http://dnb.d-nb.de/.

The paper in this book meets the guidelines for permanence and durability
of the Committee on Production Guidelines for Book Longevity
of the Council of Library Resources.

© 2018 Peter Lang Publishing, Inc., New York
29 Broadway, 18th floor, New York, NY 10006
www.peterlang.com

All rights reserved.
Reprint or reproduction, even partially, in all forms such as microfilm,
xerography, microfiche, microcard, and offset strictly prohibited.

Printed in the United States of America

Table of Contents

List of Figures, Photos, and Tables xiii
Acknowledgments xvii
Foreword xix
Introduction xxiii
      What's New in This Edition? xxv
      About the Einstein Fellowship xxv

About the Authors xxvii

Chapter One: The Search for Interdisciplinarity: Moving from Biology,
    Chemistry, and Physics to STEM and Beyond 1
    Nancy Spillane

      Introduction 1
      Background 4
      Author's Best Practices: What Do I Do? 11
      Adaptation of Best Practices 18
      Conclusion 19
      Works Cited 20

Chapter Two: Building a Foundation for Successful STEM Education at the Elementary Level 23
Carmelina O. Livingston

- Introduction 23
- Background 24
- Best Practice 29
- Conclusion 39
- Appendix 40
- Works Cited 42

Chapter Three: Engaging Girls in STEM Careers 45
Terrie Rust

- Introduction 45
- Background 46
- STEM Best Practice: Girls Exploring Technology (GET) Club 49
- How Others Can Adapt This Best Practice 59
- Conclusion 69
- Appendix 71
- Works Cited 71

Chapter Four: Teaching Mathematics to At-Risk Students 75
Brenda Gardunia

- Introduction 75
- Background 76
- My Best Practices and How Others Can Adapt Them 77
- Conclusion 86
- Appendix A 87
- Appendix B 88
- Works Cited 88

Chapter Five: The Student-Centered Sheltered Instructional Approach and Growth (SSIAG) Model 91
Eduardo Guevara

- Introduction 91
- Background: Components of the SSIAG Model 94
- Best Practice: The SSIAG 96
- Strong Evidence Supporting the SSIAG Model 100
- Conclusion 104
- Works Cited 104

Chapter Six: Putting the "Authenticity" into Science Learning 107
   Tim Spuck
   Introduction 107
   Background 109
   Coming to Know Science: My Personal Story 113
   Best Practices: Authentic Science in Action 119
   Implementing Authentic Science Outside the Regular Classroom 128
   Getting Started with the Authentic Science Rating Instrument 132
   Conclusion 135
   Appendix. 137
   Works Cited. 139

Chapter Seven: Engaging Young Minds to be Tomorrow's Innovators. 143
   Arundhati Jayarao
   Introduction 143
   Background 145
   Best Practices. 149
   Adapting Best Practices in Your Classroom 159
   Conclusion 160
   Works Cited. 161

Chapter Eight: Expand the Horizons of Your Students by Expanding Yours 165
   Jean Pennycook
   Introduction 165
   Background 166
   Best Practices. 167
   Expanding Your Horizons 175
   Conclusion 178
   Appendix. 179
   Works Cited. 179

Chapter Nine: Research Experiences for Teachers Can Enhance the
   Teaching of Science 181
   Sue Whitsett
   Introduction 181
   Background 183
   Best Practices. 187
   Adaptations of the Best Practices 193
   Conclusion 196
   Works Cited. 200

Chapter Ten: Modeling Sustainability Through STEM Service-Learning 203
Leigh Jenkins

- Introduction 203
- Background 205
- Best Practice 210
- Modeling Sustainability in Your School and Community 216
- Conclusion 219
- Appendix 220
- Works Cited 221

Chapter Eleven: Outdoor Ecological Inquiry Brings Students and Nature Together 223
Dave Oberbillig

- Introduction 223
- Background 226
- Best Practice 232
- Applying Best Practices in the Classroom 237
- Conclusion 241
- Appendix 243
- Works Cited 246

Chapter Twelve: Twenty First Century Skills Inspired Through Global STEM Projects 249
Dan Carpenter, Florentia Spires and Joseph Isaac

- Introduction 249
- Background 251
- Five Conceptions 253
- Best Practices for Global Collaborative Projects 255
- Adapting the Global STEM Education Practice 260
- Conclusion 271
- Appendix A 272
- Appendix B 272
- Appendix C 273
- Appendix D 274
- Appendix E 275
- Appendix F 276
- Appendix G 277
- Works Cited 277

Chapter Thirteen: Alternative Reality: Gamifying Your Classroom 281
Remy Dou

- Introduction 281
- Background 282
- Best Practices 287
- Structuring Classes Like a Game 289
- Additional Thoughts: MMOGs 296
- Conclusion 297
- Appendix 298
- Works Cited 298

Chapter Fourteen: Using Whiteboards to Create a Student-Centered, Collaborative Classroom 301
Buffy Cushman-Patz

- Introduction 301
- Background 303
- Best Practices: How to Adapt Whiteboarding to Your Classroom 306
- Conclusion 321
- Appendix 322
- Works Cited 322

Chapter Fifteen: Communicating Science to Public Audiences Through Media in High School: Improving Students' Attitudes and Motivations in Science 325
Bernadine Okoro

- Introduction 326
- Background 330
- Case Study 335
- Ideas for Adaptation by Others 346
- Conclusion 347
- Appendix A 348
- Appendix B 349
- Note 349
- Works Cited 350

Chapter Sixteen: Discourse Strategies for English Learners in the STEM Classroom 353
Jenay Sharp Leach

- Introduction 353
- Background 354

Persisting Opportunity Gaps 355
STEM Classroom Discourse.. 356
Best Practices: Discourse Strategies 359
Adopting Discourse Strategies in Your Classroom 367
Classroom Management Considerations 369
Conclusion 370
Works Cited... 371

Chapter Seventeen: Increasing Literacy Skills in the STEM Classroom 375
April Lanotte

Introduction.. 375
Background 376
Literacy Expectations 377
Increasing Text Understanding... 380
Best Practices.. 382
What I Learned Along the Way 383
Additional Literacy Techniques and Strategies................. 387
Adapting Your STEM Classroom to Include Literacy........ 393
Conclusion 395
Appendix.. 396
Works Cited... 396

Chapter Eighteen: Promoting Science Literate Identities Through the Use of Trade Books 399
Paulo A. Oemig

Introduction.. 399
Background 401
Misconceptions in Teaching Content Area Literacy 403
Comprehensible Input, Sociocultural Perspectives and Trade Books 404
So Why Trade Books? 406
Best Practice... 406
Adapting the Practice... 412
Conclusion 415
Appendix A.. 416
Appendix B.. 417
Appendix C.. 417
Works Cited... 417

Chapter Nineteen: Building Community Partnerships and Integrating Arts and Social Studies to Strengthen STEM Learning 421
John F. Smith and June Teisan

    Introduction 421
    Background 424
    Best Practices 429
    How Others Can Implement 436
    Conclusion 441
    Works Cited 441

Chapter Twenty: Zoology Brüt: Using Backward Design to Explore the Sixth Extinction Through Art, Architecture and Appetite 445
Melissa George

    Introduction 445
    Background 446
    Zoology Brüt: A Best Practice Design 450
    Using Backward Design for Course Planning 463
    Conclusion 471
    Works Cited 472

Chapter Twenty-One: Using Self-Regulated Learning Processes to Support Scientific Thinking 477
Erin Peters-Burton

    Introduction 477
    Background 478
    Self-Regulated Learning in Practice 482
    How Others Can Support Self-Regulated Learning in Practice 494
    Conclusion 500
    Works Cited 500

Chapter Twenty-Two: Teaching Students Metacognition Through Discipline-Based Research and Technology 503
Rebecca Vieyra

    Introduction 503
    Background 504
    Best Practice 508
    How Others Can Adapt This Best Practice 521
    Conclusion 525
    Works Cited 526

Chapter Twenty-Three: Applications of Satellite Imagery, Remote Sensing, and Computer Visualizations: Observing the Earth and Visualizing the Future 529
John D. Moore

    Introduction 529
    Background 531
    Best Practice 535
    Adoption of Best Practice 538
    Conclusion 540
    Works Cited 542

Chapter Twenty-Four: Integrating Informal STEM Learning into Your Curriculum 543
Remy Dou and Terrie Rust

    Introduction 543
    Background 544
    Best Practice 546
    Adapting the Best Practice 549
    Conclusion 553
    Appendix A 554
    Appendix B 555
    Appendix C 555
    Appendix D 562
    Works Cited 562

Index 565

Figures, Photos, AND Tables

Figures
Figure 1.1. Scaffolding, Transfer, and Application 7
Figure 3.1. GET Program Outline 51
Figure 3.2. GET Logo 52
Figure 3.3. GET Icon 52
Figure 3.4. GET Nomination Letter 53
Figure 3.5. GET Flyer 54
Figure 3.6. GET Checklist 64
Figure 3.7. GET FAQs 65
Figure 6.1. Acquisition of Tools vs. Opportunity for Creativity 108
Figure 6.2. The Authentic Science Environment Is the "Science Soup of the Day." 113
Figure 12.1. Global Collaboration Through Partnerships 250
Figure 12.2. Certificate of Recognition Sample 277
Figure 16.1. In STEM Classrooms, Students Investigate, Evaluate, and Develop Explanations and Solutions 360
Figure 16.2. Consensus Placemat 368
Figure 16.3. Graphic Organizer for Three Way Interview 368
Figure 17.1. Frayer Model 386
Figure 17.2. Graphic Word Organizer Example 390
Figure 18.1. Student-Created Trade Book 410
Figure 21.1. Phases of SRL in the Independent Research Project 489

xiv | FIGURES, PHOTOS, AND TABLES

Figure 21.2. Phases of SRL in the Knowledge Building Section of the Class. ... 490
Figure 21.3. Phases of SRL in the Citizen Science Section of the Class. 494
Figure 22.1. Diagnoser Tools Pre-test Question 510
Figure 22.2. Physics Toolbox Sensor Suite Displaying Graph of
              G-forces in an Elevator. 517

Photos

Photo 2.1. Constructing an ROV 29
Photo 3.1. GET Members with Astronaut Sally Ride at the February 2006
            Sally Ride Science Festival at ASU. 55
Photo 5.1. Family Science Night. 99
Photo 5.2. Student Field Trip to Moody Gardens in Galveston, TX. 100
Photo 6.1. Oil City High School Students Nick Kelly and Sandy Weiser
            Fill the Dewar on the Kitt Peak National Observatory's
            0.9-Meter Telescope During an Observation Run. 129
Photo 6.2. During the 2009 Winter Meeting of the American
            Astronomical Society, Oil City High School Students Discuss
            Their Research with Dr. Neil deGrasse Tyson, Astrophysicist
            and Host of the PBS Series *NOVA scienceNOW*. 131
Photo 8.1. Jean During Her Immersion Research Experience with
            Penguins in Antarctica. 171
Photo 9.1. Petals That Students Removed Covered with PVA. 199
Photo 10.1. 2009/2010 Advanced Placement Environmental Science
            Students Pose for a Group Picture After Taking Measurements
            and Making Plans for the Greenhouse Renovation. 211
Photo 11.1. Hellgate High School Students Begin the Hike up
            Specimen Ridge, Yellowstone National Park. 235
Photo 17.1. Word Wall in an 8th Grade Science Teacher's Classroom. 385
Photo 19.1. Floral Displays in June's Classroom, Donated by Local
            Funeral Homes, Ready to Enrich the Planned Botany Studies. 422
Photo 19.2. Great Lakes Studies Included Bird Watching Hikes Led by
            Staff at Wild Birds Unlimited and the Michigan Audubon
            Society on the Grounds of Edsel and Eleanor Ford Estate. 435
Photo 19.3. Students Explore the STEM and Social Justice Themes in
            Diego Rivera's "Detroit Industry" Murals at the Detroit Institute
            of Arts as Part of the "Detroit 1933/2033" Project. 437
Photo 20.1. One of Many Rain Barrels Entered in Community Water
            Conservation Contests. 463
Photo 22.1. Student Maneuvers Bowling Ball Around
            Obstacles with a Broom. 512

Tables

Table 2.1.	Prepare and Inspire K–12 Students in STEM	25
Table 2.2.	Essential Practices of STEM.	30
Table 5.1.	Hispanic Educational Attainment in the United States.	92
Table 5.2.	Hispanic Poverty Level in the United States	92
Table 5.3.	Indicator One: Class Averages in Five Texas School Districts. Academic Year 2007–2008	101
Table 5.4.	Indicator Two: Bench Mark Test Scores in Five Texas School Districts. Academic Year 2007–2008	101
Table 5.5.	Indicator Three: Timely Submissions of Assigned Work in Five. Texas School Districts. Academic Year 2007–2008.	102
Table 5.6.	Indicator Four: Number of Same-Student Discipline Referrals in Five Texas School Districts. Academic Year 2007–2008	102
Table 12.1.	Platforms That Engage Students Beyond Their Community.	272
Table 12.2.	Identifying a Global Partner	272
Table 12.3.	Part I: Timeline for Global Collaborative Partnership	273
Table 12.4.	Part II: Timeline for Global Collaborative Partnership	274
Table 12.5.	Samples of Cross Cultural Competencies to Consider	275
Table 12.6.	Use of a Checklist for Potential Collaborative Partner	276
Table 14.1.	Recommended Supplies	322
Table 15.1.	Woodrow Wilson High School Demographics.	336
Table 16.1.	Example Tiered Sentence Starters.	362
Table 16.2.	Elaboration and Clarification Questions	364
Table 16.3.	Scientists' Meeting Discourse Prompts	365
Table 16.4.	Engineers' Meeting Discourse Prompts	366
Table 17.1.	Modified KWL Chart Sample	387
Table 18.1.	Possible Features to Include in the Things That Float and Things That Don't Data Table.	411
Table 19.1.	STEM-based Acronyms	423
Table 19.2.	Possible Pedagogical Approaches to Community-Based Learning and Partnerships	427
Table 20.1.	Understanding by Design Template, Stage 1	465
Table 20.2.	Understanding by Design Template, Stage 2	467
Table 20.3.	Understanding by Design Template, Stage 3	470
Table 21.1.	Rubric for Class Discussion.	491
Table 21.2.	Intersections Between SRL Phases and 5e Model of Instruction.	499
Table 22.1.	Answer Selections and Their Corresponding Facets for the Diagnoser Pre-test Question	511
Table 22.2.	Force Diagrams Corresponding to Motion of Hover Ball	513

Table 22.3.	Force Diagrams and Corresponding Motions and Sensations of an Elevator	518
Table 22.4.	Sampling of Amusement Park Physics Assignment	519
Table 22.5.	Example Reflection Sheet with a Diagnostic Question, Pre- and Post-answer, Corresponding Learning Target, and Confidence Rating to be Completed by the Student Before Checking Answers	520
Table 22.6.	End-of-Unit Reflection Questions	520
Table 24.1.	Major Events in Informal Science Learning	554

Acknowledgments

The editors would like to express our deep gratitude to all the authors who took time from an already busy life to share their best practices with others. In addition, we would like to acknowledge educators who, on a daily basis, strive to enrich the lives of all children without bias, who uphold that responsibility with fervor, and who balance their role as parent, counselor, mentor, and friend each day in the classroom. Those educators go above and beyond the required workday to provide a whole child experience toward the goal of creating a productive populace of lifelong learners. Further, we recognize that it is through equality in education opportunity that we achieve a more fair and just society for all. Every child, regardless of economic status, gender, race, nationality, sexual orientation, religion, etc. comes to us with tremendous gifts and talents. We applaud those educators who seek constant innovation in teaching and learning: unlocking the process of discovery for all.

Foreword

The Teacher's Voice: Notes from the Frontline of Education Reform

"Education is the future." We hear that often because it is central to the knowledge economy and the success of a strong democracy in our ever-more-technical age. When finding ways to improve education, many leaders in the field draw on a plethora of books, studies, and on-line material. What we too seldom hear, however, is the teacher's voice.

Now a group of science, technology, engineering, and math (STEM) teachers have taken a big step forward: putting the teacher's voice on center stage. The essays in *Best Practices in STEM Education* are written by participants in the Albert Einstein Distinguished Educator Fellowship Program. Developed and supported by the Department of Energy, the Einstein Program brings outstanding K–12 STEM teachers from around the country to Washington, D.C., where they work for 1 and sometimes 2 years. The teachers serve in technical agencies including the Department of Energy, National Science Foundation, National Aeronautics and Space Administration (NASA), and the National Oceanic and Atmospheric Administration (NOAA). Some work in congressional offices or on congressional committees whose members can and have drawn on their experience to help draft legislation, and others have been placed at the U.S. Department of Education.

Over the past several years, the Program on America and the Global Economy of the Woodrow Wilson International Center for Scholars has held a number of events with the Einstein Fellows, who, in addition to being outstanding teachers, have all reflected on past efforts in order to help define new directions for school reform. In short, they are system thinkers about education as well as experienced educators themselves. In this book, Einstein Fellows have come together to share their thinking and insights on best practices in STEM teaching and on how to continue to hone the skills of teachers.

Several prominent themes run through the chapters. Many stress the importance of project-based learning. There is a parallel emphasis on increasing student engagement through the active solving of real-world problems. They preach the same philosophy when it comes to afterschool education opportunities: when classroom structures confine learning, afterschool programs can increase motivation and access. Some of the themes focus on honing a teacher's skills through summer sessions in a lab where science practices are applied, rekindling the spark that led the teacher into a particular STEM field to begin with.

Whether you are an administrator, teacher, or informal educator, there is something for you in *Best Practices in STEM Education*. This book offers tips on implementing project-based learning, enhancing teacher preparation and meaningful professional development, improving communication in the classroom, reaching the most challenging students, increasing female participation in STEM, using language arts to enhance learning, and using science, technology, engineering, and mathematics to improve learning for all students. You will read stories and case studies about students moving from Fs to As, growing food for their school cafeteria, and contributing to professional science through the discovery of asteroids and exploding stars.

The wealth of knowledge in this collection is seemingly endless and as diverse as the authors themselves. It is no surprise that they have received numerous local, state, national, and international awards recognizing them as outstanding STEM master teachers. The authors, too, are unique in that they maintain strong connections to their disciplines outside of education, many of them having explored different careers before coming to teaching. Such experience adds to their ability to think about education as a part of the greater picture of the American economy and American competitiveness.

In the next few years, Congress is expected to turn its attention to renewing the Elementary and Secondary Education Act, the latest version of what is commonly known as the No Child Left Behind Act. The House Science Committee is already thinking about renewing the America Competes Act. Earlier versions of the America Competes Act emphasized investments in physical science and STEM education from elementary to postgraduate levels. As Congress and the Obama administration consider renewing major legislation dealing with education, they will draw on a host of academic specialists, Washington-based think tanks, and leaders within teacher associations. Too often, however, individual teachers with recent classroom experience are absent from the witness lists. In our work on education, and STEM education in particular, we here at the Woodrow Wilson International Center for Scholars have learned a great deal from listening to and talking with the Einstein Fellows. I strongly encourage teachers, those who prepare teachers, school administrators, those who fund school reform, Congress, and the Obama administration to give this publication a thorough review as they work to prepare the next generation of STEM innovators. This collection of essays offers lessons for us all.

Kent H. Hughes, Director
Program on America and the Global Economy
Woodrow Wilson International Center for Scholars
1300 Pennsylvania Ave., NW
Washington, DC 20004–3027
June 2014

Introduction

Between the private sector and government, it is estimated that the United States spends over $400 billion annually on research and development—nearly twice that of its closest competitor, China. Investment in science, technology, engineering, and mathematics (STEM) has been identified as the critical piece necessary for the nation's economy to remain innovative and competitive, one that is crucial to improvements in the quality and longevity of human life. Throughout the 1960s and 1970s, the Space Race and other STEM initiatives brought many into STEM-related careers. Those individuals are nearing retirement age and will soon leave the field. Who will take their place? Who will be the innovators of tomorrow? Are the students we are preparing today ready to meet the current and future STEM challenges facing our planet?

There is immense concern that the United States is falling behind in its competitiveness and ability to meet the global challenges that lie ahead. Compared to students in other countries, U.S. 15-year-olds rank 20th in science, 27th in math, and 17th in reading, as measured by the 2012 Programme for International Student Assessment (PISA). When these same students arrive in our colleges, they struggle there as well. Fewer than 40% of students who enter college majoring in a STEM field complete a STEM degree. In addition, the general public seems to struggle in its understanding of STEM concepts. A recent Pew Research Poll showed that 85% of scientists view the public's lack of scientific knowledge as a major problem, and nearly 50% believe the public has unrealistic expectations of scientists. Even after years of media focus and attention, 35% of Americans do not know that carbon dioxide is a gas linked to rising global temperatures, nearly 50% do not know that stem cells can develop into many different types of cells, and more than 50% do not know that an electron is smaller than an atom. We are indeed a nation at risk.

The call for the creation of a STEM Master Teacher Corps, a team of experienced and highly vetted educators that will lead the charge to improve STEM education, has been sounded. President Barack Obama's FY2014 budget called for the creation of such a corps to be established through the U.S. Department of Education. The authors of *Best Practices in STEM Education: Innovative Approaches from Einstein Fellow Alumni* are part of an elite group of K–12 STEM educators recognized nationally as Albert Einstein Distinguished Educator Fellows. Not unlike the proposed STEM Master Teacher Corps, Einstein Fellows are recognized as some of America's most talented STEM teachers, whose expertise is leveraged by federal agencies to improve STEM education and raise the profile of the STEM teaching profession. Collectively, the authors have more than 600,000 hours of practice teaching STEM. Just as important, Einstein Fellows, while limited in number, have been offering their STEM expertise in Washington, D.C. for over 20 years!

The goal of this publication is to help improve the state of STEM education. As the body of STEM-learning research grows, this volume provides the unique perspective of nationally recognized education professionals who have spent years at the interface between teaching and learning. The chapters that follow are the product of years of practice, mistakes, reflection, and refinement. They provide the experiential pragmatism backed by research so desired by practitioners. Each chapter communicates how its author has implemented a specific STEM practice in the classroom and how the practice might be modified for use in other classrooms, schools, and learning environments. These are stories of success, as well as stories of struggle.

Although the chapter order has been given significant attention, this book may also serve as a reference guide to a variety of STEM education professionals. The chapters may be read in order, or readers may choose to skip around from one topic of interest to the next. From the benefits of interdisciplinary teaching to the role of informal education in the classroom, every topic contributes to building an effective STEM education system. More important, the methods proposed are not only supported by research, but have been tried and proven by educators in a variety of diverse STEM classrooms around the country.

In the event that you have questions about what an author has written, or if you want additional information, please do not hesitate to contact individual authors. In addition, if your school or district seeks professional development to implement practices outlined within this volume, please feel free to contact the editors or authors directly. The Einstein Fellows initiated this publication to serve as a resource for teachers and schools. This volume will be effective only if its pages become worn and tattered.

WHAT'S NEW IN THIS EDITION?

The increased interest in the topic of STEM best practices led to the request for this revised edition by the publisher, Peter Lang, who selected the first edition as their *2014 Book of the Year*. We're grateful to those who've not only praised Einstein Fellows: Best Practices in STEM Education, but who have used it in their teaching practice. We've been excited at the usage across a broad audience. We are especially pleased that the first edition has been translated into Spanish which will lead to an even broader impact.

- The book title has changed in the second edition from *Einstein Fellows: Best Practices in STEM Education* (first edition), to *Best Practices in STEM Education: Innovative Approaches from Einstein Fellow Alumni*.
- Eight new chapters have been added to the original 16 chapters. These additional chapters, representing 11 new authors, provide unique approaches to STEM learning, offering readers further ways to incorporate innovative STEM best practices into their teaching.
- All web links have been updated where necessary.
- Author bios have been updated. Many of the authors have had exceptional experiences since the last edition that are worthy of sharing.
- Nationally-recognized changes to educational terminology have been noted.

ABOUT THE EINSTEIN FELLOWSHIP

Founded in 1990, the Albert Einstein Distinguished Educator Fellowship Program is a paid fellowship for K–12 STEM educators who have demonstrated excellence in teaching and leadership in STEM. The Einstein Fellowship aims to increase understanding, communication, and cooperation between the legislative and executive branches of the government and the STEM education community. This goal is achieved by embedding experienced and highly vetted STEM educators into a variety of federal agencies, which have included the Department of Energy, the National Science Foundation, the National Aeronautics and Space Administration, the National Oceanic and Atmospheric Administration, the Department of Education, as well as in the offices of congressional leaders on Capitol Hill.

The Albert Einstein Distinguished Educator Fellowship Act, authorized by Congress in 1994, gave the Department of Energy (DOE) federal responsibility for the program. Today, the Albert Einstein Distinguished Educator Fellowship Program is managed by the DOE Office of Science's Office of Workforce Development for Teachers and Scientists in collaboration with the sponsoring agencies and the Oak Ridge Institute for Science and Education (ORISE). ORISE is a world class DOE institute designed for a variety of research and scientific workforce endeavors. Teachers interested in applying for the Einstein Fellowship can apply online at http://science.energy.gov/wdts/einstein.

> Neither the U.S. Department of Energy or the AEF Program endorse this publication or the ideas expressed in it.

About the Authors

Dan Carpenter, PhD is an Assistant Professor in STEM Education at Texas Tech University in Lubbock, Texas. Dan earned a BS in Natural Science Education, an MA in Curriculum and Instruction and a PhD in Education, all from the University of Nebraska-Lincoln. He is currently the STEM Education Program Chair and co-chairs one of the largest blended delivery PhD programs in the world. His research interests include inquiry instructional models, school improvement processes, standards-based application in instructional settings, and the development of 21st century skills in K–16 settings. Prior to working at Texas Tech, Dan served as a high school science educator for about 20 years. Dan spent most of his career working in the Midwestern United States on school culture and professional learning community models. The models served both practice and policy on shared leadership structures that promote teacher job-embedded professional development and organizational improvement through teacher-driven data systems. Dan is passionate about providing all students with high quality STEM education. Dan served as an Albert Einstein Distinguished Educator Fellow at the National Science Foundation in the Division of Graduate Education (2005–2006), where he developed program evaluation policy for education programs. (Contact Dan at daniel.carpenter@ttu.edu)

Buffy Cushman-Patz is the founder and School Leader for the School for Examining Essential Questions of Sustainability (SEEQS), a public charter secondary school in Honolulu, Hawai'i. Buffy served her Einstein Fellowship from 2010–2011 as the inaugural fellow in the Office of Legislative and Public Affairs (OLPA) at the National Science Foundation. Following her fellowship year, Buffy completed her EdM in School Leadership at the Harvard Graduate School of Education (2012), with School Development as her concentration. During the program she simultaneously wrote and submitted the charter application for SEEQS and earned her principal's license while serving as a member of the leadership team of Neighborhood House Charter School in Dorchester, Massachusetts. Prior to the fellowship, she taught math and science in public, charter, and independent schools in Hawai'i. Buffy earned an MS in Geology and Geophysics from the University of Hawai'i at M noa and a BS in Geology from the University of Florida. In early 2010, Buffy returned to the Galapagos Spreading Center, her MS thesis study area, to serve as a Teacher at Sea, sharing geologic research conducted using the Alvin submersible with students in Hawai'i and around the world. She volunteered with Teachers Without Borders (2008 and 2010), leading math and science workshops for teachers in South Africa. Buffy's exploration of teaching and learning through the lenses of theory, policy, leadership, and through her firsthand experiences as a teacher in both conventional and unconventional settings, guides her work as a School Leader. (Contact Buffy at bjc231@mail.harvard.edu)

Remy Dou, PhD grew up and taught in a richly diverse metropolis. From 2011 to 2013, as an Albert Einstein Distinguished Educator Fellow, Dou worked at the National Science Foundation on projects related to both engagement and diversity in STEM education, including the development of a design and evaluation framework for federal STEM intervention programs. This framework was used by the White House's Committee on STEM Education in the development of a five-year Federal STEM education strategic plan. Currently, he works in academia performing research investigating the affective outcomes of active-learning strategies in STEM education. His focus lies in career decision-making constructs, including self-efficacy, interest and recognition. He has presented on these topics at various organizations, including the National Science Foundation, the American Association for the Advancement of Science, and the National Association for Research in Science Teaching. In addition, he invests some of his time in curriculum development and pre-service STEM teacher training. Prior to becoming an Einstein Fellow, Dou taught high school biology, AP biology, chemistry, and physics. He also led teacher technology workshops and pre-service teacher training. As a former K–12 Science Department Director, he helped transform his school's science "culture" across all grade levels. He has received various awards, grants, and accolades for his work in K–12 education, as well as for his academic research. Remy's many hobbies include writing, both nonfiction and fiction. He is a member of the Society of Children's Book Writers and Illustrators. He also serves as a department editor for *The American Biology Teacher*. (Contact Remy at douremy@gmail.com)

Brenda Gardunia taught high school mathematics in Boise, Idaho, for 22 years, working with at-risk students. She was selected for the NASA Educator Workshop in 2003 and the Fulbright Memorial Fund program in 2006, where she spent three weeks as a guest of the Japanese government, learning about education and culture in Japan. Brenda served as an Albert Einstein Distinguished Educator Fellow at the National Science Foundation, 2010–2012, working on programs that provide research experiences to K–12 teachers and undergraduates. Brenda serves on the Council of Teachers of Mathematics Classroom Resource Committee and is a board member of Idaho Council of Teachers of Mathematics. She is currently on staff at the College of Western Idaho Mathematics Department. Her professional interests are finding ways to increase authentic learning experiences for high school mathematics students and improving teacher preparation programs to include stronger content knowledge for K–12 teachers, especially those who will be teaching elementary and middle school mathematics. Brenda has a Bachelor's degree in Secondary Mathematics Education and a Master's Degree in Curriculum and Instruction with an Endorsement in English Language Learners. (Contact Brenda at brendagardunia@gmail.com)

Melissa George, PhD is CalTeach Internship Coordinator at the University of California, Santa Cruz. As a practicing K–16 teacher-researcher in science education for nearly two decades, she uses empirically validated teaching practices, advocates discovery-based research courses, and sustains partnerships among stakeholders. Her MS (1998) and PhD (2005) in Curriculum and Instruction, Science Education, from Purdue University (IN), were earned under the advisement of the late Sandra K. Abell while teaching middle school in the Lafayette (IN) School Corporation. Following a two-year Einstein Fellowship (2011–2013) at the National Science Foundation (NSF) in the Division of Environmental Biology (DEB), she taught a stewardship-focused zoology course at Lafayette Jefferson High School, an ethnically and socio-economically diverse city school in Lafayette, Indiana. Her work focused on creating sustainable ways to situate, fund, and enhance the learning experiences of her students to best reflect the vision of the *Next Generation Science Standards*. She has several recent publications base on her work with partnerships. The first, in *Bioscience*, evaluates the grant proposals to the NSF's DEB from submission to reporting to analyze the implementation of broader impacts. One contribution to this project was her preparation of the supplemental table entitled "Annotated History of NSF's Broader Impacts Criterion." Melissa also co-authored two chapters in *The Power of Partnerships: A Guide from the NSF GK–12 Program*. (Contact Melissa at mgeorge.efemeritus@gmail.com)

Eduardo Guevara, PhD is a citizen of the United States and Colombia. His focus on improving the academic performance and educational attainment of at-risk students and English language learners led him to design the project- and inquiry-based Student-Centered Sheltered Instructional Approach and Growth (SSIAG) Model, and its accompanying SSIAG Teacher Training Modules, both successfully implemented in a number of public school districts in Texas. Eduardo received his BS in biology from Universidad del Valle in Cali, Colombia, an MS in Fisheries and Allied Cultures from Auburn University (AL), and a PhD from the University of South Carolina. He is certified in Science Composite and Spanish in Texas. His career includes leadership, applied research, and teaching in the British West Indies, Colombia, Mexico, and the United States. Guevara served his Albert Einstein Distinguished Educator Fellowship from 2009 to 2011 in offices on Capitol Hill. His awards and accolades include the Excellence in Science Teaching–Mentoring Award: Trainer of Trainers and Role Model in the Teaching Profession (2010), Outstanding Science Teacher Award from the Houston Independent School District's ASPIRE Program (2007–2008, 2008–2009), a Distinguished Teaching Award for Recognition of Exemplary Performance in Science Teaching by the Cynthia & George Mitchell Foundation (2007), the Excellence in Science Teaching Award and Science Teacher Mentor (2004), and the National Award on Applied Research by the Colombian National Science Foundation (1983). He is an active member of AAAS, NSTA, the Texas Science Teachers Association and the Texas Classroom Teachers Association. (Contact Eduardo at eguesansta08@gmail.com)

Joseph Isaac is a ten-year veteran of science education in the District of Columbia, spending the last eight of those years as Teacher and Department Chair of Biotechnology at McKinley Technology High School, the public school system's STEM-focused school. During his teaching career, Joe has taught Biotechnology, Molecular Biotechnology, Plant Biotechnology, Forensic Science and Advanced Placement Biology, among others. He also served as adjunct faculty at Fortis College in Maryland, where he taught General and

Protein Biotechnology. Joe served as an Albert Einstein Distinguished Educator Fellow from 2012 to 2014 at the National Science Foundation's Division of Molecular and Cellular Biosciences and the Division of Education and Human Resources, during which time he was able to present at several domestic and international conferences on the topics of STEM education in urban school settings and international collaboration in STEM subjects. One of the many highlights of Joe's fellowship experience was participating in NASA's Microgravity Flight experiences with "The Flying Einsteins" in the summer of 2013, during which time they conducted research on "Coacervate Formation in Variable Gravity Conditions." Issac completed his BS in Biology from Howard University (DC), his Master's in Teaching in Secondary Science from Trinity University (TX), and is now pursuing a PhD in Curriculum and Instruction from Texas Tech University's Global PRiSE (Pragmatic Research in Science Education) program. Joe is also currently a science curriculum writer for Planet3, which is developing an interactive learning platform for middle school Earth and Life Science. (Contact Joe at bootneylee2000@gmail.com)

Arundhati Jayarao, PhD is currently Chief Education and Operations Officer with BLUECUBE Aerospace, a start-up established by educators. As a 2009–2011 Albert Einstein Distinguished Educator Fellow, Arundhati was the legislative lead staffer on P-20 STEM and higher education issues at the office of Senator Kirsten E. Gillibrand (D-NY). Jayarao started her career as a theoretical physicist at the Bhabha Atomic Research Center (BARC), a premier national lab in Mumbai, India. She entered the field of education in 2002, teaching physics, AP chemistry, and chemistry to grades 10–12 at Oakcrest, an independent school for girls in Virginia. As a female physicist passionate about motivating girls to pursue STEM careers, she brought into Oakcrest a culture of science learning based on a rich curriculum supplemented by project-based learning and research methods. Within two years, Arundhati was leading the science department as department chair, serving as a role model and teacher to Oakcrest students, as well as a mentor to her departmental colleagues. Arundhati's enthusiasm for teaching is reflected in her many awards, including the Virginia Governor's 2008 Outstanding Educator Award, and the 2007 Coach for Regional Winners of NSTA-Toshiba's Exploravision Competition, and teacher appreciation awards from the American Chemical Society, the American Physical Society, and the American Institute of Aeronautics and Astronautics. Arundhati earned a doctorate in theoretical physics and mathematics from Bhabha Atomic Research Center and University of Mumbai, an

MS in Physics from Hyderabad Central University, India, and a BS in Math, Physics, and Chemistry from Nizam College, India. (Contact Arundhati at arundhati.jayarao@gmail.com)

Leigh Jenkins taught science for 18 years in rural Morgan County, West Virginia. Prior to receiving her teaching certification in biology and general science, she worked as an environmental specialist for a cement corporation in Texas. She taught biology and environmental science for 14 years at Berkeley Springs High School. In 2001, she was chosen as the Eastern Panhandle Conservation Teacher of the Year for bringing environmental awareness into her curriculum. In 2007, Leigh was a scholarship recipient through the Japan Fulbright Memorial Teachers Fund, where she studied Japanese culture and education. In 2009, Leigh and her Advanced Placement Environmental Science students received a $41,000 grant from the State Farm Youth Advisory Board to add solar upgrades to an existing campus greenhouse. In 2010–2011, she served as an Albert Einstein Distinguished Educator Fellow in the Office of Vocational and Adult Education at the U.S. Department of Education, where she assisted with the Sustainability Education Summit: Citizenship and Pathways for a Green Economy. Upon returning to her teaching position, Leigh initiated a sustainability team at her high school which, in 2016, was recognized as a U.S. Green Ribbon School for efforts to reduce environmental impacts, improve health and wellness, and provide environmental education that incorporates STEM learning, civic engagement, and promotes green career pathways. Leigh earned her Master's degree in Curriculum and Instruction from Shepherd University (WVA) and is currently a doctoral candidate at Shenandoah University (VA) in the Administrative Leadership Program. (Contact Leigh at jenkileigh@gmail.com)

April Lanotte is a Senior Instructor/Master Teacher for the University of Colorado at Colorado Springs' UCCSTeach secondary math, science, and engineering education program, and a former secondary science and English teacher with over twenty years of education experience. Her primary university responsibilities include the design and teaching of *Reading in the Content Area*, pre-teacher field supervision, and instruction of other teaching pedagogy courses. April earned a BA in English Literature from La Roche College (PA) and a MA in Curriculum and Instruction in Science Education with a Space Studies emphasis from the University of Colorado at Colorado Springs. In addition to her university duties, April is also an Instructional Designer and provides education support for NASA's Aeronautics Research Mission Directorate. Having grown up in a small, rural community, her education research and specialization has primarily focused on attrition rates of rural students in higher education, literacy and STEM, and bridging the gap between traditionally underserved populations and success in STEM. April served as an Albert Einstein Distinguished Educator Fellow for NASA's Aeronautics Research Mission Directorate from 2011 to 2013, where she helped shape education policy, created lesson materials to support NASA's *Dressing for Altitude* book about the history of pressure suits, oversaw rewrites for NASA's *Museum in a Box* education series, and provided education support for NASA's Space Shuttle requirements. (Contact April at april.lanotte@gmail.com)

Jenay Sharp Leach, PhD is a National Board Certified science teacher who loves to spark her students' curiosity about the natural world. Jenay earned a BS in Physics and General Science/Secondary Education from Grove City College (PA), a Master's degree in Educational Leadership and Administration from the George Washington University (DC), and a PhD in Education from the University of Virginia, where she also worked as a research assistant and supervisor of student teachers. Her research interests include inquiry, teacher professional development, and science for English Learners (ELs). Jenay has spent most of her career in Fairfax County Public Schools (VA), one of the nation's largest and most diverse districts, serving as a physics teacher, elementary science resource teacher, curriculum writer, and K–12 Science Coordinator. She now oversees science for the district as the K–12 Science Coordinator and is passionate about providing all students with equal access to a high quality science education. She served as an Albert Einstein Distinguished Educator Fellow at NASA in the Aeronautics Research Mission Directorate (ARMD) from 2010 to 2011, where she developed science curriculum and education policy. She received the 2011 ARMD Associate Administrator Award for Program and Mission Support. (Contact Jenay at jenaysharpleach@gmail.com)

Carmelina O. Livingston is an elementary educator with expertise in STEM education and pedagogy for K–12 education. In her career, she has focused on interdisciplinary, standards-based, and real-world instruction within formal and informal education settings in Charleston, SC. Some of her typical projects include research-based professional development opportunities for teachers and partnerships with the scientific and business communities for student programs, particularly in the ocean sciences. Livingston contributed to the creation of the Ocean Literacy Scope and Sequence Framework and the *Next Generation Science Standards* as a Critical Stakeholder. Her greatest inspiration was serving as NOAA Educator at Sea and riding in the submersible

2,000 feet under the ocean. She is a strong advocate for experiencing science and technology education early in life. She collaborates with community stakeholders to support STEM literacy and workforce development through events such as STEM festivals, symposiums and media resources. Livingston has a Masters of Education in Special Education/Learning Disabilities from The Citadel (SC) and Master Plus 30 in Science and Technology Education from the College of Charleston (SC). She served her Albert Einstein Distinguished Educator Fellowship at the National Science Foundation in the Directorate of Geosciences in 2011–2012. She's also served on the board of the National Marine Educators Association. Livingston is the recipient of the South Carolina Marine Educator President's Choice and Marine Educator of the Year awards and the Mickelson ExxonMobil Science Teacher recognition. Livingston's joys in life include her family and friends, her love for the ocean, and her disposition to have fun in life! (Contact Carmelina at carmlivingston@gmail.com)

John D. Moore was an Albert Einstein Distinguished Educator Fellow at the National Science Foundation's Directorate for Geosciences from 2009 to 2011. John is a past president of the Satellite Educators Association and chaired the American Meteorological Society's Board of Outreach and Precollege Education (2013–2015). He was the first author and draft committee chair for the AMS Policy Statement on Earth System STEM Education. John is the New Jersey State Coordinator for the Presidential Awards for Excellence in Mathematics and Science Teaching (PAEMST) and is the founder, Executive Director and Chairman of the Board for the American Council of STEM Educators. Formerly, he was the developer and instructor of Career and Technical Education (CTE) programs, which included Environmental Science, Geoscience and Remote Sensing, and Geospatial Technologies, at the Burlington County Institute of Technology, a CTE High School in New Jersey. Currently, John is the Director of Earth Observations and the NJ GLOBE and Environmental Discovery Center in New Jersey. (Contact John at mr.moore.john@gmail.com)

Dave Oberbillig teaches general biology and International Baccalaureate Biology at Hellgate High School in Missoula, Montana. Dave earned his BS in Biology from Metropolitan State College in Denver, Colorado. It was here that Dave first experienced science research, earning a first-place research award in the American Chemical Society Colorado Undergraduate Research competition for work on enzyme kinetics. Dave earned a Master's degree in Secondary Education at the University of Montana, which led to his serving as co-principal investigator on a National Science Foundation GK–12 grant at the University of Montana. He has served as a panelist and presenter at the NSF, GK–12 annual meeting and worked on the national planning team for that event. As the first high school educator to serve on the Ecological Society of America Education and Human Resources Committee, Dave helped develop strategies to recruit the next generation of ecologists. Dave served his Albert Einstein Distinguished Educator Fellowship from 2010 to 2012 at the Department of Energy's Office of Science–Workforce Development for Teachers and Scientists. Partnerships serve as an important part of Dave's pedagogy. His work with forest service ecologists and local community naturalists inspired his students to better understand and appreciate the local environment. Dave promotes the benefits of teacher-scientist partnerships at educator conferences. Currently, Dave also works with Garden City Harvest in Missoula to introduce his students to the biology of food production and organic farming. When not in the classroom, Dave can be found in the wilds of Montana and the West. (Contact Dave at daves.soccerstop@gmail.com)

Paulo A. Oemig, PhD has taught science in the Las Cruces Public School District in New Mexico for the past ten years. He studied chemistry in Argentina at the National School of Technical Education No. 1. His research thesis involved optimization processes in the production of lactic cultures. At the University of Utah in Salt Lake City (SLC), he completed studies in physical anthropology. While in SLC, Paulo worked for five years at an environmental laboratory. He also studied behavior ecology at Cambridge University in England. Looking to bridge physical and cultural anthropology brought Paulo to New Mexico State University (NMSU) in Las Cruces where he completed his Master's degree in Cultural Anthropology. Paulo is interested in the anthropology of science education with an emphasis in bilingual education and social justice. As an Albert Einstein Distinguished Educator Fellow (2012–2013) at NASA Goddard Space Flight Center and NASA Headquarters, Oemig became involved in several projects while reviewing and researching NASA's Education Portfolio and its New Lines of Business. In this fellowship role, he focused on designing professional development for pre-service and in-service teachers that is both comprehensive in its specificity and self-sustaining. He has presented numerous papers across the country and abroad. In 2012, Paulo was recognized with the Las Cruces Public Schools Teacher of the Year and the New Mexico Golden Apple Excellence in Teaching awards. In December, 2016, Paulo earned his PhD in Curriculum and Instruction with an emphasis in Science and Bilingual Education from NMSU. (Contact Paulo at poemig@gmail.com)

Bernadine Okoro taught science in Washington, DC public schools for twelve years. A trained chemical engineer, Bernadine has worked for Bethlehem Steel, Perfecseal, the U.S. Patent & Trademark Office, and BioCore Medical Technologies. Bernadine obtained both her Master's in Communications, Producing Film & Video, and her Master's in Teaching from American University (DC). In 2008, Bernadine worked with the National Institutes of Health, National Library of Medicine, where she collaborated with a team of teachers, college freshmen, and high school seniors to produce a YouTube video promoting the agency's "MedLine Plus" medical database to middle and high school audiences. In 2009, as a National Endowment for the Humanities (NEH) Scholar, she joined other teachers from across the United States in traveling to coastal New England in search of the inspiration and origins of Winslow Homer's paintings for a unique discovery about maritime history in the nineteenth- and early twentieth-centuries. In 2010–2011, she served as an Albert Einstein Distinguished Educator Fellow at the National Science Foundation in the Directorate of Engineering. Bernadine was one of the writers for the *Next Generation Science Standards* (NGSS) in engineering and physical science. Bernadine is passionate about finding ways to merge engineering, science, and the arts. Her 2009 novel, *Peculiar Treasures*, set in Washington, D.C., spins a dramatic tale about relationships. As a 2012 D.C. Humanities Council Community Heritage grant recipient, she produced *Preserving Trinidad*, a documentary about the history of the Trinidad neighborhood. Bernadine currently serves as a STEM Learning Consultant, creating professional development videos for science educators. (Contact Bernadine at Bernadine.okoro75@gmail.com)

Jean Pennycook served her multi-cultural, multi-lingual urban school district in Fresno, California, for over 20 years as a secondary science teacher providing quality classroom experiences in the sciences. Jean passionately encouraged her students to pursue careers in the STEM fields as well as promoted the next generation as life-long learners and scientifically literate adults. In 1992, Jean took her teaching overseas to the American International School of Florence, Italy, where she was challenged to provide science education across all the disciplines to students speaking several different languages. This experience provided a lifelong sensitivity to students with limited English proficiency. Jean's enthusiasm for teaching is reflected in her many awards and accomplishments as an educator, including: NASA teacher grants, Sierra Club Environmental

Educator of the Year, NSF Teachers Experiencing Antarctica grant, and district- and state-level committees and boards. Jean has been nationally recognized as she translates and repackages the science research of Antarctica for classrooms around the world. She makes the excitement of discovery available to all through an interactive website as she provides a virtual field trip to that extraordinary continent and the Adélie penguin breeding colony on Ross Island. In 2010–2012, Jean brought her dedication to quality education to the National Science Foundation as an Albert Einstein Distinguished Educator Fellow, working to make a difference in our education system. Jean holds a BS degree in Wildlife Fisheries Biology from the University of California, Davis, and a MS in Education and Curriculum from California State University, Fresno. (Contact Jean at jean.pennycook@gmail.com)

Erin Peters-Burton, PhD is a National Board Certified science and mathematics teacher who helps students who feel excluded in science become more aware of the scientific enterprise and the ways scientific knowledge is generated. Erin has a BS in Physics from the University of Illinois, a MEd in Educational Psychology and Social Foundations of Education from the University of Virginia, and a PhD from George Mason University (VA) in Educational Psychology and Education Research Methods. Her research focuses on interventions that explicitly teach science as a way of knowing, using metacognitive prompts set in a self-regulatory delivery system and teacher implementation of such methods, as well as explorations of the decisions students make when engaged in scientific inquiry and open-ended questioning and how these decisions compare to decisions that scientists make in their work. Erin served as an Einstein Fellow in the NASA Exploration Systems Mission Directorate from 2006–2007. During this time, she developed national engineering education strategic plans and reviewed engineering education policy. Her experience as a professional engineer, prior to her teaching career, helped frame contributions she made while serving on the K–12 National Engineering Education Standards Committee. Erin's experience as an Einstein Fellow has helped inform her work as an education researcher at George Mason University, where she is the Donna R. and David E. Sterling Endowed Professor in Science Education. She is currently pursuing research projects in the nexus of the nature of science, student learning, science teacher pedagogical content knowledge, and educational psychology. (Contact Erin at erin.peters1@gmail.com)

Terrie Rust, Distinguished Technology Educator (DTE) taught technology education, engineering education, and career exploration to middle school students for 18 years in Peoria, Arizona. Terrie built a program at her school that received Program Excellence and Teacher Excellence Awards from the International Technology and Engineering Educators Association (ITEEA). She developed a passion for integrative STEM education. Her concern for gender equity in STEM fields led to the creation of a Girls Exploring Technology (GET) club at her school. She was recognized with the Visible Difference Award from the Association of Career and Technical Education of Arizona (ACTEaz) for her efforts. Terrie received BA and MA degrees from San Jose State University (CA), a MEd degree from Northern Arizona University, and an EdS degree from Northcentral University (CA). Terrie served as a 2010–2011 Albert Einstein Distinguished Educator Fellow at the National Science Foundation, in the Directorate for Education and Human Resources, Lifelong Learning Cluster, during which time she created a database of the education outreach offerings of the 101 NSF National Research Centers and eighteen NSF Large Facilities, allowing the data to be shared by all NSF Directorates. In 2010, Terrie was awarded the Distinguished Technology Educator citation from ITEEA in recognition of her contributions to the field of technology education. After her fellowship, Terrie served for two years as the Director of Academics for Creya Learning in Hyderabad, India and has since worked directly with educators in China, Malaysia, The Netherlands and Saudi Arabia. (Contact Terrie at terrierust@gmail.com)

John F. Trey Smith began teaching middle school science and social studies in Philadelphia in 2007. He later taught biology and chemistry and chaired the science department at an all-boys high school where he established U.S. FIRST and SeaPerch robotics teams, monthly speaker and field trip programs, and a junior chapter of the National Society of Black Engineers (NSBE). With the NSBE junior chapter and educators from The Math Forum, Trey co-founded the Philadelphia Engineering and Math Challenge. He also taught graduate courses as an adjunct instructor in the University of Pennsylvania's Graduate School of Education. A teacher-consultant with the Philadelphia Writing Project, Trey partnered on a grant with the Academy of Natural Sciences at Drexel University to create an afterschool program for middle school youth to create science-themed games. In 2012, the Philadelphia Chapter of the Pennsylvania Society of Professional Engineers named him Philadelphia's Outstanding Science Teacher. He is a past Philadelphia Social Innovations Lab Fellow and a 2014 National Liberty Museum Teacher as Hero award recipient. Trey served as an Albert Einstein Distinguished Educator Fellow in the office of Senator Kirsten Gillibrand (D-NY) in 2014–2015. He helped secure bipartisan co-sponsors for Senator

Gillibrand's computer science, engineering, and career and technical education amendments included in the *Every Student Succeeds Act of 2015*. The following year, Trey was the Science Teacher-in-Residence at the Library of Congress and explored the use of historical primary sources in science classrooms. Trey is now a Learning Sciences PhD student at Northwestern University (IL). (Contact Trey at jftreysmith@gmail.com)

Nancy Spillane, EdD feels privileged to have spent most of her life helping others become excited about STEM. She taught chemistry in high school, life and physical sciences in middle school, physics labs and science methods courses at the university level, and has been a volunteer science teacher for elementary school. She also served as a science teacher and science department chair at the Williams School, a 7th–12th grade independent day school in New London, Connecticut. Her passion for education led her to a 2-year position (2009–2011) as the voice of the classroom teacher as an Albert Einstein Distinguished Educator Fellow at the National Science Foundation. Facilitated by this experience, Spillane earned her doctorate in Curriculum and Instruction as a George Washington University (DC) Presidential Merit Fellow. Her research focused on STEM teachers at highly-successful STEM-focused high schools. She worked tirelessly to communicate across university silos to engage in interdisciplinary research and collaboration. Nancy is now an Associate Clinical Professor in the WVUTeach Program at West Virginia University where she helps students majoring in STEM fields explore teaching as a possible career path. In addition to her EdD from George Washington University, Spillane earned a Bachelor's degree in Chemistry and a MEd in Teacher Education from the University of Vermont. She was recognized for teacher excellence through The Brian J. Carey Award and was awarded grants and scholarships from the Ford Foundation, Pfizer Inc., and the Siemens Foundation. (Contact Nancy at nks1300@gmail.com)

Florentia Spires is currently a STEM Instructional Leader in Prince George's County School System (MD), where she attended high school. Previously, she worked in Washington, DC at Howard University Middle School of Math and Science, where she developed a STEM curriculum, and at Edmund Burke College Preparatory School as a science teacher. Her teaching career began as a US Peace Corps Volunteer serving in Southern Africa. This global experience catapulted her love for science teaching. For two years, Florentia also served as a Peace Corps trainer for incoming new volunteers. Her teaching spans more than two decades. Florentia earned a BA and BS in Biology at Bennett College for Women (NC) and a MEd in Curriculum and Instruction at Loyola University (MD). She earned a NASA Endeavor Program STEM Certificate with Columbia University (NY). Florentia is currently pursuing her PhD at Texas Tech University in Lubbock, Texas. She served as an Albert Einstein Distinguished Educator Fellow in 2013–2014 at the National Science Foundation expanding computing in education. Florentia has been recognized with numerous prestigious awards. President Barack Obama honored her at the White House where she received the Presidential Award for Excellence in Mathematics and Science Teaching (2013–2014). She was also named a NASA Endeavor Fellow in 2011–2012 and a Helen DeVitt Fellow in 2013–2014. Florentia was selected in 2013–2014 as a NASA Master Teacher and most recently was selected to serve on the National Physics Master Teacher Leader Taskforce (2016). (Contact Florentia at florentia.spires@gmail.com)

Tim Spuck, EdD currently serves as the Director of Education and Public Engagement at Associated Universities Inc. (AUI) and has been a leader in STEM education for more than 20 years. He has developed and led a range of programs focusing on astronomy and STEM education. He currently serves as Principal Investigator (PI) on four NSF-supported projects including: the Chile-US Astronomy Education Outreach Summit, Astronomy in Chile Educator Ambassadors Program (ACEAP), Innovators Developing Accessible Tools for Astronomy (IDATA), and the 2018 National Society of Black Physicists National Meeting award. Prior to his role with AUI, Tim served as an Albert Einstein Distinguished Educator Fellow at the National Science Foundation in the Division of Graduate Education (2010–2012), taught astronomy and earth sciences at the high school and college levels for more than 20 years, and served as K–12 Science Coordinator for Oil City Schools in Pennsylvania. He has led numerous professional development programs throughout the United States and abroad, and has developed a variety of astronomy experiences for learners of all ages. While teaching, his students regularly engaged in authentic astronomy research, and have been recognized throughout the scientific community for their discoveries. Tim's own contributions have been recognized through the American Institute of Aeronautics & Astronautics Educator Achievement Award, Tandy Technology Scholars Award, the Pennsylvania Christa McAuliffe Fellowship, PolarTREC, NITARP, TLRBSE, and numerous other STEM education awards and programs. He earned his Master's degree in Science Education from Clarion University (PA) and an EdD in Curriculum and Instruction from West Virginia University. (Contact Tim at timspuck@gmail.com)

ABOUT THE AUTHORS | XLI

June Teisan, PhD is a National Board Certified teacher who brought science to life for her 7th graders in Harper Woods Schools on Detroit's northeast border for 27 years, and now designs and delivers vibrant learning experiences for students and fellow teachers across the country. June served as Michigan Teacher of the Year (2007–2008) and was one of four finalists for 2008 National Teacher of the Year. She received the Presidential Award for Excellence in Mathematics and Science Teaching in 2005 and was one of five inductees into the National Teacher Hall of Fame in 2016. She holds a BA in Biological Sciences, a MA in Teaching Biological Sciences, an EdS in Educational Leadership, and a PhD in Educational Leadership. She was an Albert Einstein Distinguished Educator Fellow in 2014–2015, placed in the Office of Education at the National Oceanic and Atmospheric Administration, where she continued to work as an Education Outreach and Program Specialist after her Einstein Fellowship. June is passionately committed to widening opportunities for under-represented and under-served students in the STEM fields, and is committed to supporting urban and early-career educators with rich, innovative professional development. (Contact June at jLteisan@gmail.com)

Rebecca Vieyra is the K–12 Program Manager at the American Association of Physics Teachers outside of Washington, DC. She taught high school physics, physical science, and engineering at public schools in both central and northern Illinois for seven years. Rebecca earned a BS in Physics Education from Illinois State University in 2007 and a MAS in Science Education from the Illinois Institute of Technology in 2010. In 2011, she became National Board Certified (AYA/Science-Physics), and in 2013 she was honored by President Obama with the Presidential Award for Excellence in Mathematics and Science Teaching. Rebecca served as an Albert Einstein Distinguished Educator Fellow in the Aeronautics Research Mission Directorate at NASA Headquarters during 2014–2015. During her fellowship, she co-authored, along with her father, a three-volume book, *Teaching High School Physics* with an emphasis on physics teacher preparation and professional development. Rebecca also started an educational technology company with her husband, Vieyra Software, to develop free mobile sensor apps for STEM Education. Vieyra Software is the producer of Physics Toolbox apps, used by over half a million students, teachers, and researchers in every country where Google Play is available. Her work in physics education, research-based teaching practices, and educational technology has led to her being invited as a guest speaker in Mexico, Italy and Indonesia. Rebecca currently resides with her family near Washington, DC. (Contact Rebecca at rebecca.elizabeth.vieyra@gmail.com)

Sue Whitsett knew in third grade that she wanted to be a teacher, and she followed her dream. Sue began her career teaching grades 7–8 in a Catholic school in Oshkosh, Wisconsin, for 5 years with a BS degree in Secondary Education (major in biology and minor in chemistry). She continued her own education and obtained her MS in Curriculum and Supervision with a science emphasis. Sue returned to the high school classroom to teach all levels of biology for 25 years, the last 24 in Fond du Lac, Wisconsin. While in Fond du Lac, Sue received awards for Teacher of the Year, was a state finalist for the Presidential Award for Excellence in Mathematics and Science Teaching three times, and became a National Board Certified Teacher. Sue was awarded a NSTA Toyota TAPESTRY grant, which allowed her students to "do science" by researching pond ecology on newly formed retention ponds. Sue's love of research led her to participate for six summers in Research Experiences for Teachers programs at the University of Rochester (NY) and the University of Wisconsin-Madison, which allowed her to bring authentic scientific research into her classroom. Sue was selected as an Einstein Fellow in 2009 and served her fellowship for two years at the National Science Foundation in the Directorate for Biological Sciences, Division of Molecular and Cellular Biosciences. Sue continues to promote science education through her work at the National Science Teachers Association as the NSTA Director of Army Educational Outreach Programs. (Contact Sue at whitsetts54935@gmail.com)

CHAPTER ONE

The Search FOR Interdisciplinarity

Moving from Biology, Chemistry, and Physics to STEM and Beyond

NANCY SPILLANE

INTRODUCTION

Since the Industrial Revolution, worldwide development and use of technology has been advancing at an exponential rate. The human capital required to adequately meet the needs of this changing infrastructure involve greater understanding and capacity in STEM (science, technology, engineering, and mathematics). In a recent address, President Barack Obama stated, "We must educate our children to compete in an age where knowledge is capital, and the marketplace is global" (President's Council of Advisors [PCAST], 2012b, p. v). To meet increasing needs for a population that is literate in STEM, capable of informed decision making, and adequately prepared for the workforce, our educational system must change. Global problems such as climate change, food security, healthcare, and energy are going to require interdisciplinary solutions. In order for our students to be prepared to participate in these solutions, we must help them to see the connections between the subject matter they learn in school and the broader applications in society.

There are movements at all levels of education that address ideas of interdisciplinary learning and application of knowledge (National Research Council, 2012; Sanders, 2012). National initiatives support interdisciplinary research (*Introduction to interdisciplinary research*, 2013). Attention is given to the need for twenty-first-century skills that focus on "the 4Cs," including critical thinking and problem solving, communication, collaboration, and creativity and innovation (*Framework for 21st century learning*, 2013). On the state and local levels, funding has enabled the design and implementation of STEM-focused schools (T-STEM Network, 2004–2010; NC STEM Learning Network, n.d.) and project-based learning models (New Tech Network, 2013). Within schools, administrative support for collaboration among interdisciplinary teams of educators has provided space and time for cooperation at the most local levels.

Change of this kind, however, is not occurring in every school and in every district. In some schools, the disciplinary boundaries are seemingly impermeable and impenetrable. There is resistance at the administrative level, and an interdisciplinary philosophy is not supported. Even in cases such as these, however, a single teacher can make a difference. Working from inside the classroom and reaching out to the department and beyond, interdisciplinary inroads can be made and can be effective. As classroom teachers, connecting across disciplinary lines allows each one of us to think more broadly about our subject's place within the universe of knowledge in exciting and transformative ways while also advantaging our students, helping them to scaffold their learning across separate classrooms, and situating content knowledge into the larger world.

Who Am I?

I entered the teaching profession as a science content area major in college and was very much embedded in the world of science. I started teaching in my classroom the way I was taught, primarily through lecture and lab. I improved as a teacher through trial and error, a little bit of informal coaching, some professional development, and regular response to student evaluations, concerns, and recommendations. My undergraduate science learning experiences were primarily theoretical—a consequence of the post-Sputnik-era philosophies of teaching—and my own coursework did not emphasize practical applications of the science content. However, through continuing education and select professional development after I started teaching, I slowly came to understand how all of the sciences interrelated. I learned how having this knowledge, as well as an even broader understanding of how science connected to subjects beyond the sciences, enabled me to expand my learning in ways I had never attempted. It only made sense that if this interdisciplinary understanding helped me, it would also help to contextualize science for my students. What could I do within my own chemistry classroom to help give this scientific content knowledge a solid place in my students' academic lives and their experiences beyond the classroom?

What Is STEM?

In the mid-1990s, under the acronym "SMET" (science, mathematics, engineering, and technology), seeds for connections among these four disciplines were sown through funding initiatives from the National Science Foundation. Later, in the 2000s, the acronym changed to the more familiar "STEM" that is used today. Since the beginning, there has been increasing research and education focused on how learning and knowledge in one of these disciplines supports gains in each of the others. Beyond just an acronym representing four independent subject areas, STEM has been defined as

> an interdisciplinary approach to learning where rigorous academic concepts are coupled with real-world lessons as students apply science, technology, engineering, and mathematics in contexts that make connections between school, community, work, and the global enterprise enabling the development of STEM literacy and with it the ability to compete in the new economy. (Tsupros, Kohler, & Hallinen, 2009, slide #10)

In this definition, we see a focus not only on the acronym's four content areas, but also on the ideas that tie them to each other and how their interrelationship results in valuable new knowledge, the grand hope being that through learning, integrating, and applying the STEM disciplines to the problems of the world, we can find solutions.

What Is Meant by "Interdisciplinary"?

Many terms have been used to describe learning that spans more than one subject area. Interdisciplinary, multi-disciplinary, cross-disciplinary, integrated or fused curriculum, and core content are among them. These are used to talk about crossing the borders between academic subject areas and allowing or enabling content to infuse, merge, blend, or support. Although some research literature intends specific meaning through the use of one or more of the terms listed, they have all been used interchangeably.

What Are the Conversations in Science Education?

In September 2010, the President's Council of Advisors on Science and Technology (PCAST), in its report to the president, *Prepare and Inspire: K–12 Education in Science, Technology, Engineering and Mathematics (STEM) for America's Future*, concluded:

> To meet our needs for a STEM-capable citizenry, a STEM-proficient workforce, and future STEM experts, the Nation must focus on two complementary goals: We must prepare all students, including girls and minorities who are underrepresented in these fields. And we must inspire all students to learn STEM and, in the process, motivate many of them to pursue STEM careers. (PCAST, 2012b, p. 11)

In February 2012, PCAST's report on higher education, *Engage to Excel: Producing One Million Additional College Graduates with Degrees in Science, Technology, Engineering and Mathematics* (2012a), included the statistic that "fewer than 40% of students who enter college intending to major in a STEM field complete a STEM degree" (p. 1) and cited the following as reasons: uninspiring introductory-level classes, difficulty in math with little assistance available, and, in particular, "members of groups underrepresented in STEM fields, cite[d] an unwelcoming atmosphere from faculty in STEM courses" (PCAST, 2012a, p. 1). Research data show that "evidence-based teaching methods are more effective in reaching all students—especially the 'underrepresented majority'—the women and members of minority groups" who, while representing 70% of all college students, only make up approximately 45% of all STEM graduates (PCAST 2012a, p. 1). Clearly, the STEM education field needs to do a better job of educating all members of society—particularly those populations formerly neglected by the STEM world—if we are going to be adequately prepared to meet future global workforce and intellectual needs.

BACKGROUND

What Is Already Happening in K–12 Science Education?

There is research evidence to support the connecting of student learning across subject areas to students' prior knowledge and to their lives beyond the classroom. Bransford, Brown, and Cocking's *How People Learn* (2000) describes learning requiring a "network of connections" (p. 129) among the objectives in a lesson, in addition to the relevant uses of new knowledge to be gained. In *The Liberal Art of Science: Agenda for Action, a 1990 report by the American Association for the Advancement of Science*, the importance of integrating science into the human experience was highlighted with recommendations to incorporate "philosophy, values and methods of science into instruction in the natural sciences" (p. viii). These recommendations were further reinforced by A. Truman Schwartz in a speech where he lamented the vast divide between the sciences and the humanities and suggested that concerns of scientific illiteracy could be better addressed by working to actively embed science within "the liberal arts tradition" (Schwartz, 2007). More recently, the National Research Council (NRC) released two documents that pay particular attention to interdisciplinary connections. *A Framework for K–12 Science Education* (hereafter referred to as the *Framework*) introduces the ideas of "cross-cutting concepts" (NRC, 2012, p. 83) and "practices of science" (p. 41) that represent knowledge and understanding permeating and connecting all sciences with technology, engineering, and mathematics as important areas of focus in writing new K–12 science standards. And moving beyond the integration across STEM fields alone, the NRC's *Facilitating Interdisciplinary Research* (2005) suggests that "interdisciplinary studies could help to increase the coherence of students' learning experience across disciplines ... and could facilitate an understanding of how to promote the transfer of knowledge from one setting to another" (p. 169).

We are hearing the calls for a STEM-literate citizenry and are made aware that specific populations are glaringly absent from our current STEM workforce. STEM literacy requires interdisciplinary learning. Integrating STEM into the broader array of high school subjects and applying this knowledge to problems beyond the classroom makes STEM knowledge more accessible to all students. It is sensible that we, as STEM classroom teachers, do our part to make these connections explicit for our students.

Current Ideas on Interdisciplinary Learning

The *Framework* represents the dedicated work of the Committee on a Conceptual Framework for New K–12 Science Education Standards on which the *Next Generation Science Standards* (2013) have been constructed. The *Framework* focuses on three key aspects of science education: (1) core ideas, (2) science and engineering practices, and (3) crosscutting concepts.

Historically, core ideas have been the primary focus of any secondary science discipline generally representing the science content. Increasingly, we have seen the *process* or *practice of science* articulated as intentional and foundational to all sciences, including but not limited to concepts such as the scientific method, inquiry method, nature of science, hands-on science, and active learning. In the *Framework*, these ideas are expanded and more clearly articulated under the umbrella of science and engineering practices, but it is in the crosscutting concepts that targeted attention and emphasis are given to those ideas that span the breadth of the disciplinary subject areas. In particular, there is an effort to focus on common themes and vocabulary that can be reinforced throughout science learning in the K–12 continuum.

On Crosscutting Concepts

In the *Framework,* crosscutting concepts are identified as those that "transcend disciplinary boundaries and prove fruitful in explanation in theory, in observation and in design" (2013, p. 83). They are foundational ideas that take root in kindergarten (or before), extend beyond high school, and permeate knowledge and understanding in science, engineering, and technology. The use of common language and reinforcement of the seven crosscutting concepts (i.e., patterns, cause and effect, scale proportion and quantity, system and system models, energy and matter, structure and function, and stability and change) provide a link to all areas of science: life sciences, physical sciences, earth and space sciences, and the fields of engineering and technology. The *Framework* highlights how these concepts flow through and among the STEM disciplines to provide coherence of thought and language in the establishment of scientific understanding throughout the K–12 continuum. Of significant importance is the concern that these crosscutting concepts are not stand-alone ideas, but rather provide common themes that connect all STEM content areas. They should not be taught in isolation from the applicable core content in each subject specific area.

A Brief History of Interdisciplinary Learning

Thoughts about the importance of interconnectedness are not new in science education or in education in general. Even in their earliest efforts to define the American curriculum, educators sought to divide learning content into manageable chunks, and then find ways to link the chunks together (Kliebard, 2004). These chunks have most often been the subject areas we are accustomed to (e.g., science, history, language arts, etc.). The links between these subject areas have taken a variety of forms, including projects, activities, themes, integrated content, and cross-disciplinary learning. This connectedness is supported by education research literature as a way to provide *scaffolding* for future learning, to aid *transfer* of knowledge and skills, and to demonstrate to students the *applicability* of their learning in the real world.

Scaffolding

Learning that has been provided a structure that supports its acquisition is considered to be "scaffolded." In the early development of the language arts curriculum, there were efforts to have English serve as a foundation for learning in social studies, science, and the arts with the idea that commonalities among these subjects could provide a support structure for student learning (Kliebard, 2004). Other approaches, termed *core*, *broad fields*, and *needs-based* curricula, searched for common ground or common themes among diverse subjects to enable structural support. The more recent curriculum theorists Wiggins and McTighe, authors of *Understanding by Design* (2005), which focuses on utilizing the desired educational outcomes and working backward to design curriculum, highlight a continuing search for connections within and among subject matter. Through their *big ideas*, they describe the "umbrella concepts" or "conceptual velcro—that help the facts and skills stick together" (2005, pp. 66, 67), and in ferreting out the big ideas, *essential questions* are formulated that "often jump curricular boundaries" (2005,

p. 281). Education theorists point to the value of making connections between content in one subject and content in another, with the idea that the commonalities among them will provide an easier, quicker, or more sustained level of learning (Figure 1.1).

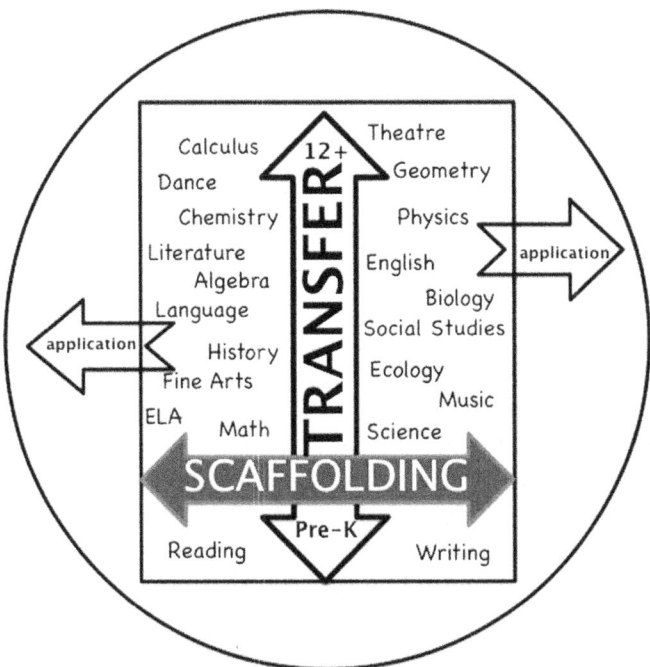

Figure 1.1. Scaffolding, Transfer, and Application.
(Source: Author.)

Transfer

Loosely speaking, transfer relates to the facilitation of new learning by previous learning. The early *mental disciplinarians*, those who believed that training the "faculties" would enable future learning by developing the "powers of the mind" (Kliebard, 2004, p. 4), felt that a curriculum of very specific coursework would lead to the mental development necessary to facilitate the acquisition of new knowledge. In the mid-twentieth century, Ralph Tyler, an American educator and author of research on developing and evaluating curriculum, highlighted "evidence that learnings which are consistent with each other, which are in that sense integrated and coherent, reinforce each other" (Tyler, 1950, p. 41) and are advantageous in student learning. Jerome Bruner, an American cognitive psychologist, suggested that the "structure of a subject" (Bruner, 1960, p. 7) provided students with interrelationships among content that could increase a student's intuitive thinking.

He proposed a spiral curriculum that clearly articulated the advantages of providing a young child with foundational conceptual knowledge early and reinforcing it through subsequent years of education. Transfer, according to Wiggins and McTighe (2005) is the ultimate evidence of understanding, representing application of previously learned knowledge, skills, and ideas to novel experiences and problems, whether they are classroom situations or something in the child's life outside of school.

Each theorist articulates the idea of transfer in a different way, but fundamental is the thought that there is learning that lays a foundation for the construction of subsequent learning. Theorists might argue about what exactly that foundational learning looks like, the best way to orchestrate its study in the school, or the absolute outcome and its measure, but they all appear to agree that aspects of prior learning enable learning that follows.

Applicability

Real world applications have been used either to directly teach subject area content, to reinforce classroom teaching, or to represent the ultimate goal of an American education. John Dewey, a leader in educational reform in the early to mid-1900s, advocated the use of *social occupations* to tie activities in school to those in everyday life in order to provide children with a context into which to fit learning (Kliebard, 2004). Both Dewey and William Heard Kilpatrick, an education philosopher who followed in Dewey's footsteps, thought learning could be made more interesting and applicable through *project organization,* also known as *activity curriculum* or *experience curriculum.* This type of education was inspired by Rufus W. Stimson, who conceived the "home project," an activity designed to help vocational education students in "applying the teachings of the school in their home farm work" (Stimson; cited in Kliebard, 2004, p. 131). During the early 1900s, when much of the motivation for a public education was directed toward efficiency, the applicability of an education was pushed almost to an extreme. At this time, the primary outcome of curriculum was social utility, and classrooms focused almost exclusively on training students for specific jobs in mills and industry. Later, in the mid-1900s, Tyler (1950) suggested that curriculum design should facilitate students' practice of what they learn in school.

All of these examples are broad-brush connections to applicability, but each theorist makes the case that curricular connections between classroom learning and the world at large are important. Students learn from their environments and can build on that knowledge in the classroom; in turn, they can take knowledge from the classroom and apply it to their real lives.

Since the beginnings of the struggle to determine the content of the American curriculum, there have been efforts to divide content and efforts to synthesize it.

While it is likely that no perfect segregation of concepts and no ideal organizational route through them exists, the easy way is the one that has dominated: subject-area content taught in isolation from other courses. This approach requires a teacher to be prepared in only one discipline, allows a classroom full of children to be treated as a single unit, and requires only minimal communication between these components. Since theorists have been recording their ideas about curriculum, the advantages and processes of teaching across, between, and among subject lines have been studied, espoused, championed, and supported by research. Connections can be made from year to year (transfer), from course to course within a single year of study (scaffolding), and from the course out into the student's world at large (application). These three facets of learning are not isolated from each other; they intermingle and constructively interfere in the grand search for understanding.

Wiggins and McTighe speak of *understanding* as the ability to "perform effectively with knowledge" (2005, p. 82), and they identify six facets that make up this understanding: (1) can explain, (2) can interpret, (3) can apply, (4) have perspective, (5) can empathize, and (6) have self-knowledge (2005, p. 84). The crosscutting concepts outlined in the *Framework* assist in enabling the first three facets of Wiggins and McTighe's *understanding*: can explain, can interpret, and can apply.

Vertical alignment (i.e., the flow of specific learning throughout the K–12 continuum) and curriculum spiraling (i.e., revisiting concepts at increasingly higher levels of learning) provide structure and enable transfer, assisting the student in *explaining* through the generalization and articulation of recurring principles. Conscious, intentional, horizontal alignment (i.e., connections across concurrently learned content) scaffolds learning between different subjects, providing new scientific ideas with a context in history, literature, or art. This helps the student *interpret* the new data to establish its significance and relevance. Correlations of course content to students' experiences beyond the classroom walls help students *apply* and use the theory in practice, and to identify and clarify the relevance of learning. Through all of these connections, the crosscutting concepts are what link and reinforce. In Jerome Bruner's words, when speaking of student learning, "the more fundamental or basic is the idea he has learned, almost by definition, the greater will be its breadth of applicability to new problems" (1960, p. 18). We see crosscutting concepts again in Wiggins and McTighe's (2005) *essential questions*. They are the glue that holds subject area knowledge together and the understanding that enables transfer to subsequent learning and to the problems our students will be responsible for solving in the world they encounter when they leave the classroom.

John Dewey and Interdisciplinary Learning

John Dewey formulated many ideas about science teaching and learning that still resonate in today's climate of STEM integration. He spoke to the need for connections between student learning, both in and out of school, and between subject area learning and its application to students' lives and experiences. Dewey felt that students must be actively engaged in their learning, proposing that "methods of instruction and administration be modified to allow and to secure direct and continuous occupations with things" (1916, p. 38) if the learning is to be significant. He suggested "that education is not an affair of 'telling' and being told, but an active and constructive process" (Dewey, 1916, p. 38). Dewey also raised concerns about dividing the world into in-school and out-of-school learning, suggesting that "when the schools depart from the educational conditions effective in the out-of-school environment, they necessarily substitute a bookish, and pseudo-intellectual spirit for a social spirit" (1916, p. 46). Making these connections is of utmost importance in enabling students to learn in ways that will help them to be informed, productive citizens in a constantly changing world.

Dewey recognized that there are different ways of thinking in each of the disciplines. To become an expert necessarily requires a narrowing of one's field of study in the pursuit of depth of knowledge in a single discipline. The languages spoken in the unique disciplines become refined in ways that are no longer universal, and it becomes easy to lose track of related fields and how they are interconnected. Dewey spoke to the need for each of us, within our own disciplines, to remember this when we find conflicts in our thinking with other disciplines. We must "discover some more comprehensive point of view from which the divergencies may be brought together, and consistency or continuity of experience recovered" (1916, p. 326).

As educators, we must look for the connections between and among subjects that students take in a single year to better scaffold student learning, to help our students see the ties that bind these subject areas rather than allowing students to assume they are unrelated because terminology is unique. As Dewey says, we must help our students see the "essential unity of method and subject matter; [and] the intrinsic continuity of ends and means" (1916, p. 323). In addition, even though the majority of our students (based on current statistics) will not go on to become scientists, scientific understanding is important. Even in the 1900s, Dewey cited this significance when he stated:

> Since the mass of pupils are never going to become scientific specialists, it is much more important that they should get some insight into what scientific method means than that they should copy at long range and second hand the results which scientific men [sic] have reached. Students will not go so far, perhaps, in the "ground covered," but they will be sure and intelligent as far as they do go. (1916, p. 221)

Given that the majority of students who enter our science classrooms may not go on to study a science, what can we, as science teachers within our own science disciplines, offer them that will increase both their scientific literacy and the likelihood that the scientific knowledge will be both useful and perceived to be of value in their lives?

AUTHOR'S BEST PRACTICES: WHAT DO I DO?

With the above ideas solidly in mind, I would like to describe my own journey in teaching science at the secondary level and consider what I have done to facilitate my own and my students' understanding of the place of chemistry within the full science spectrum. No doubt, my ideas are not unique and others have found their way to a similar end through a different pathway, but my student gains are marked, and the process is worthy of note.

There are two experiences that significantly impacted my teaching. The first was the result of a series of nominally unrelated summer experiences that exposed me to content areas beyond my field of chemistry, and the second was my effort to incorporate a theatrical play into learning in my classroom.

Through the first experience, I found that every time I learned something that was not specifically related to chemistry, I became a better chemistry teacher. Through the second experience, I learned that taking a very different approach to teaching resulted in different effects on different students, sometimes the greatest impacts being on the very students who seemed to be making the least connection with the content in my classroom. I will address these in two separate sections below, even though their respective impacts resulted in similar conclusions and effects on my teaching.

Non-Chemistry Professional Development Experience

After the first few years of my teaching career, I started searching for summer professional development opportunities that would come at no cost to me. There were a multitude of offerings, provided that I was willing to expand my vision of what a summer experience for a chemistry teacher might look like. I stopped seeking out experiences that were targeted specifically for chemistry and discovered others in material science, paleontology, microbiology, summer internships with pharmaceutical companies, experiences that introduced me to the whole of scientific and engineering research and discovery, fuel cell research and design, and countless others. Each time I learned something that did not seem to be targeted specifically to chemistry, my understanding of chemistry's place within the sciences—and more broadly within learning and discovery—expanded in ways I never anticipated. I was better able to help my students understand these connections and better able to help them think about how my course might fit into their future studies and careers. In addition, each of these experiences brought together teachers, scientists, and engineers in learning environments that enabled productive, interdisciplinary conversations about student learning.

Communicating across boundaries

During these multi-disciplinary professional development experiences, educators from a variety of backgrounds engaged in discussions about the different uses of terms in our respective subject areas, as well as the different ways we introduced students to similar ideas and concepts. We found that while we were teaching our students the same concepts, we were not using consistent language to describe similar ideas. Minimally, this enabled me to go back into my classroom and tell students, "When you learn this in biology, it's called ___, and in physics you'll speak of it as ___." I have found that when I do this and make *explicit* these differences, the students can make the connections more easily, and the content is no longer separate and unique from course to course. When I do not, there are often no connections made. The students simply think these concepts and ideas are different and unrelated to each other.

These experiences caused me to constantly search for overlapping ideas and concepts in my students' learning that I could reinforce, troubleshoot, or presage. My students usually come to me having spent a year in biology. When they leave me, they usually go on to physics. I have actively sought to highlight the connections and conflicts between these subject areas to help students see these as part of a larger whole—a continuum of learning—rather than independent and unique ways of viewing the world.

Some Examples

Physical change/chemical change

When I first start talking about physical change, I reference the water cycle. Instead of starting directly with typical chemistry examples, I start with diagrams from biology books, using similar terms, and talking very specifically on the molecular level about how they are related to our lab experiments. When we first learn of chemical change, we use examples from the nitrogen and carbon cycles, as well as photosynthesis and respiration. These are already familiar concepts, and it makes sense to build on this previous knowledge.

Chemical reactions

When introducing chemical reactions, one of my favorite discussions is helping the students "discover" the products and reactants involved in simple respiration. I start by noting that to stay alive we breathe in and we breathe out. I then ask students to name what is going in and out of the lungs when we breathe. This seems like a simple and silly question, but it is surprising how little thought students have actually given to the overall process. Most students will say that we inhale oxygen and exhale carbon dioxide. But when I ask how this can happen, it gets them thinking, and they have to remember what they learned in biology. Students usually remember that air is composed of oxygen and nitrogen, and some will remember that the percentage composition is approximately 22% oxygen and 78% nitrogen. They usually know that our bodies use the oxygen and extract it from the air, but they really have not given much thought to what happens with the nitrogen that is taken into the lungs. If they think about what they know from the nitrogen cycle, they may remember that while our bodies need nitrogen to build proteins and other macromolecules, we cannot directly extract nitrogen from the air. They may remember that we need to consume proteins (which are nitrogen-containing molecules) from other sources in order to build our own proteins. What this tells them is that humans cannot take nitrogen directly from the air and use it. (Neither can plants, for that matter.) When we breathe, the nitrogen is inert; we breathe it in, and breathe it back out unchanged.

The students have learned that oxygen dissolves in the blood by passing across the lungs' surface and is carried by the hemoglobin in red blood cells from the lungs to the body cells. (This same circumstance can later be used to discuss diffusion, solutions, equilibria, and so forth.) Once the oxygen is in the body's cells, the mitochondria facilitate the reaction between oxygen and glucose in cellular respiration. At this point, I ask students: "Where does the glucose come from, and how does it get to the cells?" We do not get too involved with these reactions, but rather try to stay focused on the cellular respiration reaction. They usually remember that one of the products of cellular respiration is carbon dioxide, that it is transported in the blood back to the lungs and expelled when we breathe out. For some reason, they often forget that a second product of cellular respiration is water. To prompt them (since I have usually taught in the northeast and it gets cold in the winter), I ask what they see when they breathe out in the winter (i.e., condensing water vapor), and where that comes from. Of course, this water vapor comes in part from the moist linings of the lungs, but it is also a product of cellular respiration. It is interesting that they do not necessarily think of this process as sets of chemical reactions when they learn them in biology class. There is so much chemistry in this "simple" process that when students connect these new descriptions to material they learned the year before, it does two things: (1) it reminds them of what they already know and allows them to revisit that knowledge, and (2) it provides scaffolding for new knowledge. Even this kind of interdisciplinary learning helps students to realize that these concepts are not isolated within a single discipline—that biology and chemistry are different ways to look at similar problems and represent interconnected and interwoven material that can reinforce and support each other in the greater understanding of the world.

More connections

Math is an easy common denominator when teaching chemistry, and something that is integrated by its very nature across all science disciplines, but I have found that students learn much more readily if I have taken the time to talk with their math teachers to find out what they already know and how they have been instructed. I do not have to use the same processes or even the same words, but it certainly helps if I can make ties between how I am teaching something and how they learned it, or will be learning it, in their math classes.

After a fairly lengthy discussion between the math and science departments several years ago at our school, we found that we were not using even simple terms such as *proportion*, *ratio*, or *significant figures* in the same ways. The similarities and differences in the ways we use language in different classes is important to know, even if we continue to use terms in ways unique to our subjects. It can be confusing to students to hear different terms used to identify the same thing, or similar terms used differently in the different subjects. It gives the impression that the content in one subject is simply different from the content in another. Being explicit, and making these differences very clear to our students—and, when possible, explaining *why* we use these words differently—helps students compartmentalize and clarify ideas. Even better outcomes result when we work across departments to arrive at a common language. When students can see the similarities among the content in these classes, when teachers can actively highlight these similarities, it connects rather than separates the learning.

Another example comes from physics. When talking about frequency, I found that the physics teacher in our school always used the variable f, and yet, in chemistry, we almost always used the Greek character **v**. Telling students to expect this difference, that both f and **v** can be used to mean the same thing (not a different kind of frequency, in this case), helps bring these ideas to a common foundation. Even though this seems like it might be obvious and simple, it does make a difference in students' understanding.

I have found over the years that there are times when I am talking with other teachers about concepts that are familiar to math, physics, biology, earth and space sciences, and chemistry, and realize that we use different language, different terms, and different approaches to teaching these subjects. Using different language to talk about the same thing often leads students to think that the material is different and unrelated. As teachers, we need to make sure we are aware of the different conceptualizations and make sure we explicitly clarify connections; students will not easily make them on their own.

A Story About a Play

This is a story about a classroom lesson—an evolutionary tale, but not about Darwin—a story of change over time, but not about climate change. It all started in Manhattan.

Spring 2001; The TKTS office (vendor of day-of-performance tickets) in Times Square, New York City

Scanning the list of available plays and musicals, I noticed a play called *Copenhagen*, with a science theme—a definite rarity in my performance-going experience. What could be better than a theater performance that is also an academic experience? A science teacher's dream! Three tickets: one for me and one for each of my school-age daughters. Stage seats facing the rest of the audience. It was as if the play were being performed just for us. The cast: three performers. The set: an empty ellipsoid floor with one simple chair. The performance: spectacular. I was in heaven.

I had been teaching science at a small school in Connecticut for a few years, a school that prided itself on its arts curriculum as well as its academics, but rarely did the boundaries blur. Theater, dance, music—the performances were often classical or contemporary presentations, sometimes based on history or biography or literary beauty alone, but never science, and here was a play with science at its very core!

The play finished and I had laughed, cried, and pondered moral questions with the characters. My elder daughter, a physics student, was similarly enthralled; my younger one, in eighth grade physical science, was less so. The language in the play was scientifically sophisticated, and an understanding of nuclear science, as well as a basic familiarity with the history of the time period, made it easier to understand, and to laugh with. I *needed* to figure out a way to share this play, this experience, with the students in my chemistry classes.

Spring 2002; The touring company

A weekday evening performance of the same play was being presented in my hometown, a mile from our school. I wangled department funds to pay for 15 tickets, which I offered to the first 15 students who were interested in going. I had talked up the play in class, and there was definite enthusiasm, but a weeknight performance beginning at 8:00 P.M. can be a tough sell to a commuter crowd with a long trip home at the end of the day. Nevertheless, 15 tickets were gone in record time. Again, a fabulous performance. I was thrilled. My students with stronger science backgrounds and those with an interest in theater found the greatest enjoyment, but no one was quite as excited as I.

Spring 2003; The local repertory company

A local college theater company was putting on weekday, daytime performances specifically for school groups. My school granted us permission to take all of our chemistry students on a field trip, and I could not wait to share this play with them. We prepared by learning about nuclear chemistry and the historical background of the play. All students had taken some American History, another plus. The repertory company did a fantastic job of staging the production to make it more student-friendly by adding furniture and props to describe time and place, and visual descriptions to support the science being addressed. But even this time, there were not many students who loved it the way I did – just the same ones: the stronger science students and some of the theater students. The rest could find a few good things to say about the experience, but not many. There had to be a way to make this experience as significant for my students as for me. How could I help them appreciate it—enjoy it—as much as I did?

A caveat

One student, Sophie, was unable to join us on the field trip, so I gave her my copy of the play to read and asked her to meet with me one week later to talk about it. Two days later, well before our scheduled meeting time, we passed in the hallway. With an excitement I did not often see in her, Sophie said, "Ms. Spillane, this is incredible! I loved this play!" I was dumbfounded. Why did she, an athlete, a fairly disinterested science student, and someone not involved in theater, love this play? What was it that spoke to her in ways that my other students did not hear? Her comments: "This play made me see scientists who were human, who had families: spouses and children. They worked together and argued with each other; they played jokes and hiked and skied with friends; they struggled with moral dilemmas and had to make decisions about their work based on far more than just the science they researched. They were real." I was astounded. This is what I loved about the play, and what she experienced by reading, that my other students did not when they watched a live performance. What could I do to help the rest of my students have the experience that Sophie did?

The Play: *Copenhagen* (Frayn, 1998) is a challenging play. It is both hilarious and profound, and much of its entertainment value requires a sophisticated scientific vocabulary and working knowledge of the scientific concepts, scientists, geographical locations, and historical references. It is about Niels Bohr and Werner Heisenberg—colleagues, collaborators, and family friends throughout most of their academic lives. They met in 1941. The content of the meeting is not known, but the two scientists never spoke again: a friendship lost. This play explores three possibilities of what might have happened during that fateful encounter, and in the process covers vast historical and scientific territory.

The Plan: After several years of experimentation, I finally settled on a process that incorporated *Copenhagen* as a spring project in all of my chemistry classes. I committed 4 or 5 class days specifically to the play, but the assignment spanned 6 weeks overall. The students read the play outside of class over the course of 5 weeks. When they finished the reading, they were assigned in groups of four students to different sections of the play to collectively discuss and dissect in order to promote deeper understanding. They then performed the play for their classmates as a series of dramatic readings. Three of the students performed the roles of the three characters in the play: Niels Bohr, Werner Heisenberg, and Margarethe Bohr. The fourth, a role I added, served as the narrator to describe and explain what was going on to an audience that might not have fully understood the science or the history. The details of this classroom project are fully described in another publication (Spillane, 2013). What I found through this experience was that this project reached students in very different ways than much of the chemistry-as-usual in my class. It provided a context for their learning that brought the scientists and their research to life. It introduced emotional, moral, and ethical conflict to decisions the scientists made, all the while documenting the process of a scientific endeavor. It provided a different perspective of World War II than students learned in their American History classes, and either built upon a previous foundation or provided a more solid foundation for future learning in European History, depending on the grade level of my students.

This project provided me with an opportunity to work with teachers in my school from a variety of disciplines. In particular, I engaged with history, English, and theater teachers. I had never used a play in my classroom, and while I was comfortable with discussions of science content, I was not sure how to approach a discussion that was more literary, historical, and ethical in nature. I was not sure how our history teachers taught the history of this time period and needed to learn what ideas were taught in which courses, and what I might expect my students to already know. The English teachers helped me brainstorm ways for my students to engage in discussion about the play, and the theater teacher gave me ideas about the play's staging and performance. Each subject area teacher saw the experience from a different perspective and helped me better understand how my students might engage differently with the play to get more out of it.

What this one experience made me realize was that I can (and should) try to meet my students on different levels and tie learning in my class to any other hook I can find. Not everyone will be as intrigued by the mathematical or abstract explanations of phenomena as I am, and it is my job to make connections to as many aspects of a student's experience as I can. Tying course content across disciplinary boundaries enabled me to see different abilities in my own students and helped me experience some of the different approaches to teaching and learning that are taken in different subject areas.

ADAPTATION OF BEST PRACTICES

Some schools are STEM schools and by their very nature support a broad and integrated student experience in science, technology, engineering, and math. Other schools focus more narrowly on one or more fields of STEM such as engineering or health sciences, and still others prepare for specific career trajectories like aviation or nursing. Still other schools focus heavily on different ways of presenting course content, such as project, or problem-based learning. Schools such as these often provide professional development to help teachers facilitate changes in the classroom to meet the school's mission of integrating STEM across the curriculum. Time is provided for interdisciplinary planning, and necessary conversations are initiated to talk across disciplinary lines to enable useful and positive scholastic activity. Many of us, however, are not in schools like these. Even if a school has a name that reflects this kind of integration and forward thinking, the realities of time constraints and financial limitations may not enable the kinds of cross-disciplinary conversation and integration that can truly make a difference for students. So what can an individual teacher do to implement change?

Some First Steps

Walk next door, or across the hall, and start a conversation with the teachers in your school who are teaching in the other science subject areas. Really engage and seek out those "crosscutting concepts" where your subjects overlap. Make sure you are speaking the same language or talk about why it should or must be different. Engage with each other about what you actually say to students. Consider pulling out the standards for your different subjects such as *A Framework for K–12 Science Education* or the *Next Generation Science Standards*, and look for areas where you can support and enhance learning for each other.

1. Walk down another hallway and find teachers who teach the students of the same grade level you are teaching. Start asking them about standards they are trying to meet. Are there ways that you can work together to satisfy different standards through a collaborative lesson? Is there something that you are struggling to teach your students in your class that another subject-area teacher might be able to help with? Ask English teachers how they help students learn to write essays and engage in discussion of literature, or how they work with students to edit their written work. Can you use any of these techniques in your class? If the students are learning to peer-edit work in their English classes, can you ask the English teachers to help you understand what the students know so you can ask them to do the same thing with their introductory paragraphs for their lab reports or

their essay responses on exams? These collaborations should help students see that what they are learning in English class, for example, can be applied to other subjects.
2. Walk to the math department and find out how the math teachers teach students to work with fractions or ratios. Do they use mathematical terms the same way you do? How do they treat units in their classes? How do the math teachers teach students about exponents or scientific notation? Can you use the same process so that students see you are both providing the same information?
3. Use the *Next Generation Science Standards* (2013) to look across the science and engineering fields to find connections. Also, look for the links to the Common Core math and English language arts standards (Common Core State Standards Initiative, 2012a, 2012b) that are included within these standards, and bring these teachers into the conversation.
4. Give yourself the opportunity to attend a talk, a conference, or a workshop that you usually would not. If it seems relevant, go. If it does not initially seem relevant to your field of study, go anyway, and try to think how it can be made relevant or how it can be connected to some aspect of your teaching. Figure out how you can use these experiences to better understand how your students think about their futures and how your classroom learning will inform their future fields of study or career paths.
5. In every case where you are integrating content or trying to connect to a different subject area, make connections explicit. Do not assume that just because you are talking about something the students have learned in another class, or are using language that is common to both, that your students will remember what they learned in a different subject or in a different year. *Actively and regularly remind them how these concepts and ideas are connected to each other.*

CONCLUSION

Having taught in the classroom over many years, I have come to appreciate how connections—those threads of understanding that bleed from one apparently isolated concept to another—help students not just understand content, but retain it, apply it, and even come to appreciate it. I think back on much of my own learning and realize that most of my experience was in learning through lecture. When I first started teaching, I taught the way I had learned. But each year, each day, through each lesson, I made small changes—tiny modifications—as I noticed my students picking up different understandings in different ways. Like any evolutionary process, my methods changed and my students' learning changed; each reinforced the other for positive gain. The process, however, was slow. In some cases it was trial and error, and in others, it was almost accidental in discovery. At times I came up against roadblocks and had to figure out why my students were struggling. I would then think about what might help them learn more effectively and more efficiently, and seek a different approach to facilitate their learning.

There is much research data validating the benefits of an integrated curriculum. Yet our classrooms have been slow to change. In many college classrooms today, even those that are teaching our future teachers, the lecture format is status quo—designed for teacher input and student absorption. It appears that it is easiest and most efficient to teach through lecture, in the hope that students will get it. Certainly some students, the successful ones, do get it. Students who can figure out how to learn from the prevailing methods do learn. In fact, we judge student success this way: "I teach the way I teach; some students learn as a result; these students are successful." But what if *I* change? What if I teach in a way that many can learn, or many more can learn? What if, through my teaching methods, a student who was not "successful" under my previous method becomes successful, and a student who previously believed him or herself to be a failure is not? How is this not something worth striving for? With national support for interdisciplinary learning and increasing state and local support for STEM schools, now is as good a time as any to do what we can to implement change, to talk with one another across disciplinary boundaries, and to seek commonality of purpose.

WORKS CITED

American Association for the Advancement of Science. (1990). *The liberal art of science: Agenda for action: Report of the project on liberal education and the sciences.* Washington, DC: American Association for the Advancement of Science.

Bransford, J. D., Brown, A. L., & Cocking, R. R. (Eds.). (2000). *How people learn: Brain, mind, experience, and school.* National Research Council, Committee on Developments in the Science of Learning. Washington, DC: National Academy Press.

Bruner, J. S. (1960). *The process of education.* Cambridge: Harvard University Press.

Colwell, R. (1998). The national science foundation's role in the Arctic. *Paper presented at the Opportunities in Arctic Research: A Community Workshop.* Arlington, VA: National Science Foundation. September 3–4, 1998.

Common Core State Standards Initiative. (2012a). *The common core state standards for English language arts & literacy in history/social studies, science, and technical subjects.* Retrieved March 19, 2013, from http://www.corestandards.org/ELA-Literacy

Common Core State Standards Initiative. (2012b). *Standards for mathematical practice.* Retrieved March 19, 2013, from http://www.corestandards.org/Math/Practice

Dewey, J. (1916). *Democracy and education.* New York: The Free Press.

Framework for 21st century learning. Washington, DC: Partnership for 21st Century Skills. Retrieved April 19, 2013, from http://www.p21.org

Frayn, M. (1998). *Copenhagen*. New York: Anchor Books.
Introduction to interdisciplinary research. (2013) Arlington, VA: National Science Foundation. Retrieved March 19, 2013, from http://www.nsf.gov/od/oia/additional_resources/interdisciplinary_research/
Jones, R. (2003). NIH roadmap for medical research calls for interdisciplinary research. *FYI: The AIP Bulletin of Science Policy News.* September 2013.
Kliebard, H. M. (2004). *The struggle for the American curriculum, 1893–1958* (3rd ed.). New York: RoutledgeFalmer.
National Research Council. (2005). *Facilitating interdisciplinary research.* Washington, DC: National Academies Press.
National Research Council. (2012). *A framework for K–12 science education: Practices, crosscutting concepts, and core ideas.* Washington, DC: National Academies Press.
NC STEM Learning Network. Research Triangle Park, NC. Retrieved April 19, 2013, from https://www.ncstem.org/
New Tech Network. Napa, CA. (2013). Retrieved April 19, 2013, from http://www.newtechnetwork.org/
Next generation science standards. (2013). Achieve, Inc. Retrieved April 19, 2013, from http://www.nextgenscience.org/next-generation-science-standards
President's Council of Advisors on Science and Technology (PCAST). (2012a). *Engage to excel: Producing one million additional college graduates with degrees in science, technology, engineering and mathematics.* Retrieved September 28, 2013, from http://www.whitehouse.gov/sites/default/files/microsites/ostp/pcast-engage-to-excel-final_2-25-12.pdf
President's Council of Advisors on Science and Technology (PCAST). (2012b). *Prepare and inspire: A framework for K–12 science education: Practices, crosscutting concepts, and core ideas.* Washington DC: National Academies Press. Retrieved September 28, 2013, from http://www.whitehouse.gov/sites/default/files/microsites/ostp/pcast-stem-ed-final.pdf
Sanders, M., (2012, December). Integrative STEM education as *"best practice."* Paper presented at 7th Biennial International Technology Education Research Conference, Queensland, Australia.
Schwartz, A. T. (2007). Chemistry education, science literacy, and the liberal arts: George C. Pimental Award, sponsored by Rohm and Haas Co. *Journal of Chemical Education, 84,* 1750–1756.
Spillane, N. K. (2013). What's Copenhagen got to do with chemistry class? Using a play to teach the history and practice of science. *Journal of Chemical Education, 90*(2), 219–223. doi: 10.1021/ed2007058
T-STEM Network. (2004–2010). Texas High School Project. Retrieved April 19, 2013, from http://tstem.thsp.org/index.htm
Tsupros, N., Kohler, R., & Hallinen, J. (2009). *STEM education in southwestern Pennsylvania: Report of a project to identify the missing components.* Leonard Gelfand Center for Service Learning and Outreach at Carnegie Mellon University and The Intermediate Unit 1 Center for STEM Education. Retrieved March 19, 2013, from http://www.iu1stemcenter.org/links.php
Tyler, R. W. (1950). *Basic principles of curriculum and instruction: Syllabus for Education 360.* Chicago: University of Chicago Press.
Wiggins, G. P., & McTighe, J. (2005). *Understanding by design* (2nd enl. ed.). Upper Saddle River, NJ: Pearson Education.

CHAPTER TWO

Building a Foundation for Successful STEM Education at the Elementary Level

CARMELINA O. LIVINGSTON

INTRODUCTION

Most teachers in the elementary grades bear the responsibility of teaching all subject areas. Since reading and mathematics are tested on state-standardized exams, more instruction and effort are dedicated to these two disciplines, leaving little time for classroom experiences in science and technology. How can elementary teachers integrate content learning and experiences for students in the STEM (science, technology, engineering, math) disciplines on a daily basis? What teacher support system can be provided for the purpose of implementing best practices in STEM education? What resources can district officials, leaders in business and industry, parents, and community provide our students that will encourage them to pursue STEM fields in college or acquire STEM skills for future jobs? How do we get all stakeholders to support, partner, and advocate for STEM experiences early in life for students?

To prepare our future workforce in STEM careers or twenty-first-century skills-based jobs, our nation will need to be unified and consistent in providing the best approaches to STEM education for all students. It is important to build a strong foundation for STEM experiences during the formative years of cognitive and social development of a student's life. With successful, research-based learning experiences at the elementary level, students will progress to middle and high school with a groundwork in STEM learning and practice. Building a competent pipeline of future STEM innovators will drastically improve our nation's effort to stay competitive in the global economy and remain the world leader in science and technology.

BACKGROUND

What Is STEM and Why STEM Education?

Many educators and professionals know that the acronym STEM stands for science, technology, engineering, and mathematics, but do they really know what STEM means? STEM has different meanings for people in various professional settings. Educators, business and industry professionals, health practitioners, and political leaders all have a different interpretation of STEM. A common definition presents STEM as "an interdisciplinary approach to learning where rigorous academic concepts are coupled with real-world lessons as students apply science, technology, engineering and mathematics in contexts that make connections between school, community, work and the global enterprise" (Tsupros, Kohler, & Hallinen, 2009).

STEM is "more than just S+T+E+M. It is a way of learning about the world that reflects the best of each discipline and brings those bests together in entirely new ways" (S^2TEM Centers SC, 2012). Whether educators view STEM as four separate disciplinary silos or view these disciplines as connected by an interdisciplinary approach to learning, STEM education is crucial to the next generation of skilled workers. Before implementing any STEM lesson, project, or experience, one must have in mind a definition or interpretation of STEM and how one can inspire the next generation of students with strong foundational skills for innovation in the twenty-first century.

According to several research studies, the United States lags behind as a world leader in science and technology education. The 2009 *Programme for International Student Assessment* (PISA) (National Center for Educational Statistics, 2010) compares educational performance among countries. The National Science Foundation (NSF) and the National Science Board reported that PISA "emphasizes students' ability to apply skills and information learned in school to solve problems or make decisions within their work setting" (NSF, 2010a). The 2009 PISA scores showed that the United States ranked 17th in science and 25th in math out of 33 industrialized countries. Our students are not performing as well in mathematics and science compared to Nordic, Baltic, European, and Asian countries, and our nation is failing to produce a sufficient number of workers to meet the growing needs of STEM and non-STEM employers (National Center for Education Statistics, 2010). The United States produces the greatest number of STEM degree recipients, yet when compared to other nations in the overall number of degrees for the country's population, the United States ranks 23rd. The President's Council of Advisors on Science and Technology (PCAST) developed specific recommendations to ensure that the United States is a leader in K–12 STEM education. Table 2.1 below identifies the nation's important goals and recommendations (PCAST, 2010).

Table 2.1. Prepare and Inspire K–12 Students in STEM.

	President's Council of Advisors on Science and Technology (PCAST) 2010 Report Prepare and Inspire: K–12 Education in Science, Technology, Engineering and Mathematics (STEM) for American's Future
Goals	1. To prepare all students to be proficient in STEM subjects. 2. To inspire students to learn STEM and motivate them to pursue STEM careers.
Recommendations	1. Support the state-led effort to develop common standards in math and science. 2. Recruit and train 100,000 STEM teachers over the next decade. 3. Recognize and reward the nation's STEM teachers by creating a STEM master teacher corps. 4. Use technology to drive innovation by creating an advance research projects agency for education. 5. Create opportunities for individual and group experiences outside the classroom. 6. Create 1,000 new STEM-focused schools over the next decade. 7. Ensure strong and strategic national leadership.

(Source: President's Council of Advisors on Science and Technology (PCAST) 2010 Report.)

Despite its history as a world leader in research and development, the United States falls short in the competitive global economy (Task Force on American Innovation, 2012). All countries seek to expedite more innovation for economic and social gains, boost competitiveness in international trade, develop new and more effective ways to meet societal needs, and improve the standard of living for their citizens. For science- and technology-based innovation to occur, the United States must invest in a workforce educated in STEM (Atkinson et al., 2010). In the 2011 report produced by the Information Technology and Innovation Foundation, *The Atlantic Century II: Benchmarking E.U. and U.S. Innovation and Competitiveness* (Atkinson & Andes, 2011), the United States ranks fourth in categories of human capital, innovation capacity, entrepreneurship, IT infrastructure, economic policy and economic performance among 44 countries and second to last on progress in improving the country's innovation capacity and competitiveness over the last decade. The report further explains that the United States has suffered a Great Recession because of America's declining innovation performance. This is at the heart of the current slow economic recovery (Atkinson & Andes, 2011). The National Science Foundation's report mentioned that 50% of all economic growth in the United States can be attributed to scientific innovation (NSF, 2010b).

Despite these troubling signs, the United States has made some recent progress toward improving STEM education. The nation has the most productive STEM workforce in the world (PCAST, 2010). A growing body of research has supported how students learn STEM and has made it possible to devise more effective instructional materials and teaching strategies in mathematics and science. A bipartisan consensus emerged for the importance of STEM education with the reauthorization of the Elementary and Secondary Education Act in 2010. In addition, a state-led initiative by the nation's governors and leading state school officials forged consistent and higher standards for mathematics and English in grades K–12 in a document known as the Common Core State Standards (National Governors Association, 2010). The recent development of the *Next Generation Science Standards* will change the way science is taught in classrooms around the nation.

In order to reclaim or revitalize our nation's leadership position in the global economy, the United States must strive to adopt domestic policies to invest in innovation, productivity, and competitiveness, and to prevent the growth of innovation mercantilism among many other countries (Atkinson & Andes, 2011). The bottom line is that it is crucial for our nation to become a leader in the global economy by investing in a workforce educated in STEM for science- and technology-based innovation. The nation will need to provide support and resources to education systems including elementary, secondary, and higher education. The education systems can help create a STEM-literate society by encouraging students to pursue careers in STEM fields or to provide experiences and opportunities practicing twenty-first-century innovative skills in everyday life and work-related situations. Our nation's workforce is directly linked to how well we educate skilled workers to learn content, practice effective pedagogy, and become critical, creative, collaborative, and communicative in everyday life.

Why Is Elementary STEM Learning So Crucial to STEM Education?

Elementary STEM learning lays the foundation for STEM literacy and proficiency because students' interest in science and math typically develops early in life. "All young children have the intellectual capability to learn science … young children have rich knowledge of the natural world, demonstrate causal reasoning, and are able to discriminate between reliable and unreliable sources of knowledge" (Michaels, Shouse, & Schweingruber, 2008 pp. 37–38). In other words, children come to school with the cognitive capacity to engage in science. At an early age, students' interest in STEM disciplines naturally occurs because of their innate sense of wonder and ability to explore and acquire knowledge. Since this occurs early in life, the need to build solid skills is essential for successful progression to higher-level subjects. Early in life, children spend most of their time systematically exploring the world through trial and error, and they grasp what seem like complex concepts very quickly (Gopnik, 2012). Further, a group of scientists and graduate students reported that interest in science began before middle school; males recalled interest in science through self-initiated activities, while females recalled interest in science through school-related activities (Maltese & Tai, 2010).

Recent efforts to support the preparation of STEM-literate students and teachers have focused mostly on the high school and middle school settings. STEM-centered programs usually concentrate on career building and encourage high school students to learn more about future STEM career possibilities. Forty percent of high school students plan to major in a STEM field in college, but only 16% graduate with a STEM degree (U.S. Department of Education, 2007). In the next several years, nearly 50% of STEM professionals will be retiring from their profession, and our nation will have a difficult time filling those STEM jobs. The U.S. Bureau of Labor Statistics (2009) reported that 80% of future jobs will require professionals to be STEM literate and STEM skilled by 2018.

Realizing the need to get more students interested in STEM majors, many formal and informal education systems facilitate professional development opportunities for teachers. Both teacher and student programs and opportunities are usually offered at the high school level. Statistically, many states in the nation also have been seeing a decline in the number of college students graduating with STEM majors, so programs and initiatives throughout the country now bring STEM experiences to middle schools. The research shows that "by the time students reach fourth grade, a third of boys and girls have lost an interest in science. By eighth grade, almost 50 percent have lost interest or deemed it irrelevant to their education or future plans" (Murphy, 2011, p. 1). Since research has shown that 50% of eighth graders lose interest in science, efforts to expose students to STEM classroom and afterschool experiences prior to middle and high school may improve interest and skills. Students' learning and thinking in elementary grades are similar to learning and thinking in science. Tremendous attention and support should be directed at the elementary level.

Elementary teachers are certified as education generalists and teach all subjects. Since elementary teachers are accountable for achievement targets in the state-tested disciplines of English/language arts and mathematics, little or no time is given to teaching concepts in the disciplines of science, social studies, and health.

In the context of science education, we cannot expect students to catch up on 6 years of missed science by the time they get to middle school. The STEM disciplines are a cumulative process. Students' proficiency in science develops through detailed learning progressions of practices, concepts, and ideas (National Research Council, 2012). The K–12 education system needs to advocate for high-quality STEM programs for elementary teachers and students, increase time spent on teaching the STEM disciplines, and consistently provide professional development opportunities for elementary teachers to teach these subjects well (Keeley, 2009).

Many elementary educators lack confidence in teaching content in STEM disciplines, especially science. According to a 2011 report entitled *High Hopes—Few Opportunities: The Status of Elementary Science Education in California*, 40% of all elementary teachers spend less than 1 hour per week on science, and 13% spend 30 minutes or less (Dorph, Shields, Tiffany-Morales, Hartry, & McCaffrey, 2011). Nationally, elementary teachers spend an average of 2.3 hours per week on science education. The U.S. Department of Education noted that 62% of elementary teachers stated that they get little or no professional development in science education and that many districts do not provide a science specialist or coach to assist teachers with student science achievement levels (U.S. Department of Education, 2007).

The elementary grades are part of the K–12 STEM pipeline, and STEM initiatives should emphasize the importance of consistent elementary-level opportunities and experiences for teachers and students. Teacher preparation and professional development in content and pedagogy should occur frequently and should include the best practices and experiences in STEM fields with sponsoring research scientists, engineers, and mathematicians (Epstein & Miller, 2011). School districts should invest in elementary STEM coaches or specialists to work closely with elementary teachers. Public support from parents, community business partners, and industry should contribute support services to promote STEM literacy and include programs and opportunities for the elementary-level gamut. Educational leaders should collaborate with the many politicians and informed citizens who will be making critical decisions that affect our nation. STEM education drives innovation, and it is up to all stakeholders involved to create a foundation for an ecosystem of innovation that allows the United States to be a leader in the global market (Change the Equation, 2012).

Standards-Based, Interdisciplinary, and Real-World Instruction and Learning

Over the past 25 years, I have had the privilege of teaching all grade levels in elementary education. I have taught in a self-contained setting (i.e., teaching all subjects) for kindergarten through fifth grades. I have also taught K–12 science in both formal and informal environments. Even before the term "STEM" was coined, recommendations on best practices in these disciplines were encouraged (Sanders, 2009). Believe it or not, if you are an elementary teacher and enjoy teaching science, technology, engineering, and mathematics concepts, by challenging your students to think critically and work collaboratively, you are a contributor to the vast world of STEM education.

I did not realize I was teaching integrative STEM lessons when I first started, because the term or idea of STEM education was in its infancy and not yet part of mainstream education. The idea of standards-based, integrated, and real-world education and learning has been around for many years. Some practices have been identified differently, but most are consistent with today's methods. When best practices are implemented, it is important to understand that the pedagogy of each practice should be considered at all times during implementation. Some educators may argue that certain best practices "don't work in my classroom" for demographical reasons. I tend to disagree. If the research-based practice is implemented according to the pedagogy in mind, your students will improve academically.

BEST PRACTICE

Photo 2.1. Constructing an ROV. Fifth-grade students plan, design, and construct a submersible remotely operated vehicle (ROV).
(Photo by C. Livingston, 2010.)

All students can learn! Teaching content in STEM disciplines is only one facet of instruction. Best practices in STEM education include standards-based, interdisciplinary, and real-world instruction and learning. Emphasizing twenty-first-century skills for innovation will strengthen learning and help create STEM-literate and STEM-skilled students. A brief overview of these best practices highlights important components for implementation. To illustrate these practices, I will be referring to a fifth-grade course I taught where students were engaged in the creation of ROVs.

The Practice of Standards-Based Instruction and Learning

- Understand what students should know and be able to accomplish in a given period of time.
- Provide the basis for content in instruction and assessment.
- Ensure that students learn what is important.
- Aim for a high and deep level of student understanding.
- Specify how teachers and students will meet their education goals.
- Learn content, concepts, and practice skills at grade-specific levels.
- Influence the quality of teaching and learning.
- Identify successes and problems in student performance.
- Recognize and strengthen academic weaknesses.
- Determine educational growth through assessment of standards. (Texley, 2005; Ohio Department of Education, 2012)

In Table 2.2 eight essential practices of science, engineering, and mathematics are identified.

Table 2.2. Essential Practices of STEM.

Science and Engineering Practices	Mathematical Practices
1. Ask questions (for science) and define problems (for engineering).	1. Make sense of problems and persevere in solving them.
2. Develop and use models.	2. Reason abstractly and quantitatively.
3. Plan and carry out investigations.	3. Construct viable arguments and critique the reasoning of others.
4. Analyze and interpret data.	4. Model with mathematics.
5. Use mathematics and computational thinking.	5. Use appropriate tools strategically.
6. Construct explanations (for science) and design solutions (for engineering).	6. Attend to precision.

Science and Engineering Practices	Mathematical Practices
7. Engage in argument from evidence.	7. Look for and make use of structure.
8. Obtain, evaluate, and communicate information.	8. Look for and express regularity in repeated reasoning.
(Source: National Research Council, 2012.)	(Source: National Governors Association Center for Best Practices, 2010.)

The importance of standards-based instruction in my classroom is crucial to learning. When my students are engaged in learning, they are interested and motivated to learn concepts in depth. While creating an ROV, my fifth-grade students quickly realized they needed the knowledge of ocean science concepts and the precision of mathematical information to develop a model and explain the reasoning behind their design plans.

The Practice of Interdisciplinary Instruction and Learning

- Captures intellectual interests
- Connects information, knowledge, and practice across disciplines
- Uses a transverse process that unifies knowledge beyond disciplines by developing concepts and skills through real-life context
- Develops higher-order cognitive skills such as problem solving and critical thinking
- Employs multiple perspectives
- Evaluates information, data, and perspectives (Drake & Burns, 2004; Nicolescu, 2002)

For me, an elementary teacher, the implementation of interdisciplinary instruction is a common practice in my classroom. Integrating concepts and practices across disciplines promotes deep conceptual learning and understanding. My fifth-grade students also noticed the integration of concepts and skills while working on their ROVs. They researched ocean science content, analyzed and interpreted data from various research-based sources, and used mathematical and computational thinking with their ROV designs.

The Practice of Real-World Instruction and Learning (Emphasis on Project-Based Learning and Real-World Scientific Research)

- Promotes lifelong learning
- Engages students to learn about various research
- Focus on conducting actual research by practicing research skills and methods

- Contributes to meaningful scientific data input that scientists can really use
- Helps students to value information gathered about the world in one's own backyard and in faraway countries
- Uses students' data to research many topics
- Helps students develop twenty-first-century skills (critical thinking, communication, creativity, and collaboration)
- Encourages students to take responsibility for their own learning
- Views knowledge holistically by investigating real-world issues and topics while integrating many disciplines
- Allows the teacher to evaluate and assess the students' progress on cognitive and social development skills

In my classroom, the implementation of real-world learning includes project-based learning (PBL) and project-based science (PBS). These practices facilitate inquiry-based learning in science education but differ in pedagogical emphasis. The former originated in medical education reform, and the latter originated in progressive and constructivist science education reform (Berglund, 1999; Edutopia, 2012; Colley, 2008). My fifth-grade curriculum incorporated these practices when students were introduced to research conducted during the 2010 Gulf oil spill. They analyzed video clips of ROVs capping the drilled hole in the ocean floor. They compiled and communicated data found in various environmental analyses. Using a real-world application of the Gulf tragedy enhanced instruction and sparked interest and concern among students, thus motivating them to persevere with their ROV projects.

Emphasizing Twenty-First-Century Skills for Innovation with Instruction and Learning

Creativity and Innovation

- Think creatively
- Work creatively with others
- Implement innovation

Critical Thinking and Problem Solving

- Reason effectively
- Use systems thinking
- Make judgments and decisions
- Solve problems

Communication and Collaborations

- Communicate clearly
- Collaborate with others

(Route 21, 2012)

When implementing best practices, it is important to emphasize twenty-first-century skills throughout instruction. My students worked together in teams to plan, design, and construct their ROVs. A requirement for each ROV team was to communicate information among all team members, discuss and collaborate ideas, and solve issues and problems. Students quickly learned to capitalize on other students' ideas and skills. All team members contributed to the completion of the project.

Incorporating best practices with twenty-first-century skills will help create a strong foundation for STEM learning. Why should I, as a K–12 teacher, include these skills with instruction? Students will need to work collaboratively, make decisions by critically and creatively thinking of possible solutions, and communicate effectively by using various forms of technology for future employment success. I use many interactive technological tools while integrating science, engineering, and math content with the arts, social studies, and real-world applications. My students use Google Earth and Skype to gather, analyze, and communicate information during instruction. These two programs help bring real-world application to content learning while strengthening twenty-first-century skills.

During the Gulf oil spill and prior to building the ROVs, I challenged my students by asking them, "Would you need to know the weather and physical ocean conditions to launch your ROV?" My fifth graders used Google Earth to determine oceanographic and meteorological conditions in the Gulf of Mexico by using near-time and real-time imaging and data. Google Earth is a user-friendly virtual globe, map, and geographic information program that can be downloaded free right onto your computer. I simply used the basic functions of Google Earth, which include a search panel, 3D viewer of the Earth, and a layer panel.

I divided my class of 25 students into 6 research teams. Since I did not have enough computers in my classroom, my students used our school's computer lab. Using the program, each team entered "Gulf of Mexico" into the search panel. The Gulf of Mexico's image quickly came into focus on the 3D viewer of the Earth. Next, I provided the latitude and longitude of the oil spill to enter into the search panel. Students then clicked on specific layer panels to find their assigned dataset. Teams One and Two were assigned to gather cloud information. Teams Three and Four were instructed to gather the daily weather conditions at the oil spill location. Teams Five and Six were asked to gather ocean observation data of sea surface temperatures, currents, and wind patterns from buoys. My students were amazed to see real-time data available in various forms and were fascinated by the visualizations of current cloud patterns and radar and projected images of rain actually falling into the ocean. All teams came together to have a class discussion of the analyzed data and to decide—based on ocean and weather conditions—if their team's ROV was safe to launch.

Using Skype as a teleconferencing tool to communicate with scientists, engineers, and other STEM professionals can impact learning. Students can meet new people, talk to experts, share ideas, and create learning experiences with teachers and students from around the world. Skype is a software application that allows users to make video calls over the Internet. Skype can be downloaded free of charge right onto one's computer. My class had a Skype session with the education and research crew on an oceangoing drilling research vessel named the *JOIDES Resolution* (JR). This research vessel goes out for frequent expeditions throughout the year to gather core samples of the Earth's crust. Teachers can sign up online for a live ship-to-shore broadcast with the crew. Refer to the Appendix of this chapter for the web address.

During the Skype session, my students went on a live tour of the ship, learned content information of the Earth's crust, saw core samples, and had a question-and-answer session with an educator and scientist. The students were fascinated by the fact that they were communicating with a scientific crew out at sea on a ship. Bringing real-world experiences such as the *JOIDES Resolution* ship-to-shore live broadcast helps students to see the importance of learning concepts and skills, and how these skills are applicable to everyday life problems and issues in our homes, communities, and nation.

Using the best practices of standards-based, interdisciplinary, and real-world instruction and learning with emphasis on twenty-first-century skills has inspired me to grow professionally and implement research-based teaching strategies. By using these practices over the years, I have learned from experience that students tend to be more engaged in lessons, collaborate with peers, think more critically about problems or issues, and have fun while learning. Students have displayed an understanding of in-depth, integrated knowledge that is applicable to real-world situations. Students have often asked me to help replicate experiments and projects in order for them to conduct learning experiences at home.

Using the Best Practices with Ocean Science and Engineering

I am an advocate of ocean literacy at the state and national levels. I believe ocean science education and literacy are important channels to the STEM arena. You do not have to live by a coast to understand and advocate the importance of the ocean on our planet. Ocean literacy curricula, lessons, teacher workshops, and research experiences for teachers are available to assist classroom instruction, and much of it is based on the *Next Generation Science Standards* and Ocean Literacy Scope and Sequence Framework. Refer to the resources section of the Appendix for more information.

As an elementary teacher, I especially enjoy teaching ocean science. My interest was generated by several stellar professional development experiences. Many of these experiences were tailored to middle and high school teachers by providing content and implementation strategies at their grade level. Of the professional development sessions I attended, the content, resources, and experiences provided were adaptable to the K–5 grade level. In fact, this is true of similar professional development opportunities. Adapting strategies to the K–5 level can be achieved, provided one feels comfortable with the content and pedagogy. Elementary teachers can invest in learning content, networking, and communicating with STEM professionals and middle and high school teachers, and adapting learning experiences to the needs and developmental levels of their students.

My first professional development encounter with ocean engineering and technology was a teacher workshop sponsored by the Center for Ocean Sciences and Education Excellence SouthEast (COSEE SE) using underwater, remotely operated vehicles (ROVs) to teach science concepts and expose students to marine technology and engineering. This workshop was offered to middle and high school teachers. I strongly believed this opportunity could be adapted to elementary-level students with few modifications.

The workshop extended the Massachusetts Institute of Technology's "Sea Perch" program. The Sea Perch program encouraged teachers to create an underwater ROV by using a step-by-step manual and working with engineers. The final products of the workshop included a wealth of resources for ocean engineering, real-world application ideas that included regional and national ROV competition, and an ROV to keep and use in your classroom. Throughout the workshop, I realized my fifth-grade students would be able to construct the ROV but would have difficulty constructing the remote control boxes and motors.

Another opportunity to attend an ROV teacher workshop came in the summer of 2009 at a National Marine Educators Association conference held in Monterey Bay, California. Marine Advanced Technology Education (MATE) sponsored this workshop. The "MATE ROV" program encouraged teachers to build an underwater ROV with materials provided in a kit. I really liked the versatility of the program. The re-usability of the ROV kit can be a repeatable classroom activity year after year. The MATE program gives students the flexibility to plan and design their own ROVs without following a manual.

After the workshop, I did a bit of homework back at school. I needed funding to purchase 15 ROV kits for my students, so I checked with my building principal and the school's Parent-Teacher Association (PTA) to see what kind of funds might be available for STEM initiatives. I discovered that the PTA had set aside

$2,500 for innovative math and science projects. My school is a math and science magnet school, which may not be typical of elementary PTAs across the United States. But do not underestimate the potential of working closely with your local PTA. Many times, if they don't have the funds, they can help raise them for special projects that benefit students.

I wrote a brief proposal to the PTA board requesting the purchase of 15 kits for the first year of implementation and attended a PTA board meeting to justify my request. The request was approved, and just like that I had ROV kits for my students. Realizing how important it was for the PTA to understand the impact of their support, I invited its board members to the fifth grade's concluding ROV event on World Ocean Day. The event included fifth-grade students operating their ROVs and teaching K–4 students how to operate the ROV in kiddie pools on the school grounds. Fifth-grade students served as role models for the younger students. Students were excited and engaged throughout the event, and parents, teachers, PTA board members, and administrators were impressed. One parent commented, "I can't believe fifth graders designed these ROVs!" A kindergarten teacher commented, "It's good to see the older children teaching what they have learned to the younger students. What a powerful learning experience!"

As we all know, elementary students are very active and full of energy. To begin the ROV lessons with my students, I used my favorite strategy related to learning concepts or vocabulary: "Move, Listen, and Stop." This sort of activity helps build vocabulary and concepts while children keep active when learning. An example that has worked for me in the past is an activity I put together on biotic and abiotic factors in ocean ecosystems. This is a strategy that will fit into most elementary school science curricula.

Here's a roadmap you can follow:

Focus Standard of Learning: Biotic and abiotic factors of an ocean ecosystem
Vocabulary: biotic, abiotic, ecosystem
Concepts: biotic—living, abiotic—nonliving
Directions:

1. Ask the children to begin a specific type of motion, such as march, hop on one foot, or swing arms. Let them know they cannot stop the movement until they hear an abiotic factor stated out loud by the teacher.
2. Once the children are in motion for a few seconds, start shouting out some words and eventually say "water." Students should stop their movement. When they've stopped, briefly discuss that water is an abiotic factor and why.
3. Ask the children to begin movement again, whether it is the previous movement or a different one. This time, shout some other words like fish, whale, seaweed, and so on. Because these are not abiotic factors of the ocean

ecosystem, students should continue movement. While children continue to move, tell them, "you're correct, fish are living organisms and they are not considered abiotic factors."
4. Continue to change vocabulary or concepts as you continue to implement "Move, Listen, and Stop."
5. Repeat the first step, but this time focus on biotic factors of an ocean ecosystem.

Connecting learning with physical motion does increase attention, accommodates kinesthetic learners, and gets rid of excess energy. My children enjoy learning concepts and practicing skills through movement because they are having fun while learning. This strategy can be implemented anytime with K–5 students.

Providing experiences in ocean science and engineering brings awareness of STEM careers and skills for students. Students benefit from learning the essential skills and concepts through real-world, project-based experiences by using real data and real results. The NOAA Teacher at Sea program and the Sea Perch and MATE underwater robotics programs allowed me to capitalize on my own experiences and integrate my knowledge and skills with engaging science and engineering programs for students.

In our school's science enrichment program, I taught 125 fifth graders the importance of ocean exploration and why the ocean is so important to everyday living. I shared my experience of riding in a submersible 2,000 feet below the ocean surface off the coast of South Carolina during my Teacher at Sea expedition, which was called NOAA Ocean Explorer Islands in the Stream. The research, logs, images, and video clips of the expedition are available online at the NOAA Ocean Explorer website. Using actual video footage and other online resources, I presented my submersible dive experience and took my students on a virtual dive to the unexplored ocean floor. Students witnessed their teacher communicating with the sub pilot while using a robotic arm to gather sediments and specimens of living animals. My students were fascinated to see the actual experiences of their teacher. Many students in class remarked how cool it was to see their teacher on a ship working with scientists and conducting various investigations. They asked to see the images and video clips of the expedition over and over again.

After the presentation of my Teacher at Sea expedition, I used real-world materials gathered from the research expedition to begin the ROV lesson. Students used tools such as microscopes and magnifiers to observe sediments and specimens taken from the submersible dive. To supplement the virtual dive experience, students were grouped in teams as "real scientists" to construct an ROV. Students selected their role from a list of engineering positions and researched how to plan, design, and construct the best ROV by using the engineering design process and their creation of a blueprint or model of their design.

Scientists and engineers from various local agencies were scheduled to come and share their expertise with the students. I had established a network of local scientists and engineers during the NOAA Explorer expedition and ROV teacher workshops. All I had to do was call them to present their research from the expedition or share their professional expertise with my students. A presentation day was planned on World Ocean Day for fifth-grade students to work closely with K–4 students. The purpose of the presentation day was twofold: students teaching students and students presenting their experiences. All students presented their designs and demonstrated the ROV's movement capabilities in a child-size pool. The students illustrated how submersible pilots communicate with scientists and other researchers while operating an ocean exploration tool. Even after the ROV experience, students (both girls and boys) asked where to purchase ROV kits, and several students mentioned wanting to have a career in ocean science or engineering.

Implementing Best Practices with Limited Resources

How can these "best practices" be adapted and implemented by other formal and informal educators with limited resources? Collaboration and partnerships between educators and STEM professionals are key to implementation of best practices, even when resources are limited. Start small and work collaboratively through partnerships with educators and stakeholders. Stakeholders in your area can include business and industry partners who will benefit from future STEM-literate and STEM-skilled employees, workforce-related community organizations and associations, parent associations, formal and informal educators from institutions, local and state agencies and organizations, and national and state public and policy representatives.

After attending my first ROV teacher workshop, I asked the PTA at my school to purchase ROV design materials. I partnered with a high school marine science teacher who also received funding for ROV materials. We partnered to provide an ROV learning experience with high school and fifth-grade students. Both schools benefitted by providing these experiences within our school district.

Starting small and collaborating with other stakeholders does not have to be a full-blown event. Partner with the teacher across the hall from you to implement a group project. Your class can be responsible for one part of the project while the other class works on different parts, or you can have both classes work together as individual teams, each with specific responsibilities and duties. Invite scientists, engineers, and other STEM educators to be an active part of your lessons. You will be surprised at how their expertise and resources add a real-world dimension to your lesson, experiment, or project. Invite the parents of your students who are scientists, engineers, or STEM-skilled professionals. I take full advantage of incorporating my parents during class lessons and school-wide science, engineering, or math events.

Parents are often too shy to come forward and share their expertise because they do not know how their skills can assist teachers and students. Educators have the ability to translate information from STEM professionals using best pedagogical practices. I highly recommend that you collaborate or partner with them in enhancing student experiences. Many STEM professionals will have an outreach component to their employment and will often search for opportunities to share their expertise with students in their community.

Engaging elementary students to participate in real-time data and technology projects can be accomplished in a variety of ways. Local or online citizen science projects are examples of how to get students to make observations and record data in a collaborative database. There are many citizen science projects available for teachers and students online. Examples of projects can include recording the number of fireflies seen on a specific night or monitoring water quality in your local river.

A variety of additional collaborations can be created. Working with other elementary schools in the same district can help bring attention and support to a project. Middle or high school students can mentor elementary students on career-building skills. Students can compete in local, regional, and national competitions to heighten their knowledge in the content area and engage in both science and engineering practices. Teachers can team up with informal educators from museums and aquariums, or scientists and engineers from local agencies and universities during after school programs that reinforce activities and/or projects using a project-based, real-world application.

How can you find resources? Start within your district and seek recommendations from the science, math, engineering, or curriculum specialists, local teacher organizations, and social networks. Local informal education programs such as aquariums, parks, and museums are another asset for fostering partnerships on projects and events. Informal educators have expertise in educating and informing the general public. Many informal education programs include children's after school programs. Educators have the opportunity to share their expertise about education and mesh the expertise of various other stakeholders to provide the best learning experience for the students. A list of organizations and resources directed at real-world issues in elementary STEM education and ocean sciences is provided in the Appendix. The winners in these experiences are always the students!

CONCLUSION

Whether you are a rookie or a seasoned teacher, this is an exciting time in our country for STEM education. Now is the time to make a difference in your classroom instruction by facilitating and implementing STEM education best practices. The United States faces the ultimate challenge of regaining its competitive edge as an innovative world leader in the global economy. Another "Sputnik moment" in the history of our nation has sparked an unbelievable amount of research, national discussion, legislation considerations, state-led alliances, and partnerships with business and industry to cultivate a STEM-literate citizenry and STEM-proficient workforce. Our country will need a STEM-literate and STEM-skilled workforce to lead the nation in accelerating innovation. A key factor in the challenge is that change is required in the K–12 education system in order to provide a foundational preparation in STEM disciplines. The preparation involves all stakeholders, including teachers, students, and district and state education specialist leaders. Investment and support from parents, business and industry, and legislators will be critical to the STEM system.

Whether you are an elementary, middle, or high school teacher, STEM best practices will help provide a strong foundation of STEM skills in the lives of students. It is crucial to build a stable infrastructure of STEM experiences during the formative years of students' development. Having taught elementary science for many years, and knowing that students come to school with the cognitive capacity to engage in the top-level domain of critical and creative thinking in STEM disciplines, the implementation of standards-based, interdisciplinary, and real-world instruction reflects the best practices of STEM education at the elementary level. You can implement these practices today when teaching content. With fifth graders, I chose ocean science and engineering experiences to help students learn content, think innovatively, and apply their skills to a future career. With successful experiences at the elementary level, students will progress to middle and high school as STEM proficient and might contribute to the pipeline in STEM fields. Our nation's STEM workforce will be the next generation of innovators and world leaders.

APPENDIX

Here is a list of resources and professional organizations to assist you with developing and implementing STEM instruction at the elementary level. The organizations below offer many online professional development opportunities through webinars, graduate-level courses, and video streaming sessions. For further information on STEM education in ocean science and engineering, refer to the resources in this section.

Organizations

- Alliance for Science & Technology Research in America, U. S. Innovation, http://www.usinnovation.org/

- American Association for the Advancement of Science, http://www.aaas.org/
- American Council of STEM Educators, http://www.acstemed.org/
- Association for Supervision and Curriculum Development, http://www.ascd.org/
- International Technology and Engineering Educators Association, http://www.iteea.org/
- National Council of Teachers in Mathematics, http://www.nctm.org/
- National Marine Educators Association, http://www.marine-ed.org/
- National Science Teachers Association, http://www.nsta.org/

Resources

- American Geophysical Union (AGU) GIFT Workshop, https://education.agu.org/education-activities-at-agu-meetings/gift/
- Bridge, http://web.vims.edu/bridge/
- Center for Ocean Sciences and Education Excellence SouthEast (COSEE SE), http://www.cosee-se.org
- Change the Equation, http://changetheequation.org/
- Engineering Is Elementary, http://www.eie.org
- JOIDES Resolution Educator at Sea Program, http://joidesresolution.org/
- Journal of STEM Education, http://ojs.jstem.org/
- Marine Advanced Technology Education (MATE), http://www.marinetech.org/
- Mickelson ExxonMobil Teachers Academy, http://mickelson.nsta.org/
- MIT Sea Grant Sea Perch Program, http://seaperch.mit.edu/
- Monterey Bay Aquarium Research Institute EARTH, http://www.mbari.org/earth/
- NASA Education Resources, http://www.nasa.gov/education/resources
- National Center for Technological Literacy, http://legacy.mos.org/
- National Lab Network, http://www.nationallabnetwork.org/
- National Math + Science Initiative, http://www.nms.org/
- National Science Digital Library, https://nscl.oercommons.org
- National Science Foundation, http://www.nsf.gov/
- *Next Generation Science Standards*, http://www.nextgenscience.org/
- NOAA Education Resources, http://www.education.noaa.gov/
- NOAA Ocean Explorer, http://oceanexplorer.noaa.gov/
- NOAA Teacher at Sea Program, http://teacheratsea.noaa.gov/
- Ocean Literacy Framework, http://oceanliteracy.wp2.coexploration.org/
- Office of Naval Research Sea Perch Program, http://www.seaperch.org/
- STEM Connector, http://www.stemconnector.org/
- STEM Education Coalition, http://www.stemedcoalition.org/
- The Connectory, http://theconnectory.org

WORKS CITED

Atkinson, R., & Andes, S. M. (2011). *The Atlantic century II: Benchmarking EU and U.S. innovation and competitiveness.* Washington, DC: Information Technology and Innovation Foundation.

Atkinson, R., et al. (2010). *Innovation policy on a budget: Driving innovation in a time of fiscal constraint.* Washington, DC: Information Technology and Innovation Foundation. Retrieved from http://www.itif.org/files/2010-innovation-budget.pdf

Berglund, K. (1999, November 1). World wide weather: Involving students in GLOBE's real-life scientific research. *Science and Children,* pp. 31–35.

Change the Equation. (2012). *Up with the system.* Retrieved November 23, 2012, from http://www.att.com/Common/about_us/files/pdf/120827_stem.pdf

Colley, K. (2008). Project-based science instruction: A primer. *Science Teacher, 75*(8), 23–28.

Dorph, R., Shields, P., Tiffany-Morales, J., Hartry, A., & McCaffrey, T. (2011). *High hopes—few opportunities: The status of elementary science education in California.* Sacramento, CA: Center for the Future of Teaching and Learning at WestEd.

Drake, S., & Burns, R. (2004). *Meeting standards through integrated curriculum.* Alexandria, VA: Association for Supervision and Curriculum Development.

Edutopia. (2012). *Project based learning professional development guide.* Retrieved November 27, 2012, from http://www.edutopia.org/project-based-learning-guide

Epstein, D., & Miller, R. T. (2011). *Slow off the mark: Elementary school teachers and the crisis in science, technology, engineering, and math education.* Washington, DC: Center for American Progress. Retrieved from http://www.americanprogress.org/wpcontent/uploads/issues/2011/04/pdf/stem_paper.pdf

Gopnik, A. (2012). Scientific thinking in young children: Theoretical advances, empirical research, and policy implications. *Science, 337*(6102), 1623–1627. doi: 1126/science. 1223416

Keeley, P. (2009, June). Elementary science education in the K–12 system. *Science and Children,* pp. 8–9.

Maltese, A. V., & Tai, R. H. (2010). Eyeballs in the fridge: Sources of early interest in science. *International Journal of Science Education, 32*(5), 669–685.

Michaels, S., Shouse, A. W., & Schweingruber, H. A. (2008). *Ready, set, science! Putting research to work in K–8 science classrooms.* Board on Science Education, Center for Education, Division of Behavioral and Social Sciences and Education. Washington, DC: National Academies Press.

Murphy, T. (2011, August 29). STEM education, it's elementary. *U.S. News & World Report.* Retrieved from http://www.usnews.com/news/articles/2011/08/29/stem-education--its-elementary

National Center for Education Statistics. Programme for International Student Assessment (PISA). (2010, December). *Highlights from PISA 2009: Performance of U.S. 15-year-old students in reading, mathematics, and science literacy in an international context.* Retrieved from http://nces.ed.gov/pubs2011/2011004.pdf

National Governors Association Center for Best Practices, Council of Chief State School Officers. (2010). *Mathematics common core state standards.* Washington, DC: National Governors Association Center for Best Practices.

National Research Council. (2012). *A framework for K–12 science education: Practices, crosscutting concepts, and core ideas.* Committee on a Conceptual Framework for New K–12 Science Education Standards, Board on Science Education, Division of Behavioral and Social Science and Education. Washington, DC: National Academies Press.

National Science Foundation, National Science Board. (2010a). *Preparing the next generation of stem innovators: Identifying and developing our nation's human capital.* Retrieved from http://www.nsf.gov/nsb/publications/2010/nsb1033.pdf

National Science Foundation, National Science Board. (2010b). *Science and engineering indicators 2010.* Retrieved from http://www.nsf.gov/statistics/seind10/pdf/seind10.pdf

Nicolescu, B. (2002). *Manifesto of transdisciplinarity.* Albany: State University of New York Press.

Ohio Department of Education. (2012). *Standards-based education.* Retrieved from http://ims.ode.state.oh.us/ODE/IMS/SBE/default.asp

President's Council of Advisors on Science and Technology. (2010). *Prepare and inspire: K–12 education in science, technology, engineering, and mathematics (STEM) for America's future.* Washington, DC: Board on Science Education and President's Council of Advisors on Science and Technology. Retrieved from http://www.whitehouse.gov/sites/default/files/microsites/ostp/pcast-energy-tech-report.pdf

Route 21. (2012). *Partnerships for 21st century skills.* Retrieved December 1, 2012, from http://route21.p21.org

Sanders, M. (2009). STEM, STEM Education and STEMmania. *Technology Teacher, 68*(4), 2–26.

S²TEM Centers SC. (2012). *A theory of action for STEM education.* Retrieved September 24, 2013, from http://www.stemreports.com/2011/a-theory-of-action-for-stem-education

Task Force on American Innovation. (2012). *American exceptionalism, American decline.* Retrieved from http://www.innovationtaskforce.org

Texley, J. (2005). *Implementing the national science standards.* Retrieved August 22, 2005, from http://go.hrw.com/resources/go_sc/gen/HSTPR116.PDF

Tsupros, N., Kohler, R., & Hallinen, J. (2009). *STEM education in southwestern Pennsylvania: A project to identify the missing components, Intermediate Unit 1.* Pittsburgh, PA: Carnegie Mellon. Retrieved from http://www.scribd.com/doc/22053137/Science-Technology-Engineering-Math-Education-Survey-Report-Pennsylvania

U.S. Bureau of Labor Statistics. (2009, November). *Occupational employment projections to 2018.* Retrieved from http://www.bls.gov/opub/mlr/2009/11/art5full.pdf

U.S. Department of Education. National Center for Education Statistics. (2007). *Changes in instructional hours in four subjects by public school teachers of grades 1 through 4.* Retrieved from http://nces.ed.gov/pubs2007/2007305.pdf

CHAPTER THREE

Engaging Girls IN STEM Careers

TERRIE RUST

INTRODUCTION

The U.S. government's call for increasing the number of graduates in science, technology, engineering, and math (STEM) careers is particularly critical for women, since they make up a relatively small number of STEM graduates, especially in the technology and engineering fields. In our service-focused economy, careers in business and law and other service-related areas are a much higher draw for women than STEM careers. How to reach out to girls to study STEM has been a much-discussed challenge for over 20 years. Schools must now take up the call to promote female engagement in STEM courses. However, equally critical is the need for schools and districts to look at the bigger picture of gender equity in STEM education. Simply encouraging girls to take these courses has not been effective in increasing their numbers in high school and college courses, nor in the STEM workforce.

Research has shown that attracting girls prior to high school is key to introducing them to technology and engineering, and it is the determining factor in whether they take these courses in high school. It's also been shown that several other key factors are important to girls in making the decision to study STEM fields: strong mentoring during these critical years, an engaging curriculum that focuses on areas of importance to girls, and building strong math and science backgrounds, which are vital to all STEM careers.

This chapter will focus on one school district's attempt to increase female interest in STEM careers at the junior high level through the creation of a Girls Exploring Technology (GET) club and provide suggestions for starting a similar program.

BACKGROUND

Throughout history, there has been what societies have considered "woman's work" and what they have considered "man's work." At various points in time, women by necessity took on some male-associated tasks. Innovations in technology, the Industrial Revolution, and wartime requirements brought more and more women into the workforce—into what some considered "male territory." Changes in the role of women, the economy, and the acceptance of women into the permanent workforce during the twentieth century created opportunities never before available to women. Yet even with all the changes in the structure of the workplace, educational opportunities, and Title IX, women today still face myriad intentional and unintentional discriminatory practices when it comes to education and the labor market (Pickert, 2009).

To this mix we add another component: the decline of the United States as the leading innovator and creative body among the world's larger economies. To counteract this decline, especially within the scientific and engineering areas (which are most noted for providing innovation), government agencies and business and industry have realized that their past practices of importing talent will not meet their long-term needs. They now understand that the situation demands a different approach, because without more women entering the fields of science, technology, engineering, and math, the challenge of regaining our innovative capacity and global competitiveness will not be met (Committee on Science, 2007). Women currently fill close to half the jobs in the U.S. economy, but less than 25% of STEM jobs (Beede, 2011). They make up half of the college-educated workforce yet are underrepresented as STEM degree holders (Beede, 2011). "Women with a STEM degree are less likely than their male counterparts to work in a STEM occupation; they are more likely to work in education or health care" (Beede, 2011, p. 6).

Underrepresentation of women in the sciences, engineering, and the information technology (IT) sector persist. "Girls continue to lose interest and abandon plans for careers in the [science] field at a higher rate than boys" (Mitchell, 2006, p. 10). Girls account for approximately 50% of the school population, yet they make up only 10% of those enrolled in most technology education programs (Phelps, 2010). Women engineers account for a paltry 14% of all engineers, even though engineers make up the largest STEM occupational group (Beede, 2011). "In 2009, women comprised 27% of the computer and math workforce (the largest of the four STEM components), a drop of 3 percentage points since 2000" (Beede, 2011, p. 3). These statistics are even lower for minority women. Only one in ten minority women is a STEM professional (Writers, 2012). Rebecca Lucore, Executive Director of the Bayer USA Foundation, noted: "We need to be aware of the impact that peer pressure and societal concerns have on girls and minorities, resulting in them falling into familiar roles and patterns" (Bayer Corporation, 2010, p. 9).

It may take decades for women to attain the same numbers as men in STEM jobs, but there is a genuine concern that without more women taking up the call to enter these much-needed STEM career fields, our nation's economic future may be in jeopardy. More positions exist than can be filled by qualified men. In addition, women bring a unique perspective that cannot be matched by their male counterparts. "A 2007 study conducted by the National Center for Women and Information Technology indicates that mixed male and female teams were 26 percent to 46 percent more likely than normal to produce patents. A male only team produced the least amount of patents" (Kabir, 2012, p. 1). "Attracting and retaining more women in the STEM workforce will maximize innovation, creativity, and competitiveness" (Hill, 2010, p. 21) and bring greater diversity in ideas and development (Kabir, 2012).

Attempts to help girls become more aware of STEM career fields and to provide greater access have not had the desired effect of increased enrollment in high school STEM courses or in declaring STEM college majors. Other factors have been identified in helping to explain this apparent lack of interest, including lack of female role models, gender stereotyping, teaching methods that are not reaching girls, low feelings of acceptance, less family support to pursue a STEM career, and applying girls' career interests to real-life situations. Research confirms this. Mitchell and Hoff (2006) noted that discriminatory treatment of females is evidenced in many science and math classrooms by instructors asking questions differently of boys and girls (lower-level factual questions to girls vs. open-ended questions to boys), doubting girls' work and accomplishments, deliberately grading girls' work more severely than boys' work, and creating what girls perceive as a "chilly climate" (an indicator of a low level of socioemotional warmth).

Indirect discrimination occurs in the ways male teachers interact with their female students, often not realizing that their teaching methods are less effective at reaching girls than boys (Mitchell & Hoff, 2006). Weber and Custer noted that "the pedagogical approach may have a significant impact on student interest beyond the inherent interest in any particular activity" (2005, p. 62). Girls and boys see, hear, and process information differently. Understanding these differences requires differentiated instruction (Kovalik, 2008).

Females also receive a tacit message from a world that continues to produce mixed messages on the topic of gender and achievement (Lips, 1999; Hill,

2010). The scientists portrayed in science textbooks are predominantly male. The achievements of female scientists are infrequently highlighted (Mitchell & Hoff, 2006). The media depict female scientists as less than feminine. Role models help to dispel stereotypes and instill passion in young girls (Bayer Corporation, 2010).

Most girls, including many minority students, find a strong connection to STEM subjects only when what they are learning is tied to real-life situations, especially those that are humanitarian in nature (McNees, 2004) or socially relevant (Weber & Custer, 2005). According to research conducted by the Girl Scout Institute, two-thirds of girls who have expressed an interest in STEM careers choose medicine or healthcare as their primary career over any other career (Modi, Schoenberg, & Salmond, 2012).

It was traditionally assumed that girls are not interested in computers because the demographics have shown a huge gender gap in participation between males and females in playing computer games. The reality is that females account for almost 40% of all computer/video-game players and more than 50% for online games (Sweedyk & de Laet, 2005). Perceptions were based on participation observations rather than looking at the reality, which is that the gaming culture dismissed the games females liked to play as "real." It assumed females wanted "touchy-feely" games, when they simply wanted less violence and fewer hyper-sexualized female characters. These dichotomies only served to reinforce gender stereotypes.

Female students need broad exposure to mentors from a variety of STEM fields. Mentors play a critical role in encouraging girls to become interested in STEM fields. "Women were almost twice as likely as men to have chosen a science and engineering major through the active influence of someone important to them, such as a relative, teacher, or close friend" (Committee on Science, 2007, p. 65). Mentors have been shown to be a primary determining factor in whether females pursue and persevere in STEM careers and are able to visualize themselves in those careers (Hopkins, 2012). Maria Klawe, president of Harvey Mudd College and former dean of engineering and professor of computer science at Princeton University, said this about the importance of mentors to women of all ages: "When students see successful women working in a wide variety of technology fields and enjoying those fields, they begin to understand who they can become and how STEM can help them get there" (Coger, 2012, p. 3).

There is still resistance on the part of some parents who see various areas of STEM as male oriented and discourage their daughters from pursuing careers in those fields. These parents are usually unaware of the changing face of the twenty-first-century workforce and may see the educational needs of their daughters as different from those of their sons. Teachers still hear (and sometimes repeat) these references to older stereotypes about girls doing poorly in math. "Girls who hear their mothers or female teachers saying boys are better at math and science are more likely to give up sooner" (Moret, 2012, p. 1). These adults would be surprised to learn that girls perform nearly as well as boys on standardized math tests at the eighth-grade level (Hill, 2010).

Research has shown that middle school is the most suitable time in a young girl's education in which to foster interest in STEM careers (Hopkins, 2012; Mahoney, 2011; Modi, Schoenberg, & Salmond, 2012). During middle school is the ideal time to be introducing career options to all students, and focusing on the nation's critical need for STEM workers is one way to begin exposing girls to the vast opportunities available in STEM fields. Student clubs, workshops, and camps, which provide opportunities for girls to get hands-on experience and exposure to STEM careers, are also helpful (Devaney, 2007). Many organizations that host such events and clubs target middle school girls (Mahoney, 2011; Perales, 2012). "Participation in clubs [like this] has a positive impact on girls' enrollment in advanced math, science, and technology classes in high school" (The GEMS Club, 2010). Girls need to see that STEM is "cool, fun, and pays well" (Sheehy, 2012, p. 1). Although the concept of one-shot workshops is useful in providing awareness of STEM careers to girls, year-round programs provide a sustained focus and have a greater influence on girls' career decisions (Phillips, 2002).

And what about the gender differences that account for the subtle measures of increased male abilities in cognition that are critical for success in the science and engineering fields? Several studies indicate that the differences lie in mental rotation and spatial reasoning skills (Moret, 2012; Perales, 2012; Hill, 2010). Even though males have a head start in these areas, girls can quickly catch up. Girls in programs that help them to build spatial reasoning see a dramatic change in a short time. "If girls grow up in an environment that enhances their success in science and math with spatial skills training, they are more likely to develop their skills as well as their confidence and consider a future in a STEM field" (Hill, 2010, p. 45).

Although the studies listed above are by no means exhaustive, they do provide insight into the problems that have existed and, in some cases, still exist for girls in relation to enrolling in STEM courses and pursuing STEM careers.

STEM BEST PRACTICE: GIRLS EXPLORING TECHNOLOGY (GET) CLUB

In early 2005, at a Peoria (Arizona) Unified School District (PUSD) Communities and Schools Committee meeting, Career and Technical Education (CTE) Academy Director Gay Evangelista shared concerns with CTE Consultant Katreena Hayes-Wood about being overwhelmed with developing ideas to address girls and non-traditional careers. Gay had received a small grant for this work but was not sure how best to use it. Amazingly, at about the same time, PUSD's Centennial High School (CHS) senior Erin Cammarata reached a point where she felt something had to be done about her being the only female in most of her computer classes. She wanted more girls to become interested in computer courses but didn't know how to go about it. Erin approached her computer teachers with a concept: high school girls mentoring younger girls so that the younger girls would feel comfortable in what many of them perceived as a male-only field. Erin's teachers brought her idea to Gay at the district level (Rust, 2006). Gay had also been receiving pleas from Dennis Kuhn, one of the computer science teachers at CHS, for intervention to increase female enrollment in computer classes. Kuhn had observed how a class with even a few girls improved the learning environment. As a result, a model for reaching out to girls began to take shape.

During the spring of 2005, a committee was formed, and organization meetings were held to lay the framework for the program and to plan a kickoff event. Members of the committee included Gay, Katreena, Erin, Diana Denton (one of Erin's computer teachers), a National Academy Foundation advisor, instructors from Glendale (Arizona) Community College and DeVry University, four girls from grades 10 and 11 enrolled in the CHS computer classes, several eighth-grade girls representing their various schools, and a Technology, Life, Careers teacher (me) from Oasis Elementary (one of CHS's feeder schools)

Early in 2002, I had read some research about results garnered from single-gender classrooms (Streitmatter, 1999). The focus of those classrooms had been to see whether middle school girls would fare better if separated from boys, since it was at this point that girls began to distance themselves from boys specifically in the areas of math and science. I had noticed some of the same characteristics in the girls whom I taught in my Technology, Life, Careers classes, so I formally proposed to my principal what I called a "gender equity class." This class would consist of eighth-grade girls (voluntary) who would have their math, technology, and science classes together, but mix with the other students (male and female) in the rest of their classes. I sensed my female students were holding back and felt that the idea of this all-girl class might just show my hunch to be true. To my surprise, my principal, who was also female, rejected the gender-equity classroom concept! When the idea for Girls Exploring Technology (GET) was proposed 3 years later, I realized that it would allow me to do some of the things I'd previously hoped to achieve. I wholeheartedly endorsed GET and was excited to become involved.

Katreena, the District CTE Consultant, conceived the concept of Girls Exploring Technology and used a track-and-field model of "get ready ... get set ... go" for establishing the scaffold of the program (Figure 3.1).

The primary concept behind the program was derived from Erin's original idea: high school girls acting as mentors to younger girls, taking control of the active teaching and leading. Using an industry model, they were given the title of

ENGAGING GIRLS IN STEM CAREERS | 51

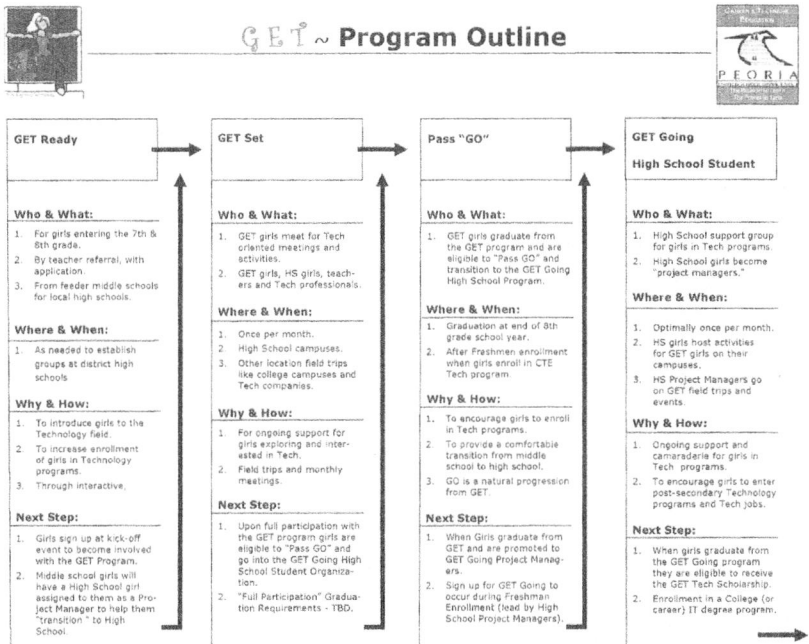

Figure 3.1. GET Program Outline.
(Source: Author.)

Project Managers. Since the seventh- and eighth-grade Technology, Life, Careers program in the district included extensive career study and computer skills development, this age group was determined to be the primary target for the GET program. Encouraging girls of this age had also been seen to spur math and science interest in high school. Furthermore, establishing confidence in this group prior to high school course selection was deemed important, since the main reason for starting the GET program in the first place was to encourage more girls to enroll in computer classes once they reached high school.

We drafted a mission statement to guide our actions: "The mission of Girls Exploring Technology is to provide young girls an opportunity to explore areas of technology in an environment that is safe, comfortable and encouraging."

One of the project managers created a GET logo: a girl emerging from a computer screen (Figure 3.2). An exclamation point was our marketing icon (Figure 3.3).

A Get Ready kickoff event was scheduled for June 9, 2005. The Get Ready step in the program was held during the summer to introduce girls to the GET program prior to the start of the school year. This event would be open to any incoming seventh- or eighth-grade girl attending any of CHS's feeder schools.

Figure 3.2. GET Logo.
(Source: Author.)

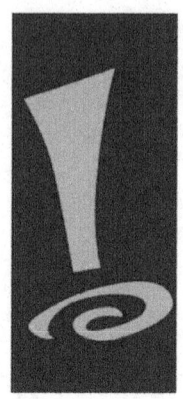

Figure: 3.3. GET Icon.
(Source: Author.)

Sixth-grade teachers and the Gifted Education teacher were asked to nominate girls with strong science, math, and technology interests (Figure 3.4).

Invitation letters were then sent to these nominees along with a flyer introducing the GET program (Figure 3.5).

Nineteen girls attended the event, which was hosted by Glendale Community College (GCC). One of the advantages of having a community college partner was the variety of opportunities the school could provide for facilities and mentors at that level. This first Get Ready event included activities introducing website development, creating animated graphics, exploring how computer networks work, touring the college IT facilities, and hearing firsthand from the high school project managers about how to get involved in this exciting career field during high school. All of the activities were led by the project managers with assistance from the advisors (the adult members of the organizing committee). Each girl received a GET folder that not only included information on that day's activities, but also contained reprinted articles on successful women in IT careers, web links to a variety of technology sites designed to appeal to girls' interests, and, most important, the pathway of courses in pursuit of computer science as a career focus during high school. The press was invited and took pictures and conducted interviews of the girls. GET t-shirts had been designed in bright pink with the lime green exclamation point logo and were worn by all the project managers and adult advisors, sending a message that GET was, indeed, something special. Erin, the high school senior who originally had the idea for high school girls acting as mentors for middle school girls, was actively involved in the event and was able to see her idea come to fruition.

Once the kickoff event was over, planning began for the "Get Set" phase: field trips, in-school and after-school meetings, and periodic activities that would

Date

Do you have 7th and 8th grade female students who have strong math, science and problem solving skills?
(Note: Include 8th grade if letter is send in fall and 6th grade if letter is send in spring)

We hope you do!

Let us introduce you to GET! **G**irls **E**xploring **T**echnology – It's a new pilot program being launched by the district's CTE department focused entirely on helping girls explore the vast and interesting field of Information Technology. (IT).

On date of event goes here we will host the "GET Ready" kick-off event at a local college and or company. Girls will be invited, through an application process, to participate in this event. This is where we need your help.

We want to encourage girls with the natural skill set required for IT to participate in this fun and interactive event. We'd like you to help us identify the girls who have shown an interest in IT, or perhaps the female students who you think may benefit from further exploring the IT arena.

The event begins at time goes here in the morning and ends at time goes here . We are providing transportation to and from location goes here from school location goes here on the day of the event.

During the course of the day girls will have a hands-on experience using various forms of technology in programs being led by IT Directors, women in the IT industry. Assisting our industry leaders will be the IT Project Managers, the high school students currently enrolled in our CTE Information Technology program. Girls attending this event will Technical Activities planned for the event go here.

This event will be the kick-off to the GET Program and the girls you recommend will pave the way for others interested in the field of I Won't you help us by giving applications to the girls you believe would enjoy this opportunity?

We appreciate your time, effort and recommendations. If you have any questions, or are interested in attending our kick-off event, please let us know by deadline to receive applications goes here.

Sincerely,

GET Representative Name
Enclosures

Figure 3.4. GET Nomination Letter.
(Source: Author.)

provide the girls with additional opportunities to explore the many areas of information technology. Any seventh- or eighth-grade girl could join GET from that point on and participate in any or all of the provided activities. Oasis Elementary, the school where I taught, began a GET club that grew to 40 members. (PUSD's elementary schools are K–8; the junior high population totaled about 185.) Although Oasis was the only school with an established GET club, any girl

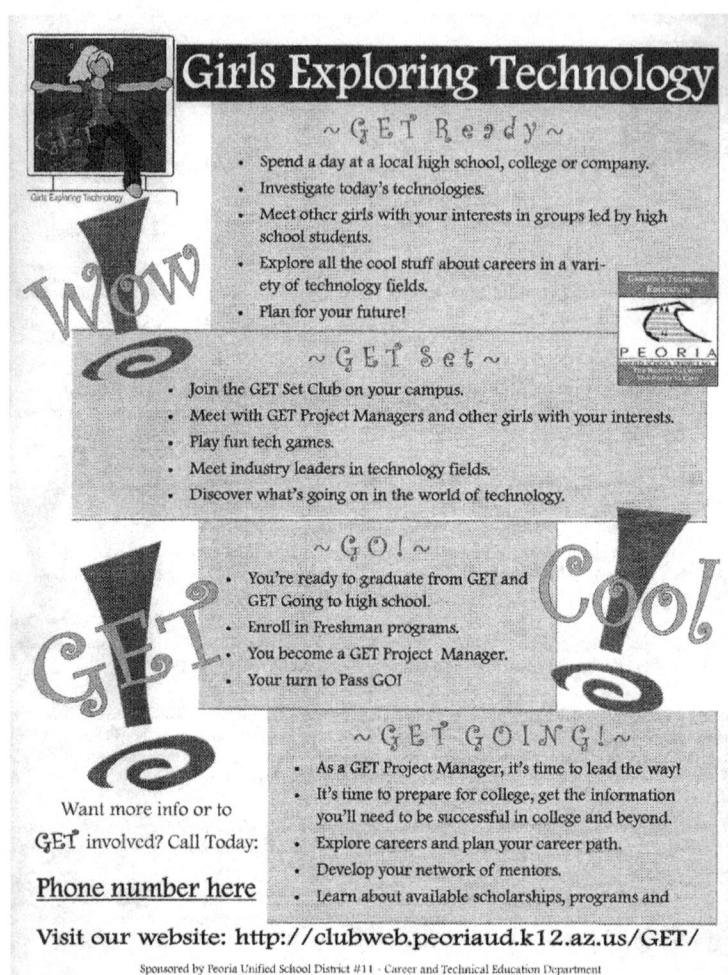

Figure 3.5. GET Flyer.
(Source: Author.)

attending one of the CHS feeder schools could attend our GET events. There were several girls who participated in this way.

Formal Get Set events held during the 2005–2006 school year included a trip to DeVry University (a GET partner) to attend a *her world* event in which the girls were able to learn about careers, participate in hands-on electronic projects, and view the senior students' final projects. They also attended a presentation by a former CHS student who now worked for Cisco Systems and was able to share her experiences and describe opportunities for women in her field; a Saturday workshop held at CHS to hear from women in the IT field and to participate in activities teaching binary code and computer graphics; and, a trip to the Sally Ride Science Festival held on the campus of Arizona State University (ASU) that featured not only a formal presentation by astronaut Sally Ride, but also workshops and vendor booths to engage the girls. Another workshop, hosted by Mesa Community College's School of Business for middle school girls, introduced them to technology careers.

Photo 3.1. GET Members with Astronaut Sally Ride at the February 2006 Sally Ride Science Festival at ASU.
(Photo courtesy of Deb Jones.)

Informal Get Set events included monthly meetings held during lunch and occasional after-school workshops. During the brief lunch meetings, the girls might be introduced to a particular career, learn a new tech skill, view a video highlighting young women in tech careers, or plan upcoming field trips. The purpose of the after-school workshops was to allow the girls an opportunity to collaborate on a problem to solve, to bond as a group, and to interact with the project managers. These informal meetings were designed to quickly engage the girls and have them leave with something new to think and talk about.

Funding for our events that first year came partly from the original grant but was also supplied by our business and education partners, from district CTE funds, and tax credit funds specifically donated by parents for our club at Oasis Elementary. GET girls purchased their own GET t-shirts, which were offered in two colors: the original bright pink with a lime green exclamation point logo, and a black shirt with a glow-in-the-dark lime green exclamation point. The girls wore these shirts to all GET events. The unique logo often attracted the attention of other event attendees, providing the girls with many opportunities to share with others the concept of GET. These shirts had the added effect of uniting the group.

Also, during the first year, the project managers designed and launched the first GET website (hosted on our district server) and a monthly newsletter that was emailed to the GET members. They also created GET clocks to be used as raffle prizes. These clocks used the GET computer girl logo as the backing of the clock face. I had one of these clocks in my classroom, proudly displayed above the GET events board.

As the school year was drawing to a close, it was time to plan for our first "Go Promotion" ceremony. This ceremony would provide a time to recognize the GET activities held during the year and to promote those eighth-grade girls who had been most active in the year's activities to the role of Assistant Project Managers in the "Get Going" phase at the high school. Assistant project managers would be freshman girls who would assist the project managers at GET events.

The Go Promotion was held one evening in early May at our school district's Administration Center. Formal invitations were sent to the girls, their parents, and school and district administrators. The project managers introduced the GET program to the audience with a PowerPoint presentation highlighting some of the year's events. I announced each of the honorees and explained the importance of their personal involvement in GET. The girls were each presented with a certificate and a pink carnation. Refreshments were served. The event was well attended, and the parents expressed excitement about what their daughters had accomplished. The promotion event was made possible through parent donations.

Summer was fast approaching, and the advisory team and project managers began planning for the 2006 Get Ready event. Sixth-grade teachers were again asked to recommend girls. A new group of girls would be introduced to the GET program, and the project managers and assistant project managers would have an opportunity to build confidence in their new charges. Because of the success of the first year's program, the advisory team decided to enlist the help of other GCC instructors who wanted to see more girls in their programs. So it was that attendees at the June 12th Get Ready event at Glendale Community College participated in sessions covering computer programming, web design, computer-aided design and drafting (CADD), computer networking, and automotive technology (to learn the role computers play in today's automobiles). Presentations on the GET program and website were included. The 2006–2007 GET program year was off to a great start.

Formal and informal events were held that embraced a broader spectrum of technology-related careers and skills: a trip to the Sally Ride Science Festival at ASU; a tour of Honeywell's avionics facility that included a discussion about careers; a day at Universal Technical Institute for electronics activities and a facilities tour; and a half-day workshop focusing on women in aviation careers held at the Challenger Space Center. The girls attended monthly lunch meetings and periodic after-school workshops as well as a Saturday workshop at CHS sponsored by the automotive and building trades teachers. They also hosted a GET booth at the regional career fair attended by eighth-grade students from several cities.

The second Go Promotion event was held in May 2007. A video of the year's GET events, which included interviews and comments from each of the girls being promoted, was a highlight. Two of the project managers were graduating from high school, and their contributions to the GET program were recognized by CHS GET advisor Diana Denton.

Gay's original plan for the program was to establish GET groups at each of the district's high schools, eventually reaching out to all the girls in their feeder schools. Three district high schools established GET groups. The CHS GET group held few of their own activities on their campus, choosing instead to focus on mentoring the junior high girls. A GET group begun at Peoria High School (PHS) focused on business technology. The Sunrise Mountain High School (SMHS) GET group focused on automotive technology. Both PHS and SMHS groups invited junior high school girls from their feeder schools to various events, but, of the two, only Sunrise Mountain's GET girls planned to host a Get Ready event. Members of these two new GET groups attended the Honeywell avionics event along with the CHS GET group and participated in staffing the GET booth at the regional career fair.

GET Faces New Challenges

The 2007–2008 school year saw many changes that directly impacted the GET program. Two of the CHS project managers had graduated in May (they had been the backbone of that program, and there was no one to replace them); funding for GET activities was minimal; district busing schedules made it nearly impossible to go on any field trips; Gay's replacement (she retired in 2006) was capable but not as engaged in program participation; DeVry University stopped including junior high girls in their *her world* event, which reduced involvement by GET's DeVry Advisory Committee member; and our GET Advisory Committee member from GCC could no longer participate. In addition, Diana Denton, the high school GET advisor, was unable to attend as many events or help mentor at CHS, and the district's high school course scheduling (which prevented many freshmen from signing up for electives) hit computer science enrollments hard. Another factor affecting girls at the high school level was the availability of other CTE clubs to participate in, forcing girls to make a choice as to which one to join. The GET website, which had been designed and updated by a project manager who had graduated, was redesigned by Katreena's company. But since there were few events, there was little new input that year. I was able to continue the GET club on my campus, but the number of participants dropped. Without the high school girls as mentors and the events to generate enthusiasm, the concept was losing momentum. The GET groups on the other two campuses experienced similar roadblocks but kept together during that school year.

By the end of the 2007–2008 school year, it was clear that the GET program was not going to be able to continue on the Oasis campus. The GET groups at Peoria High School and Sunrise Mountain High School soon followed suit, disbanding for lack of an advisor to take them forward.

The GET program at Oasis Elementary held steady at a membership of between 32 and 40 girls during the first 2 years. That was about 34% of the entire Oasis female junior high population. The club offered a great opportunity to those girls who participated in it. The number of participants dropped by half during the third and final year. All the girls who joined GET were richer for the experience.

In a video filmed for the May 2007 Go Promotion ceremony, the Oasis girls said that GET had made them aware of the opportunities in the field of technology, expanded their choices for future jobs, and provided them with a chance to meet women working in engineering, information technology, aerospace, and more. The GET group activities allowed them to learn and be challenged in a fun setting because their friends (old and new) were there, too. Several girls said they were now considering careers in fields they didn't know existed prior to their exposure through GET, or careers they now felt confident to pursue. Even though the GET program was only in existence for a brief time, the impact on the girls would be long lasting. Involvement with the GET program was definitely worth the time and effort.

As previously mentioned, many factors led to the GET group's dissolution. If other advisors at the high school and community college could have been found to carry the program forward, momentum might have continued. The high school advisor positions were critical in taking the assistant project managers and building them into strong mentors for the middle school girls. Without this, the girls who were promoted to the high school were not able to fulfill that GET program function. Perhaps locating grant funds could have offset the program costs. There might have been other possibilities, but—since at the time I was the one holding the program together—I felt overwhelmed and unable to tackle many of these issues.

One of the intended consequences of the GET program was to encourage girls to enter careers traditionally underrepresented by women. The two GET project managers who graduated in May 2007, as well as Erin, who started it all, majored in computer science at college. Although I was unable to follow my middle school students through the years to see where their careers led, I know that at least one majored in computer science, too. Several more pursued other STEM careers.

Throughout my experiences with GET, I became a greater champion for girls and STEM gender equity. In my personal life, even I had to deal with negative comments about my involvement with technology. That negativity led to an even stronger resolve to make a difference, even if for only one girl. I believe the GET program allowed me to be a conduit to accomplish that.

HOW OTHERS CAN ADAPT THIS BEST PRACTICE

Engaging girls in STEM activities/career exploration can be accomplished in diverse ways. This section will detail some of the ways in which classrooms, schools, informal STEM education organizations, and corporations can, have, or are addressing the need. It is my desire that among these examples, readers might identify one method that could work for them as a first step toward reaching girls in their locale.

My focus is on programs and program ideas reaching middle school girls. Programs and program ideas designed to reach high school girls will be addressed briefly later in this section.

For the STEM Educator: Getting Started

Professional development and teacher training in strategies to expand options for women and girls in STEM may be required to provide the teacher a firm footing as he or she travels this new path. Discussed here are several organizations that offer exceptional resources for teachers to gain beneficial knowledge and skills.

As you begin to think about ways to address the need for girls to get exposure to STEM activities and careers, I highly encourage you to watch this video blog by technology teacher Dave Janosz on the topic of gender equity (http://www.youtube.com/watch?v=cB49Fhlb-hY). The video blog explains what teachers can do to attract girls to STEM programs and includes interviews with professionals who share their viewpoint on successfully reaching out to girls.

The National Alliance for Partnerships in Equity (NAPE) developed the STEM Equity Pipeline Project (http://www.napequity.org/stem/stem-equity-project/), a site offering webinars and online courses designed to create gender-equitable classrooms. The site also contains a collection of STEM education resources for students and teachers.

The National Center for Women in Information Technology (NCWIT) provides free resources to formal and informal educators to help provide strategies, tools, and practices to attract more girls and underrepresented groups to computing and information technology (https://www.ncwit.org/resources/alliance/47). A free White Paper entitled *How to Recruit Women and Girls to the Science, Technology, Engineering and Math (STEM) Classroom* is available from the ncwit.org site as well.

The Women in Education Proactive Network (WEPAN) Knowledge Center (http://www.wskc.org/) provides free access to information related to women in STEM.

The Society of Women Engineers (SWE) K–12 Outreach website (http://societyofwomenengineers.swe.org/k-12-outreach) hosts a special page for educators. A sampling from the resources provided includes an Outreach Tool Kit and details on how to engage girls in grades 6–12 through the *Invent It. Build It.* Expo.

The National Girls Collaborative Project (NGCP) Resources page (http://ngcproject.org/resources) includes archived webinars, statistics, stories, exemplary practices, and relevant links.

Familiarize yourself with the STEM-related courses offered at the high school your students will attend. Meet some of the teachers; showing up at their parent night event is an easy way to circulate and introduce yourself. Get to know at least one of the guidance counselors at the high school. Once you are armed with the tools necessary to move forward, you can consider your options.

Classroom-Level Program

One of the easiest ways of addressing the growing need for introducing middle school girls to STEM careers is at the classroom level. Teachers of science, technology (technology education and computer courses), pre-engineering, and math can promote gender equity and introduce STEM careers easily through poster displays. Girls are impacted by visual stimuli, and posters are effective tools. As this first step in generating interest takes hold, teachers of STEM subjects should also locate interesting stories about women in their course's STEM field to share with students. (Boys are just as interested in these as girls!) These stories should be introduced along with the subject matter rather than separate from it. In this way, students can see how what they are learning not only applies to real life, but also can lead to engaging careers. An added benefit is that girls get introduced to role models. A great resource is http://stemstories.org/. Stories are categorized by field of study, birthdays of those profiled, and topics.

In my classroom, I set up a book cart providing my students with a selection of books on technological advancements, invention stories, profiles of famous women inventors and scientists, profiles of black inventors and scientists, and career information. Many companies provide free career materials, including the armed forces, NASA, and most of the local community colleges and trade schools. Some books I ordered through my school librarian, some were donated by parents through my "teacher wish list" at our school book fair, and some I purchased.

Reading time must be scheduled to give students an opportunity to access your reading materials. Downtime after tests or when in-class work is completed are both good opportunities for reading time. This is another subtle method of sparking interest.

Another effective tool is inviting guest speakers to your classroom, especially women who can serve as role models for the girls. At the beginning of each school year, I asked the eighth-grade students to provide information on their parents' jobs. When I was looking for a guest speaker, I would go to this list first. Parents were glad to talk to students about their careers. Whether parents or professionals, speakers were briefed prior to their presentation about topics, time limitations, and addressing specifics about their career with students, especially the girls.

If GET had extended beyond the 2007–2008 school year, the introduction of online chats or Skype sessions with women in STEM careers would have taken place through the newly designed website. The possibility of students attending webinars was also being considered. These opportunities are easily accommodated through the Internet.

I mention all of these resources first because only after girls generate an interest in STEM fields can you think of moving to another level such as forming an interest group for girls who are enrolled in that class. As an example, even though all the middle school girls on my campus were required to take my class, not all of them expressed interest in knowing more than what was covered in the curriculum. How many girls are enough to start a classroom-based interest group? Numbers are not as relevant as the desire and initiative of the classroom teacher, whether they be male or female.

One important aspect to mention here is that a classroom-level group will most likely meet no resistance from an administrator, since it is subject-matter based and the teacher has volunteered to take charge. There is also little if any monetary expense associated with this type of group. Activities such as introducing a concept extension from your curriculum, watching a brief online video, or bringing in a guest speaker could easily be held during lunch or after school at zero cost. Even this minimal exposure by middle school girls, no matter its depth or breadth, could serve as a launching pad for developing and furthering their interest in STEM activities and careers.

An example of a successful classroom program is the one led by technology education teacher Kenny Zushma. His group of budding female engineers are students at Heritage Middle School in Livingston, New Jersey. The FemGineers (http://www.livingston.org/Page/29642) focus on engineering through participation in the National Engineers Week Future City Competition (http://futurecity.org/). The Future City Competition is inexpensive, engages the student teams with an engineer mentor, and integrates all the STEM areas. Girls especially love the challenge of the human element built into the competition criteria. (Note: The

Future City Competition is not just for girls. It does, however, lend itself well to all-girl teams.)

School-Level Program

If you analyze GET, you discover that it was actually a hybrid of the classroom and school models: members were all students from the Technology, Life, Careers course I taught, but we were a school group under the umbrella of the high school CTE program, since we worked in conjunction with our feeder high school.

School-level programs and clubs can reach girls from a variety of STEM areas. For instance, a technology teacher can begin a group that might include members who are not enrolled in that course. School-level involvement usually requires approval from the administration and, one hopes, energetic participation from multiple STEM teachers who would be willing to share the workload. The endorsement of the school principal becomes important in selling your program to parents and for generating funds when necessary. My principal was extremely supportive, arranging for substitutes to cover my class so the GET girls could attend events, being present at the Go Promotion ceremony, helping me locate funding, making contacts, and much more.

Starting a school-level program involves many of the details mentioned in the section describing the GET program. What will be the name of your group/club be? What is your target audience? (Although we had sixth graders on my campus, they were considered part of the elementary, not the junior high scheduling block, and were therefore not included in GET.) Have you thought about which higher education, community, and business partners you would like to approach to work with you? What activities will you focus on (field trips, after-school meetings and workshops, and so on)? Dangling some of those planned events in front of girls increases their excitement about joining. Do you plan to have group t-shirts or other identifying clothing or accessory item? Girls like to be identified in a special way. Will you have a kickoff event? The GET Checklist (Figure 3.6) and GET FAQs (Figure 3.7) provide important items we needed to consider. These lists are a good starting point, no matter what kind of STEM club might be planned.

A school-level program that started at one school in 1994 and is now in 20 schools nationwide is the GEMS Club: Girls Excelling in Math and Science. The purpose of this club is to expose fifth- and sixth-grade girls to the fun and wonder of the fields of math, science, and technology. The organization's website (http://www.gemsclub.org/) provides many valuable resources, including a free 71-page toolkit with all you need to know to start a GEMS Club at your school.

Forming Partnerships

At the ideation stage, consideration of whom you might approach as partners in your endeavor is of primary importance. These partners can provide advice, facilities at which to hold events, monetary or in-kind donations, role models or mentors, guest speakers, and a variety of other offerings. As partners, you share common goals and outcomes that build on each other's strengths. Partners share clearly defined visions. Forming partnerships is not the same as asking for a donation. Partnerships can leverage resources in ways a donation cannot (Bayer Corporation, 2010).

Make contact first through a formal letter. Introduce yourself, the program you are developing, and your goals, and extend an invitation to partner with you. Next, schedule a face-to-face meeting to discuss whether you would be a good match. Have clearly defined parameters for what you expect from this partnership. The discussion should also identify how the organization would benefit from partnering with you.

Community colleges are usually looking for ways to connect with schools. Reach out to department heads and instructors in the STEM disciplines. Some high school teachers teach an evening course at a community college. Check with teachers you know at your local high school for help in identifying those teachers. You may find they teach a STEM subject and would be interested in working with you, or they may be able to connect you with the right person at the college.

Four-year colleges and universities often have a Department of Gender Studies. They might be interested in supporting a gender-based group. Some university STEM-related departments or STEM centers host annual events/workshops to encourage girls' interest in those fields. If a university close to you does this, it might be a good partner candidate.

STEM-based businesses and corporations are often on school district advisory committees. Check with your school district's CTE department for a list of those businesses as a starting point. Contact your local Chamber of Commerce for listings of member businesses. Those members are often interested in community engagement. Conduct a search to see if there are female CEOs or female business owners in your area who would be good prospects for you to contact. Review other business–education partnership models to learn what their experience has been and to help you understand how these types of partnerships can be successful.

Funding

Funding is always going to be an issue. District budgets are squeezed tight. Economic forces impact businesses. Expecting the girls to pay for the program

> **Welcome to the World of G E T** — Page 3
>
> ## The G E T Checklist
>
> Below is a list of things to do to as you prepare to bring GET to your school:
>
> ____ Get approval and support from school administration, high school and middle school principals.
>
> ____ Establish a planning committee, then schedule first meeting to include:
>
> ____ Executive Director (program facilitator, create a GET Program Planning Binder)
>
> ____ GET Director/s (Technology Teacher and Industry Sponsor/s)
>
> ____ College Representative interested in hosting kick-off event
>
> ____ High School Project Managers (high school girls currently enrolled in Technology classes)
>
> ____ Middle School Teachers interested in organizing GET on their campus
>
> ____ Determine Committees and Assign Committee Responsibilities
>
> ____ Executive Director (keeps GET Program Planning Binder)
>
> ____ Secretary (takes notes, sends copies of notes after each meeting to planning committee members)
>
> ____ Other committees to be determined by planning committee
>
> ____ Determine Date and Location for **GET Ready** kick-off events (Saturday seems to work best for kick-off events). See Sample Run Book.
>
> ____ Plan **GET Ready** event schedule and related details, to include:
>
> ____ School Transportation
>
> ____ Delivery and return of GET Interest Inventory & Application
>
> ____ Start and end times for **GET Ready** event
>
> ____ Follow up meeting to evaluate **GET Ready** event and begin planning for **GET Set** Event/s (we recommend at least one for every quarter—nine week period)
>
> ____ Planning for **GO!** Graduation for 8th grade girls graduating to high school and promotion to **GET Going** as Junior Project Managers
>
> ____ Get Project Managers develop a GET Website
>
> *GET Girls attend a Technology Fair*
>
> *"There are no shortcuts to any place worth going."*
>
> ~ Beverly Sills
>
> *Opera Singer*
>
> CAREER & TECHNICAL EDUCATION
>
> PEORIA UNIFIED SCHOOL DISTRICT
> The Reason to Learn
> The Power to Earn

Figure 3.6. GET Checklist.
(Source: Author.)

is self-defeating. GET was fortunate in that we had grant funds to help us get underway. I mentioned several other financial resources that we utilized in the sections on GET and Partnerships. I would like to add a couple of other suggestions, which I hope may be useful. (As mentioned in the Class-Level section, funding requirements depend upon the extent to which the planned club engages the girls.)

Grants are available from a variety of sources. Some grants, such as those offered through the Wells Fargo Teacher's Partner Program, are available in amounts up

| Welcome to the World of G E T | Page 5 |

G E T FAQ's

Here are some of the questions we frequently get:

1. How many girls do you generally invite to a GET Event? *Depending on the facility, but generally we plan for between 20-30 girls per GET Ready event. For the smaller GET Set Events we plan for between 10-20.*
2. How long do your GET Events last, and on what day of the week? *For the GET Ready Events we plan for about six hours on a Saturday. For a GET Set Event we plan for 2-3 hours on a Saturday morning, after school, or a day during a school break.*
3. How can we pay for GET Events? *Often you can get companies to provide giveaways and cash donations, or offer the use of their facility to help with expenses. Use of tax credit funds in states which offer them is another funding source. Also try mini grants for transportation or facilities, or supplies.*
4. How do you find the high school and middle school girls that might be interested in starting a GET Program? *One of the best places to recruit high school girls is from your technology classes. For middle school recruitment talk to your district middle school math and science teachers. Generally girls who excel in those subjects are more interested and gifted in the field of technology.*

GET Girls working on web pages

GET Project Managers teach a GET Activity

Figure 3.7. GET FAQs.
(Source: Author.)

to $500. Applications are evaluated based on overall quality of the project, educational benefits to students, integration of the proposed project with school curriculum, and number of students who will benefit. Wells Fargo Teacher's Partner grant applications are reviewed frequently, so you'll get a response one way or the other within a short time. I was the recipient of two of these grants: one helped me pay expenses to a conference, and the other provided classroom equipment. These grants are available in many states with Wells Fargo branches. For details, Google "Wells Fargo Teacher's Partner Program" to see if these grants are available in your locale.

Kids in Need Teacher Grants provide K–12 educators with funding to provide innovative learning opportunities for their students. Individual Teacher Grant awards range from $100 to $500 and are used to finance creative classroom projects (the Future City Competition project would qualify in this case). The number of grants awarded varies from year to year, depending on the total funds being requested. Typically, 300–600 grants are awarded annually. Applications are accepted during the back-to-school season. For an application, visit http://www.kinf.org/grants/.

A number of grants directed toward gender equity programs are available in selected states. The National Girls Collaborative Project (NGCP) is one group that offers mini-grants. Mini-grants are awarded to girl-serving, STEM-focused programs to support collaboration, address gaps and overlaps in service, and share promising practices. Details can be found at http://www.ngcproject.org/resources/mini-grants/schedule. Google "gender equity grants" to locate similar resources.

Contact your school district's CTE Director to learn if there are any CTE grant funds that might be available for girls' programs.

One-time donations can be solicited from the school community, businesses, and interested parties. In your quest for business partnerships, you may encounter many turn-downs. A business may decline for various reasons but still share your vision. Asking for a one-time donation would be appropriate at this juncture. Have a reasonable figure in mind if you are asked "How much?" You don't want to shortchange your opportunity, but you also don't want to appear unreasonable. That particular business may decide to donate again if it is so inclined. Asking for donations from your school community will depend upon school and district policies regarding such requests. Approaching individuals for a donation only requires making the ask. A formal note of thanks should be written to donors. Publicly acknowledge the donor in any publicity connected to the event ("This event was made possible through generous contributions by …"). Provide donors with copies of clippings containing that acknowledgment. They may choose to make another donation during their next funding cycle.

Publicity

Publicity is important to get the word out about your group and its activities. Media coverage multiplies the impact of your event. The more people are aware of your group, the more likely it is that your school, parents, community members, business community, and school board will take notice. Getting noticed provides openings for dialogue that may create interest in partnering or supporting other events/activities or for making a donation to your group.

Become friends with the education writers covering education in your community. How do you find them? They are always in attendance at your school board meetings. You can identify them because they'll be writing notes (usually in a small notebook!) during the board meeting. Approach them before or after the meeting and ask them to cover your event. Have all the details ready: when, where, time, who will be involved, and so on. Get their contact information so you can email them a reminder. If a photographer will not accompany them to the event, ask if you can provide a photo to accompany the article. Space limitations may prohibit this, but it doesn't hurt to ask. If they can't (or won't) cover your event, ask if you can submit an event photo with a caption (and photographer's name). Newspapers often need to fill awkward spaces in their pages, and in such cases a captioned photo of this type can be useful.

Education sections of news publications usually include the name of education staff writers. Contact them directly by email or phone. Many television and radio stations also have someone who is designated to cover education events. When contacting the media, ask for these designees specifically.

School-Level Specialized Workshops

School programs need not be just about formal clubs. Specialized workshops can provide a short-term solution that will garner lasting memories and encourage girls' interests in STEM. These types of workshops can be planned as after-school events, daylong events, or 1- or 2-week summer camps. The length and breadth depend on availability of space, time, funding, and planned activities.

One example of a specialized school camp was held at Northern Valley Regional High School in New Jersey for girls in grades five to eight. The Girls in Engineering and Technology camp was led by high school technology instructor Nicole Hodgson and was designed to give young girls added skills and confidence in math and science by experiencing a variety of hands-on projects. Lessons and projects covering civil, electrical, and biomedical engineering exposed these viable career options to the participants (Albrizio, 2012). Nicole holds 1-day workshops as well.

School-level clubs and workshops can be rewarding experiences for both the teacher (as I can attest) and the participants. Even more important, the girls you reach will see opportunities in STEM they may never have known existed.

Informal STEM Education Organizations

It may not be possible to start a class-level or school-level girls program at your school. There are myriad opportunities for you to engage middle school-aged girls in STEM activities offered through informal STEM education venues and organizations. The listings here will provide teachers with ideas to serve as a starting point.

Many cities have interactive science centers. These centers provide a host of programs, some of which are directed to girls. The Carnegie Science Center in Pittsburgh, Pennsylvania, established the Girls, Math & Science Partnership for girls ages 11–17 to encourage them to become the female science and math leaders of tomorrow (Gormly, 2010). Women in Science Programs offered at the Connecticut Science Center deliver programming aimed at engaging girls and young women in STEM fields. The Center offers Saturday Science workshops and a #DaVinci Coder Girls-only Hackathon competition. The website highlights women in STEM leadership and includes a page of valuable links aimed at showing how equity is being tackled on a variety of levels. Visit https://www.ctsciencecenter.org/programs/wis for details. Chabot Space and Science Center in Oakland, California, offers a program called Techbridge. Started locally in 2000, the program is now offered nationally. Techbridge presents a multi-faceted approach that introduces hands-on activities while also emphasizing career exploration to inspire girls to develop a passion for technology, science and engineering. Details can be found at http://techbridgegirls.org/.

One of the most highly profiled organizations committed to gender equity and providing programs for girls is the American Association of University Women (AAUW). Their Tech Trek program is a weeklong science and math summer camp designed to develop interest, excitement, and self-confidence in young women who are entering the eighth grade. Tech Trek events are held around the country. AAUW also sponsors a Tech Savvy event, a daylong conference focusing on STEM careers for sixth- to ninth-grade girls. For details on these opportunities, visit http://www.aauw.org/what-we-do/stem-education/tech-trek/ and http://www.aauw.org/what-we-do/stem-education/tech-savvy/.

The National Girls Collaborative Project (NGCP) provides an online state-by-state program directory (http://www.ngcproject.org/find). Locating programs in your state is a click away!

ExxonMobil hosts annual "Introduce a Girl to Engineering" events at company sites across the country, with activities designed to encourage careers in engineering. The program seeks to promote curiosity among middle school students and, in particular, to help shrink the gender gap in STEM fields (Brice, 2010).

Sally Ride Science holds Sally Ride Science Festivals for fifth- to eighth-grade girls. The astronaut speaker, workshop sessions, and vendors make this an exciting event. Also offered are Sally Ride Science Camps for girls entering grades 4–9. These camps are held for 2 weeks during the summer. Both the festivals and camps are held on university campuses; multiple festivals and camps are offered annually. The GET girls visited the festival held at ASU twice. Learn more at https://www.sallyridescience.com/.

Honeywell sponsors the Fiesta Bowl Aerospace Challenge for middle school students in Arizona. Even though this event is open to both boys and girls and is limited to Arizona, it is a prime example of a way to engage girls in science and engineering (https://fiestabowl.org/events/aerospace-challenge). The event is held in collaboration with the Challenger Space Center in Peoria, Arizona. Challenger Space Centers are located across the United States and offer great science and space programs that girls will find engaging. To locate a center near you and to access resources, check out http://www.challenger.org/.

Various corporations sponsor their own events. Local corporations who sponsor such events in your community would make excellent partners for your program.

Focus on High School Girls

Some programs focus on reaching high school girls only. Those listed below are especially noteworthy.

As part of its Global Diversity and Inclusion Program, Microsoft offers DigiGirlz Day and DigiGirlz High Tech Camp events, held at Microsoft corporate or retail locations from coast to coast (and at a few international locations). The DigiGirlz Day event introduces attendees to careers at Microsoft (taught by employees), career planning assistance, and Microsoft product demonstrations. At Microsoft's DigiGirlz High Tech Camp, girls spend the day interacting with Microsoft employees, attending technology tours and demonstrations, and participating in hands-on workshops. Details about the program and locations of the camps can be found at https://www.microsoft.com/en-us/diversity/programs/digigirlz/default.aspx.

Cisco has long-supported girls through its Girls in ICT (formerly Girls in IT) program. It offers many workshops, and its website provides access to videos and career information that could be useful to girls of any age. Cisco Networking Academies are available in many high schools and provide training for girls to prepare to enter this rewarding career field. The Girls in ICT Portal is located at http://girlsinict.org/.

Career and Technical Education programs at high schools often sponsor clubs that provide opportunities for girls to examine STEM careers in depth.

No matter what level of participation you can provide, the most important consideration is the engagement that girls will be provided by those offerings.

CONCLUSION

In the twenty-first century, a woman's place can be anywhere she'd like it to be: in a lab, in front of a classroom, in a business, in a professional capacity, as an entrepreneur, in space, and, yes, even in the home. The differences that set twenty-first-century women apart from their predecessors are the myriad choices and opportunities available to them through globalization and technological advancements. It's clear, though, that directing female students to career areas of greatest need to our country's economic stability (in STEM) remains a challenge. This task will continue to fall on the shoulders of individual parents and educators as long as the educational system doesn't automatically support it. A September 12, 2011, Brookings Institution forum in Washington, D.C., reported that the main reason the United States is falling short in producing the number of STEM graduates needed to fill twenty-first-century jobs is "a lack of support for teachers" (EducationViews.org).

One key to an increased interest in STEM careers for girls is engaging them at an early age. Girls relate to STEM topics best when those topics are tied to their areas of interest. All STEM areas have fields that interest girls. A broad exposure to multiple STEM areas gives girls a deeper knowledge base upon which to make informed choices.

Girls also need strong mentors to encourage them and role models for them to emulate—mentors having been identified above as a determining factor in choosing a career, and role models an imperative for girls to see that a particular career road had been successfully traveled before.

Meeting the above-mentioned needs can be successfully accomplished by engaging girls through school STEM clubs beginning at the junior high level. Teachers and administrators should be encouraged to develop such groups as a way to address the U.S. workforce's need for women to enter STEM career fields. It takes dedicated leaders who are willing to invest the time to work with girls in the first place and who are committed to carrying the program forward from year to year to make such programs successful.

Educating parents, teachers, and school guidance counselors (Bayer Corporation, 2010) to the fallacies of women's shortcomings in science and math—and to the countless career opportunities that exist in STEM—should be a high priority in helping to encourage more girls to enroll in STEM courses and pursue STEM careers. In addition, gender bias issues should be identified and addressed by textbook publishers, principals (as they observe classroom teachers), business and industry leaders, colleges and university faculties, and the media. If girls are to see themselves in the roles society needs them to play in the development of our STEM workforce, they need to see beyond the limits currently being imposed upon them by society. "By emulating successful educational practices, and eliminating both overt gender discrimination and subtle gender bias, we can make substantial progress in closing the STEM gender gap" (Coger, 2012, p. 5).

Classroom-level and school-level clubs, single- or multiple-day workshops or camps, week long summer camps, guest speakers, videos, STEM stories, and gender equity posters are all successful ways to engage girls and introduce them to the array of STEM careers open to them. "Providing the necessary resources, exposure, and encouragement would help young women understand that their gender shouldn't determine the career path they choose, and that pursuing a STEM career does not make them less feminine" (Coger, 2012, p. 4).

In 2002, when I first began researching how I could interest my female students in pursuing careers in technology, the pressure for inclusion coming from the agencies in Washington, D.C., was a whisper. Since then, the voices have become louder. The call for increasing the STEM workforce and, specifically, the number of female STEM workers, is being hailed loudly by government and industry. Their message now is this: "Women are necessary in STEM fields to sustain, or even exceed, our innovative potential as a nation" (Kabir, 2012, para. 4). STEM courses in schools are broadening their appeal to reach out to females, and corporations and businesses are working with schools and women's organizations to help identify areas of need to prepare women for careers in their organizations. Is this enough? No, but it is a good start.

You can make a huge difference in attracting young girls to the possibilities that exist in STEM fields. Engaging girls at the middle school level through a club at your school provides sustained results and is an investment not just in the lives of your students, but that of our nation as well.

APPENDIX

AAUW Online STEM Resources for Women and Girls
http://www.statelibraryofiowa.org/ld/c-d/continuing-ed/conted-ials/archives-webinar/online-roundtables2013/stem1.
This document provides six pages of links to websites, organizations, and resources.

How to find female role models in STEM [tutorial].
http://www.iwitts.org/tutorial
IWITTS also provides access to a variety of gender equity posters and other resources.

Infographic: What girls say about science, technology, engineering, and math.
http://www.transportationyou.org/wp-content/uploads/InfoGraphic-STEM-report.pdf

WORKS CITED

Albrizio, L. (2012, July 26). News: At a camp in Old Tappan, young girls exposed to engineering. *NorthJersey.com: News.* Retrieved from http://www.northjersey.com/news/163812686_Engineering_camp_exposes_young_girls_to_the_field.html

Bayer Corporation. (2010). *Building a diverse United States STEM workforce: Perspectives on creating successful business partnerships.* Pittsburgh, PA: Bayer Corporation.

Beede, D. J. (2011). *Women in STEM: A gender gap to innovation.* Washington, DC: U.S. Department of Commerce.

Brice, A. (2010, February 19). ExxonMobil "Introduce a Girl to Engineering Day." Retrieved from http://www.icis.com/blogs/recruitment-campaign/2010/02/exxonmobil-introduce-a-girl-to-engineering-day.php

Coger, R. (2012, October 29). Diversity in academe 2012: Why STEM fields still don't draw more women. *Chronicle of Higher Education.* Retrieved from http://chronicle.com/article/Why-STEM-Fields-Still-Dont/135302/?goback=%2Egde_3391879_member_180439874

Committee on Science, Engineering, and Public Policy. (2007). *Beyond bias and barriers: Fulfilling the potential of women in academic science and engineering.* Committee on Maximizing the Potential of Women in Academic Science and Engineering. Committee on Science, Engineering, and Public Policy. National Academy of Sciences, National Academy of Engineering, and Institute of Medicine of the National Academies. Washington, DC: National Academies Press.

Devaney, L. (2007, July 17). Wanted: More IT workers. Retrieved from http://www.eschoolnews.com/news/showStory.cfm?ArticleID=7256

EducationViews.org. (2011, September 14). Panel: STEM education crisis stems from unsupported teachers. *Education News*. Retrieved from http://educationviews.org/panel-stem-education-crisis-stems-from-unsupported-teachers/

The GEMS Club: Girls Excelling in Math and Science. (2010) *Research*. Retrieved from http://www.gemsclub.org/research

Gormly, K. (2010, December 14). Science Center program focuses on girls as math, science leaders. Retrieved from http://triblive.com/x/pittsburghtrib/ae/more/s_713584.html#axzz2C6hWnJ9y

Hill, C. C. (2010). *Why so few?* Washington, DC: AAUW.

Hopkins, K. (2012, May 31). Education: College mentors key to prospective female STEM majors. *U.S. News and World Report*. Retrieved from http://www.usnews.com/education/high-schools/articles/2012/05/31/college-mentors-key-to-prospective-female-stem-majors

Kabir, F. (2012, October 31). Opinion: Broken STEMs. *Cavalier Daily*. Retrieved from http://www.cavalierdaily.com/article/2012/10/broken-stems

Kovalik, S. (2008). *Gender differences and student engagement*. Rexford, NY: International Center for Leadership in Education.

Lips, H. (1999, March). *Women, education, and economic participation: Keynote address presented at the Northern Region Seminar, National Council of Women of New Zealand*. Auckland, New Zealand.

Mahoney, G. (2011, December 6). Interactive workshop and career expo brings STEM, industry to middle school girls. *Free Press Release*. Retrieved from http://www.free-press-release.com/news-interactive-workshop-and-career-expo-brings-stem-industry-to-middle-school-girls-1323208785.html

McNees, P. (2004, January 6). Why Janie can't engineer raising girls to succeed. *The Washington Post*, p. C09.

Mitchell, S., & Hoff, D. L. (2006). (Dis)interest in science: How perceptions about grades may be discouraging girls. *Electronic Journal of Science Education, 11(1)*.

Modi, K., Schoenberg, J., & Salmond, K. (2012). *Generation STEM: What girls say about science, technology, engineering, and math*. New York: Girl Scouts of the USA. Retrieved October 2, 2013, from http://www.girlscouts.org/research/pdf/generation_stem_full_report.pdf

Moret, B. (2012, April 20). Communities: Book review. *Pink brain, blue brain: How small differences grow into troublesome gaps and what we can do about it* by Lise Eliot, Ph.D. *The Washington Times*. Retrieved from http://communities.washingtontimes.com/neighborhood/parenting-first-time-through/2012/apr/20/booK-review-pinK-brain-blue-brain-how-small-differ/

Perales, L. (2012, September 7). Girls learn STEM skills through Pearl Project. *Herndon Patch*. Retrieved September 24, 2013, from http://herndon.patch.com/groups/schools/p/girls-learn-stem-skills-through-pearl-project

Phelps, M. (2010, March 19). *Green technology improves classroom gender equity*. ITEEA Conference: Conference Presentation. Charlotte, NC.

Phillips, K. B. (2002). Science career interests among high school females one year after participation in a summer science program. *Journal of Women and Minorities in Science and Engineering*, 235–246.

Pickert, K. (2009, January 29). *Lilly Ledbetter*. Retrieved from http://www.time.com/time/nation/article/0,8599,1874954,00.html

Rust, T. (2006). Girls exploring technology: A program to interest girls in non-traditional occupations through career and technical education. *Illinois Journal of Technology Education*, 5–8.

Sheehy, K. (2012, June 28). STEM education. *U.S. News and World Report*. Retrieved from http://www.usnews.com/news/blogs/stem-education/2012/06/28/stem-disconnect-leaves-women-minorities-behind

Streitmatter, J. (1999). *For girls only: Making a case for single-sex schooling*. Albany: State University of New York Press.

Sweedyk, E., & de Laet, M. (2005, Summer). Women, games, and women's games. *Phi Kappa Phi Forum*, 25–28.

Weber, K., & Custer, R. (2005). Gender-based preferences towards technology education content, activities, and instructional methods. *Journal of Technology Education, 16* (2), 55–71.

Writers, S. (2012, June 17). *Blog: Ten startling stats about minorities in STEM*. Retrieved from http://www.onlineuniversities.com/blog/2012/06/10-startling-stats-about-minorities-stem/

CHAPTER FOUR

Teaching Mathematics to At-Risk Students

BRENDA GARDUNIA

INTRODUCTION

As a 20-year veteran teacher, I returned to the classroom following a 2-year leave of absence to complete an Albert Einstein Distinguished Educator Fellowship in Washington, D.C., at the National Science Foundation. My time as an Einstein Fellow was inspiring and invigorating, and I was ready to get back to teaching, thinking I could just pick up where I left off 2 years earlier—only with more enthusiasm. This was far from my new reality. I returned to the same school district but was assigned to a different school. I went from an alternative high school of 450 students to a city high school of 1,400 students. Whereas my previous assignment was four class periods with an average classroom size of 15 students, I now had five class periods of 30 students each. Besides being in a new school with new people and new preps, I was a traveling teacher and moved among five different classrooms every day. The math department at my new school had different expectations for students and teachers than where I taught previously and was also a pilot school for the new Common Core State Standards for Mathematics (CCSSM); we were redesigning the curriculum as we went, as well. It was like being a brand new teacher again. I quickly felt overwhelmed by preparing lessons, creating presentations, grading papers, going to meetings, and communicating with parents.

One evening, while I bemoaned the difficulty of meeting all of the expectations of the administration, students, parents, and the math department at my new school, my son gave me some advice: "Mom, stop trying to look and be like everyone else. Just remember to do what you do best, teach students, your way. You know how to get struggling students to learn math." I took his advice to heart and became less worried about conforming and more concerned with using the tricks and tools of teaching that I had learned (and relearned) over the years.

This chapter is about what I know. During the past 20 years, including my 2-year experience as an Einstein Fellow, I learned many things about teaching mathematics to at-risk students—what makes an "at-risk" student, what practices and strategies are successful when teaching at-risk students, and how they fit within the new CCSSM.

BACKGROUND

Who Is an At-Risk Student?

Students who are not experiencing success in school and are potential dropouts are considered at-risk students (Donnelly, 1987). Many factors have been associated with increased dropout rates, including race/ethnicity, gender, economic status, and parental education. Black, Hispanic, and Native American students, and students from low socioeconomic backgrounds are more likely to be deficient in basic math and reading skills (National Research Council, 2011). Independent of sex, race/ethnicity, and socioeconomic background, students in single-parent families and students who have frequently changed schools are also more likely to drop out (National Research Council, 2011). Even though these factors are strongly related to student graduation rates, they do not predetermine whether a student graduates. There are many students who share this background who still succeed in school, and many students free of these disadvantages who do not succeed.

Research shows that nearly all students who drop out of school are disengaged in school. Engagement activities include arriving at school and class on time, being attentive in class, coming to class prepared, and completing assignments. Students with a higher level of engagement initiate questions during class and seek help when they don't understand (Finn & Rock, 1997). Failing mathematics and/or English classes and poor attendance are variables that can help in the early identification of students who may not graduate on time (Balfanz, 2008). While gender, race/ethnicity, and socioeconomic status cannot be controlled, school engagement can be influenced. To help at-risk students become engaged in their learning, the teacher's emphasis should be placed on learning and progress over performance and ability (Alderman, 1990). What happens in the classroom is critical to helping students experience success in school and progress toward on-time graduation. Teachers have a stronger effect on student learning than the influences of poverty and other socioeconomic factors (Darling-Hammond, 2000; Good & Brophy, 1997). Teachers can—and do—make a difference.

An important feature of the classroom is that it is an environment in which all students feel safe: safe to be right, safe to make mistakes, and safe to try new things. Teachers must build respect in the classroom (Payne, 2008). One long-standing tradition in secondary education is to not allow students to retake assessments. But as schools and districts are attempting to implement standards-based learning, many educators are re-thinking learning and testing. Students—especially at-risk students—benefit from being allowed additional opportunities to learn and master content (Marzano & Heflebower, 2011; Wormeli, 2011). At-risk students might need multiple chances. Giving a student more than one opportunity to show mastery is different from not holding a student accountable. In fact, it is teaching accountability for learning. At-risk students already know how to fail. They have done that before. They may not know how to succeed. These students may need a second or third chance before they experience success.

MY BEST PRACTICES AND HOW OTHERS CAN ADAPT THEM

Background Information

My 20 years of teaching were spent at an alternative high school for at-risk students. In my state, an at-risk student is defined as one who meets three or more of the following criteria:

- Has substance abuse behavior
- Is pregnant or a parent
- Is an emancipated youth
- Is a previous dropout
- Has serious personal, emotional, or medical problems
- Is a court or agency referral
- Has repeated at least one grade
- Absenteeism was greater than 10% during the preceding semester
- Overall grade point average (GPA) is less than 1.5 (4.0 scale)
- Failed one or more academic subjects
- Is two or more semester credits per year behind the rate required to graduate (Idaho State Department of Education, 2013)

When my school first opened, the idea of having an alternative high school for at-risk students was a new idea for my district. I was one of six teachers. No one had any specific training for working with at-risk students. Over the next 20 years, our student population grew from 65 students to nearly 500 students, with a staff of over 40. I want to share with you some of the most important lessons I learned.

Create a Safe Environment

To create a safe environment for learning, it is important to put the emphasis on learning and not on just getting the right answer. Asking questions is encouraged, and making sure that everyone participates is important. At-risk students are very skilled at hiding what they do not know. Many of my students are English language learners (ELLs) and are very reluctant to participate. One technique I have found that works, even for high school students, is to write each student's name on a Popsicle stick. During discussions, I draw a stick out of a can I store them in to see whose turn it is to voice his or her opinion or answer a question. I also use these name sticks to identify working groups. When getting the right answer or solving a problem the "right way" is no longer the goal, students are more willing to try. When they see that everyone gets a turn, and that it is okay to not have all the answers all of the time, they are more willing to take risks.

Students often come into a math class with holes in their learning sequence. Things may have happened in their lives that made learning math more difficult or next to impossible for a time. This was never brought to my attention more pointedly than when I was tutoring at the Booth Memorial Home for pregnant and parenting teens. I was attempting to teach a young woman algebra, but she was really struggling with some of the basic concepts of arithmetic. She didn't seem to have any concept of how to add, subtract, multiply, or divide fractions. I was frustrated with her at first, as it appeared to me that she was not even trying. Then I realized that fractions are normally taught to students when they are in the fifth grade. This young girl had missed a good part of that year because she was pregnant and had a baby when she was 12 years old.

You cannot tell what challenges your students have in their lives. Math is a subject that is very sequential, with new concepts building on previously taught concepts. If you are a teacher of at-risk students, it is important to create an environment in your classroom in which they feel safe to admit that they have holes in their mathematics knowledge without feeling that they are stupid. I tell my students that they do not need to fake it or pretend that these holes don't exist. It can be a lot of work trying to hide the fact that you never learned something that you think everyone else already knows.

Another student, John, was 17 when he came to my class. He had earned just 12 credits toward graduation. In Idaho, students need 44 credits to graduate. John had dropped out of school two years earlier and spent most of that time using drugs and hanging out with friends who had also dropped out of school. He now wanted to quit the drugs and come back to school. Since he was so far behind, the counselors wanted him to give up on the idea of getting a high school diploma and to work toward getting his GED (General Education Development Test certification) instead. But John was determined to come back to high school and walk across the stage on graduation night. I knew his math skills were very poor, but I had no idea how poor until he shared with me that he didn't know how to multiply two-digit numbers. I never once did or said anything that made him think that I judged him for not knowing something that every 17-year-old should know. Making him feel embarrassed or ashamed for not knowing math would not change the fact that he had never learned how to multiply multiple-digit numbers, a skill that he needed to master before he could progress further.

We started that very day going back to basic arithmetic skills. He also did not know how to subtract numbers where renaming (borrowing) was required. We worked every day after school. Even though it seemed like an impossible goal for John to catch up and master the algebra requirement for high school graduation, he didn't give up, and as long as he was willing to work, I didn't give up on him, either. Once he felt safe enough to admit to me, his teacher, what he did not know or understand, we were able to make progress and not waste time pretending that he understood what he did not know. John never gave up and continued to work toward his goal of high school graduation. He was nearly 21 years of age, but he stood tall as he walked across the stage in his graduation robe and flat hat with the tassel tickling his nose to receive what many people once told him was impossible: a high school diploma.

Multiple Chances to Show Mastery

Michael was 15 when he first came to my class. He had been "kicked out" of his prior school for behavior issues and had a long-standing pot-smoking habit. He and his dad lived in a pickup camper. He was disengaged in school. He had made a lot of mistakes along the way and was behind in credits, with a GPA near 1.0. We chose not to judge him by his past mistakes and gave him a chance to start again.

At first his progress was rocky, but as time went on he started experiencing some success. Drug use was still a problem, but he was coming to school and learning math. In my class he was not allowed to go on until he had shown mastery of the current topic. Sometimes that meant going back and relearning and then retesting. He moved through Algebra 1, Geometry, Algebra 2, and Pre-Calculus. When I last saw him, he excitedly told me he had graduated from high school and was enrolled in a community college. He had kicked his drug habit and was studying to be a drug counselor. He wanted to let me know how much he appreciated me for never giving up on him and for helping him to believe that he could do better and learn more.

At many schools, students are not allowed to retake tests if they do poorly the first time. When I question the rationale for this policy, the response is that if students are allowed to retake tests, they have an unfair advantage over students who studied and did well the first time, and that we are encouraging bad habits and irresponsibility. Students could delay studying and then simply retake the test later if they did poorly. This may be true for the general or advanced student, but the typical at-risk student has had lots of experience with failure. It does no good for them to fail again and again. They need to know that failure is not an option. They need to know that if they do fail, they can try again as long as they don't give up on themselves (Wormeli, 2011).

Be Yourself

You, the teacher, are one of the most important factors in your students' success. Teachers come in all shapes and sizes. I have seen great teachers who were loud and theatrical, others who were calm and quiet. Some teachers are very organized and procedure-driven, while others are random and more spontaneous. Aspiring artists study and copy the techniques of master artists, experiencing and practicing different styles. But eventually they must take all that they have learned and find a style that is comfortable and unique to them. Teaching is an art. It is important that you study and learn from master teachers. Read and be informed of what the research says. Eventually you will find the teaching style that works for your personality and your particular set of students.

Let your students see you as a real person—a person who has challenges and celebrations, just as they do. My undergraduate degree is in secondary mathematics education, and I have a master's degree in education. I have done some exciting things in my teaching career. As a Fulbright Memorial Fund participant, I spent three weeks in Japan as a guest of the Japanese government, and I was selected to be an Einstein Fellow. But this is not the beginning of my story. My story started many years earlier when I was born the third of four children to parents who dropped out of high school at 16 and 17 years of age. By the time I was eight years old, my family had moved 11 times, following the jobs available to a father with only an eighth-grade education. I did well in school, but—in keeping with what seemed to be a family tradition—I married right after high school and had my first baby a year later. By the time I was 34, I was a single parent with five young children, no child support, no job skills, and on welfare. The future seemed bleak for my family and me. What changed my story? Education.

I soon realized that I would never be able to support my family with the jobs I qualified for at that time. I decided to go back to school. I moved my young family to the university campus and spent the next three years doing everything I could to reach my goal of getting a degree in secondary mathematics education and to become a teacher. Having an education gave me options that I could never have had otherwise, and it allowed me to give my children opportunities they would not have known. It was education that got me away from a minimum wage job, welfare, and poverty. My students need to know where my passion for learning and education comes from.

Everyone has a story. Listen to your students' stories and find an appropriate time to share your own story. You are a very important role model. As a teacher, you may spend more hours with a student each day than any other adult. They do not need for you to be perfect, but they do expect you to be real. Share your story.

Common Core State Standards

In 2009, a state-led effort in Idaho coordinated by the National Governors Association Center for Best Practices and the Council of Chief State School Officers and Governors initiated the development of Common Core State Standards in Mathematics (CCSSM). Besides benchmarks and standards for mathematical learning for each grade level, the CCSSM includes Eight Standards for Mathematical Practice that describe the expertise that mathematics educators should seek to develop in their students (Idaho State Department of Education, 2013). These standards of mathematical practice are:

1. Make sense of problems and persevere in solving them.
2. Reason abstractly and quantitatively.
3. Construct viable arguments and critique the reasoning of others.
4. Model with mathematics.
5. Use appropriate tools strategically.
6. Attend to precision.
7. Look for and make use of structure.
8. Look for and express regularity in repeated reasoning.

As teachers design lessons that keep these eight standards of mathematical practice in mind, students will be required to go beyond learning algorithms to solving problems, and they will begin to gain a deeper understanding of the processes of mathematics and the use of mathematics to explain the world around them. For at-risk students, these mathematical practices are especially important. It cannot be assumed that students are learning and applying these practices as they learn and do the math. They must be explicitly taught. My students often tell me that they never understood the "why" before. Math never made sense. It just seemed to be a bunch of rules and procedures they had to memorize. They want it to make sense. The Eight Standards for Mathematical Practice can help students make sense of the mathematics they are learning. For teachers, that may mean getting away from the textbook and looking for multiple ways to help students learn a concept. It will challenge you as a teacher, but the rewards will be worth the effort.

My Favorite Lesson

All teachers have a lesson they can point to that is a favorite for various reasons: it may be particularly engaging, it may bring student learning to new levels, or it may allow students to experience learning in a whole new way. Here is an example of one of my favorite "away-from-the-textbook" lessons. It's hands-on, visual, and does not require access to technology. It does require students to use mathematics to help make sense of their world by modeling with mathematics and using appropriate tools. The lesson incorporates several of the Eight Standards for Mathematical Practice contained in CCSSM. The lesson is appropriate for an Algebra 1 course, though I have used a variation of this lesson for Algebra 2 students with equal success.

Lesson Title: Writing Linear Functions

Learning Objective: Through this activity, students will

- Generate, organize, and record data
- Graph the data and find the line of best fit
- Verbalize the meaning of the intercepts and slope (as rate of change)
- Relate the data to concept of function
- Write functions that describe linear relationships between two variables.

Teacher Background Information

This lesson is divided into three sessions of approximately 50 minutes each. Functions will be represented in four ways: a table, a graph, an equation, and using function notation. Part 1 of this activity has students collect data for a linear function with a negative slope and represent it using a table and a graph. They will then explain what is happening in words, which will lead them to write an equation using function notation. Part 2 will look at a function with a positive slope in a similar manner. During Part 3, students will, with teacher guidance, practice writing functions in function notation for various linear relationships. The lesson ends with an assessment activity in which students apply what they have learned and mathematically model and then generalize a real-world situation.

Part 1: Modeling a Function with a Negative Slope
Materials needed: pieces of cord about 50 cm long, cloth measuring tapes, rulers, masking tape, graph paper

Procedure

Students can work in teams of two. Each team needs to secure the cloth tape to a flat surface with a couple of pieces of masking tape, so that they can measure from 0 to 50+ cm. Teams will need a piece of cord that is approximately 50 cm in length and graph paper. Students measure the piece of cord, tie a knot in the cord, and then measure the length of the cord again. They continue tying knots and measuring the length of the cord after each knot until they have eight or more knots. This data is recorded using a T-chart with the number of knots in the left column and the length of the cord after each knot is tied in the right column. Remind students to be sure to measure and record the length of cord with zero knots. After all students have recorded their data, collect the pieces of cord. I ask them to untie all of the knots before returning the cords. I also collect the measuring tapes. This makes for fewer distractions.

Using the data they collected, students next create a *scatter plot* on the graph paper. This is a good time to talk about *dependent and independent variables*. The independent variable (number of knots) is on the horizontal axis. The dependent variable (length of cord) is on the vertical axis. We also discuss the scale or interval length needed on each axis so that it is appropriate for the data collected. Remind students to add a title and label axis. We then talk about trends and *line of best fit*. For this introductory lesson, students will use an "eyeball fit." Using a ruler, they find the line that best fits the points. Some points may be slightly above or below this line. We then talk about the *x-intercept* and *y-intercept*. The y-intercept is where their line intercepts the vertical axis. I ask, "What was this point?" (The point represents the initial length of string before any knots.) The x-intercept is where their line intercepts the horizontal axis. I ask, "What is this point?" (That point represents when the length of cord is zero.) We then talk about the constraints in the real world, because they can never tie enough knots so that the cord has zero length. We also talk about *rate of change* or slope of the line. The students are now ready to start writing a function to describe the length of the cord with any given number of knots. First they write it out in words.

- Length of cord = initial length of cord − number of knots x length of cord used for one knot

Then I introduce *function notation*.

- Let l = length of cord. Let k = number of knots

We are looking for a function of the length of cord as we tie knots, so the length depends on the number of knots.

- Length of cord = initial length of cord − rate of change x number of knots
- l(k) = 50 − 2.5 • k

Part 1 ends with a student review of the terms and functions covered during the class.

Part 2: Modeling a Function with a Positive Slope
Materials needed: various jars and lids, rulers, containers to hold water, paper towels, graph paper, worksheet for practice writing functions, assessment

Procedure

Again, students work in teams. Each team has an empty jar, a jar lid, and a container holding about four or five cups of water. They will be recording data on a sheet of graph paper in a T-chart in the upper left corner. First, they record the height of water in the jar after zero lids of water have been added (this will be 0). Then students add one lid of water to the jar and measure the height of the water and record. They continue to add a lid of water and measure the height, recording the number of lids of water added and the height of the water in the jar after each addition of water on the T-chart until they have at least eight data points. Clean up all water and put away jars and lids before proceeding to the next step.

Students create a scatter plot of the data, independent variable (number of lids) on the horizontal axis, and dependent variable (height of water) on the vertical axis, using rulers to draw line of best fit. This is a good time to talk about positive slope and average rate of change.

Have students explain the graph using a word sentence.

- Height of water = the initial height of water + height of water after one lid x number of lids

Then write using function notation

- h = height of water
- l = number of lids
- h(l) = 0 + 3l

Discuss what each part of the function sentence tells. What is the dependent variable? What is the independent variable? What does the 0 represent? What does the 3 represent? How high is the water after you add five lids to the jar?

Part 2 ends with students discussing the process and making predictions for the height of the water with an increase in lids.

Part 3: Practice Writing Functions
Materials Needed: *Handout: Practice Writing Functions* (Appendix A), and *Function Lab Assessment* (Appendix B)

Procedure

Review the concepts of functions, independent and dependent variables, and writing function rules. Guide students, as needed, to identify independent and dependent variables, and then write a function rule for each example on *Handout: Practice Writing Functions*.

After completing the *Handout*, students will be ready to complete the *Function Lab Assessment*.

During previous work in this lab, students worked in small groups, but I ask students to complete the *Function Lab Assessment* individually. At first, many of my students find it frustrating that I don't tell them exactly how to answer the assessment questions. I explain to them that there are multiple ways of finding the answers. I want to know what they think. I do tell them to consider what we have been doing in the Function Lab as they complete the tasks in the assessment. Students turn in everything they have completed during the three days as a portfolio of their work.

How Do Students Benefit from This Type of Lesson?

The lesson in my current Algebra 1 textbook on functions and function notation introduces vocabulary and mechanics of evaluating functions. Students would memorize the process but not understand the concept of functions. Independent and dependent variables meant nothing to them. But after this lesson, it all starts to make sense. They can see that the length of cord depends on the number of knots tied. It is easy to see the relationship between slope of the line and the rate of change. Intercepts are more than just points on the graph. Students become excited when they see how easy it is to write their own functions for given situations. As students complete the assessment, it is fun for me to see the "aha" moment when they realize that they can write a function and use it to make a prediction.

I included this lesson here as an example of creating lessons that enhance and enrich textbook lessons. Developing and writing this type of lesson takes time and thought on the part of the teacher, but it is time well spent. I also wanted to show that creating effective lessons does not always require expensive technology or supplies. Once students start to understand that math is more than just numbers and operations, they start to gain the deeper understanding asked for in the Eight Standards for Mathematical Practice in the CCSSM.

Professional Development

In order for students to achieve the higher level of thinking and problem-solving abilities outlined in the CCSSM, teachers need strong content knowledge of mathematics in their pre-service preparation and rich professional development experiences throughout their teaching careers. Professional development might be offered at the building, district, or state level, but often teachers must seek out opportunities on their own.

While I was an Einstein Fellow, I worked with the Research Experiences for Teachers (RET) Program. This is a program in which K–12 teachers come to a university during the summer to work on real research projects under the mentorship of research professors. During my time at NSF, I had the opportunity to talk with many teachers and researchers involved in this program. Their research experiences ranged from nanotechnology to creating apps for smart phones. All of them shared stories of how this professional development experience changed them as teachers. Their desire to be lifelong learners was renewed. They were able to take their newfound excitement back to the classroom. There are many other opportunities for teachers and scientists to work together that are described in other chapters of this book. Search them and find something that works for you and your students. Something powerful happens when teachers and students work side by side with scientists, engineers, and mathematicians for extended periods of time. Students come to understand why they learn mathematics. They gain insight into the usefulness and beauty of mathematics. Mathematics becomes relevant to their world. Students begin to see themselves as mathematicians, scientists, and engineers. I have discovered some of the best and most rewarding professional development activities by doing an Internet search and by visiting the exhibitors' hall during math conferences.

CONCLUSION

The art of teaching mathematics is changing. Adoption of the Common Core State Standards in Mathematics is challenging educators to look at how they teach mathematics and to search for better ways to raise the standards of mathematical education for all students. At-risk students are an especially vulnerable population but also stand to gain the most. Teaching strategies that are effective for at-risk students are often described by teachers as just "good teaching" and can increase learning for all students (Ladson-Billings, 1995). Teaching strategies and professional development that have proven successful in my classroom, especially as I work with at-risk students, are:

- Create opportunities for students to try again if they are not successful the first time.

- Make your classroom a safe environment for students to make mistakes and to take risks.
- Find ways to include all students, even the reluctant learners.
- Develop the teaching style that works for you.
- Share your story with your students and listen to their stories.
- Become familiar with the Common Core State Standards' Eight Standards for Mathematical Practice and start to find and develop lessons "outside of the textbook."
- Improved teacher preparation is a must, with increased emphasis on strong content knowledge.
- In-service teachers need rich, authentic professional development opportunities that can inspire and excite them to be lifelong learners. Excited and inspired teachers return to the classroom ready to prepare and inspire their students.

Never take for granted your influence as a teacher on the lives of your students. Hone your craft carefully. Learn from the best and then make it your own. "It is the supreme art of the teacher to awaken joy in creative expression and knowledge" (Einstein, n.d.). As you teach at-risk students, you are helping them learn more than facts and skills. "A good teacher can inspire hope, ignite the imagination, and instill a love of learning" (Henry, n.d.). A good teacher can change lives.

APPENDIX A

Handout: Practice Writing Functions   Name_____

Read each scenario. Define your variables. Indicate independent variable and dependent variable. Write a function using function notation that models the situation.

1. You are driving a car at an average speed of 65 miles per hour. Write a function rule for how far the car has traveled after h hours.
2. You are going on a trip to your grandmother's house, which is 350 miles from your house. You will drive at an average of 60 miles per hour. Write a function rule for how far you will still need to travel to reach grandmother's house.
3. You are going on a trip. You have already driven 105 miles when you stop to eat. You will average 65 miles per hour for the remainder of the trip. Write a function rule for how many total miles you have driven after t more hours.

4. Due to drought, the water in the Payette River is dropping at an alarming rate. Soon there won't be enough water left to safely raft the river. The water level is 7,500 cfs (cubic feet per second), and it is dropping at a rate of 800 cfs each day. Write a function rule to determine how much water is left after x days.
5. It has been a heavy snow year. There is already 2 feet of snow on the ground, and it has snowed an average of 3 inches every day for several days. (Hint: think about what unit of measure you should use.) Write a function rule to describe the height of snow on the ground after d days.

APPENDIX B

Function Lab Assessment   Name_____

During the spring, the Snake River swelled to 18 feet over flood stage. After cresting, the river fell at an average rate of 5 inches per hour.

1. Use a graph, table, or drawing to show the height of the river above flood stage after each drop during the first 10 hours after cresting.
2. How long will it take for the river to drop below flood stage? Show how you determined your answer.
3. Develop a method to find the height of the river above flood stage at any given hour after it crests. Use your method to find the height after 17 hours. Show all work.

WORKS CITED

Alderman, M. K. (1990). Motivation for at-risk students. *Educational Leadership, 48*(1), 27–30.
Balfanz, R. (2008). *Early warning and intervention systems: Promise and challenges for policy and practice.* Washington, DC: National Academies Press.
Darling-Hammond, L. (2000). Teacher quality and student achievement: A review of state policy evidence. *Educational Policy Analysis Archives, 8*(1).
Donnelly, M. (1987). *At-risk students.* ERIC Digest Series Number 21. Retrieved January 21, 2013, from http://www.ericdigests.org
Einstein, A. (n.d.). Untitled. Retrieved May 13, 2013, from http://www.brainyquote.com/quotes/quotes/alberteins122147.html
Finn, J. D., & Rock, D. A. (1997, April). Academic success among students at risk for school behavior. *Journal of Applied Psychology, 32*(2), 221–234.
Good, T. L., & Brophy, J. E. (1997). *Looking in classrooms* (7th ed.). New York: Longman.
Henry, B. (n.d.). Untitled. Retrieved May 13, 2013, from http://www.brainyquote.com/quotes/quotes/b/bradhenry167806.htm

Idaho State Department of Education. (2013). *Common core state standards for mathematics.* Retrieved from http://www.sde.idaho.gov/site/common/math/docs/CCSSI_Math_Standards.pdf

Ladson-Billings, G. (1995). But that's just good teaching! *Theory into Practice,* 34(3), 159–165.

Marzano, R. J., & Heflebower, T. (2011). Grades that show what students know. *Educational Leadership,* 69(3), 34–39.

National Research Council. (2011). *High school dropout, graduation, and completion rates: Better data, better measures, better decisions.* Committee for Improved Measurement of High School Dropout and Completion Rates: Expert Guidance on Next Steps for Research and Policy Workshop. Washington, DC: National Academies Press.

Payne, R. (2008). Nine powerful practices. *Educational Leadership,* 65(7), 48–42.

Wormeli, R. (2011). Redos and retakes done right. *Educational Leadership,* 69(3), 22–26.

CHAPTER FIVE

The Student-Centered Sheltered Instructional Approach AND Growth (SSIAG) Model

EDUARDO GUEVARA

INTRODUCTION

Demographic trends in the United States present several challenges to society, especially when it comes to education (U.S. Census Bureau, 2011a; McKeon, 1994). The number of Hispanic students and the socioeconomic level of their families, as well as their placement regarding poverty level, increase the challenges to the nation's educational system. The challenge is multiplied when schools must enroll students requiring additional services such as bilingual support staff, special education, and so on, while at the same time facing a national trend to slash public education budgets (Cardenas & Kerby, 2012, p. 4; La Fe Policy Research & Education Center, 2012, p. 2). Texas, the second-largest state in the United States and second only to California in number of Hispanic residents (Cardenas & Kerby, 2012, p. 1; U.S. Census Bureau, 2011a, Table No. 5.1), is an example of insufficient funding for education and limited language proficiency in English coming together to create many challenges. The trend is reflected in the number of Hispanics enrolled in grades K–12 and their corresponding low educational attainment.

When Hispanics with limited English proficiency, referred to as English language learners (ELLs), do not understand academic instruction in English, they feel frustrated and disenfranchised and tend to disengage from the educational process. Frustration leads to disruptive behaviors, low self-esteem, and low academic performance (Zemelman, Daniels, & Hyde, 1998) and high dropout rates. The resulting low academic performance and low educational achievement (LEA)

Table 5.1. Hispanic Educational Attainment in the United States.

Educational Attainment	US Population (× 1000)		Hispanics (× 1000)	
Males and Females	201,543	100%	26,672	100%
Less than 9th grade	10,277	5.10%	5,510	20.70%
9th to 12th grade (no diploma)	14,763	7.30%	4,403	15.00%
High school graduate	61,911	30.70%	7,957	29.80%
Some college or associate's degree	53,249	26.40%	5,442	20.40%
Bachelor's degree	39,286	19.50%	2,669	10.00%

(Source: U.S. Census Bureau, 2011a.)

increase their risk for dropping out before graduation and the unlikelihood of their pursuing future career training (AAMC, 2004). LEA increases the dropout rate, limiting career options and compromising level of qualification within the labor force. LEA has a negative impact upon productivity at large and upon the tax base, augmented by a pervasive perception on the part of parents and students about schools' role in "warehousing."

Table 5.2. Hispanic Poverty Level in the United States.

Poverty Level	US Population (× 1000)		Hispanic Population (× 1000)	
Both genders	305,688	100%	49,869	100%
Below Poverty level	46,180	15%	13,243	26%
At or above Poverty level	259,508	85%	36,626	73%
Under 18 years	74,494	100%	17,435	100%
Below Poverty level	16,401	22%	6,110	35%
At or above Poverty level	58,093	78%	11,325	65%
18 to 64 years	192,015	100%	29,576	100%
Below Poverty Level	26,258	14%	6,619	22%
At or above Poverty level	165,757	86%	22,957	78%
65 years and over	33,179	100%	2,857	100%
Below Poverty Level	3,520	9%	514	18%
At or above Poverty level	33,658	91%	2,343	82%

(Source: U.S. Census Bureau, 2011b.)

Moreover, the high number of low socioeconomic-status Hispanic families with limited English proficiency moving into Texas school districts during any academic year has a major impact on the educational system (Table 5.2). Newcomer ELLs completing their English-language survival skills are expected to perform at the same level as their English-speaking classmates in non- English as a Second Language (ESL) classes. Efforts to increase the educational level of Hispanics are desperately needed, and intervention is urgently required to provide teachers with tools that facilitate the transition from Spanish to English for the ELL student.

Two of the main factors associated with low educational attainment in Hispanic ELLs are the relatively small number of educators qualified as bilingual teachers and the lack of parental involvement in the educational process. Cultural and language barriers increase the difficulty of communication between parents and schools and require additional training and multicultural awareness skills in teachers (Zemelman et al., 1998; Guevara, 2006, 2007, 2008a, 2008b). School districts provide a wide variety of professional development (PD) opportunities to their faculty and staff in order to meet the educational needs of their enrollment. Teachers receive different types of training based on different criteria and circumstances. School districts focus their professional development activities on best practices to provide better teaching tools and teaching support to teachers. To date, a wide range of strategies and policies have been implemented with mixed results and outcomes (Hull, 2007). Hull (2007) points out two "perceived needs" to increase educational attainment in ELLs and at-risk students: the need to increase teacher preparation and the need to diversify instructional delivery and classroom management. In order to address these perceived needs, over the last 7 years I designed, fine-tuned, and implemented the Student-Centered Sheltered Instructional Approach and Growth (SSIAG) Model.

SSIAG is both an instructional strategy and a growth model. It has a definitive structure based on a sequence of strategies and techniques that are continuously implemented throughout the semester in a sequential series of phases:

- Phase 1: Setting up the stage;
- Phase 2: Hands-on, inquiry-based, exploratory activities;
- Phase 3: Project-based learning;
- Phase 4: Instructional delivery; and
- Phase 5: Growth model: impact assessment of the instructional approach.

In the spring of 2007, the Office of Community Outreach and the South East STEM (Science, Technology, Engineering, Math) Center at the University of Texas Medical Branch (UTMB) in Galveston, Texas, provided funding to train a pilot group of 19 schoolteachers from five districts in Texas in the SSIAG Model. The 80-contact-hour training program began with a 5-day intensive Summer Institute, followed by monthly daylong, follow-up sessions throughout the fall of 2007. Over this period trainees began to adopt and implement their newly acquired skills in their classroom.

At the end of the 2007–2008 academic year, the pilot teachers implementing the SSIAG Model validated the aforementioned assessment criteria and reported major improvements in their practice as educators, improved student academic performance, and the development of a strong parental support structure.

After 2008, the implementation of the model continued in the five districts, and in January 2010, the Texas Education Agency and the Texas Regional Collaboratives awarded these educators the Excellence in Science Teaching Award for their instructional performance.

BACKGROUND: COMPONENTS OF THE SSIAG MODEL

As a student-centered instructional approach diverging from the traditional teacher-centered instructional approach, the SSIAG Model uses problem-based learning (PBL) strategies, the first component of the Model, to engage the learner in critical thinking and team cooperation without time constraints and rigid boundaries. The teacher becomes a facilitator in the learning process. Also, the PBL strategy allows for flexible assessment of learning beyond multiple choice testing.

Sungur and Tekkaya (2006) compared the efficacy of PBL with a teacher-centered, textbook-oriented traditional instructional approach on various aspects of students' self-regulated learning. The authors found that when compared with those students who received the traditional instructional approach (control group), "PBL students had higher levels of: intrinsic goal orientation, task value, use of elaboration learning strategies, critical thinking, metacognitive self-regulation, effort regulation, and peer learning" (p. 307).

Ferreira and Trudel (2012) examined the impact of PBL on student attitudes toward science, their problem-solving skills, and their perceptions of the learning environment, and reported a, "significant increase in student attitudes toward science, problem-solving skills, and positive views of the learning environment." Also, the authors found that PBL "facilitated the development of a sense of community in the classroom" (p. 7). This finding is in agreement with what I have found to be true with the implementation of the SSIAG Model in my classroom (Guevara, 2008b).

The second component of the SSIAG Model is the open-ended question used to check for understanding and to facilitate reasoning and reflection on the part of the learner. During assessments, I ask the students to draw what they perceive as the question so as to clarify their understanding before answering. With several possible answers, open-ended questions allow students to participate and explore the topic more deeply in that the answer can be an assumption, an opinion, a thought, an idea, or a series of research-based facts. It is not the answer itself, but the process by which the answer is determined that is important when responding to an open-ended question.

Aguirre (1996) examined the effect of two assessment modes on ELL Spanish-dominant students' performance on in-depth concept development and scientific vocabulary use. The author asked students to answer the same open-ended question on electricity and magnetism either by drawing or by writing a paragraph. The study found that the open-ended drawing mode worked better for below-average students, who expressed a preference for the drawing mode of assessment.

Parental involvement is the third component of the SSIAG Model. Rather than just involving parents when there is a problem with a student's grades or behavior, SSIAG focuses on involving parents throughout the learning experience. The model entices parents to ask students about "what they are learning in the classroom" on a regular basis, preferably at least once a week. During the first week of each semester, parents receive an information packet with documents to be signed and returned to the classroom teacher by the parent who enrolled the student in the school. During an open house organized by the school, I provide parents with additional information and make them aware of the Science Night event, in which self-selected students demonstrate what they have learned so far. The Science Night event includes a potluck-style dinner as a community-building activity.

Parent-teacher-student conferences are scheduled in extreme situations, that is, only after previous attempts to increase academic performance or classroom discipline have had limited outcomes, or when immediate action and change must be accomplished in the short term.

The literature provides numerous examples of the beneficial impact of parental involvement in students' academic performance. Some studies are based on self-reported data either by teachers or by parents. The age of the learner and whether or not parents belong to the PTA are important factors to consider. Most empirical data indicate a positive association between parental involvement and students' performance (Nyarko & Vorgelegt, 2007; Topor, Keane, Shelton, & Calkins, 2010). Additional research (Schimpl-Neimans, 2000; Schmitt & Kleine, 2010) consistently reports associations between a wide range of sociodemographic factors (e.g., personal priorities, self-interest, child's and parents' level of education, family economics, structure, and family size), and the nature, extent, and educational outcomes of parental involvement.

Topor and colleagues (2010) studied two potential mechanisms of association between parent involvement and academic performance in elementary school children. The authors compared the child's perception of cognitive competence and the quality of the student-teacher relationship and found a statistically significant association between parent involvement and a child's academic performance— over and above the impact of the child's intelligence.

The ongoing Youth Save Ghana Experiment, a four-country study that targets young people ages 12 to 18 living in predominantly low-income households, looked at the extent of parental involvement in children's schooling and whether or not sociodemographic factors are associated with parental involvement (Chowa, Ansong, & Osei-Akoto, 2012). Preliminary results report that only one measure of parental involvement—"talking to children about what they learn in school"— is significantly and positively associated with academic performance (p. 6). The SSIAG Model entices parents to ask the question daily: "What did you learn today?" In response, students are expected to engage in conversation with their parents and to explain the learning that is taking place in the SSIAG classroom.

Finally, in its fourth component, the SSIAG Model uses a multidimensional growth model approach to measure student success. O'Malley and fellow researchers (2011) classify growth models according to three general types: (a) Growth to Proficiency (yearly changes in the number of students meeting proficiency), (b) Value-Added/Transition (score changes made by students over two or more occasions), and (c) Projections (measures that predict a student's future scores).

As a career-long educator I have found that all students learn, even if the learner does not reach what is considered "proficiency level." Growth models, rather than a single snapshot "high-stakes" test, provide a way to quantify growth in individuals. Instead taking a "snapshot" once a year (Adequate Yearly Progress, or AYP), growth models provide several points of reference, with specific information on student progress beyond the concept of "proficiency level" currently in use in most states. Instead of focusing on the "school performance" AYP criteria, "growth models" focus on the educational development of individual students. In *Growth Models and Accountability: A Recipe for Remaking ESEA* (Carey & Manwaring, 2011), a report sponsored by the Stuart Foundation, Education Sector's Policy Director Kevin Carey and Robert Manwaring, a fiscal and policy consultant, argue that Congress should incorporate some measurement of student growth into any accountability system: "Educational accountability isn't just a matter of identifying which schools have the most failing students …" (p. 2).

The SSIAG is grounded in a value-added growth model measuring annual student growth in four dimensions: (1) class average, (2) benchmark test scores, (3) timely submission of assigned work, and (4) same-student discipline referrals.

BEST PRACTICE: THE SSIAG

The framework of the Student-Centered Sheltered Instructional Approach and Growth Model is defined by the educational principles and ethical framework of progressive educators such as John Dewey and Albert Einstein, who were committed to nurturing creative talent, proactive thinking, and lifelong learning. The framework integrates the concepts of project-based learning (PBL), "learning by doing," and consistent structure with parental involvement and teacher education. Conventional educational tools (e.g., 5-E Model, note-taking, and handouts), as well as computer technology (e.g., PowerPoint presentations, videos, and Internet access), are widely included. PBL in the SSIAG Model is structured around open-ended driving questions, tasks, and products that focus on integration, synthesis, and critical evaluation of information arising from real-world situations. We strive toward excellence in the attainment of the following objectives:

- Development of superior thinking processes such as critical thinking and objective reasoning;
- Mastery of analytical skills and balanced reasoning;
- Acquisition of thorough and effective work ethics and judgment;
- Development of strong consciousness of the historic dimension of human ideas and activities;
- Recognition of the capacity of the individual to assume responsibility for the consequences of his or her own decisions; and
- Development of an understanding that the transformation process involves the political, institutional, and individual willingness to accept change as a constant in our lives and that one should address challenges with an open mind, trusting our ability to conquer the impossible.

In order to achieve the objectives outlined above, the following five SSIAG Model Operational Principles were developed:

1. Create an environment with Consistent Structure that focuses on student needs, learning styles, classroom management, product, and outcomes, and at the same time reinforces parental involvement and parent education.
2. Include Active Learning supported by learning by doing coupled with Self-Reflection to engage the students in a series of activities as follows: (a) the monthly project whose outcome is a 3-D Working Model; (b) weekly Inquiry-Based, Team-Oriented Labs; (c) instructional delivery is driven by the "open-ended question" concept; and (d) biweekly Computer Labs, where individual search leads to a Class Presentation.
3. Implement Project-Based Learning (PBL) strategies as the core of the SSIAG Model because they allow the students to: (a) engage in learning knowledge and skills to improve their critical thinking abilities; (b) focus on inquiry and autonomous learning; (c) learn within the context of culture, community, and past experience; (d) use what they know to explore, negotiate, interpret, and create to *construct solutions*; and (e) learn by means of projects the same essential information teachers might teach through traditional lecture and discussion.

4. Project-Based Learning is crucial in transforming old teaching techniques because teachers: (a) engage in considerably fewer "busy work" activities in the classroom; (b) do not cover a "laundry list" of topics; and (c) identify topics reflecting core curriculum ideas and concepts and incorporate those topics into projects.
5. PBL uses open-ended driving questions on a daily basis in attempting to achieve the following: (a) focus students on analysis, integration, synthesis, and application of gained knowledge to real-world situations; (b) challenge and engage students' interest, leading to investigating and learning criteria, confronting difficult issues, and trying out unfamiliar behaviors; and (c) provide consistency with curricular standards and frameworks. This leads students to master skills, knowledge, and processes that define a course of study.

Five Strategies—The Core of the SSIAG Model—That Can Be Implemented By Teachers

Innovative teaching of the prescribed curriculum: Instructional delivery should be focused on open-ended questions to engage students' interest, leading them to investigate and confront difficult issues. Open-ended questions are coupled with teamwork activities, leading to daily formative assessment represented by products such as exit tickets and self-reflection entries in the student notebook to answer the question "What did I learn today?"

Project-based learning: During common lesson planning, the teacher identifies topics reflecting the most important ideas and concepts in the curriculum and incorporates these topics into research project assignments. By working on 3-D projects (e.g., building working prototypes) and written projects (e.g., short "How Things Work" assignments, biographies, and article critiques), students associate and relate their classroom learning with out-of-class learning. The students can learn the same essential information teachers might teach through lectures, labs, and classroom discussions, with less "busy work" activities (e.g. fill-in worksheets) focused on products and outcomes, which are presented to the class using Power Point, theater sketches, or dramatizations.

"Learning by doing": Tasks and products (e.g., lab reports, class presentations, and research projects) lead to "learning by doing" wherein the students apply and build upon acquired knowledge and skills. In "learning by doing," the main instructional strategies include, but are not limited to, procedural scaffolding, hands-on cooperative learning, and development and delivery of multimedia presentations and acquisition of note-taking skills.

Classroom management: At the beginning of the semester, students receive the course syllabus, including contracts, routines, expectations, procedures, and rubrics to be signed by parents. Class routines concerning sign-in sheets, lab team members, and individual student folders are established. This process is based on the pedagogical principle of Consistent Structure, in which students and parents become increasingly more responsible for their choices.

Parental involvement: This process is promoted with periodic parent-teacher contacts coupled with events such as Family Science Night, potluck dinners, and field trips. Parents are encouraged to actively participate either as judges at Science Night "Show & Tell," chaperones on field trips, or sponsors of potluck dinners with home-prepared ethnic dishes.

Feedback from parents about how the activity went, and their ideas on future activities, is acquired through voluntary anonymous surveys that are sent home to be completed by parents. The surveys are returned to the teacher's suggestion box within one week following the activity.

Photo 5.1. Family Science Night. Students share their knowledge with family members. (Photo by Eduardo Guevara, 2008.)

Parents are also contacted beyond disciplinary or low-grades situations through printed materials to be signed and returned, email (when available), and phone calls. And Family Science Night is an opportunity for students to demonstrate their newly acquired skills and working 3-D prototypes to parents and peers.

Field trips to knowledge-based locations such as the National Aeronautics and Space Administration's (NASA) Johnson Space Center in Webster, Texas, and Moody Gardens and the Oil Rig Museum, both in Galveston, Texas, are conducted periodically on weekends to facilitate the participation of parents as chaperones and to reinforce topics and instruction previously covered in the classroom.

Photo 5.2. Student Field Trip to Moody Gardens in Galveston, TX.
(Photo by Eduardo Guevara, 2008.)

STRONG EVIDENCE SUPPORTING THE SSIAG MODEL

The impact of the SSIAG Model in five Texas districts, with students enrolled in 4th to 12th grade and in subjects ranging from English as a second language/language arts to core science courses was reported by the first cohort of teachers trained in 2007 to implement the model (Guevara, 2008b). During the first academic year of implementation, these educators reported statistically significant improvement in the four criteria used as indicators of the impact of the Model, as described in the following tables (Table 5.3, Table 5.4, Table 5.5, Table 5.6).

Class averages and Bench Mark test scores were improved in most classes with two exceptions: (a) two alternative schools with students on academic probation because of repetitive non-compliance with Code of Conduct, and (b) a class of fifth graders in a school seriously damaged by Hurricane Ike, where the students were relocated to schools within Galveston Independent School District.

Table 5.3. Indicator One: Class Averages in Five Texas School Districts. Academic Year 2007–2008.

District	Course	Level	Fall 07	Spring 08	Increase (%)
Galveston	ESL-LA	4th (N = 18)	80	87	9
	ESL-LA	5th (N = 18)	80	90	11
Houston	Science	5th (N = 26)	71	78	9
Houston	Science	6th (N = 34)	75	84	11
Galveston	Science	8th (N = 106)	79	85	7
Alvin	IPC	10th (N = 55)	78	91	14
	Physics	10th 12th (N = 41)	73	88	17
Hitchcock	Spanish AP	11th (N = 94)	78	93	16
Houston	IPC	9th 10th (N = 125)	65	77	16
	IPC	9th (N = 110)	66	78	15
Waller	Biology	9th 10th (N = 140)	62	75	17
	Biology	11th 12th (N = 55)	66	75	12
	Chemistry	11th 12th (N = 60)	70	86	18
	Physics	11th 12th (N = 24)	73	78	6

(Source: Guevara, 2008b.)

Table 5.4. Indicator Two: Bench Mark Test Scores in Five Texas School Districts. Academic Year 2007–2008.

District	Course	Level	Fall 07	Spring 08	Increase (%)
Galveston	ESL-LA	4th (N = 18)	72	80	10
	ESL-LA	5th (N = 18)	75	79	5
Houston	Science	5th (N = 26)	75	79	5
Houston	Science	6th (N = 34)	68	76	11
Galveston	Science	8th (N = 106)	71	77	8
Alvin	IPC	10th (N = 55)	73	80	9
	Physics	10th 12th (N = 41)	72	79	9
Hitchcock	Spanish AP	11th (N = 94)	76	92	17
Houston	IPC	9th 10th (N = 125)	72	80	10
	IPC	9th (N = 110)	70	80	12.5
Waller	Biology	9th 10th (N = 140)	71	79	10
	Biology	11th 12th (N = 55)	71	80	11
	Chemistry	11th 12th (N = 60)	72	81	11
	Physics	11th 12th (N = 24)	67	77	13

(Source: Guevara, 2008b.)

Table 5.5. Indicator Three: Timely Submissions of Assigned Work in Five Texas School Districts. Academic Year 2007–2008. Each semester, there was an average of 50 assignments per class.

District	Course	Grade Level and #(N) of Students per Class	Timely Submissions				Overall Increase in Compliance Rate (%)
			Fall 07	%	Spring 08	%	
Galveston	ESL-LA	4th (N = 18)	21	42	41	82	62
	ESL-LA	5th (N = 18)	21	42	42	84	63
Houston	Science	5th (N = 26)	24	48	38	76	62
Houston	Science	6th (N = 134)	24	48	39	78	63
Galveston	Science	8th (N = 106)	22	44	40	80	62
Alvin	IPC	10th (N = 55)	28	56	36	72	64
	Physics	10th/12th (N = 82)	31	62	45	90	76
Hitchcock	Spanish AP	11th (N = 94)	33	66	48	96	81
Houston	IPC	9th/10th (N = 125)	29	58	46	92	75
	IPC	9th (N = 114)	28	56	45	90	73
Waller	Biology	9th/10th (N = 140)	29	58	41	82	70
	Biology	11th/12th (N = 140)	27	54	48	96	75
	Chemistry	11th/12th (N = 120)	29	58	46	92	75
	Physics	11th/12th (N = 24)	32	64	41	82	73

(Source: Guevara, 2008b.)

Table 5.6. Indicator Four: Number of Same-Student Discipline Referrals in Five Texas School Districts. Academic Year 2007–2008.

District	Course	Grade Level and #(N) of Students per Class	Fall 07	%	Spring 08	%	Overall Reduction (%)
Galveston	ESL-LA	4th (N = 18)	6	33	1	6	81
	ESL-LA	5th (N = 18)	4	22	1	6	72
Houston	Science	5th (N = 26)	11	50	5	19	62
Houston	Science	6th (N = 134)	14	10	8	6	40
Galveston	Science	8th (N = 106)	12	11	7	7	36
Alvin	IPC	10th (N = 110)	20	18	6	5	72
	Physics	10th/12th (N = 82)	20	24	5	6	75
Hitchcock	Spanish AP	11th (N = 94)	6	6	1	1	83
Houston	IPC	9th/10th (N = 125)	16	13	5	4	69
	IPC	9th (N = 114)	12	11	6	5	54

District	Course	Grade Level and #(N) of Students per Class	Fall 07	%	Spring 08	%	Overall Reduction (%)
Waller	Biology	9th/10th (N = 140)	20	14	2	1	92
	Biology	11th/12th (N = 140)	11	8	1	1	87.5
	Chemistry	11th/12th (N = 120)	18	15	1	1	93
	Physics	11th/12th (N = 24)	5	21	1	1	95

(Source: Guevara, 2008b.)

The "Number of Same-Student Discipline Referrals" was influenced by high student mobility throughout Texas school districts. Hull (2007) and Carey and Manwaring (2011) cite student mobility as one of the limitations of growth models.

Between 2009 and 2011, nine of the original group of teachers trained in the model continued working in Texas schools. Four of the original group relocated to different districts within the state, and five remained in the districts where they had been implementing the model since 2007. I was selected an Albert Einstein Distinguished Educator Fellow and, with a leave of absence, moved to Washington, D.C., to serve my fellowship in the U.S. Congress. The remaining ten educators retired or changed career paths. In 2010, the Texas Education Agency (TEA) and the Texas Regional Collaboratives (TRC) honored the nine teachers of the original SSIAG group for their contributions to the academic performance of their at-risk students and awarded them with the "Excellence in Science Teaching" TEA-TRC award. At the same time, I was recognized as "Trainer of Trainers and Role Model in the Teaching Profession," a TEA-TRC mentoring award.

After serving 2 years as Einstein Fellow, I returned to Texas for the Fall 2012 semester to continue implementing the SSIAG Model at Galena Park ISD. At Galena Park I began implementation of the model in the high school physics classes. Physics is taught as a two-semester course for 11th and 12th graders. Every teacher at NSSH teaches seven periods, each with an average of 28 students. During the Fall 2012 semester, a total of 168 11th graders were assigned to me, and during the Spring 2013 semester, I taught another 161 12th graders (seniors aspiring to graduate at the end of the semester). I implemented the SSIAG Model starting at the beginning of each semester.

It's important to note that during the first 6-week grading period, all classes experienced a period of adjustment to the implementation of the SSIAG Model. There were initially many questions from students, school administrators, parents, and colleagues, especially when the Introductory Course Packet was sent home to be read and signed by parents. Most parents and students were not used to the structure of the model, and it took numerous phone calls focusing on differentiated instruction to parents, who were not familiar with the approach of replacing the traditional district-sanctioned instructional delivery with the problem-based learning and student-centered instructional strategies. However, the SSIAG Model again proved successful. As both semesters progressed, my students consistently ranked in third place in both the science department-developed Common Assessment (CA) tests, as well as in the district-developed Common Based Assessment (CBA) tests. Data from the four impact indicators of the SSIAG Model were consistent with previous reports, and they will be included in future presentations at local, regional, and national conferences.

CONCLUSION

The Student-Centered Sheltered Instructional Approach and Growth (SSIAG) Model is a compendium of instructional and assessment tools that can produce dramatic results in high-need school environments. Through personal experience and from data collected by other teachers, the SSIAG Model demonstrates the powerful impact it can have on student success and a teacher's ability to manage pressures from within and outside the classroom. Broad implementation of SSIAG can contribute significantly to school reform efforts around the nation, in particular those serving underrepresented minorities.

Albert Einstein, a great scientist and educator, once said: "I never teach my pupils: I only attempt to provide the conditions in which they learn. ..." The SSIAG Model can help you provide the right conditions for learning in your classroom.

Educators interested in learning more about the SSIAG Model are welcome to contact the author at rsedci@gmail.com or visit the website www.rsedci.com.

WORKS CITED

Aguirre, M. (1996). Effect of open-ended questions on Spanish-dominant LEP students' ability to demonstrate in-depth science concept development and use scientific vocabulary. *Journal of Multilingual Education Research*, 11, 46–69.

Association of American Medical Colleges (2004). *More apply to U.S. medical schools*. http:/www.aamc.org/newsroom/pressrel/2004/041020.htm

Cardenas, V., & Kerby, S. (2012). *The state of Latinos in the United States: Although this growing population has experienced marked success, barriers remain*. Washington, DC: Center for American Progress.

Carey, K., & Manwaring, R. (2011). *Growth models and accountability: A recipe for remaking ESEA*. Education Sector Reports. Retrieved October 5, 2013, from http://www.educationsector.org/sites/default/files/publications/GrowthModelsAndAccountability_Release%20.pdf

Chowa, G. A. N., Ansong, D., & Osei-Akoto, I. (2012). *Parental involvement and academic performance in Ghana*. Youth Save Research Brief 12-42. St. Louis, MO: Washington University, Center for Social Development.

Ferreira, M. M., & Trudel, A. R. (2012). The impact of problem based learning (PBL) on student attitudes toward science, problem-solving skills, and sense of community in the classroom. *Journal of Classroom Interaction*, 47(1).

Guevara, E. (2006) *Successful student-centered strategies to meet educational challenges in Texas schools: The SSIAG model*. Paper presented at the 7th Annual UTMB Regional Conference, Galveston, TX.

Guevara, E. (2007). *The Student-Centered Sheltered Instructional Approach Model (SSIAG, Guevara, 2006) improves academic performance in ELLs*. Paper presented at the Annual Conference for the Advancement of Science Teaching (CAST), Austin, TX, November 16–17, 2007.

Guevara, E. (2008a). *The Student-Centered Sheltered Instructional Approach Model (SSIAG, Guevara, 2006) improves academic performance in ELLs*. Paper presented at the Spring Conference of the Metropolitan Association of Teachers of Science (MATS), Houston, TX, February 2, 2008.

Guevara, E. (2008b). *The Student-Centered Sheltered Instructional Approach Model (SSIAG™, Guevara, 2006) improves academic performance in ELLs*. Paper presented at the National Science Teachers Association, Conference in Science Education, Boston, MA, March, 26–30, 2008.

Hull, J. C. (2007). *Measuring student growth: A guide to informed decision making*. Center for Public Education. Retrieved October 5, 2013, from http://www.centerforpubliceducation.org/MainMenu/Policies/Measuring-student-growth-At-aglance/Measuring- student-growth-A-guide-to-informed-decision-making.html

La Fe Policy Research & Education Center. (2012). *Texas budget crisis: Impact on Latinos*. San Antonio, TX. Retrieved October 5, 2013, from http://www.lafepolicycenter.org/documents/TEXAS-BUDGET-CRISIS-IMPACT-ON-LATINOS-4-11.pdf

McKeon, D. (1994). When meeting common standards is uncommonly difficult. *Educational Leadership*, 51(8), 45–49.

Nyarko, K., & Vorgelegt, V. (2007). *Parental involvement: A sine qua non in adolescents' educational achievement*. Unpublished doctoral dissertation, Ludwig-Maximilians University, Munich, Germany. Retrieved October 5, 2013, from http://edoc.ub.uni-muenchen.de/8384/1/Nyarko_Kingsley.pdf

O'Malley, K. J., Murphy, S., McClarty, K. L., Murphy, D., & McBride, Y. (2011, September). *Overview of student growth models: White paper*. New York: Pearson Education. Retrieved June 18, 2012, from www.pearsonassessments.com/.../Student_Growth_WP_083111_FIN

Schimpl-Neimanns, B. (2000). Social origins and educational participation: Empirical analysis of relationship between educational inequality between 1950 and 1989. *Cologne Journal of Sociology and Social Psychology*, 52(4), 636–669.

Schmitt, M., & Kleine, L. (2010). The influence of family school relations on academic success. *Journal for Educational Research Online*, 2(1), 145–167.

Sungur, S., & Tekkaya, C. (2006). Effects of problem-based learning and traditional instruction on self-regulated learning. [Abstract]. *Journal of Educational Research*, 99(5), 307–317.

Topor, D. R., Keane, S. P., Shelton, T. L., & Calkins, S. D. (2010). Parental involvement and student academic performance: A multiple mediational analysis. *Journal of Prevention & Intervention in the Community*, 38(3), 183–197.

U.S. Census Bureau (2011a). *Demographic profile*. Retrieved December 2012 from http://www.census.gov/population/www/cen2010/.atlas.html/

U.S. Census Bureau (2011b). *Demographic profile*. Retrieved December 2012 from http://www.census.gov/population/www/socdemo/educ-attn.html

Zemelman, S., Daniels, H., & Hyde, A. (1998). *Best practice: New standards for teaching and learning in America's schools*. Portsmouth, NH: Heinemann.

CHAPTER SIX

Putting THE "Authenticity" INTO Science Learning

TIM SPUCK

INTRODUCTION

Have you ever heard someone say, "We need to teach kids that science is fun"? When I hear it, I always cringe a bit. From my personal experience, doing science is much more. Science is frustrating, time consuming, hard work, exciting, motivating, powerful, rewarding—and yes, fun—all rolled into a package of authenticity. When learners experience science as anything else, failing to negotiate the struggles as well as the excitement that practitioners of science face daily, they likely end up making important decisions about careers and science beliefs based on a misunderstanding of science. Through science you can come to know the unknown, use data and evidence to predict the future, or share knowledge that may one day improve life for all humanity. I find these facts utterly amazing. This is the power of science within each of us.

As a K–12 science department chairperson, I often hear high school teachers refer to science taught in the elementary or middle school as "playing around." On the other hand, elementary schoolteachers describe high school science as "boring." Students often see science similarly. As an educator, what I've observed in schools across the United States (see Figure 6.1) is a growing focus on the development of tools and, at the same time, decreasing opportunities for students to use and further develop their creativity through authentic problem-solving experiences. Is it any wonder that one of the most commonly asked questions by students is "I'm never going to use this; why do I need to know it"?

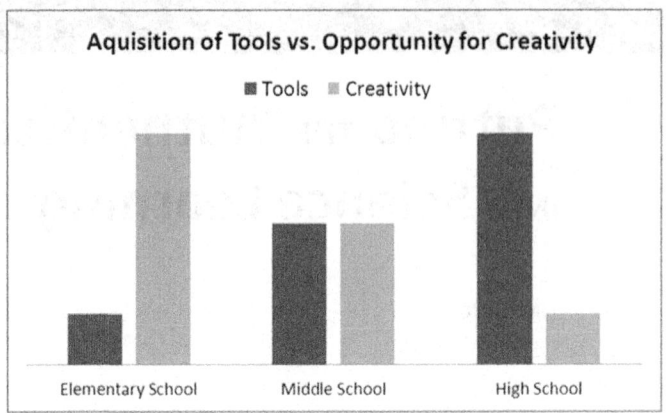

Figure 6.1. Acquisition of Tools vs. Opportunity for Creativity. The focus in student learning shifts from a world of questions and wonderment in elementary school to acquisition of new tools as students proceed toward high school. Examples of "tools" include information, use of science instruments, and so on. These tools are often taught using prescriptive (recipe)-type methods in isolation from a real-world application.
(Source: Author.)

When learners ask this question, that's a positive. It means they're struggling to find purpose and meaning in what they are doing. We should encourage students to not only question the universe, but to question what is being done in the classroom. As a teacher, I want my students to ask, "Why do I need to know what you want me to learn?" Not only does it bring context to student learning, but, in formulating my response, I too gain a deeper understanding of the useful application of the material I'm teaching.

In an effort to bring purpose and meaning to learning, teachers seek out opportunities to help students understand science by understanding what scientists do. Schools often declare "career week" or ask teachers to focus on science careers in their classes. As educators, we want to expose learners to career options, but that begs the question: "Why aren't students coming to know what scientists do from the science they are already doing in school?" Why, when we think of "career week," do many of us think about bringing someone into the classroom who can tell students what it's like to be a scientist, or sharing an article about what scientists do that they can read? If learners in our schools experienced science in an authentic way, wouldn't they already know the answer?

As a science educator with more than 50,000 hours of practice, I can say with a high level of confidence that engaging young learners in authentic science experiences not only benefits those who choose science careers, but it also holds great benefit for those who do not. So, how do we embed authentic science experiences into learning environments for all students? Authentic science is not a prescriptive program that has a clear beginning and end. It's not an activity you engage students in, but rather an environment you create together. Authentic science is a journey you choose to embark on both to create and sustain an authentic environment for all learners: empowering them to actively participate in the community that science is.

A primary goal for educators should be to communicate truth to all learners. In achieving this goal, we must first understand the roots of our personal understanding of science and the misconceptions we might hold, and then work to embed ourselves and our students in experiences that reflect the current practices of science. As you read on, think about all your days and nights spent on planet Earth. How have you come to know what science is and what scientists do? Is the picture you are painting for the learners in your classroom an authentic one?

BACKGROUND

The proposed 2013 research and development budget for the United States totals $142.2 billion (*Research and development FY 2013*, 2012, p. 3). Between the private sector and government, it is estimated that the United States spends over $400 billion annually on research and development—nearly twice that of its closest competitor, China (Battelle, 2011; Thibodeau, 2012).

Over the past decade, science, technology, engineering, and mathematics (STEM) jobs grew at three times the rate of non-STEM jobs (Schiavelli, 2011). And, looking at the job market, it would appear that research and development will be a booming area of job growth for years to come. By 2018, STEM jobs are expected to grow by 17%, nearly double the rate of growth in non-STEM fields (Casey, 2012; Schiavelli, 2011).

Scientists are held in high regard in our society too. A recent study shows that 70% of Americans believe scientists contribute "a lot" to society's well being (Pew Research Center, 2009). In addition, 54% of American students say they plan to pursue a career in STEM, and the number of college freshmen interested in a STEM major is on the rise (Drew, 2011; Koebler, 2011). So what's the big deal about science? There seems to be plenty of interest and support, so our work is done, right?

Far from it. The problem is, we need more than science fans seated in the stands. We need a strong, knowledgeable science workforce that understands science in an authentic way, with the tools necessary to address the problems of today as well as the challenges of tomorrow. While the public shows strong support for science, scientists give the public a rather poor rating. Eighty-five percent of scientists view the public's lack of scientific knowledge as a major problem, and nearly 50% believe the public has unrealistic expectations of scientists (Pew Research Center, 2009). Alarming, too, is the apparent lack of understanding of basic science knowledge. Even after years of media focus and attention, 35% of Americans do not know that carbon dioxide is a gas linked to rising global temperatures, nearly 50% don't know that stem cells can develop into many different types of cells, and over 50% don't know that an electron is smaller than an atom (Pew Research Center, 2009).

Our schools fare no better than the general public. The Programme for International Student Assessment (PISA) looked at students in 60 countries. The PISA showed that U.S. 15-year-olds score significantly lower in science than their peers from 18 other nations and institutions, and significantly lower in math than their peers from 23 other nations and institutions (Fleischman, Hopstock, Pelczar, & Shelley, 2010). When these same students arrive at our colleges, they struggle there as well. Fewer than 40% of students who enter college majoring in a STEM field complete a STEM degree (Drew, 2011; President's Council of Advisors on Science and Technology, 2012). Throughout the 1960s and 1970s, the Space Race and other science initiatives brought many into science-related careers. Those individuals are nearing retirement age and will soon leave (Barton, 2003). Who will fill the void? Without fundamental science knowledge and knowing what scientists do, our students find themselves at a significant disadvantage.

While the problem may seem insurmountable, it's not. There is evidence that students who have an opportunity to participate in original scientific research (i.e., authentic science) while in high school are more likely both to enter and to maintain a career in science compared to students who do not have these experiences (Markowitz, 2004; Roberts & Wassersug, 2009). In addition, students who participate in authentic science experiences show significant increases in conceptual knowledge, question and hypothesis development, modeling, and logical argumentation (Charney et al., 2007). And there is no need to limit these research experiences to high school or even middle school students. While it can be a bit more challenging to engage elementary school students in real-world research, a recent paper published in the British journal *Biology Letters* shows it's quite feasible. In a project designed and implemented by a group of 8–10-year-olds in England, children helped scientists better understand how bumblebees select flowers to forage from (Blackawton et al., 2011). While some may consider such ideas contrary to the findings of Piaget, it's important to keep in mind that Piaget's work was restricted to the thinking of the child working alone. There is evidence that the use of collaborative cognition can enable the child to engage successfully in more complex intellectual tasks, and children can understand the goal of hypothesis testing by age seven (Metz, 2003).

Defining Authentic Science

In defining the phrase "authentic science," let's first explore the meaning of the word "authentic." Authentic practices engage learners in real-world situations where their meanings and purposes are socially constructed through negotiation among members of the community (Cronin, 1993; Lee & Songer, 2003). In order for a learning exercise to be considered authentic, it must meet five criteria: (1) the learner must engage in higher-order thinking, (2) the learner must demonstrate depth of knowledge, (3) the learner must connect to the real world beyond the classroom, (4) the learner must promote substantive dialogue and seeking of understanding through collaboration, and (5) the exercise must lead to social support or encouragement and recognition of the learner's achievement by leaders and peers (Newmann & Wellage, 1993). In addition, authenticity is emergent in nature (Rahm, Miller, Hartley, & Moore, 2003). Individuals need time to learn and reflect as they come to know an experience and enter into a relationship with it. In science, authenticity can emerge as a result of learners asking their own questions and developing their own methods of finding answers, or they can engage in the questions and research of others for sufficient time: developing an identity as a "scientist" working on the research project (Rahm et al., 2003).

The other part of the equation, "science," is often characterized as the processes scientists exercise throughout their daily lives, or simply what scientists do (Chinn & Malhotra, 2002; Mathews, 2003; Rahm et al., 2003). This practice of science (and engineering) includes:

1. Asking questions (for science) and defining problems (for engineering)
2. Developing and using models
3. Planning and carrying out investigations
4. Analyzing and interpreting data
5. Using mathematics and computational thinking
6. Constructing explanations (for science) and designing solutions (for engineering)
7. Engaging in argument from evidence
8. Obtaining, evaluating, and communicating information (National Research Council, 2011)

Science also incorporates human values, curiosity, frustrations, surprises, and disappointments that occur, especially with unexpected results (Zion et al., 2004). Social interaction is another key feature of scientific practice. Social interaction among scientists includes the same mix of cooperation and competition, agreement and argumentation that accompanies all human social activity (Edelson, 2003). The practice of science, it would appear, is a mix of many activities, each with the potential to impact the other.

It is difficult to discuss science without also addressing inquiry. Although *A Framework for K–12 Science Education: Practices, Crosscutting Concepts, and Core Ideas* (National Research Council, 2011) and the *Next Generation Science Standards* abandon the use of the word inquiry, it continues to be used throughout science education and therefore warrants attention. Inquiry is a tool or a form of scientific practice (Michaels, Shouse, & Schweingruber, 2007) but one that in and of itself does not encompass science. For example, scientists work in teams and collaborate with others in completing their research. Yet one can quite easily conceive of engaging in inquiry in isolation from others. As much as collaboration might enhance the experience and the outcome, it's simply not a necessary part of inquiry. For example, if I were stranded alone on an island with no way of collaborating with others, would I be able to engage in inquiry? The answer is yes. And it may well be that my life depends on it. However, I would not be practicing science the way it is practiced in our society today.

Based on what we know about authentic practices and science practices, it appears that authentic science requires: (1) asking question(s) relevant to a real-world issue, (2) using the tools of scientists and engineers, (3) seeking out and evaluating evidence, (4) using that evidence to make a claim, (5) sharing that claim with others in a way that it can be verified, critiqued, and used, and (6) engaging in activities in constant dialogue with colleagues and within a real-world setting.

There are as many ways to achieve authenticity in science as there are questions in science. Authentic science does not simply happen in a prescribed or static manner. It emerges, it is negotiated, and it depends on how all of the components of the system come to interact with each other over time (Rahm et al., 2003). An authentic science experience provides an opportunity for the learner to take part in the development of the research question (Trumbull, Bonney, Bascom, & Cabral, 2000), or to adopt questions that represent true uncertainty in their world (Edelson, 2003). Science also takes on tremendous personal and professional commitment. Time is an essential component for authentic science (Anderson, 2002; Bencze, 2000; Chinn & Malhotra, 2002; Rahm et al., 2003; Robinson, 2004; Zion et al., 2004). Scientists will often spend years investigating a single question. Science learners need time to reflect on their world, ask questions, discover what may already be known, probe the problem at hand, reflect on their findings, reflect on their interpretations, and share their interpretations for critical review.

As you think about what authentic science is, consider what it takes to create a good pot of soup. You add a variety of meats (tofu for vegetarians), spices, vegetables, and so forth, to water. As the mix simmers over time, the flavors interact, and each individual item is impacted by the flavor of the others (see Figure 6.2). The broth acts as a medium through which the parts influence one another to create something different, something greater, than each part could possibly be alone. This "something" is the total environment that is necessary for authentic science to emerge and take form. Just as the soup of the day changes over time, so does authentic science.

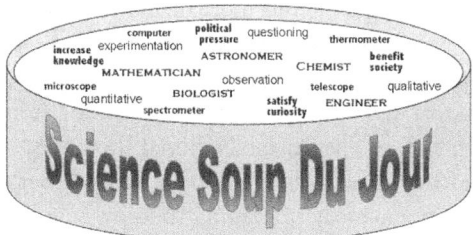

Figure 6.2. The Authentic Science Environment Is the "Science Soup of the Day." The authentic science environment is the "science soup of the day." The graphic represents how those who do science, the reasons they do science, the tools used for science, and the methods of doing science all interact in a sort of "soup." Each item in the soup has unique properties that impact other parts of the soup in unique ways.
(Source: Author.)

The "science soup du jour" can also be thought of as STEM. Ask yourself this: "Is it possible to do real science today without using technology to collect and analyze data, engineering practices to design and redesign the investigation, or mathematics to make sense of the data?" While many consider STEM an acronym for science, technology, engineering, and mathematics, my experience leads me to think of it as much more. Like authentic science, STEM is the "science soup du jour." The people who engage in the practice of science (i.e., scientists, technicians, engineers, and mathematicians) engage in research—often collaboratively, across disciplines. These collaborations are impacted by the tools currently available, political pressures of the day (e.g., the Space Race, climate change, stem cell research), the quest to improve society, and so on. The factors that influence practitioners, like the soup itself, also evolve over time. Therefore, it seems clear that one cannot engage in authentic science without the integration of technology, engineering, mathematics, and societal influences. And one cannot claim to be doing authentic science, or authentic STEM, if his or her work has no potential to impact society's understanding of the past, present, or future in some meaningful way. Authentic science and authentic STEM are at the very least quite similar, but are more likely one and the same.

COMING TO KNOW SCIENCE: MY PERSONAL STORY

Like science itself, the journey toward understanding science evolves over time, and in most cases begins years earlier, as children. To claim an authentic knowledge of science without first considering the roots of that understanding would be similar to claiming to know your family without consideration of your grandparents or great grandparents. While on the surface it may seem possible to "know" your family without knowledge of your more distant relatives, one trip to the doctor's office requiring information on family history might prove otherwise. So, how did I come to know what science is and what it is not?

As you read my story, consider this question: "Are the stories of my father, my quest for flight, and my experience at the National Radio Astronomy Observatory (NRAO) examples of authentic science in action, or is something missing?" Also, what has been your personal story of coming to know science? In what ways is it similar to or different from mine? And, most important, has your "story" given you an authentic understanding of science?

It was not until several years into my own teaching career, after I completed my master's degree in science education, that I had my first formal experience at a science research institution doing "real science." But my story begins years earlier. I grew up in what, by today's standards, would be a large family: my mother, father, four sisters, and two brothers. Home was rural northwestern Pennsylvania. My mom and dad both placed great value on the quest for new knowledge. My mother was the first in her family to graduate from high school, and I can still remember my father going to night classes in his mid-50s and proudly walking across the stage to receive his GED. Although both my parents strongly supported education, it was my dad who was the epitome of what we refer to as the "lifelong learner." He was fond of reciting the words of an unknown author: "If I had my life to live again, I would double my effort to learn all that I could, for there is much to be discovered, but too few willing to do the work necessary to seek it out."

Not only did he talk about lifelong learning; he also practiced it on a daily basis. By trade my father was a male nurse. However, his childhood quest to find a cure for all diseases led him to become one of the early homeopathic specialists. Homeopathic care is much more common in medical science today, but in the 1960s, 1970s, and 1980s, that was not the case. My father was a bit of an outlier, but he never allowed that to influence the pursuit of his lifelong dream or the care he gave to those who came to him. Nor did he limit his quest for knowledge to homeopathy. I walked into the house one day, and my dad—then in his mid-70s—was standing there with his new key grinder.

I commented, "What are you doing?"

He smiled and replied, "Always wanted to learn to be a locksmith, so I signed up for a distance learning course, and right now I'm learning to duplicate keys. Do you need one made?"

Laughing a bit I said, "Dad you're 77 years old. You're not going to start a locksmith business, are you?"

"No," he replied, "but you just never know when this could come in handy."

It was against this sort of backdrop that I grew up. As a kid I was on my own quest, but I guess you could say curiosity got the best of me on a regular basis. I loved to take apart old radios, clocks, machinery, and so on, to see what they looked like inside. Of course, the plan was to try and figure out how they worked and fix them, but let's just say my success rate was less than stellar. My dad, looking over my shoulder one day, shook his head and said, "Tim, you're really good at tearing things apart. Not so much at putting them back together, though." And he was right. But I learned a lot about systems and how things work together rather than in isolation from one another to make something happen.

Throughout my youth I discovered and deepened my interest in flying and space-related science. When I was 7 years old my father showed me the constellations and the planets and gave me my first taste of the night sky. I was hooked. Every visit to the library meant a new book about the science of flight, the planets, or the stars. I spent a great deal of time outside exploring nature, but I especially loved spending nights looking up at those points of light "hanging" in the blackness of space. The fact that I was growing up during the Space Race between the United States and the Soviet Union and was a faithful watcher of *Star Trek* probably helped, but there was more to this quest than the excitement of potentially becoming an astronaut. Flying was about getting a new perspective. Seeing something from above gave me a different, broader view and understanding.

I grew up bound and determined to unlock the secret of flying—a daunting task for a family that hovered near and sometimes under the poverty line. But what I didn't have in dollars I like to think I made up for in determination, or as my parents referred to it, "stubbornness." I tried jumping from the porch roof with an old umbrella. We all know how that worked out. Umbrellas really do turn inside out. Then came the Cardboard Wing Program (CWP). I figured the wings were lightweight and sturdy enough that if I jumped off the roof and flapped my arms, I'd be flying with the birds in no time. After dropping to the ground like a rock, I spent some time reassessing my design. I had an idea: perhaps I was just too big, but with someone smaller, like my little brother, it just might work. One broken collarbone later, and a significant amount of trouble, I had to scrap CWP.

With CWP outlawed, I turned to something a bit safer. I recalled mowing grass and the significant volume of air coming out from under the mower deck. I read a bit about hovercrafts and helicopters at the library, so the vision in my mind seemed logical: I just needed to build a lightweight wooden propeller, attach it to the lawnmower, mount the engine upside down on an aluminum frame, strap the thing to my upper body, have someone pull the cord to start the engine, and with the throttle in hand I'd be off the ground in no time. I started working on the wooden propeller and began investigating the lawnmower engine. My dad walked into the garage just as I was about to pull the second bolt. Needless to say, he was not happy. I quickly replaced the engine bolts. After my father got over the initial shock of what I was planning, he pointed out the old rusty mower sitting in the corner of the garage. I thought, *Well, I'll find a way to get it running.* After pulling the old motor off the deck, I turned it upside down, trying to determine the best way to attach the blade on my new Personal Heli-Pack (PHP). It didn't take long for the "black syrup" to soak into my jeans. *Oil, damn it!* I thought. Apparently you can't run a lawnmower engine upside down without the oil running out or up into the cylinders. The PHP had hit a roadblock. There was no money to buy something different, and I hadn't thought about a pulley with a belt to drive a vertical shaft at the time, so the PHP project was mothballed.

There were additional attempts throughout my youth and teenage years to achieve flight (some of them better left a secret), and many other self-directed attempts at understanding the mysteries of the universe.

After ninth grade, I left my high school academic program to attend vocational school with a focus on electronics technology. Vocational school brought meaning and greater authenticity to my learning. I could finally understand the application of science, technology, engineering, and the language of mathematics because I was using it in my electronics technology program. I was now able to fix those clocks and radios I had been tearing apart! On the other hand, the academic resources and courses available to me were limited. When I started college in electrical engineering, on my way to becoming a NASA engineer building satellites and other spacecraft, I again hit a brick wall. Like the "oil" in the lawnmower engine, I ran into calculus, and without the background or knowledge of where to turn, I changed direction. Lucky me!

Now I'm a science teacher specializing in earth and space sciences. I've led my students on numerous quests discovering exploding stars, asteroids, and young, sun-like stars in their early stages of development. I've been on research teams with top scientists and been given time on some of the most sophisticated scientific instruments in the world. I've worked in some of the best research labs in the nation, been to the most remote parts of planet Earth, and logged nearly a million miles in the air, never tiring of the magnificent view from above. Like my students, I'm a science teacher and a learner.

So what prepared me as an educator to engage my students in these real-world science practices? Sure, I think as a child I engaged in a great deal of scientific practice, and I think I had a great role model in my father, but was it authentic? Authentic science requires dialog and argument. As a child building my flying machines, who was I in dialogue with? No one. Had I been presenting my research to others who shared a similar interest and insight, I might have used pulleys, belts, and a long shaft to significantly improve the design of the PHP. I might have learned about the need for grant writing. My first experience practicing science research in a team did not come until after I completed my MEd in science education and after 4 years of teaching.

The art of authentic science never really struck home until 1992, when I had my first "real" science research experience at NRAO in Green Bank, West Virginia. I had attended a 2-week summer institute designed to introduce authentic science inquiry to teachers. While there, I was placed in a research team along with three other teachers, a mentor teacher, and a scientist. We were given a crash course in radio astronomy and embedded in the research process. Between lecture, labs, and our research projects, the days blended one into the next, often providing less than 5 or 6 hours of sleep per night. But together, in that short but concentrated period of time, we completed a science research project using the facility's 40-foot radio telescope. The focus of our research was to push the instrument to its limits and determine the faintest detectable object in space. My team beat the previous record, and more than 20 years later the memories of that experience are vivid in my mind.

I believe I received both a quality undergraduate and graduate education experience, but as in my academic high school, the *practice* part of doing *real science* was not part of teacher preparation. I find it ironic that there is virtually no high school in the United States that would hire someone to coach a sport who did not have years of experience playing and practicing the game. Yet schools hire many people, myself included, with very little or no experience doing real science to teach (i.e., coach) students the game of science.

My experience at NRAO-Green Bank came at a point in time when it made a significant difference in my teaching career. I returned home that summer to revamp my curriculum to be more reflective of my experience as a scientist. In the 20 years that followed, I actively looked for additional opportunities to engage with the professional science community. I joined the Hands-On Universe (HOU) project, a national effort to give high school teachers and students access to training and telescopes for astronomy research. I spent a summer at Lawrence Berkeley National Labs (LBNL) through the Department of Energy's Teacher Resource Associate program. While at LBNL, I worked with HOU and the Supernova Cosmology Project (SCP) to develop an asteroid search project for teachers and students. The very same images used in the HOU asteroid search were used by the SCP team to discover that the expansion rate of the universe is accelerating, an accomplishment that earned them the 2011 Nobel Prize in physics. Five years after my LBNL experience, a Research Experiences for Teachers (RET) program at NRAO provided me an opportunity to work with radio astronomer Dr. J. Richard Fisher, whom I had connected with years earlier. In the years that followed, I participated in the Teacher Leaders in Research-Based Science Education at the National Optical Astronomy Observatory, the Spitzer Space Telescope Research Program for Teachers and Students, the NASA-IPAC Teacher Archive Research Program, and Polar Teachers and Researchers Exploring and Collaborating (PolarTREC). Not only have these experiences taken me on incredible learning adventures to Antarctica, Greenland, Japan, Kitt Peak National Observatory, and Palomar Observatory; they have also provided me the opportunity to experience real science and engineering in action. And, every step of the way, I expanded the network of STEM professionals that my students and I had access to for future research projects.

As you can see, my journey to understanding what science is began many years ago when I was a child. In many ways my father was my initial role model, but over the years the pursuits I've engaged in have provided more and more people I could look to for guidance. So, what's your story? How have you come to know what real science is? As you consider this question, think about how you can take part in a teacher-scientist partnership experience. This type of experience is an excellent way to get started creating an authentic learning environment for you and your students.

Teacher Tip—Finding Scientist Partners you can Collaborate with

- Apply for an RET. The National Science Foundation supports a variety of research experiences for STEM teachers. Check out the website at http://www.pathwaystoscience.org/programs.aspx?descriptorhub=RET_NSF
- Visit the Graduate STEM Fellows in K–12 Education website at http://www.gk12.org/. If there's not an active site near you, do a search using the Alumni Center link to see if a past participant is located near you.
- Check out the Math Science Partnership program at http://hub.mspnet.org/index.cfm and find a project near you.
- The Science Education Partnership Award (SEPA) program is supported by the National Institute of Health. Visit the website at http://nihsepa.org.
- Visit PolarTREC at https://www.polartrec.com.
- Check out the NASA/IPAC Teacher Archive Research Program (NITARP) at http://nitarp.ipac.caltech.edu/.
- Visit the website of your local university or STEM research institutions or business, and look at the personal websites for faculty or other persons in the areas of STEM that interest you most. Call or send an email to see if there might be an opportunity for you to work with them over the summer. In some cases they might have funding to support you. If the project is funded by NSF, the Principal Investigator for the project can request an RET supplement to provide financial support for you.
- Get your students involved with regional science & engineering competitions. Visit http://www.sciencebuddies.org/ or https://www.googlesciencefair.com/en/competition. Use this as an opportunity to have STEM professionals come to your school to help mentor your students.

BEST PRACTICES: AUTHENTIC SCIENCE IN ACTION

So, where and what do I teach? I teach in Oil City, a small rural community in northwestern Pennsylvania, population ~10,000. In the late 1800s and early 1900s, as the local oil industry peaked, Oil City was a thriving metropolis. Major oil company corporate headquarters for Penzoil and Quaker State were located here, and money and jobs seemed to flow as easily as the oil itself. By the year 2000, all that had changed. Today, the Oil City School District finds itself labeled as one of the 10 poorest school districts in the state, and all the challenges that come with poverty are all too present, both in and out of our schools.

Within this system I teach a variety of earth and space science courses and from time to time have taught physical science, computer science, and general science. In my 25 years of teaching, I've taught grades 8 through 16, served as the K–12 science department chairperson, and sponsored an afterschool astronomy club and science fair team. It's also important to note that the science classes I teach at Oil City have taken place in 90-minute blocks for one semester, as well as 45 minutes per day over the course of an entire school year. I have found that authentic science practices can be implemented regardless of schedule design.

Setting the Stage for Authentic Science in the Classroom

Just like anywhere else in the world, students at Oil City are inherently curious. But as in other communities, curiosity seems to be relatively high in elementary school but decreases as learners move into higher grades (see Figure 6.1). As a science teacher, I look for ways to help students reconnect with their curious side. Curiosity, after all, is an essential component of authentic science, so from the first few days of class I work to engage them.

Early in the process (three or four days into the school year), I present my science classes with the following assignment:

> "This evening I want you to find someone elementary school age or younger. The important thing is that they understand what you're about to ask of them. It could be your brother or sister, or the sibling of one of your friends. Make sure you have a way of keeping track of time. Counting one Mississippi, two Mississippi, etc., will work fine for this particular investigation. Find out how old the child is and record it. Ask them to take a few seconds to look around. After they've had an opportunity to check things out, ask them to come up with five questions as quickly as they can about what they see in their environment. Record (in seconds) how long it takes them to come up with five questions. Bring your results with you to class tomorrow."

The next day students walk into the classroom, take their seat, and the first thing out of my mouth is, "Let's start off today with a short experiment. I need you to grab a sheet of paper and something to write with. Take a few moments and look around and come up with five questions about what YOU observe in the world around you, and write down the questions. Turn your paper over when you finish so I can record your time."

My students are in grades 10–12, so you might be able to guess the results. I'm merciful and rarely let the suffering go on for more than a few minutes. There are a few students who finish up in a minute or two and a fair number who finish in three minutes. But the number of students who struggle to come up with five questions always amazes me. And it's the college prep students who sometime struggle the most.

From there we jump right into the observations my students made the previous evening in their interviews with younger children. We chat a bit about their experience and their results, then take the data (i.e., age and time needed to ask five questions) they collected and add it to the data I just collected in class. With all the data in hand, we do what any good scientist would do: we graph it, placing age of person on the X-axis and time needed to ask five questions on the Y-axis. And, of course, we add a trend line. If I feel ambitious with a group, we sometimes head to the computers and plot the data in Excel.

With the graph in front of us, I ask the most important scientific question: "What does it all mean? What do the data tell us?" I get a variety of responses, including "Guess we're really not smarter than a 5th grader" and "We had to write the questions down and the younger kids just had to say it." I should point out that the latter statement is important because, while they may not realize it, learners are identifying a possible source of error in the experiment. As a teacher, I like to use these opportunities to help students recognize what good practice in science looks like. But after the discussion, the reality remains right there for all to see and confront for themselves. As high school students who are supposed to be more educated than their significantly younger peers, they find it more difficult to ask questions. Why?

As a class we can't escape this fact, or perhaps it's me who won't let my students escape. I keep hammering on it, asking them, "What class are you in? What do scientists do? What are the first steps of scientific processes?" We go on to discuss how at one point in their lives they were likely similar to the younger kids surveyed, and the reasons why their ability to ask questions seems to have diminished.

From there I look for a concrete example of something they've learned and are actually quite good at. For example, I'll say, "I've observed something about you in the past few days. Most of you walk into class and sit down prior to me asking you to do so. Why is that?" The discussion moves forward, and my students come to a realization that the act of walking into a classroom and taking their seats without being told is something that they've learned to do. And for a variety of reasons, although perhaps not intentional, the reluctance to ask questions is something they have learned over time as well. Perhaps they've learned somewhere along the way that if they ask the wrong question they might be ridiculed or embarrassed, or may be labeled as someone who doesn't pay attention. Perhaps they've learned that it's safer to remain quiet and less conspicuous. Whatever the reasons, something tremendously valuable that learners are born with—something that separates our children from other species on this planet—is taken away from them somewhere along the way. And, as part of an education community, we desperately need to change that. Instead of helping them find their desks, we must help learners of all ages discover the gift of curiosity and questioning. The art of questioning is foundational to authentic science, and it is from this point that the door to the authentic science environment is opened.

I do this exercise year after year because the looks on the faces of my students and the conversations taking place among high school students as they walk out of class that day are precious. If you want to bring about new learning, you need to move people off center. This activity helps learners step outside their minds and begin to look at themselves scientifically. They begin to more fully understand how others and their environment influence their thinking and behavior. This is an essential starting point if I am to respond to the question my students are sure to ask: "Why do we need to do research projects in this class? Why can't we just do bookwork and take quizzes and tests?" And yes, as hard as it is to believe, when students are pulled away from the traditional classroom—from what they've learned is the norm, from their comfort zone—they will crave the way it used to be. It's my job to help them experience and understand for themselves the significant gratification, personal fulfillment, and power that can come from doing real (authentic) science. On average, I've found it can take many students 6 to 9 weeks to begin feeling comfortable with this new approach to science.

Keeping the Pump Primed

In an effort to continuously help my students feel comfortable asking questions and sharing ideas, I build in little surprises throughout my courses that keep them off center and wondering what's next, exposing them to authentic practices in science. One example from my Space Science class takes place just before we study images from different types of telescopes (e.g. radio, infrared, optical, ultraviolet). In Earth Science, we use the same investigation to introduce the greenhouse effect and the role of light of different wavelengths. I start off having students construct a piece of jewelry using UV-sensitive beads. However, at the time, I don't inform them that the beads are sensitive to ultraviolet light. I'll typically choose a day when they're in research teams and ask that the team design and construct a piece of jewelry that is symbolic of their group or research project. Each person in the group must construct the same piece and wear it where it is readily visible (e.g., finger, arm, pinned on their shirt) for 24 hours. The assignment is worth ten points, but if I see a student within the 24-hour window without his or her piece being visible, that student is docked two points each time.

The next day, I start off with a quick check to make sure everyone is wearing his or her jewelry. Inevitably students start asking each other if their beads changed color. If the conversation is a little slow in coming, I indirectly prompt them. We discuss what could have caused the beads to change color. Temperature, moisture, and sunlight are all typical responses put forward as potential causes. With all the ideas on the table, students are asked to select an idea they'd like to test, design an experiment to test their idea (hypothesis), collect the data that evening, and come back the next day and report their findings. The following day yields all sorts of interesting stories. I've had students place beads in refrigerators (some even think to unscrew the light bulb first), freezers, swimming pools, bathtubs, and tanning beds. Pretty quickly they identify that it's light that causes the change. So I close the blinds and hold the beads up to the fluorescent light fixtures, which results in little color change. When we place them next to the window, there's more change, but when we hold them outside the window, the change in color is most noticeable. As a class, we eventually get to the idea that there's something about sunlight that causes the beads to change color, and that usually leads someone to say, "Isn't there UV protection on windows?" We pull out the UV lights and expose the beads, and sure enough, the mystery is solved. Not only does this activity help in the development of skills learners will need for their research projects; it also serves as an excellent way to introduce the science content needed to discuss multi-wavelength astronomy and the greenhouse effect. Although this activity is not fully authentic science, these types of exercises prepare the mind and set the stage for deeper engagement in authentic practices.

Authentic Science: Project-Based Learning

The time to ask five questions and the UV beads activities help to create an environment where authentic science practices can be more deeply explored. This deeper exploration can be achieved by engaging students in science research projects or project-based learning (PBL) methods. You may be saying to yourself, "I don't have time for research projects in my science classes. I have too many standards I need to address for the test!" I would argue that, as STEM educators, we can't afford to not take the time for PBL. Over traditional methods of instruction, project-based learning is shown to improve academic achievement, student attendance and self-reliance, problem-solving skills, and attitudes toward learning, all while showing no negative impact in performance on high-stakes tests (Bell, 2010; Schneider, Krajcik, Marx, & Soloway, 2002; Thomas, 2000). The benefit of embedding long-term science research projects into your classroom is that the projects actually get you to your goal, and often much further.

My student Dani is a perfect example of the benefits of PBL. She took a project initiated in class and turned it into a science fair winner. As Dani describes it, she was on a "downward spiral" when she entered my Space Science class in 11th grade. She had lots of negatives going on in her life, both in and out of school. She actually failed the first two marking periods. Her research project turned her around. One of the projects in Space Science is to design and construct a 3-D model of a constellation. Just imagine all of the science and math standards we are addressing with this one project. Learners are paired in an effort to promote collaboration. They design and construct their model, and present it to the rest of the class for critical review. But it doesn't end there. Following the critical review, the pair must revise the model based on feedback from the class and present the model a second time. Dani continued to work on revisions to her model well after my project requirements were met. One day I finally said to her, "Why don't you enter the regional science fair?" Her grades did a 180-degree turn, from F's in my class to A's. Dani invented the Constellation Cube, a clear Plexiglas cube with clear plastic rods running from side to side. Precisely placed on the rods were glow-in-the-dark beads. A person could take the cube into a dark room and see what the constellation Orion would look like from different points within the Milky Way galaxy. Dani went on that year to win first place at the Pittsburgh Regional Science & Engineering Fair. Today, nearly 5 years later, Dani states, "If it wasn't for you and those experiences I never would have graduated." Dani is not alone in her story. In my 25 years as an educator, I've observed the positive impact that authentic science experiences can have on students of all ages and abilities.

Creating an atmosphere of exploration in the classroom where students like Dani feel comfortable and motivated to explore ideas on their own begins on Day One and extends into each and every day that follows. Establishing a foundation where learners feel comfortable sharing their ideas and asking questions is an essential part of the authentic science environment.

Embedding Research Projects into the Regular Classroom

Not only do I implement smaller short-term projects on a weekly or bi-weekly basis; I also require my students to take on a self-directed team research project that spans the entire year. These yearlong projects begin early in the school year. Sometime around week 4 or 5 we complete a survey of astronomy or earth science topics, depending on the course. This is a quick look at the areas of current scientific research taking place in each field. Students are asked to spend some time thinking about the areas of research they find most interesting, and as a class we compile a list of the top five or six areas of interest. Students fill out an index card ranking their personal interest in each of the research areas, their favorite subject in school, and list the name(s) of anyone in the classroom they do not feel they could work with. Using that information, I place students into teams of two to four, trying to ensure that they get one of their top two research areas and are not placed with individuals they cannot work with. Because it typically indicates personal strength, when placing students in teams, I also try to take into account their favorite subject area in school. Placing students with different strengths on the same team helps ensure that the team has a variety of skills and can enhance the experience for all. With the teams established, they meet and decide which research area the group plans to focus on and who will be my point of contact for the team.

From there we're off and running. The teams typically work together at least one day a week. On occasion, I'll plan a structured activity for the day, but most of the time is spent working toward their research goals.

Goals for Student Research Teams

- **Goal 1**: Explore the area of research your team has selected and develop a question that is of interest to you and relevant to the scientific community. Present the question and your justification to the class for constructive critique.
- **Goal 2**: Explore what scientists currently understand about your research topic. Modify your research question(s) as needed, and prepare a summary of what scientists currently know relevant to your research question. Present the summary to the class along with any changes you made to your research question for constructive critique.
- **Goal 3**: Identify what type of data you can use to help inform the research question. Determine if this data is currently available or if it needs to be acquired through additional data collection measures. Revise the research question and the summary as needed.
- **Goal 4**: Develop a data plan. The plan should identify what data is needed, how the data will be acquired, when and where the data will be acquired, who will acquire the data, how that data will be stored, potential sources of error, and how the data will be analyzed. Revise the research question and the summary as needed. Present the data collection plan to the class along with any changes you made to your research question for constructive critique.
- **Goal 5**: Modify the data plan based on feedback from presentation.
- **Goal 6**: Collect the data. Modify data collection plan as needed, but be sure to make note of ANY changes made during data collection.
- **Goal 7**: Analyze the data the team has collected according to your data plan. Be sure to note any changes in the data plan.
- **Goal 8**: Consider the data you have acquired and the research question. If necessary, modify the data plan and research question. Collect and analyze any data as needed.
- **Goal 9**: Draw conclusions from your data. Identify potential sources of error and statistical significance.
- **Goal 10**: Prepare and deliver a written and oral presentation of your research project to the class. Write a formal research paper, and prepare a PowerPoint presentation.

Working through these ten goals takes time, patience, and guidance, but over the course of the school year—or semester, in some cases—I've enjoyed watching students grow and begin to think more deeply. In addition, their research often takes me outside my own comfort zone and pushes me to learn something new. And then there are those times when I just have to say, "This is outside my expertise and we need help." In that case, my students figure out who the experts are in that particular field of research, and they give them a call or send an email. We have rarely approached a professional scientist who is not willing to answer questions and provide assistance. In cases where scientists have helped, I try to make sure students send them a copy of their final research paper. I want learners to understand they are part of the scientific community. Even if they never engage in another research project, they walk away with an understanding of what it means to be a scientist. They understand what scientists do, making them more informed consumers of science.

There have been some great success stories at Oil City High School (OCHS) as a result of implementing authentic science practices in the form of research projects. In 1994, using an HOU telescope, my students Heather Tartara and Melody Spence provided professional astronomers with some of the earliest supernovae light curve data in history by capturing first light of SN1994I in the Whirlpool Galaxy (Pennypacker, Deustua, Perlmutter, Goldhaber, & Arsem, 1994; Richmond et al., 1996). Following the SN1994I, we moved on to hunting asteroids. Through the HOU Asteroid Search project, my students assisted in the discovery of 1998 FS144, one of the first 100 Trans-Neptunian objects (Pack, 2000). Over the years, Oil City students working with the project discovered a number of main-belt asteroids as well (Morelli, 2006).

While it's true that the network of scientists I've built over the years has provided opportunities for my students to engage in advanced research, authentic science projects need not be highly sophisticated. Sometimes they can focus on dispelling myths. My student Kim, for example, wanted to see if student behavior was really impacted by phases of the moon. She often heard the comment "It must be a full moon, because students are really acting up," so she secured 2 years of discipline report data (minus the names, of course) from the high school office. In her investigation, she used Excel to plot the number of student discipline referrals per day vs. phase of the moon. The graph was great. The data points were all over the place, with a correlation coefficient of +0.08, indicating no relationship whatsoever between student discipline and phase of the moon. Now, every time I walk into the office and hear the secretaries say, "The kids are acting crazy; it must be a full moon," I take pleasure in reminding them of the study completed by Kim that clearly demonstrated otherwise.

Authentic science isn't just for college-bound students, either. I implement research projects with non-academic students as well, but I often add a bit of a twist. Rather than a research paper and formal presentation, students create Myth Buster episodes similar to those on the popular television program. The teams still identify the question (myth), complete background research, design and conduct the investigation, collect data, and draw conclusions. The twist is that they use iMovie or Windows Movie Maker to present their study, just like it's done on the Myth Busters television program. Even the most complacent students get energized, and I've seen some great stuff come from kids who have traditionally been underachievers. One of my favorites over the years has been "Does the Five-Second Rule Work?"

Authentic science practices can be implemented through various projects and content focuses. But how do you handle planning, grades, state standards, and the like? As far as planning, you do what scientists do: you plan for the unexpected. That means being flexible with goals and expectations of the learners in the classroom, including yourself. Just like in any scientific study, things come up that you may not expect. As teachers, we are often quick to isolate ourselves. We think we need to have all the answers, but in reality, we don't, and no one else does, either. So think more globally. Don't just think, "How can I solve this problem?" Ask "Who out there can I contact to help solve the problem?" Perhaps it's a teacher down the hall, or a curriculum specialist, or a scientist at the local university or halfway around the world. Don't be reluctant to get on the phone or send an email; you'll be surprised at the responses and the support you'll get. Keep a list of these individuals and their contact information. Once they've made contact with you or your students, these individuals become part of your STEM community and can be valuable resources in the future as you build a network to support new and richer authentic science experiences for yourself and your students.

As far as grades go, I've always believed that everyone, regardless of background and ability, can—and should—contribute significantly to a project. Any successful science research team I've worked with has included individuals with different strengths who contributed in different ways. Each student receives a grade for his or her progress on the research project every 9 weeks, and the culminating presentation at the end of the year makes up 50% of the final exam grade for the course. Throughout the year, I ask my students to keep a research journal, and every time they do something for their project they are to record it. In addition, during the weekly research day, I meet with each group to discuss progress and make recommendations. I also observe individuals in the classroom and make notes for future reference. Near the end of each grading period, I evaluate the progress the team has made toward the goals established for that marking period. But giving everyone on the team the same grade has never seemed fair to me. So I also ask the researchers (students) to privately evaluate the participation of each team member, as well as themselves. Students are told they have 100 points to work with, and they must distribute them to each team member based on that student's level of contribution during the grading period. The more they feel the individual contributed, the more points they assign to that individual. In addition, they must justify in writing (to me) their distribution of points. The scores are averaged for each person on the team, indicating the percent of work contributed, and that percentage is multiplied by the total team points I award the team. This value then serves as the individual's grade on the research project. If I feel the distribution is unfair (based on my notes and what I've observed), I'll speak with each member of the team privately before I assign grades to individual students. In addition, students have the right to appeal their grades to their team and to me.

Using this grading approach provides an opportunity for all stakeholders to provide significant and useful input. Students do not hold back, and it ensures both collaboration and a competitive spirit within the group, similar to what I've experienced with professional science research teams. In my many years of engaging in this practice, I have not had a single case in which the grade distribution has not been amicably resolved. Simply put, it works. The key to this process is that students understand the system up front, and the teacher maintains awareness of what is happening within each research team.

When it comes to connecting learning to science, math, or language arts standards, why should I have all the fun? Authentic science takes place within a community of learners. As much as I want to understand the world of my students, I want them to understand my world. It helps if students understand the challenges facing teachers, too. So I point my students to the standards documents and ask them to identify which of the science, math, and language arts standards they are meeting with their research. Some people are surprised by my request, but with a little practice, students can be quite effective. Asking students to identify standards does two things: (1) it gives me a starting point from which I can add or delete standards being addressed, saving significant time, and (2) it makes learners aware of what the education institution itself expects from them. Often students are quite surprised by what they find in state and national standards documents. It's amazing how much easier life can be when teachers and students gain an understanding of each other's worlds.

Changing the Landscape to Facilitate Authentic Science in Your School

Let me be the first to say that making a change in how we approach teaching and learning is always a challenge. While making changes at your education institution can be difficult at times, here are a few tips that can help:
1. Pull together a collection of research/evidence that supports the change you wish to make. Just like knowing how to respond to your students when they ask why they need to know something, your administrators and colleagues want to understand why changes are being made.

2. Don't be afraid to discuss with your students why you're doing things differently. Share some of the research with them as well.
3. Be patient with yourself and your students. Over the years I discovered that shifting students from a traditional classroom to a more authentic science-based environment took between 6 and 9 weeks before learners started feeling comfortable.
4. Find an administrator who is supportive of your initiative and communicate with that individual on a regular basis. It's preferable that this person be your building principal, but it's not necessary. You may choose a department chair, curriculum specialist, or superintendent. Having one or two administrators on board can make all the difference in the world.
5. Explore the ideas of implementing authentic science practices with your colleagues. Is there another teacher in your building with whom you might collaborate? Having a "partner in crime" always makes life a bit easier.
6. Work to find opportunities for you and your students to work with scientists and engineers. Explore summer professional development opportunities as well as working with your local university. It's clear to me that teacher-scientist partnerships are perhaps the single most effective way of bringing authenticity to the practice of science in our schools.
7. When you and your students do something special (e.g., enter a science fair, hold a Skype meeting with a scientist working with students on their research, collect data for a project that will benefit the community, and so on), prepare a press release and send it to the local news media. Follow up the press release with a phone call to those at the news agency just to make sure they received it, and see if they have questions. This often leads to the reporter coming to your school to do a more in-depth interview with you and your students. You may need to work this through your media office at school, if the school has one.

IMPLEMENTING AUTHENTIC SCIENCE OUTSIDE THE REGULAR CLASSROOM

Inevitably, one of the consequences of implementing authentic science in your classes is that student interest will increase. What can you do with all that interest? It breaks my heart to see students with interest and motivation but nowhere to channel it. So I make sure students know that if they want to come to my room during their spare time to work on their research, they are welcome to do so. This can take place during lunch or study hall, but it typically occurs after school. Since the early 1990s, I've maintained an astronomy research team as part of the Astronomy Club at Oil City High School. Participation is not based on academic ability but on interest and motivation. And it's not always easy work.

I learned over the years to establish expectations early on. If students want to pop in after school every now and then to work with the astronomy research group, that's fine. But if they want to be part of the team that travels to NRAO, or the Annual Meeting of the American Astronomical Society (AAS) to present their research, they owe me time, and lots of it. Students understand from Day One that they are part of a team, and just like with the football or swim team, if they don't make practice, they don't play. I'm willing to share time with another sport, and from time to time I will speak directly with the other coaches to explain what students are doing, but I refuse to allow learners to think their participation in the science research team is any less important than their participation in another activity. Sometimes that means students have to make a choice, and that, in and of itself, is a great lesson in life. No matter who we are, we can't do it all. Just like in authentic science, in life we have to assess the situation, weigh the evidence, and make a decision.

Over the years, these teams have participated in a wide variety of research, the vast majority of which has an astronomy focus. Sometimes they're extensions of projects students are working on in class. Other times they're projects I'm working on, and occasionally they are student-developed projects. Early in my teaching career, the Astronomy Research Team designed and conducted investigations using the 40-Foot Radio Telescope at NRAO Green Bank. The team spent months planning the projects, then each year traveled 6 hours to the facility, spent several days collecting data, and upon their return home spent countless more hours reducing the data and summarizing and sharing their findings with astronomers at Green Bank. We also incorporated a number of projects using HOU optical telescopes.

Photo 6.1. Oil City High School Students Nick Kelly and Sandy Weiser Fill the Dewar on the Kitt Peak National Observatory's 0.9-Meter Telescope During an Observation Run. Students received support for observing time through the Kitt Peak Teacher Observing Program (TOP).
(Photo by Tim Spuck, January 29, 2007.)

In 2004, I spent time at the Kitt Peak National Observatory in Tucson, Arizona, and shortly thereafter joined the Spitzer Space Telescope Teacher/Student Research Program (now called NITARP). These professional development experiences brought new opportunities for students. Each year, thanks in part to support from these programs, my students travel to the AAS annual meeting to present their research. This is the largest annual meeting in the United States for professional astronomers, and Oil City High School students are there presenting and defending their research posters alongside graduate students and other professional scientists. It's amazing to see the transformation in maturity and confidence that takes place in these high school students between the morning of their presentation and the evening that follows. They come to realize they are doing real and meaningful science, and there are others out there who want to know what they know. These high school students see themselves as part of something much larger, a community of learners.

Neil deGrasse Tyson, world-famous astrophysicist and host of the PBS series *NOVA scienceNOW*, put it best during a radio interview with OCHS students at the January 2009 AAS meeting in Long Beach, California:

> There are a lot of projects in the universe that can serve the backgrounds of people that are still in high school. And somebody's got to do the work, and here we are and there they've done it. And so, that's part of what makes the community of astrophysicists so vibrant, and the participation of high school students so important. I see it as a kind of cog in this huge wheel that turns. They [high school students] become college students, then graduate students and postdocs, and senior faculty, and part of this research organism: this research organism that is the national community of astrophysicists. (*Interview with scientists and Oil City students*, 2009)

Both in and out of the classroom, there is only one true litmus test used prior to involving students in my research or approving their own. I ask myself, "Is there a possibility that the proposed research could enhance the body of scientific knowledge out there today?" If the answer is yes, then the rest of the equation is simply a matter of using solid scientific practices throughout the process. And as Dr. Tyson indicated, with a bit of training and mentoring, there's a lot of science that high school kids can do. One clear example was the development of a new way to identify potential T-Tauri stars using low-cost optical telescopes. I especially love this because it's an Oil City High School original and demonstrates what a mature authentic science program can do.

Over a 3-year period, working with several different groups of students, our team, with guidance from our mentor scientist Dr. Luisa Rebull, conceptualized and developed a method to identify young sun-like stars in their early stages of development. Using this new method, it's now possible for scientists to conduct a low-cost, all-sky survey and produce a catalog of potential T-Tauri star candidates.

Photo 6.2. During the 2009 Winter Meeting of the American Astronomical Society, Oil City High School Students Discuss Their Research with Dr. Neil deGrasse Tyson, Astrophysicist and Host of the PBS Series *NOVA scienceNOW*. Left to right are Matt Walentosky, Alex Holcomb, Samantha Wheeler, Neil deGrasse Tyson, Jennifer Butchart, Shana Kennedy, and Rachel Siegel. (Photo by Tim Spuck, January 7, 2009.)

Once identified, these candidates can then be followed up with observations using more sophisticated instruments.

Time outside the regular classroom provides unique opportunities for students. Not everyone wants to be a scientist, nor should they. These additional opportunities to explore science subjects provide valuable experiences for students who may want to pursue careers in science, or may not be part of your regular classes. Due to scheduling issues, Rachel Siegel never took a class from me, but she was an active participant in my after-school astronomy research group. Rachel was never really interested in being an astronomer, but she had given some thought to a possible science career. This is what Rachel has to say more than 4 years after her experience in astronomy research:

> My experience with the Spitzer Space Telescope Research Program served as the introduction to a world of science beyond the pages of high school-level textbooks. As I immersed myself in research, I felt empowered by small milestones I accomplished with the help of teachers and peers who shared my same passion for discovery. My participation in the Spitzer Program fueled my desire to become involved in research at the undergraduate level in my chosen discipline of science: biology. Currently, I am collaborating with an optometrist to study the effectiveness of several contact lens solutions on biofilm removal from silicone hydrogel and hydrogel contact lenses. In retrospect, without my early exposure to

science beyond the classroom, I cannot say for certain that I would have pursued my chosen career path. I owe a huge THANK YOU to my teachers and mentors who encouraged me. (R. Siegel, personal communication, January 20, 2013)

Another student, Matt Walentosky, is finishing his undergraduate degree in physics and plans to pursue a PhD in astronomy following graduation. He initially had no interest in astronomy. For that matter, when Matt first came to me at the end of ninth grade asking to be part of a team going to California for research, he wasn't selected. The following year Matt would not be denied. He became a leader in the astronomy research team each year thereafter. At the 2008 International Science & Engineering Fair (ISEF) in Atlanta, Matt took second place with his research on cataclysmic variable stars. He had his first summer research internship at the National Radio Astronomy Observatory prior to his freshman year in college and has secured similar internships every summer since. Matt has an amazing story, and has this to say today:

> I would say more than anything, attending the AAS meetings and different scientific endeavors in high school opened my eyes to the many different possibilities and careers available in science. Growing up in a smaller town kind of limits the different career opportunities most students are exposed too. When I first became involved in astronomy research the idea of a career in the pure sciences seemed very unrealistic and outlandish, but the more I became involved the more I could visualize myself doing it. (M. Walentosky, personal communication, January 3, 2013)

The list of students attending Oil City High School and benefiting from authentic science experiences is long. Throughout their high school careers many have been recognized at regional and international science competitions. Their names have risen to the top, in part of their own accord, but also thanks in significant part to the people from previous years that helped create the authentic science environment at OCHS. Each year I became a better mentor because of the students and scientists I worked with in the previous year. As my support network of STEM professionals grew, I gained confidence in my ability to enter the unknown and come out on the other side a better person than when I first entered. I grew from that boy with an inquisitive mind, trying to build a flying machine, but lacking the discipline and knowledge of the practice of science, into a person who can both understand the nature of science and communicate it to others. It was not a quick journey, but it was certainly a worthy one.

GETTING STARTED WITH THE AUTHENTIC SCIENCE RATING INSTRUMENT

In addition to embedding yourself in science research experiences with real scientists and engineers and getting your students involved with these projects, what else can you do to turn your science classroom into an authentic science-learning environment? Through years of experience engaging both myself and my students in authentic science practices, working directly with scientists and engineers, and analyzing related research, I've been able to identify common characteristics of authentic practices across disciplines. Using these characteristics, I worked to develop useful criteria to measure the level of authenticity of activities my students were engaged in. The Authentic Science Rating Instrument (ASRI) is the result of that effort (see Appendix to this chapter). ASRI allows the user to more objectively assess student engagement and provides a baseline from which you can quantify the improvement made over time. ASRI is easy to use and a great way to begin your journey to create and/or improve the authentic science-learning environment in your school.

An Example of the ASRI in Action

In order for you to get started, think of a lesson, lab, or other activity you do with your students, and let's put the ASRI to work for you. My example here is Galaxy Zoo, but as I go through and rate Galaxy Zoo, think about how you might score each of the 10 ASRI questions for the activity you've selected.

Galaxy Zoo is an online citizen science project. Participants log on and visually analyze galaxy images from the Sloan Telescope. They assess galaxy features and, based on those features, classify the galaxy. For more information on Galaxy Zoo and lots of great citizen science projects you can engage your students in, visit https://www.zooniverse.org. It's also important to note that the developers at Zooniverse are constantly making improvements to their web interfaces. It is likely that Galaxy Zoo will have been significantly improved by the time this document makes it into print. The rating below is based on my prior experience and does not include recent modifications that may have been made.

In providing a rating for this activity, let's answer each of the 10 questions from the ASRI measure (see Appendix).

Question 1: The analysis of each galaxy completed by the learner is entered into a database. This data is being used by the professional astronomy community in an effort to better understand the nature of galaxies and their distribution within the universe, so participants are contributing to the body of scientific knowledge. As a result the answer is "Yes" (1.5 points).

Question 2: Although learners do engage in some reading and exploration of background information before getting started, they are not required to "thoroughly" explore or summarize what scientists currently know about galaxy classification and evolution. As a result the answer would be "In Part" (0.5 points). (Note:

A teacher could engage students in a more thorough review and summary of the literature on galaxy classification and evolution, increasing the score.)

Question 3: The learner does use an interactive website to view and visually compare galaxies, but tools or instruments are not used to make direct measurements. As a result the answer would be "In Part" (0.5 points).

Question 4: Mathematics is not used by the learner in the analysis process, so the answer is "No" (0 points).

Question 5: The learner does use a variety of visual evidence to classify the galaxies, and as a result the answer is "Yes" (1.5 points).

Question 6: The learner does not have an opportunity to develop or refine the question. As a result the answer is "No" (0 points). Again, if the teacher could, in this case, modify the structure in a way that students could ask their own meaningful questions, the score could increase.

Question 7: The learner does not have the opportunity to develop or refine the procedures being used. For good reason, the web page structure controls all experimental methods. So, again, the answer is "No" (0 points).

Question 8: Participants in Galaxy Zoo do communicate, although indirectly, the results to their peers/colleagues. In addition, there is little if any opportunity to critique the work of their peers or provide feedback to the learner. The answer is "In Part" (0.5 points).

Question 9: By design, participants typically work by themselves at their own computer in isolation from others. In a classroom setting, a teacher could establish teams to promote an increased level of collaboration. In addition, there is an online discussion link for each galaxy analyzed. The discussion link does provide an opportunity to engage in some dialogue about the galaxy being classified. As a result, the answer is "In Part" (0.5 points).

Question 10: The result of their work is added to the database and recorded in a place where it is accessible and used by other scientists. So the answer here is "Yes" (0.5 points).

Based on my assessment, Galaxy Zoo has an ASRI score of 5.5 on a scale of 0–10. So you may be thinking, "Only 5.5 out of 10? That's not great." I use this example specifically because Galaxy Zoo is an excellent science resource for teachers and students. I use it on a regular basis to introduce galaxies and to engage learners in doing meaningful science that benefits the scientific community. Galaxy Zoo is not designed to provide the full "scientist" experience, but it does engage the learner in many aspects of authentic science. As teachers we can—if we so choose—make changes in how the activity is implemented with our students in a way that increases the ASRI score. So how about you: what is the ASRI score for the activity you selected?

It's important to note that while we may all desire a fully authentic experience for our students 100% of the time, the structure of schools themselves can prevent such a goal from being realized. For example, if you only have access to your students for 45 minutes a day, or one semester, or in almost all cases a single year, it may limit the experiences you can provide for your students. Further, limitations in resources can also be restrictive. So the idea here is to first evaluate your lessons, labs, and activities against the ASRI criteria. Once you have a baseline established—whether for a single activity, a course, or an entire science program—teachers and administrators can work together to improve the ASRI score over time, providing more authentic science experiences for learners of all ages.

CONCLUSION

My hope is that you've gained insight into what authentic science is and how I've worked over the years to create a more authentic learning environment for students. I hope you have also become motivated to use the ideas and tools in this chapter to help create, or improve, the learning environment you provide for your students. Sometimes when we try putting "real science" into a box it looks confusing or even overwhelming. As teachers we're used to being the "go-to person" with all the right answers, but in science there's only the best answer supported by current evidence. And as science and the tools scientist use advance, the body of evidence changes. That can be a scary fact to accept. It places both teacher and learner in a state of uncertainty—not a place either has traditionally been comfortable being in.

A key to understanding and accepting authentic science is recognizing that there are lots of pieces interacting within a changing environment. We can think of the authentic science environment as the "science soup du jour" where, through a medium, different pieces interact and influence each other over time. When we try to learn about carrots in isolation from the rest of the soup, we might increase our knowledge about the carrot, but what if, as a learner, I don't like carrots? I don't see a use for them, so I lose interest and stop paying attention. Isn't this what happens in our schools every day? We teach subjects, vocabulary, or concepts in isolation from each other, and students fail to see their meaning or how it relates to them. If I came to understand "carrots" in the context of how their flavor interacted with other spices, vegetables, and meats, do you think I might be a better soup maker? If our students are to learn science, then they need to practice science not in pieces, but as a whole, or as authentic science.

Fortunately, it's not that difficult to turn your science classroom into a more authentic learning environment. You don't need expensive equipment or vast STEM expertise. You simply need a real-world problem and problem solvers. Take a walk around your school or local community, read the newspaper, talk to locals working in STEM-related fields. There are science-related problems at every corner that you and your students can help solve. Implement yearlong research projects in your classroom with your students. Think about the roots of your understanding of science and potential misconceptions you might hold and ask yourself: "When was the last time I was a scientist myself?" If it's been more than 5 years, get out there and get another research experience. Start applying for programs that give you an opportunity to work directly with scientists and engineers. When you run into a roadblock, pick up the phone or send an email, and find someone to help out. Take two or three of your favorite science lessons and determine their ASRI score, think about how you might make them more authentic, and then give it a shot with your students. The important thing for all educators is to start somewhere and then keep moving forward.

To the scientists and engineers, and to the funding agencies, I say this: You have a vested interest in doing all that you can to make the learning our students experience more authentic. The health not only of our nation, but the entire planet, depends on it. If we desire a population that supports science, we must educate a population to authentically understand science. This cannot happen without real, living role models that both teachers and students can look to for guidance. Lasting partnerships between those doing research in science, technology, engineering, and mathematics, and those teaching these subjects, are essential. You make a difference!

As stated earlier, authentic science is not a prescriptive program that has a clear beginning and end. Nor is it a series of activities you engage students in, but rather an evolving environment you facilitate for both yourself and your students. Authentic science requires meaningful interactions between learners on a quest to solve real-world issues of the day. If you want tomorrow to bring something better, then you must begin to make changes today. Begin by empowering yourself and your students to do real science. I wish you well on your journey toward a more authentic science-learning environment.

APPENDIX

Authentic Science Rating Instrument Developed by Tim Spuck

The Authentic Science Rating Instrument (ASRI) is a simple tool that can be used to rate individual learning activities, or it can be used more broadly to examine a science course or an entire program. By asking a series of 10 questions relative to a proposed student experience, a level of science authenticity can be established as a quantitative value. The ASRI instrument is grade independent, and it can be used in the evaluation of activities and programs pre-K through grey. It is important to consider that the focus here is not to ensure a rating of 10, but rather to evaluate the implementation of current activities and programs and find ways to modify or replace activities so that the overall ASRI score is increased. Users of ASRI should be aware that the instrument is an early attempt at quantifying the authenticity of an experience in science, and improvements to the instrument may be made over time.

Using the ASRI

Individual Lessons, Labs, and Activities: Using the ASRI score sheet, ask the 10 questions relative to the planned lesson, lab, or activity, scoring each of the questions. The sum of the scores for questions 1 through 10 is the Individual Activity Authentic Science Rating (IAASR) for the individual lesson, lab, or activity. To improve the lesson or lab, modify the activity in a way that increases the overall IAASR score for the lesson, lab, or other student experience.

Individual Course Authentic Science Rating (ICASR): For each lesson, lab, or activity in the course, estimate the number, or fraction, of hours, both in and out of class, students are expected to engage in the activity. Refer to this value as the Number of Hours per Activity (NHA). Multiply the score for each individual activity (the IAASR) by the NHA for the activity. We will refer to the result as the Activity Course Rating (ACR). Calculate the ACR for all lessons, labs, and activities in the science course. Determine the sum of all ACR values for the course. Then determine the total hours students are expected to engage in all course-related activities, both in and out of the classroom. Refer to this value as the Total Course Hours (TCH).

ICASR = Sum of all ACR/TCH

As you modify individual course activities, the authenticity value for the course, or the ICASR, will change.

Science Program Authentic Science Rating (SPASR): The ASRI can also be used to gauge authenticity of the science program in your school or school district. Simply take the sum of the ICASR scores for courses in your school, or school district, and divide by the total number of courses. The SPASR is the average of all ICASR ratings for science courses in your school or school district.

NOTE: In the event that multiple individuals are providing scores for comparison purposes, the scoring team must be standardized to ensure a fair comparison across scores.

Asri score sheet

By engaging the learner in the proposed activity, will the learner …

1) work toward a solution to a real-world problem, provide the scientific community with answers to current or new science related questions, or contribute in a meaningful way to the body of knowledge the scientific community has access to?

　　___ Yes (1.5 points)　　　　___ No (0 points)

2) thoroughly explore and summarize the current information available on the subject being studied?

　　___ Yes (1 point)　　___ In Part (0.5 points)　　___ No (0 points)

3) use science instruments and technology (e.g., rulers, thermometers, computers, digital cameras, I-phones, data analysis software, microscopes, telescopes, spectrometers, etc.) to collect and analyze data?

　　___ Yes (1 point)　　___ In Part (0.5 points)　　___ No (0 points)

4) use "grade-appropriate" mathematics (e.g., math functions, graphing, plotting coordinates on map, derive equations, etc.) in the analysis of data?

　　___ Yes (0.5 points)　　　　___ No (0 points)

5) analyze evidence and use the analysis as a basis for drawing conclusions?

　　___ Yes (1.5 point)　　___ In Part (0.75 points)　___ No (0 points)

6) have the opportunity to develop or refine the question driving the activity, and to present new questions that come about as a result of their work?

　　___ Yes (1 point)　　___ In Part (0.5 points)　　___ No (0 points)

7) have the opportunity to develop and/or refine procedures or methods being used?

 ___ Yes (1 point)   ___ In Part (0.5 points)   ___ No (0 points)

8) communicate the methods used and results of their work to their peers/colleagues for review and critique, and engage in the review and critique of the work of their peers/colleagues?

 ___ Yes (1 point)   ___ In Part (0.5 points)   ___ No (0 points)

9) collaborate with others in meaningful ways throughout the process?

 ___ Yes (1 point)   ___ In Part (0.5 points)   ___ No (0 points)

10) record the results of their work where it is accessible to the broader scientific community?

 ___ Yes (0.5 points)   ___ No (0 points)

TOTAL SCORE ___/10.0

WORKS CITED

Anderson, R. D. (2002). Reforming science teaching: What research says about inquiry. *Journal of Science Teacher Education, 13*(1), 1–12.

Barton, P. E. (2003). *Hispanics in science and engineering: A matter of assistance and persistence.* Policy Information Report, Educational Testing Services. Retrieved October 6, 2013, from http://www.ets.org/Media/Research/pdf/PICHISPANIC.pdf

Battelle. (2011). *Battelle–R&D magazine annual global funding forecast predicts R&D spending growth will continue while globalization accelerates.* Retrieved December 16, 2013, from http://www.battelle.org/media/press-releases/battelle-r-d-magazine-annual-global-funding-forecast-predicts-r-d-spending-growth-will-continue-while-globalization-accelerates

Bell, S. (2010). Project-based learning for the 21st century: Skills for the future. *The Clearing House, 83*(2), 39–43.

Bencze, J. L. (2000). Procedural apprenticeship in school science: Constructivist enabling of connoisseurship. *Science Education, 84*(6), 727–739.

Blackawton, P., Airzee, S., Allen, A., Baker, S., Berrow, A., Blair, C., Fraquelli, L., et al. (2011). Blackawton bees. *Biology Letters, 7*(2), 168–172.

Casey, R. (2012). *STEM education: Preparing for the jobs of the future.* Washington, DC: U.S. Congress Joint Economic Committee.

Charney, J., Hmelo-Silver, C. E., Sofer, W., Neigeborn, L., Coletta, S., & Nemeroff, M. (2007). Cognitive apprenticeship in science through immersion in laboratory practices. *International Journal of Science Education, 29*(2), 195–213.

Chinn, C. A., & Malhotra, B. A. (2002). Epistemologically authentic inquiry in schools: A theoretical framework for evaluating inquiry tasks. *Science Education, 86*(2), 175–218.

Cronin, J. F. (1993). Four misconceptions about authentic learning. *Educational Leadership*, 50(7), 78–80.
Drew, C. (2011, November 4). Why science majors change their minds (it's just so darn hard). *The New York Times*. Retrieved October 6, 2013, from http://www.nytimes.com/2011/11/06/education/edlife/why-science-majors-change-their-mind-its-just-so-darn-hard.html
Edelson, D. C. (2003). Realizing authentic science learning through the adaptation of scientific practice. In B. J. Fraser, & K. G. Tobin (Eds.), *International handbook of science education* (pp. 317–331). Dordrecht: Kluwer Academic Publishers.
Fleischman, H. L., Hopstock, P. J., Pelczar, M. P., & Shelley, B. E. (2010). *Highlights from PISA 2009: Performance of US 15-year-old students in reading, mathematics, and science literacy in an international context*. NCES 2011-004. Washington, DC: U.S. Department of Education, National Center for Education Statistics.
Interview with scientists and Oil City students [Video/DVD]. (2009, January 7, 2009). Heim, M. (Director). Oil City, PA: WKQW-96.3 Radio.
Koebler, J. (2011). *Mexican, Indian students hungrier for STEM*. Retrieved January 10, 2013, from http://www.usnews.com/news/blogs/stem-education/2011/12/01/mexican-indian-students-hungrier-for-stem
Lee, H., & Songer, N. B. (2003). Making authentic science accessible to students. *International Journal of Science Education*, 25(8), 923–948.
Markowitz, D. G. (2004). Evaluation of the long-term impact of a university high school summer science program on students' interest and perceived abilities in science. *Journal of Science Education and Technology*, 13(3), 395–407.
Mathews, M. R. (2003). The nature of science and science teaching. In B. J. Fraser & K. G. Tobin (Eds.), *International handbook of science education* (pp. 981–999). Dordrecht: Kluwer Academic Publishers.
Metz, K. E. (2003). Scientific inquiry within reach of young children. In B. J. Fraser & K. G. Tobin (Eds.), *International handbook of science education* (pp. 81–96). Dordrecht: Kluwer Academic Publishers.
Michaels, S., Shouse, A. W., & Schweingruber, H. A. (2007). *Ready, set, science! Putting research to work in K–8 science classrooms*. Washington, DC: National Academies Press.
Morelli, B. N. (2006). *Using astronomical databases in the search for minor planets*. Unpublished manuscript.
National Research Council. (2011). *A Framework for K-12 Science Education: Practices, Crosscutting, Concepts, and Core Ideas*. Committee on a Conceptual Framework for New K-12 Science Education Standards. Board on Science Education, Division of Behavioral and Social Science Education. Washington, DC: The National Academies Press.
Newmann, F. M., & Wellage, G. G. (1993). Five standards of authentic instruction. *Educational Leadership*, 50(7), 8–12.
Pack, H. (2000). *The hands-on universe web-based asteroid school project and discovery of trans-Neptunian object 1998 FS144*. Paper presented at the Amateur-Professional Partnerships in Astronomy, 220 303.
Pennypacker, C., Deustua, S., Perlmutter, S., Goldhaber, G., & Arsem, E. (1994). Successful operation of remote telescopes for education and research. *Bulletin of the American Astronomical Society*, 26, 1423.
Pew Research Center. (2009). *Scientific achievements less prominent than a decade ago*. Washington, DC: Pew Research Center for the People & the Press. Retrieved October 6, 2013, from http://www.people-press.org/files/legacy-pdf/528.pdf

President's Council of Advisors on Science and Technology. (2012). *Engage to excel: Producing one million additional college graduates with degrees in science, technology, engineering, and mathematics.* Report to the President. Washington, DC: Executive Office of the President.

Rahm, J., Miller, H. C., Hartley, L., & Moore, J. C. (2003). The value of an emergent notion of authenticity: Examples from two student/teacher-scientist partnership programs. *Journal of Research in Science Teaching, 40*(8), 737–756.

Research and development FY 2013. (2012). No. 37. Washington, DC: American Association for the Advancement of Science.

Richmond, M. W., Filippenko, A. V., Bustamante-Donas, J., Moeller, M., Pawellek, C., Tartara, H., & Spence, M. (1996, January). UBV RI photometry of the type ic SN 1994I in M51. *Astronomical Journal, 111,* 327–339.

Roberts, L. F., & Wassersug, R. J. (2009). Does doing scientific research in high school correlate with students staying in science? A half-century retrospective study. *Research in Science Education, 39*(2), 251–256.

Robinson, G. (2004). Replicating a successful authentic science research program: An interview with Dr. Robert Pavlica. *Journal of Secondary Gifted Education, 15*(4), 148–154.

Schiavelli, M. (2011, November 3). STEM jobs outlook strong, but collaboration needed to fill jobs. *U.S. News and World Report,* p. 1.

Schneider, R. M., Krajcik, J., Marx, R. W., & Soloway, E. (2002). Performance of students in project-based science classrooms on a national measure of science achievement. *Journal of Research in Science Teaching, 39*(5), 410–422.

Thibodeau, P. (2012). *China set to surpass U.S. in R&D spending in 10 years.* Retrieved December 24, 2012, from http://www.computerworld.com/s/article/9234976/China_set_to_surpass_U.S._in_R_D_spending_in_10_years_

Thomas, J. W. (2000). *A review of research on project-based learning.* San Rafael, CA: Autodesk Foundation. Retrieved October 6, 2013, from http://www.bobpearlman.org/BestPractices/PBL_Research.pdf

Trumbull, D. J., Bonney, R., Bascom, D., & Cabral, A. (2000). Thinking scientifically during participation in a citizen-science project. *Science Education, 84*(2), 265–275.

Zion, M., Slezak, M., Shapira, D., Link, E., Bashan, N., Brumer, M., Valanides, N., et al. (2004). Dynamic, open inquiry in biology learning. *Science Education, 88*(5), 728–753.

CHAPTER SEVEN

Engaging Young Minds TO BE Tomorrow's Innovators

ARUNDHATI JAYARAO

INTRODUCTION

I am enough of the artist to draw freely upon my imagination. Imagination is more important than knowledge. Knowledge is limited. Imagination encircles the world.
—ALBERT EINSTEIN

Innovation, inspiration, communication, collaboration, creativity, problem solving, and twenty-first-century skills are some of the buzzwords being tossed around by policymakers, news reporters, education researchers, economists, scientists, industrialists, CEOs, hiring managers, and many others these days. They are seemingly the links between K–12 education and U.S. global competitiveness. In September 2010, the President's Council of Advisors on Science and Technology (PCAST) released a 142-page report entitled *Prepare and Inspire: K–12 Science, Technology, Engineering, and Math (STEM) Education for America's Future*, outlining "the road ahead to regain and maintain U.S. competitiveness" (President's Council of Advisors on Science and Technology, 2010). The PCAST report made several recommendations, including one asking that students be given opportunities for inspiration through individual and group experiences inside and outside the classroom. It calls upon the federal government to develop a coordinated initiative that PCAST terms "INSPIRE, to support the development of a wide range of high-quality STEM-based after-school and extended day activities (such as STEM contests, fabrication laboratories, summer and afterschool programs, and similar activities" (p. ix).

PCAST purports that such activities, as well as the presence of STEM mentors and role models, would inspire students, particularly girls and members of minority groups, who are underrepresented in science and engineering, to study STEM subjects and enter STEM professions. The PCAST recommendations follow several reports that speak to U.S. global competitiveness and its link to U.S. economic prosperity. A clarion call initiated by the 1983 report *A Nation at Risk: The Imperative for Educational Reform* (National Commission on Excellence in Education, 1983), revived by the 2007 report *Rising Above the Gathering Storm: Energizing and Employing America for a Brighter Economic Future* (Committee on Science, Engineering, and Public Policy, 2007), and reinforced by the 2010 update, *Rising Above the Gathering Storm, Revisited: Rapidly Approaching Category 5* (National Research Council, 2010), underlines the importance of a robust STEM education, a workforce requirement and an urgent need for action to produce innovators and entrepreneurs.

Innovation, inspiration, and creativity are not unfamiliar concepts in American history. In fact, a recent exhibit, "The Great American Hall of Wonders," held from July 2011 to January 2012 at the Smithsonian American Art Museum in Washington, D.C., in collaboration with the U.S. Patent and Trademark Office, examined

> the nineteenth-century American belief that the people of the United States share a special genius for innovation. It explores this belief though works of art, mechanical inventions and scientific discoveries, and captures the excitement of citizens who define their nation as a "Great Experiment" sustained by the inventive energies of Americans in every walk of life. (Smithsonian Newsdesk, 2011)

So has the education fabric changed so dramatically from the nineteenth to the twenty-first centuries for our nation to focus attention on the exigency of devising novel methods of teaching and fresh policies of education reform to produce innovators? Who are innovators? Is one born an innovator? What is innovation, anyway? Is innovation a teachable skill? Can innovation be part of a regular K–12 classroom? Can one inspire to innovate? Is imagination a learned skill? What of creativity? What roles do problem solving, collaboration, and communication play in the teaching of innovation? Can a classroom be designed as a hub of inspiration, imagination, and creativity that spins the wheel in the making of innovators? Possibly you have pondered many of these questions yourself.

As a non-traditional entrant from a full-fledged physics research career into the field of education, I have dared to ask these questions myself. Science and mathematics fascinated me; even as a young girl, I was drawn to the fields of physical sciences and mathematics, charmed by the sheer logic of the subject matter. As I forayed into the field of education, donning the role of a high school physics and chemistry teacher at an independent girl's school, I craved communicating my passion for the physical sciences to the young ladies in my classroom. Easier said than done. As I taught the classes, I recognized that problem solving, thinking outside the box, and extending classroom concepts to the real world posed major hurdles to science appreciation. What I found logical as a young girl (and continued to find) seemed to drive my girls to the edge of a precipice. I had to improvise and think creatively to excite my students, or I knew I would lose them! To my good fortune, I had the support of my school administrator, who permitted me a certain amount of leeway in my teaching methods that I might not have had in a traditional public school setup. Armed with some freedom to experiment in lesson delivery, I introduced several programs that engaged students in active learning and infused the delight of project-based learning in my classroom.

This chapter will illustrate how an educator can employ science competitions, engineering challenges, and project-based learning to successfully promote student interest, increase student engagement, and inspire young minds toward innovation. It will also discuss effective means to increase students' confidence in their ability to participate in engineering activities, motivate students to consider physical science and engineering majors and careers, and promote peer-to-peer learning in collaborative environments.

BACKGROUND

I've missed over 9,000 shots in my career. I've lost almost 300 games. 26 times I've been trusted to take the game-winning shot and missed. I've failed over and over and over again in my life. And that is why I succeed.
—MICHAEL JORDAN (GOLDMAN & PAPSON, 1998)

The Role of Competitions in Education

During the academic year, students can be exposed to several extracurricular opportunities related to science challenges or team competitions at local, regional, and national levels that induce independent or collaborative learning outside the classroom. Conversely, goals and driving principles from the competitions can be suitably modified to fit a classroom environment or serve as alternative assessments or evaluative instruments for classroom learning.

The role of competitions in education has been studied extensively and has its share of proponents (Verhoeff, 1997; Fulu, 2007; Lawrence, 2004) and opponents (Lam, Yim, Law, & Cheung, 2001). Verhoeff (1997) argues that well-organized competitions are a source of motivation for students to perform at their best and are excellent motivators for learning. Fulu's (2007) research presents evidence of several positive outcomes, including recognition of student learning by external bodies, student motivation and engagement, and gain in self-esteem and skills in cooperative learning (Johnson & Johnson, 1988). Lawrence (2004) discusses competitive programming and concludes that a combination of game development and friendly student competition is a significant motivator for increased student performance.

Lam and colleagues (2001) studied two groups of 7th-grade students that were randomly assigned competitive and non-competitive conditions in a Chinese classroom setting. They concluded that the competitive group was more prone to negative self-evaluations after failure at a task, and performance oriented to the extent of sacrificing learning opportunities for better performance. There are other studies that point to the efficacy of cooperative learning. In a study involving 80 engineering students, the authors Cantador and Bellogín (2012) organized a competition where students had to collaborate, resolve, and evaluate questions and exercises about specific topics on the subject. The study was further tweaked to incorporate a team-based competition with characteristics and elements of cooperative learning. Cantador and Bellogín concluded that competitions promoting positive interdependence of team members, a balanced workload, face-to-face interactions and social skills, and individual accountability were successful in helping students focus on the learning instead of the victory and prizes associated with the competition.

Teaching and Learning Innovation

The latest mantra chanted in education circles today is the urgent need to teach students to be innovative and creative. Be it at the K–12, postsecondary, or graduate research level, educational and research bodies across the nation are experimenting with revolutionary models of teaching and learning that focus on harnessing the creative and collaborative ability of all learners. In his book *Creating Innovators: The Making of Young People Who Will Change the World* (2012a), Tony Wagner, a high school teacher, K–8 principal, and Innovation Education Fellow at the Technology & Entrepreneurship Center at Harvard, promotes the theory that with the abundance of information and knowledge available at the click of a mouse, the only skill that sets apart a class of learners today is their capacity to innovate. Wagner's book itself incorporates innovation in that it uses Microsoft's tag technology to embed video links to live interviews throughout the book. Wagner, in his online essay "Educating the Next Steve Jobs (2012b)", underscores the importance of allowing students to take risks, encouraging perseverance and not penalizing failure. He concludes:

> My most important research finding is that young innovators are intrinsically motivated. The culture of learning in programs that excel at educating for innovation emphasize what

I call the three P's—play, passion and purpose. The play is discovery-based learning that leads young people to find and pursue a passion, which evolves, over time, into a deeper sense of purpose. (Wagner, 2012b, p. 1)

A growing number of higher education establishments, including Olin College of Engineering (*About Olin College*, 2012), the M.I.T. Media Lab (MIT Media Lab, 2012) and the D-school at Stanford (d.school, 2012), are developing new approaches to teaching innovation and fostering creativity through collaborative and cooperative learning. Stanford design school's K12 lab, an educational research program, works on transforming teachers into designers. The K12 lab's flagship project is the creation of a design thinking space for a K–8 school, a 3,500-square-foot Innovation Lab at the Nueva School in Hillsborough, California, that applies design processes to teaching and learning. Piggybacking off concepts of product design, the K12 lab developed a design-thinking methodology for education that approaches learning as an iterative process wherein each failure leads to the exploration and discovery of new ideas. Through teacher workshops (K12 Laboratory *Workshops*, n.d.), graduate classes (K12 Laboratory *Classes*, n.d.), and their work with industry, public, and nonprofit partners (K12 Lab Partners, n.d.), the K12 lab has documented evidence in producing teacher innovators (K12 Laboratory, *Teachers as Innovators*, n.d.) and young student designers (*Designers at the Lab*, n.d.) who can transform the classroom into a palpably vibrant and ingenious platform for teaching and learning innovation.

Several foundations and organizations have formed professional learning communities (PLCs) to strategize new approaches to teaching twenty-first-century skills that combine the rigor of the 3Rs—reading, writing, and arithmetic—with the urgently needed 4Cs—critical thinking, communication, collaboration, and creativity. Some examples of successful implementations of such PLCs are the Hewlett Foundation's Deeper Learning Initiative (Deeper Learning, 2012) with its affiliations with over 500 K–12 schools, EdLeader21 (EdLeader21, 2012) with its consortium of 100 school districts and the Partnership for 21st Century Skills, P21 (Partnership for 21st Century Skills, n.d.), a coalition of education nonprofits, foundations, and businesses. Under its Innovative Schools Program (Microsoft in Education, 2013), Microsoft has forged partnerships with schools globally and in the United States to achieve a record of innovation and successful change implementation. Microsoft asserts that its partnerships have resulted in innovation in curriculum design, professional collaboration aided by technological advancement between teachers across the world (Partners in Learning Network, 2013), digital-age literacy, and inventive thinking. The Partners in Learning program has also contributed information and policy insight on effective education transformation that is taking place around the world.

Project-Based Learning (PBL)

The notion of projects in a classroom is definitely not new. Educators of all disciplines have been employing this tool for several decades now. There are two other methodologies—project-based learning (PBL) and problem-based learning—and some educators have interchanged the two freely. In fact, I would go a step further and also throw in two other terms: inquiry-based learning and the more recent game-based learning. There is a clear distinction between just doing projects and employing any of the other four learning methodologies. Other authors have addressed inquiry-based and game-based learning in this book; hence I will not delve into those details except to say that there are clear distinctions between inquiry-based, game-based, and project-based learning methodologies.

As far as a project and PBL are concerned, the distinction (*The Difference Between Projects and Project-Based Learning*, 2012) is very clear. A project is generally assigned with specific directions and questions, has a rubric, and is not always relatable to either the real world or a student's world. Project-based learning, on the other hand, is far more complex in that it completely organizes learning around the project and grants the student independence of thought and execution.

While project-based learning and problem-based learning share some characteristics, a major difference between the two is that in problem-based learning, the problem is well defined by the educator, and students work in teams or individually over a period of time to solve it. This is an approach frequently and aptly used in medical, architecture, and business education. In project-based learning, there is no project or problem specified. Student voice and choice is very important, and students control the project they will be working on. Moursund (2007) has delineated the differences between the two learning methodologies as applicable in mathematics education.

There are some detractors (Kirschner, Sweller, & Clark, 2006) that argue against PBL as well as the other learning methodologies listed above as being minimally guided and not conducive to learners with little prior knowledge of the subject. However, there is a vast body of research (Buck Institute for Education, 2013) on the efficacy of project-based learning and a pedagogy that supports the notion that students learn best by tackling and solving real-world problems, and that students have increased control over their learning. One of the leaders in project-based learning is the Buck Institute for Education (BIE) (2013). BIE is a not-for-profit organization that has dedicated efforts toward education instructional reform by creating and disseminating products, practices, and knowledge for effective project-based learning. The Buck Institute website has extensive literature, research, and implementation modules. A good place to start is the paper on the fundamentals for PBL (Larmer & Mergendoller, 2010) that describe the seven essentials of project-based learning. Larmer and Mergendoller emphasize the importance of a good driving question or meaningful inquiry that is compelling, open-ended, and captures the essence and purpose of the project. An excellent review paper by Thomas (2000) discusses the success of PBL when it first started 2 decades ago.

BEST PRACTICES

Engaging Young Minds at Oakcrest School

"I respect your scientific background but do realize that we are certainly not a TJ (Thomas Jefferson High School for Science and Technology, based in Alexandria, VA) ... science and technology is not the top agenda or mission of our school," cautioned the Head of Oakcrest School in response to my statement that I am passionate about science, and enthusing girls about science would be my top priority at Oakcrest. I was interviewing for a science teacher position at Oakcrest School in July 2002. I took the challenge and plunged headlong into a teaching career after years of research in the field of quantum optics. What followed over the next 7 years was a discovery of the joys of teaching. I owe a ton of gratitude to the head of my school for investing confidence in me and allowing me the latitude to experiment with my teaching philosophy—a philosophy laced with a driving motivation to produce young minds that might revel in the delight and wonder of the natural and physical world, learn out of curiosity, and be empowered to innovate.

I embraced a teaching career with two primary goals: (1) to inculcate a passion for STEM in young minds, and (2) to increase the number of girls in the fields of physics and engineering. The experience and training I underwent as a scientist showed me a logical path to learning, appreciating, and communicating science. To my teaching advantage, the path I naturally adapted and applied to induce children into science—whether they were my own children, the children I helped out as a PTA volunteer, or the students in my classroom—helped me to be successful at fostering children to be innovative thinkers and learners. It was only much later that I discovered that what I was pursuing successfully in my classroom modeled the underlying philosophy of project-based learning!

At the outset, I would like to respond to a few questions that other teachers have asked me. I was asked about content delivery, completion of syllabus, and validity of grading of group or individual projects. I have never compromised on what curriculum needs to be covered for a grade level and what depth I need to go into for the curriculum. All the classes at Oakcrest School were at honors level. That does not imply that non-honors students cannot engage in the projects listed. In addition to the detailed laboratory-oriented and problem-based course curriculum, I involved all my classes in a variety of learning activities, including these:

- Research projects involving:
  - Writing papers
  - Collaborative & independent learning
  - Oral presentations
- Open-ended projects on
  - "Chemistry in Everyday Life"
  - "Chemical Innovations & Applications"
- Participation in National Mole Day, National Chemistry Week, World Water Monitoring Day, and related activities
- Participation in regional and national science competitions:
  - American Chemical Society's Chemagination Challenge
  - Dupont Science Essay Challenge
  - National Science Teachers Association (NSTA)/Toshiba Exploravision Contest

To teach a student how to innovate, one has to allow the student to initiate the learning process; to think "outside the box"; to stick with a problem but also realize when a new approach needs to be taken; to be very comfortable with failure, but also ready to learn from it and continue moving forward; and to know how to work on a team and collaborate to produce an end product that might really not have a single owner. Some of the key principles I used in my classroom projects are listed below. One will find any project-based learning authority delineating similar principles (Project Based Learning for the 21st Century, 2013), for instance, Buck Institute for Education.

1. Engage student in active learning: Underscore the importance of student voice and choice. The student owns the project and learning process.
2. Ask a driving question: Prod the student to question why, where, and what: why does the student need to know this information, where is the relevance in real life to the knowledge acquired, and what will I do with this knowledge? The student asks the driving question/s; critical thinking begins with asking really good questions. Forming a problem is as important as solving it. An average 4-year-old asks hundreds of questions, but as the child grows up, this skill seems to be either schooled out or parented out or both. It is important for a teacher to realize that asking questions is neither disruptive nor annoying.
3. Persevere: In science, answers are never straightforward. Many of us know that experiments never seem to work out the way we would like them to.

Scientists try several different approaches before being successful. The crux of innovation is perseverance. For a student to understand that failure is a stepping-stone to success, that a solution to a problem or even a design for a question does not immediately take place, is as important as finding a solution to a problem. Genius is indeed 99% perspiration and 1% inspiration. Most innovators have some common strands: they are adaptable and receptive, do not penalize failure, and are cognizant that taking risks is an important lesson in learning to be innovative.

4. Team work: Brainstorm, collaborate, and communicate. Being able to brainstorm, piggyback on ideas, and learn to pull together disparate thoughts to form a solution to a problem is another important skill for an innovator. Thinking "outside the box" is a skill that can be cultivated. It is a result of combining and practicing these three processes. This involves teamwork—understanding how to collaborate, cooperate, and work on a team to produce a product that is an amalgamation of thoughts. This is the kind of creative collaboration that innovators need to engage in to hone their skills. Innovation is most often a team sport.
5. Understand how to seek feedback and revise one's work: Again, a skill that can be taught by a combination of points 3 and 4.
6. Publicly present the product: Any innovator has to be able to successfully present the idea and product. Unless you understand completely what it is that you have tried to solve and innovate, it is hard to speak about it. If a student owns a process, he or she is much more able to communicate the solution or innovation.

Teaching Innovation to Oakcrest Sophomores

The first thing I decided to do was to introduce the students to a sampler of assignments to come. I would start the students on a weeklong, or couple of days, or overnight-deliverable sort of assignment to introduce them to thinking "outside the box." This is what I call "the appetizer project."

At the start of the year I would introduce a mini project titled "My Chemistry World" or "Chemical Limericks" or "It's Elemental, Oakcrest." I offered a different one each year. The purpose of this mini project was to set the tone for learning for the rest of the year. The project delivery coincided with our back-to-school nights, so I was able to introduce parents to my teaching philosophy, expectations in the classroom, and where it would be leading. The first few weeks of chemistry revolved around measurements, lab safety, lab apparatus, chemical and physical changes, and an introduction to the periodic table. For instance, students could create limericks related to any of the topics we had covered, or they would be asked to look around to describe what reminded them of chemistry class in their homes. This nanoscale project, as I referred to it, was a very open-ended project worth a quiz grade. The end result I was looking for was for students to take back what they had learned in class and see it being applied in the real world. They were required to draft and submit a visual on a piece of paper. Sometimes the class display would form the slogan of the project, with each student assigned an alphabet as a visual promoting his or her chemical curiosity.

As we moved into October, I made full use of National Chemistry Week, National Mole Day, and World Water Monitoring Day to introduce a major project and several associated activities.

Start with a Bang and Definitely Do Not End with a Whimper

Do celebrate the science. I cannot stress enough the importance of both the introduction to the project and the project culmination events. Projects should be promoted by novel opening events and end with a public presentation by the students that recognizes their efforts.

National Chemistry Week was celebrated in a big way at Oakcrest. The sophomores, or Sopho-moles, as they started calling themselves, were engaged in various activities. October began with my 10th graders being asked to come up with chemistry experiments that they could devise and lead 6th graders in. The students worked in teams of three to search the Internet and come up with age-appropriate experiments that could hook younger children on chemistry. Although the sophomores were still in the initial stages of their chemistry course, they could still lead 6th graders through lab safety and simple chemistry experiments that mainly involved household chemicals. The idea was to have a lab that was mentored by 10th graders and offered about six to eight activities, each designed to be completed in 10 minutes, and for 10th graders to feel the thrill of communicating the joy of science experimentation. Chemistry was and always is a great subject to get young ones hooked. It was colorful, it foamed, it fizzed and popped, and it produced vivid visible changes, all of which had the students ooh-ing and aah-ing.

This would take place during the double-period lab days, so some amount of collaboration between teachers of both grades was required, as well as schedule matching. I had two sections of 10th graders lead two sections of 6th graders. Here is a sample list of the activities that the 10th graders came up with:

1. Cabbage magic: use the red cabbage juice as a universal indicator and test household products for acid or base properties
2. Elephant's toothpaste
3. Baking soda volcano experiment
4. Density of copper pennies
5. Tensile strength of spaghetti and comparison of different types of spaghetti

6. Hydrophilic qualities of milk
7. Acids and bases to write secret messages

The 10th graders were allowed to freely search the Internet and laboratory websites to come up with experiments or modifications of experiments that they could understand and be excited enough about to want to lead young ones through. My experience has been that it empowers the 10th graders when they set up experiments themselves, leads to pride of ownership, and builds a capacity to communicate knowledge, thereby building a better understanding of the subject. In turn, the 6th graders loved the peer learning process and viewed the whole experience as an enjoyable field trip.

This would qualify as a 2-week project, since students had to formulate an inquiry-based science experiment that they could lead and mentor. It had the students' voice and choice. A bonus for me from this project was motivation for 10th graders as mentors to take lab safety seriously. The assessment for this short project was based on the lab activity, the originality, the mentoring, the follow-up lab report, including the scientific details about the experiment, plus:

- any questions asked by 6th graders that the student was able to clarify,
- any questions raised by the 10th grader herself, and
- any modifications or new activity related to the experiment.

Of course, there are always students who will take the shortcut by a wholesale copy and paste, but this can be circumvented by using a turn-it-in option.

As part of the National Chemistry Week celebrations, I also celebrated Mole Day. This started out on a small scale within the classroom but turned into a school-wide celebration within a couple of years.

I subscribed to the National Mole Day Foundation (National Mole Day Foundation, 2013) and obtained the Mole Day theme song and Mole Day paraphernalia. The AP chemistry students would lead the entire school body with a pledge to the mole. With Mole Day being so close to Halloween, they would demonstrate Halloween chemistry magic activities. The sophomores followed up with a Mole Day song and dressed up to relate to the Mole Day theme. At the end of the assembly, enthusiastic 9th graders would yell out that they could barely wait for 10th-grade chemistry. Later that year, I bumped into a 6th grader at a pet store and was pleasantly surprised when she reeled off "Hi Dr J, the mole is 6.0×10^{23}"! I knew then that science fun had arrived at Oakcrest!

My longer projects involved the American Chemical Society's Chemagination Contest (ACS Chemagination, 2012) and the Toshiba/NSTA Exploravision Contest. The introduction to these long-term projects, as I mentioned, started with a bang. After the Mole Day assembly, I would introduce two main projects,

ACS's Chemagination Contest and the Exploravision Challenge to both sophomores and AP students. The Chemagination contest was mandatory for all 10th graders, whereas the NSTA Exploravision project was optional and open to any willing science students in grades 10–12.

The American Chemical Society's Chemagination Contest (ACS Chemagination, 2012) asks high school students to imagine that they are living 25 years in the future and have been invited to write an article for ACS's *ChemMatters*, a magazine for high school students that focuses on the role of chemistry in everyday life. The object of the article is to "Describe a recent breakthrough or innovation in chemistry (and/or its applications) that has improved the quality of people's lives today" (ACS Chemagination, 2012). In addition to the article, students are asked to design a cover for the magazine. The article must be written as if the students are living 25 years in the future, looking back at innovations that have occurred since the present time. The innovation must fall under one of four categories: Alternative Energy Sources, Medicine/Health, Environment, or New Materials. As the topics are very broad and open ended, students have the freedom to select a category of their choice. The students are required to work in teams of two or three. Evaluation of the entry is based upon:

- the written article, which is submitted in advance, and
- the presentation of the innovation on a self-standing display and interviews with judges (much like science fair judging).

This contest started out small and grew to be large scale, with Oakcrest actually hosting the private schools in the Maryland-Virginia area. Students presented their work to scientists who came in as judges, and the winning teams were sent to present their innovation at the Mid Atlantic Regional Meeting of ACS in New York. The project involved all the elements of innovative thinking, learning, and teaching. The project was launched in October, with the final presentation in May.

The NSTA/Toshiba Exploravision Contest (Exploravision: Today's young minds drive tomorrow's innovation, 2012) was the other long-term project. Each year that I taught, I offered this as an optional, extra-credit project and encouraged all my chemistry, AP chemistry, and physics students to participate. Exploravision asks students to imagine what technology might be like in the future. The competition was another excellent way for students to learn how to work in collaborative learning groups on an interdisciplinary project. It taught them how to innovate and work effectively in teams of two to four students, and stick with specific deadlines as laid out by the competition organizers. In 2007, one of my teams was declared regional winner. "Oakcrest Sophomores Explore the Future of Technology" screamed the *Arlington Catholic Herald* (Arlington Catholic Herald, 2007) headline. "Students Named Regional Winners," announced *The Connection*

Newspaper (The Connection, 2007) that joyous morning in March 2007. It was a proud moment for me as the coach of three young women who conceived a nanotech futuristic device that notified and treated a person having a heart attack. This team exemplified innovation, perseverance, and teamwork. The team started out with several ideas, and we held periodic discussions. I would listen to them brainstorm and formulate new ideas. They researched various technology innovations. They finally zeroed in on an innovation and were working hard at it until they stumbled on a research paper that indicated the project was currently being pursued in labs. This meant the students had to scramble to get on another project and work even harder to meet the deadline. The students willingly persevered through the Christmas break and put together another project that went on to become a regional winner.

I highly recommend the Exploravision project for several reasons. This is a project that can be used in any grade from K to 12. It is a team project, is interdisciplinary, teaches students the basic principles of innovation, and fosters imagination and creativity. It is a great project that allows students to learn how to brainstorm, piggyback on ideas, and collaborate as a team while allowing them complete freedom in choice of discipline. In addition, it is one of those competitions that is all-inclusive, being open to all K–12 students without any bias toward outstanding academic achievers.

Teaching Innovation to the Oakcrest AP Class

At our school, AP Chemistry was offered only in 11th grade, and students were required to complete the 10th-grade chemistry course before they could take the AP. Because Oakcrest is a small private school, I was teaching the entire sophomore class, and my AP class consisted of students who had already been exposed to my teaching philosophy for a year. AP students were assigned a major responsibility in choosing, setting up, and demonstrating more complex experiments in the school-wide assembly. Since Mole Day came just before Halloween, there was no better bait to hook students on chemistry than Halloween-themed demonstrations. Each year the AP students had to come up with a fresh set of experiments that would have a dramatic effect. AP students made wide use of Flinn, or any other chemical company manuals used for ordering laboratory chemicals, to select experiments that the AP students were confident they could explain and demonstrate safely. We used experiments such as Witches Brew, elephant's toothpaste, Halloween "chemmystery," and the disappearing pumpkin. Experiments involved concepts of catalysts, decomposition reactions, exothermic reactions, heat-shrinking plastics, polyurethane foam, polymers, oxidation-reduction reactions, fluorescence, chemiluminescence, and triboluminescence. AP students demonstrated the experiments to the entire student body and answered any questions that students asked. This provides an example of giving students choice and voice, independence and leadership. Students also learn from this experience that experiments do not always go as planned. They learn that scientists improvise and come up with different approaches when an experiment refuses to work. They take ownership of the demo and practice it, and train themselves to anticipate failure. Lessons learned are to be innovative in the face of sudden experimental failures, and to be confident enough to explain the science and venture an explanation on why an experiment might have failed—all these delivered extempore to a school-wide assembly consisting of peers and adults.

I believe in allowing students to have a feel for real-world science and what happens in a research lab. Toward this end I have liaised with local labs and developed a network of scientists who are willing to offer summer internships to students. Students regularly intern at National Institute of Health, Cancer Research Institute, National Aeronautics and Space Administration, Udvar-Hazy Museum, Smithsonian Air and Space Museum, Walter Reed National Military Medical Center, and other labs. In addition to providing real-world experience and connections to classroom learning, the internships open up opportunities to present student research work at regional science fairs.

I was constantly engaged in opening up symposiums, workshops, science challenges, and other opportunities as they presented themselves. AP students were regularly nominated to attend junior science symposiums. Unfortunately, this could be offered to only a couple of students who expressed interest in research.

The way our academic year was structured, there was not much class time remaining after the students took the AP exam. I would arrange for open-ended lab investigations that allowed the AP students to have a more relaxed approach to class work. In addition, since the curriculum was completed, the students were in a position to appreciate lab work and better predict where their investigations might lead.

I have also tried to encourage AP students to take up the DuPont Science Challenge (The DuPont Science Challenge, 2012). The DuPont Challenge calls upon students to perform research, think critically, and write a science essay that provides innovative ideas on our most pressing global challenges, or demonstrates the application of science, technology, engineering, and mathematics (STEM) to our daily lives. The DuPont Challenge asks students to consider our most important challenges by researching and writing a 700–1000-word science essay in one of the four categories:

- Together, we can feed the world.
- Together, we can build a secure energy future.
- Together, we can protect people and the environment.
- Together, we can be innovative anywhere.

I have variously used the principles in these competitions to offer alternate assessment modules for classes. It is interesting (or should I say satisfying?) that the science competitions that I introduced to my students 10 years ago are today the highlight of White House Science Fairs!

Teaching Innovation to Oakcrest Physics Students

A physics teacher in a girl's school has to motivate students to learn in a field viewed by many as dry and male dominated. In an effort to resolve these feelings, I developed an alternative assessment that was innovative and resulted in a positive learning environment. I needed to ensure relevance to everyday life, a ton of humor and, of course, the use of physics principles! As a fan of Rueben Garret Lucius Goldberg, aka Rube Goldberg, a renowned American inventor, sculptor, author, engineer, and cartoonist extraordinaire, I thought, "What better than a Rube Goldberg Machine Challenge?" As fans of Rube Goldberg can attest, his unforgettable cartoon character and inventor, Professor Lucifer Gorgonzola Butts, illustrated inventions that are known today as the "Rube Goldberg Machine" (Rube Goldberg Contest, 2012), an extremely complicated device that executes a very simple task in a complex, multi-step, and indirect way. Many of Goldberg's illustrations portrayed complicated, almost absurd, but immensely innovative machines.

At the end of a module on mechanics, I invited the girls to use concepts of Newton's laws of momentum and energy transformation, and simple and compound machines to design and construct a Rube Goldberg machine. The machine would have to perform a simple, everyday task that could be self-selected by the girls but would have to be accomplished using at least 15 steps. It would also require the use of innovative energy transformations versus a simple battery. The machine was allowed to have only a single human intervention. The students were allowed to start the machine on its first step, and the machine would have to be so devised as to perform the remaining 14 steps without any additional nudging.

This involved finding creative solutions relying on basic principles of physics. It encouraged girls to use machine tools, gears, pulleys, and levers to build the machines. Students worked in teams, brainstormed in a lab, and willingly (or so I would like to believe!) spent several weekend hours building the machine. They rummaged through their homes for castaway items that could be innovated as gears, levers, and pulleys. My students came up with intricate ideas to perform simple tasks ranging from squeezing toothpaste, to putting sprinkles on cupcakes, to switching on a blender. Once the girls realized the significance of physics principles, they readily competed in making the coolest, weirdest, or prettiest-looking Rube Goldberg machines. On presentation day, the machines were brought out and demonstrated to the school, engaging the interest of the entire school body of

6th to 12th graders. In fact, the admissions office loved to showcase the machines as a unique feature of the school curriculum.

No doubt there was plenty of moaning and groaning (at the very least) as a crucial step would refuse to work after hours of setup, leading the girls to invent new solutions to make it work. However, the brainstorming, designing, inventing, and building of the machine was an achievement that the girls could look back on and compliment themselves about. It gave the girls an insight into physics and applied science, and a meaningful entry into engineering design. In addition, they crossed an important gender barrier by inventing a machine that would leave them with a "can-do" spirit of accomplishment, pride in using machine tools, and the delight of playing with complex mechanical objects they created—all in the name of physics. My ultimate reward was the decision of an outgoing senior to change her college options to major in physics instead of foreign languages.

A second alternate assessment project I offered was class participation in the Team America Rocketry Challenge (TARC) (Team America Rocketry Challenge, 2012). I would introduce students to principles of flight and rocket building. Students would then use licensed simulation software such as RockSim (Apogee Components, 2013), SpaceCad (SpaceCad, 2013), or the freely downloadable OpenRocket (OpenRocket, 2013) to simulate rocket designs and understand the components of building a rocket. TARC also has a list of volunteer engineers and aerospace mentors who are willing to come to schools to start students off by explaining the principles of rocket building to them. After familiarizing themselves with the simulated rockets, students had to work on a budget, raise funds (if necessary) to purchase rocketry components, and follow the competition directions to build a rocket. They would test the rocket out several times before they were ready for launching in the qualifying rounds. One of my students was selected for NASA's INSPIRE (INSPIRE, 2013) two years in a row and went on to pursue a career in aerospace engineering.

In addition to the work I did with students in grades 10–12, I also initiated a celebration by the entire school body (grades 6–12) of National Engineering Week (DiscoverE Engineers Week, 2013). A visitor entering Oakcrest on any day during this week would see students of all grades solving engineering challenges by building bridges, trebuchets, or parachutes. It was important to make science fun for students (all girls, in my case) to get them hooked into natural, physical, and engineering sciences, to teach them to think innovatively, and to make mistakes and learn from them.

Albert Einstein was right on many counts but never more so than in this quotation attributed to him: "In the matter of physics, the first lessons should contain nothing but what is experimental and interesting to see. A pretty experiment is in itself often more valuable than twenty formulae extracted from our minds."

ADAPTING BEST PRACTICES IN YOUR CLASSROOM

The practices that I employed in my classroom can be readily adapted by any teacher in a K–12 classroom. Thanks to the new emphasis on the importance of STEM, there is no dearth of resources on the web. One of the best places to search for information on science day/week celebrations is the website of the relevant professional association. Most professional association websites have a link to the educational resources they offer. Many of these websites cater to varying grade levels and offer activities for most K–12 classrooms. In addition to the list of science challenges and competitions that I have adopted, there are other competitions offered by different organizations, and they usually offer options for different grade levels. Forming an after-school club or volunteering to coach a team would help a teacher get jump-started into adapting these into his or her classroom. Here is a sampling of resources:

- Brain Awareness Week (BAW) from the Dana Foundation (Dana Foundation, n.d.)
- National Engineers Week from National Engineers Week Foundation
- Holiday Chemistry from American Chemical Society (*Holiday Chemistry*, 2012)
- Physics Quest from American Physical Society (Physics Central, n.d.)
- Six Flags America Roller Coaster Design Contest offered by American Association of Physics Teachers (AAPT, n.d.)
- FIRST Lego League (FIRST FLL, n.d.)
- FIRST Robotics Competition (FIRST FRC, n.d.)
- Junior FIRST Lego League (Junior FIRST FLL, n.d.)

Second, there are several schools that have adopted project-based learning, and I list some of those schools below. This list will give you a sense of how the curriculum has been adapted and how the school community is employing project-based learning. Educators can either adopt portions or wholesale components of project-based learning or adapt the principles of project-based learning in any K–12 classroom. I have also added a link to a source that tells how PBL can be a do-it-yourself project.

1. High Tech High Schools (HTH PBL, n.d.)
2. Illinois Mathematics and Science Academy (IMSA PBL, 1993)
3. Avalon School (Avalon PBL, n.d.)
4. Pacific Education Institute PBL Curriculum (Pacific Education Institute, n.d.)
5. Greenwich Public Schools (Greenwich Public School PBL, n.d.)

6. DaVinci Schools (Da Vinci School PBL, n.d.)
7. PBL do-it-yourself (Buck Institute for Education, 2013)
8. Replicating Success: Project Based-Learning (Edutopia PBL, n.d.)

Finally, transformative methodologies such as design thinking offered by K12 Lab and MIT Media Lab have professional training opportunities. Innovation is independent of age, grade, or country. Hence, adapting any of the new transformative practices into your classroom or school will only help you become an innovator yourself.

1. Design Thinking for Educators (IDEO Design, n.d.)
2. D-Lab MIT Youth Outreach (D-Lab MIT, n.d.)
3. K12 Lab Wiki (K-12 Lab wiki, n.d.)

CONCLUSION

Entering the education arena untutored in educational pedagogy and practice was no doubt a daunting task for me. Nevertheless, entering a science classroom as a practicing physicist, with a passion for wanting to communicate science in the most appealing way, was a big advantage to me in rapidly learning the ropes and inventing best practices as I went along. Additionally, I spent several years as a very active PTA volunteer, pitching in wherever and whenever the class teacher or school called for volunteers. Being an engaged parent also helped me a great deal in finding programs that I would have liked to see in my children's school. The 4Ps: persistence, perseverance, passion, and patience (the last one in truckloads) were crucial and extremely formative in achieving success in my classroom and my professional growth.

As someone who had to teach herself before teaching the wards under her care, I learned to be innovative, imaginative, creative, bold, and collaborative. Adoption or adaptation of new practices in one's classroom is much like taking a road less travelled. There are bound to be surprises and perhaps some shocks, so it is okay to be nervous. But, as a good teacher would encourage his or her students, remember to be kind to yourself and not be scared of failure. Of course, being responsible for your students makes one hesitate to adopt practices that might have the potential to boomerang. One way to look at adopting a new practice would be as an experimentalist in a lab. Do your background research and then start playing around with controlled variables, making sure you do not introduce too many unknown parameters into the experiment. With every new result, analyze the data and be prepared to modify. The final piece of learned wisdom I leave behind is to start small, be prepared to learn and re-learn from your errors and surprises, and incrementally boost the size of your program. Once you have the winning formula, as I often told my students, just plug and chug!

WORKS CITED

About Olin College. (2012). Retrieved from http://www.olin.edu/about_olin/

ACS Chemagination. (2012). *Chemagination description and rules.* Retrieved from http://marmacs.org/2012/Chemagination_Rules.pdf

American Association of Physics Teachers. (n.d.). *Six Flags America roller coaster design contest.* Retrieved from http://aapt.org/Programs/contests/rollercoaster.cfm

Apogee Components. (2013). *RockSim information.* Retrieved from http://www.apogeerockets.com/RockSim/RockSim_Information

Arlington Catholic Herald. (2007, March 22). Retrieved from http://catholicherald.com/stories/Oakcrest-Sophomores-Explore-the-Future-of-Technology,312?content_source=&category_id=&search_filter=oakcrest&event_mode=&event_ts_from=&list_type=&orderby=&order_sort=&content_class=&sub_type=stories&town_id=

Avalon PBL. (n.d.). *Project based learning,* Retrieved from http://www.avalonschool.org/project-based-learning-pbl/

Buck Institute for Education. (2013). *What is PBL.* Retrieved from http://www.bie.org/about/

Cantador, I., & Bellogín, A. (2012). *Healthy competitions in education through cooperative learning.* Retrieved from http://www.academia.edu/2787282/Healthy_Competitions_in_Education_through_Cooperative_Learning

Committee on Science, Engineering, and Public Policy. (2007). *Rising above the gathering storm: Energizing and employing America for a brighter economic future.* Retrieved from http://www.nap.edu/catalog.php?record_id=11463

The Connection. (2007, March 27). Retrieved from http://www.connectionnewspapers.com/news/2007/mar/27/students-named-regional-winners/

d.school. (2012). *Institute of design at Stanford,* Retrieved from http://dschool.stanford.edu/

Dana Foundation. (n.d.). *Brain Awareness Week resources.* Retrieved from http://www.dana.org/brainweek/resources/

Da Vinci School PBL. (n.d.). *Project-based Learning,* Retrieved from http://davincischools.org/projectbasedlearning.shtml

Deeper Learning. (2012). Retrieved from Hewlett: http://www.hewlett.org/deeperlearning

Designers at the lab. (n.d.). Retrieved from http://dschool.stanford.edu/k12/designers.php

The difference between projects and project-based learning. (2012, December 3). Retrieved from http://www.teachthought.com/learning/project-based-learning/difference-between-projects-and-project-based-learning/

DiscoverE Engineers Week. (2013). *Discover: let's make a difference.* Retrieved from http://www.discovere.org/our-programs/engineers-week

D-Lab MIT. (n.d.). *Youth outreach: Inspiring the next generation of innovators.* Retrieved from http://d-lab.mit.edu/youth-outreach

The DuPont Science Challenge. (2012). Retrieved from http://thechallenge.dupont.com/

EdLeader21. (2012). *Every student should be ready for the challenges of the 21st century.* Retrieved from http://www.edleader21.com/index.php?pg=2

Edutopia PBL. (n.d.). *Replicating Success: Project-Based Learning*. Retrieved from http://www.edutopia.org/stw-replicating-pbl

Exploravision. (2012). *Today's young minds drive tomorrow's innovation*. Retrieved from http://www.exploravision.org/

FIRST FLL. (n.d.). *FLL FIRST LEGO League*. Retrieved from http://firstlegoleague.org/

FIRST FRC. (n.d.). *First Robotics Competition*. Retrieved from http://www.usfirst.org/roboticsprograms/frc

Fulu, I. A. (2007). *Enhancing learning through competitions*. Retrieved from http://www.learnerstogether.net/home/2007/11/14/enhancing-learning-through-competitions.html

Goldman, R., & Papson, S. (1998). *Nike culture: The sign of the swoosh*. London: Sage Publications.

Greenwich Public School PBL. (n.d.). *Project-based learning and multimedia: designing a powerpoint project*. Retrieved from http://www.greenwichschools.org/page.cfm?p=1675

Holiday chemistry. (2012). Retrieved October 10, 2013, from http://portal.acs.org/portal/acs/corg/content?_nfpb=true&_pageLabel=PP_ARTICLEMAIN&node_id=350&content_id=CNBP_024734&use_sec=true&sec_url_var=region1&__uuid=dc8181f4-dcc1-4346-af67-a2974e22298a

HTH PBL. (n.d.). *Project based learning at HTH*. Retrieved from http://www.hightechhigh.org/projects/

IDEO Design. (n.d.). *Design thinking for educators*. Retrieved from http://www.designthinkingforeducators.com/

IMSA PBL. (1993). *PBL Network: collaborative inquiry in action*. Retrieved from http://pbln.imsa.edu/

INSPIRE. (2013). *Interdisciplinary national science project incorporating research and education experience*. Retrieved from http://www.nasa.gov/offices/education/programs/descriptions/INSPIRE_Project.html#.UnlRxXATx7Q

Johnson, D. W., & Johnson, R. T. (1988). *Introduction to cooperative learning: An overview of cooperative learning*. Cooperative Learning Institute and Interaction Book Company. Retrieved October 10, 2013, from http://www.co-operation.org/?page_id=65

Junior FIRST FLL. (n.d.). *Welcome to Junior FIRST LEGO League*. Retrieved from http://www.usfirst.org/roboticsprograms/jr.fll

K12 Lab Partners. (n.d.). Retrieved from http://dschool.stanford.edu/k12/partners.php

K12 Lab wiki. (n.d.). Retrieved from https://dschool.stanford.edu/groups/k12/

K12 Laboratory. (n.d.). *Teachers as innovators*. Retrieved from http://dschool.stanford.edu/k12/teachers.php

Kirschner, P. A., Sweller, J., & Clark, R. E. (2006). Why minimal guidance during instruction does not work. *Educational Psychologist, 41*(2), 75–86.

Lam, S.-f., Yim, P.-s., Law, J. S., & Cheung, R. W. (2001). *The effects of classroom competition on achievement motivation*. Retrieved from http://www.eric.ed.gov/ERICWebPortal/contentdelivery/servlet/ERICServlet?accno=ED458511

Larmer, J., & Mergendoller, J. R. (2010, September). Seven essentials for project-based learning. *Educational Leadership, 68*(1), 34–37. Retrieved from http://www.ascd.org/publications/educational_leadership/sept10/vol68/num01/Seven_Essentials_for_Project-Based_Learning.aspx

Lawrence, R. (2004). Teaching data structures using competitive games. *IEEE Transactions on Education, 47*(4), 459–466.

Microsoft in Education. (2013). *Microsoft innovative schools*. Retrieved from http://www.microsoft.com/education/ww/partners-in-learning/Pages/innovative-schools.aspx

MIT Media Lab. (2012). Retrieved from http://www.media.mit.edu/

Moursund, D. (2007). *Problem-based learning and project-based learning.* Retrieved from http://pages.uoregon.edu/moursund/Math/pbl.htm

National Commission on Excellence in Education. (1983). *A nation at risk: The imperative for educational reform.* Retrieved October 10, 2013, from http://datacenter.spps.org/uploads/sotw_a_nation_at_risk_1983.pdf

National Mole day Foundation. (2013). Retrieved from http://www.moleday.org/

National Research Council. (2010). *Rising above the gathering storm, revisited: Rapidly approaching category 5.* Washington, DC: National Academies Press.

OpenRocket. (2013). Retrieved from http://openrocket.sourceforge.net/

Pacific Education Institute. (n.d.). *The Pacific Education Institute helps Washington teachers at all schools levels design project-based curriculum.* Retrieved from http://www.pacificeducationinstitute.org/learning/project-based-curriculum/

Partners in Learning Network. (2013). Retrieved from http://www.microsoft.com/education/ww/partners-in-learning/Pages/network.aspx

Partnership for 21st Century Skills. (n.d.). Retrieved from http://www.p21.org/index.php

Physics Central. (n.d.). *Physics quest.* Retrieved from http://www.physicscentral.com/experiment/physicsquest/index.

President's Council of Advisors on Science and Technology. (2010, September 16). *Prepare and inspire: K–12 science, technology, engineering, and math (STEM) education for America's future.* Retrieved from http://www.whitehouse.gov/sites/default/files/microsites/ostp/pcast-stem-ed-final.pdf

Project Based Learning for the 21st Century. (2013). *What is PBL?* Retrieved from http://www.bie.org/about/what_is_pbl

Regional Winners Exploravision. (2007). Retrieved from http://www.exploravision.org/winners/regional-winners-2007.php

Rube Goldberg Contest. (2012, November). Retrieved from Argonne National Laboratory Educational Programs: http://www.dep.anl.gov/rube/

Smithsonian Newsdesk. (2011). Innovation in 19th-Century America. Retrieved from http://newsdesk.si.edu/releases/exhibition-dedicated-innovation-19th-century-america-opens-smithsonian-american-art-museum-

SpaceCAD. (2013). Retrieved from http://www.spacecad.com/

Team America Rocketry Challenge. (2012). Retrieved from http://www.rocketcontest.org/

Thomas, J. W. (2000). *A review of project based learning.* Retrieved from http://www.bie.org/research/study/review_of_project_based_learning_2000

Verhoeff, T. (1997, November). *The role of competitions in education.* Retrieved from http://olympiads.win.tue.nl/ioi/ioi97/ffutwrld/competit.html

Wagner, T. (2012a). *Creating innovators: The making of young people who will change the world.* New York: Simon & Schuster.

Wagner, T. (2012b, April 13). Educating the next Steve Jobs. *The Wall Street Journal.* Retrieved from http://online.wsj.com/article/SB10001424052702304444604577337790086673050.html

CHAPTER EIGHT

Expand the Horizons OF **Your Students** BY Expanding Yours

JEAN PENNYCOOK

INTRODUCTION

We have all attended teacher professional development seminars, institutes, workshops, and meetings. Perhaps we learned a new activity, took home an inspiring lesson plan, or won a piece of science equipment in the raffle. All fun and worthy, but is there anything else out there? You bet there is. The National Science Foundation (NSF), the National Oceanic and Atmospheric Administration (NOAA), the National Aeronautics and Space Administration (NASA), and countless other federal and private organizations offer field experiences, laboratory-based research opportunities, international travel, and more for classroom teachers during both the school year and summer release time. Extend the horizons of your students by expanding yours. This chapter offers a host of reasons why getting one foot out of the classroom will keep the other foot happy.

Teacher professional development (PD) events strive to provide experiences that improve instructional practices in the classroom. Some research has suggested that effective, quality PD may improve not only instructional quality, but also student achievement (Seymour, Hunter, Laursen, & Deantoni, 2004). The National Research Council (National Research Council, 1999) has stated that providing quality PD opportunities for teachers in grades K–12 should be an important goal of school districts and administrators.

For most teachers, however, current PD practices of schools and districts are provided as a one-shot opportunity and are not likely to improve instructional practice (Wei, Darling-Hammond, Andree, Richardson, & Orphanos, 2009). Teachers report that these workshops have little impact on their teaching practices (Garet, 2001) or content knowledge. Much of the current research states that short-term PD events—those lasting less than 30 hours—are not effective and do nothing to change the instructional practices of the teacher or the achievement of the students (Loucks-Horsley, Love, Stiles, Mundry, & Hewson, 2003; Yoon, 2007).

One type of PD, however, has been shown to have an impact and does change how teachers view their educational practices, their content, and their method of instruction. Referred to as Research Experiences for Teachers (RET), or immersion programs, these PD events provide teachers with an opportunity to experience the nature of learning as well as the practice of scientific work over an extended time frame (Dresner, 2006). Ranging from 2 to 10 weeks, covering one summer or several years, these programs provide teachers with an opportunity to immerse themselves in science research or an industry setting, stepping back from the classroom for a short time and returning with new skills.

BACKGROUND

Professional development for teachers should parallel the experience that providers want students to receive from the teachers when they return to the classroom (Loucks-Horsley et al., 2003). In an immersion PD, or RET experience, teachers become students as they acquire new content knowledge and insight into the workings, practice, and process of science or industry (Davis, 2003). During an immersion PD experience, teachers experience for themselves as their students would: through critical thinking, motivated learning, construction of new knowledge, and professional collaboration, all of which contribute to the process of a change in thinking about their teaching practices (Hashweh, 2003). After returning to the classroom, teachers who have taken part in an immersion PD or RET experience have included more inquiry-based activities in their instruction and provided more collaborative learning experiences for their students. They have also re-examined their teaching practice and instructional philosophy (Woolfolk, 2007) as they work to change what and how they teach (Loucks-Horsley et al., 2003).

Participants in immersion PD programs have indicated that this type of experience contributes to their professional growth, is a catalyst for their continued involvement in PD, encourages them to take on leadership roles in their school, promotes confidence in their content knowledge, and helps stimulate change in their teaching practice (Pop, 2010). Other valuable aspects have included building a professional network to share their learning experience and to help in translating their learning to the classroom. Research on the effects of immersion experiences for teachers have shown that teachers change how they think about teaching science, which leads to improving their instructional practice by including more inquiry-approach lessons (Hanuscin, 2007; Pop, 2010). RET and immersion PD programs provide an experience where the teacher is the learner engaged in the process of learning, which parallels the concept of constructivism (Davis, 2003). A total immersion or RET experience follows this model of PD, as teachers are working alongside their counterparts in research or industry while honing their skills and increasing their knowledge. A few studies have been conducted on RET experiences, and they credit the immersion-type programs with creating positive and sustainable changes to participants' instructional practices (Dixon & Wilke, 2007).

BEST PRACTICES

Why I Participated in an Immersion Professional Development Program

The first few years of a teacher's professional career are frequently spent developing their course curriculum, honing their classroom management skills, and navigating the waters of building and district politics. In an effort to be supportive, some districts require new teachers to participate in "new teacher" activities, providing them with workshops, structured mentoring from master teachers, and organized reflection on their practice. Instead of helping, these add-on PD requirements create a burden for new teachers, and some have reported a burnout factor that leads to their early exit from the profession. A study by the National Education Association states that half of U.S. teachers new to the profession are likely to quit within the first 5 years of teaching (Kopkowski, 2008). It is particularly hard to hold onto science, technology, engineering, and math (STEM) teachers, since the private-sector salaries for those trained in these fields far outpace those in education. For those who stay, teaching may become routine and the content stale without professional development experiences, which infuse new energy, new ideas, and fresh ways to teach.

After teachers have been established in their classroom and are comfortable in their instructional curriculum, another typical array of PD opportunities arises in the form of learning to navigate the newly adopted textbook, implementing the latest technology, infusing the latest content standards in the coursework, or developing new teaching strategies. Sometimes these bring new ideas to the teachers; other times they recycle old ones with new names. Workshops can last an afternoon, a half-day, a full day, or several days in a row.

The first 10 years of my teaching career followed the above pattern very closely. In that time I attended every form of PD, including online webinars, self-taught classes, university coursework, district- or state-level 1-hour-to-1-week workshops, "sit and get," "talk and take," "grab and go," "tips and tricks," and so on. The variety of venues and methods was limited only by the imagination of the presenter or the program budget. Sometimes PD was delivered by an outsider, other times by an in-house expert. Always well-meaning, the presenters—either paid professionals or volunteers—truly believed in the importance of their product or idea, but rarely did the experience change my teaching or how I thought science should be taught.

After logging—and in many cases donating—hundreds of hours of my life in various PD experiences and not feeling very fulfilled in any of them, I participated in an RET program working with scientists doing research and using the latest equipment, methods, and analytical tools. It changed what I teach and how I teach. Not only was I able to bring this experience back to my students, but I was also more confident and better able to provide inquiry-based activities, real-world applications to the instructional content, and insight into many educational paths that would lead to exciting career possibilities for my students in the STEM fields. Through my experience, my students' eyes were opened to the variety of opportunities in the STEM fields that they could now explore. I was able to incorporate my learning experience as a student into the teaching of my own students, and together we continued to explore the cutting-edge and continuing growth of information that is the hallmark of scientific research.

One day, while I was attending yet another professional development meeting presented by my school district personnel, I noticed a flyer in our info-stack about a National Science Foundation funded program that selected teachers to join research teams in the polar regions. The opportunity included doing "educational outreach" about the research as well as being an active part of the field team. All expenses, including travel and equipment, were paid; they even covered the substitute pay for my classroom. I could be gone between 2 weeks and 3 months, and teams were headed to the Arctic and the Antarctic. The application was competitive and lengthy. I spent 2 weeks making it perfect and sent it off with a small prayer, knowing that this would be something that would make a difference in my teaching. To my surprise, I was selected, along with about 15 other teachers from around the country. Half of them went north to the Arctic; I was sent south to the Antarctic with the remaining cohort.

The RET experience I participated in was with a research team studying an active volcano in Antarctica. After a 2-month total immersion field experience that included living in a tent on the rim of Mt. Erebus as a member of the Teachers Experiencing Antarctic and Arctic (TEA) program, I never again viewed the instruction of high school science in the same way. I became the learner, revisiting what it is like to learn new skills and new content. With a different view of the world of science, I returned to the classroom with new energy and evaluated not only everything I taught, but—more important—how I taught it.

Thrown into an environment I knew little about, I had to navigate my way through the new-to-me vocabulary of volcanoes and geology. Working with technical equipment that looked like something out of a science fiction novel, I had to master its operation so that precise data could be collected in the short window of opportunity in the field. There were no instructions on how to do this; it was the first time anyone had done it. Again, I experienced the pleasure of learning something new and completing a new task, and the feeling of accomplishment in developing new skills. It was a metacognitive journey that provided me with a needed dose of empathy for my students and a renewed respect for acquiring complex knowledge. Plus, it was fun. I wanted my students to feel that way about learning science content, which was just as new to them when they entered my classroom as the volcano was new to me. I wanted them to experience the beauty and joy of learning, the mystery of nature, and the pleasure of mastery. I totally rewrote my curriculum and reformed my teaching strategies.

Real scientists do not sit in a room and read a book, routinely perform experiments with known outcomes, or watch videos about other people doing science. Rather than depending on the textbook as the main source of information, my students and I began to look elsewhere for knowledge. We worked together to learn about volcanoes around the world, how they affect our planet, how they are monitored, what we know about them, and, most important, what we do not know. The textbook provided basic information, but we were excited to learn more.

No more "cookbook" labs with predictable outcomes; real science experiments were much more meaningful. Our hands-on, inquiry-based activities included creating models of volcanoes to predict lava flow patterns, to determine the rate of lava flow based on the shape and size of the volcano, to determine the volume of ejecta based on the range of fallout from an eruption, and to connect the age of ice cores and oceanic sediment cores to layers of volcanic ash from known eruptions. Students designed their own questions, established their own procedures, gathered and analyzed data, and presented their findings to the class.

After we discarded traditional vocabulary lists, we learned new vocabulary as we went along, adding to a "word wall" of terms that were important to our understanding. The "word wall" (our designated bulletin board) provided a visual connection to the vocabulary as the unit progressed and served as a documentation of the students' learning progression. Students took turns writing the new words on strips of paper and placing them on the bulletin board. In many cases, students provided pictures next to the word as an aid to understanding and comprehension. The "word wall" grew as the unit progressed, and I was able to refer to it visually every time I used one of the words in a class discussion. This reinforced the concept of word recognition and cognitive development of academic language.

Project-based curriculum and inquiry-based activities became the routine in my teaching. Rather than reading about the different kinds of volcanoes, students looked for patterns around the world in volcanic formations. Volcano data sets were readily available, and we spent our time looking for trends and patterns. From websites such as the one from the U.S. Geological Survey (USGS.gov), students gathered maps and data charts showing the location of volcanoes worldwide and looked for ways to predict their location. This led to the students' discovery of the link between volcanoes and earthquakes, the placement of mountain ranges on the Earth, and what led to the formation of oceanic islands and continental hot spots. Other teams looked at data sets showing recent and past activity of active volcanoes and created posters showing places, both locally and globally, that posed a risk to human habitation.

Students researched where volcanoes had impacted people's lives around the world. Teams prepared reports on the dangers of living near volcanoes and what societies have had to do to be prepared for volcanic activity. Student teams selected a specific volcanic eruption and prepared a poster on the impact of these events, both locally and worldwide. During the classroom "share-o-thon," posters were set up around the room. Students took turns standing near their poster to present to others, and also assumed the role of a visitor looking at others' posters. Students were encouraged to ask for evidence of any claim on a poster or clarification of stated facts. The "share-o-thon" was styled after the poster sessions that are often held at professional conferences and meetings wherein scientists present research findings by means of a poster. Students gained practice in communicating their ideas and defending their data. This is what real scientists do.

I was excited, and teaching became fun again. The students were motivated as well. Rather than learning the history of science, we were doing the process of science. Effort, collaboration, and a final product were the reward, rather than a score on a test. I never looked back to the old ways of instruction. I continued to keep both my knowledge and my skills current through ongoing collaboration with the research team.

The immersion with a polar science research team was life changing for me. Although I continued to teach for several years following that experience, I also continued to go to Antarctica and connect the research with classrooms around the world. After taking part in the TEA program on the volcano, I decided to continue to do education outreach in the polar regions.

In 2005 I joined the Adélie penguin team and continue to connect science research to the classroom and encourage students to follow educational pathways in STEM fields. My interactive website (www.penguinscience.com) attracts over 800 classrooms to the activities about penguins, the sights and sounds of Antarctica, and opportunities that are available for exciting careers in STEM fields. Bringing the excitement of Antarctica, penguins, and STEM careers to students continues to be my passion.

An Added Benefit of Experiences Outside the Classroom

When I was in my mid-30s, I left a professional life in the sciences (as an inspector for the California Department of Food and Agriculture) to enter the education profession. At that time, career opportunities in STEM fields were limited. Over the years, opportunities and STEM job titles have increased exponentially. How could I help my students make career choices and show them pathways to interesting lives when my own information was outdated? In addition, the tools of science were ever changing, and my students had to be more prepared to embrace new tools and technologies—but only if I was confident enough to introduce them into the classroom. A perfect example of this was a tool called Geographic Informational Systems (GIS). Widely used in many of the STEM fields, the ability to use GIS provides thousands of opportunities for exciting, interesting careers. Without my involvement in science research through an immersion PD program, I would not have developed the skills to introduce this tool to my classroom, nor would I have possessed the knowledge about career possibilities that could excite my students.

Photo 8.1. Jean During Her Immersion Research Experience with Penguins in Antarctica. (Photo by Tim Spuck.)

I have now spent 11 seasons in Antarctica doing education outreach for science teams and have discovered that for every research scientist there are approximately

8 support people who do the field work and lab work, manage the data, design, build, and maintain the equipment, do logistics, and manage the project. Each of these positions requires STEM-related background, experience, and expertise and provides an exciting career in an ever-changing and challenging field. Through my exposure to the research team and others in the RET experience, I was able to guide my students in exploring different career and educational pathways that were of interest to them and that could provide dynamic and challenging lifelong careers.

Many of my students did pursue STEM-related careers based on their introduction to these pathways in my classroom activities. Five of my students became interested in the U.S. Antarctic program and spent at least one season "on the ice." Of the five, two were on science teams and have contributed to the unraveling of many polar mysteries. The other three were in STEM-related positions supporting the science in the capacities of engineer, technician, and fireman.

Confidence in my content area, new skills with cutting-edge technologies and science tools, and knowledge about current career paths and educational opportunities for my students are all positive outcomes of my immersion PD research experience. I am not alone in knowing the value of this type of experience. What follows are three unique stories in which similar experiences changed the professional lives and careers of other teachers.

Case Studies

Case study #1

Jeff had been a biology teacher in Colorado for 15 years. Although he loved his students and enjoyed teaching biology, he had little time to read the research journals. During a classroom activity about stem cell research, Jeff discovered that several of his students knew more than he did on the topic because they had prepared a debate in their social studies class. Jeff began to feel that he was out of touch with recent developments in his field and wanted to change that. In the town where he lived there was a state university, so Jeff contacted one of the researchers in the Biology Department. With a degree in biology and several years' experience teaching laboratory procedures to his students, Jeff was familiar with basic lab equipment, so he approached the researcher to see if he could use some help in the lab. The researcher was glad to accept the help of an experienced person, and soon Jeff was assisting the researcher one afternoon a week and a couple of weeks in the summer. By working in the academic environment, Jeff was exposed to the latest in research equipment and methods, which he was able to share with his students. Additionally, he was able to share the variety of jobs associated with the research, in which many of his students showed interest and later pursued as careers.

With his exposure to the practice of science, Jeff learned a great deal about the scientific process and changed many of his labs to a more inquiry-based approach, allowing his students to create their own methods and ask their own questions as they solved real-world biological problems. While working with the scientist, Jeff used some of the scientist's data to develop an activity for his class based on the research. Students learned about what the researcher was working on, and they were able to look at some of the data from the project and then engage in an analysis using their own ideas and tools. Rather than studying the history of science, as many textbooks are designed to do, Jeff's class was working with cutting-edge ideas, content, and practices.

Jeff's students responded well to this form of teaching, welcoming the change from worksheets, vocabulary lists, and practice problems. Now they felt involved in something authentic, relevant, and real. One student later created a winning science fair project based on the class work; others decided to continue taking biology as an educational pathway. Jeff's teaching became dynamic, authentic, and exciting both for him and his students. He was excited about his role in the research and his ability to bring back to his students the ongoing story of acquiring new science knowledge as it develops. As a professional development experience, this one far surpassed any he had previously engaged in. Every year he adds another layer to the exercise as the story of the research continues to unfold. His students feel they have a part in the research as well, as each year a new set of data is added to the activity, and every year the students see progress in the new knowledge.

Case study #2

Gail lives in a desert community but teaches an environmental course that includes the role of the oceans in the earth system. Many of her students have never been to the ocean and therefore do not have firsthand experience with it. Gail took oceanography courses in college and felt her content knowledge was up to date, but she wanted her students to experience some excitement on the topic. After taking some online classes in ocean science, she decided to get some firsthand experience and applied to a program that takes teachers on research vessels. The program was very competitive, but Gail was accepted for a 2-week research cruise in the Gulf of Mexico, looking at algal blooms from the Mississippi delta region. During the cruise, she connected with her classroom through the Internet and was able to have the scientist on board speak to her students using a satellite phone. Gail did a shift on the dissolved oxygen machine and learned how to process water samples for data collection.

Gail was also able to observe the workings of the ship, the role played by each member on board (as well as the scientists) in the success of the cruise, and what career paths had led everyone to where they were at that moment. Before taking part in this cruise, Gail did not know of these careers or how many jobs were related to supporting the research on a science research vessel. There were engineers, technicians, laboratory analysts, and computer programmers, to name a few. She brought back to her school a wealth of exciting STEM career ideas for her students, along with pictures and biographies of each person she met on board.

Through this immersion PD experience, Gail brought excitement, energy, and a new perspective to her science instruction, knowledge of STEM career pathways for her students, and firsthand experience in science research, all of which translated to helping her students understand the process of science and how scientists work. Gail changed her classroom practices to include current data sets from research, inquiry-based activities using real-world ocean problems, and promoting a selection of the career pathways she was exposed to. As a direct result, many of her students decided to pursue education in STEM fields.

One project her class continued to follow on a yearly basis was the migration of whales along the east coast of the United States. Scientists tagged these creatures so that satellites could monitor their movement. As the migration routes of these extraordinary mammals became apparent every year, communities required commercial shippers to alter their routes to protect mothers and their calves as they moved north along the coast. Following this story provided Gail's students with firsthand experience in how people interact with these mammals, the political and economic pressures on oceanic navigation policy, how policies are written and implemented, and the importance of being custodians of our environment.

Case study #3

In another state, Juan had been teaching science for 12 years and felt a sense of isolation in his career. As one of three science teachers in a small, rural high school, Juan was missing the sense of collaboration he had felt with the medical staff in his previous career as a laboratory chemist at the local hospital. There he had been offered ongoing PD experiences to maintain his knowledge of new developments in technology related to his position. Now, as a teacher, PD events offered to him were directed at his new administrative responsibilities, student behavior issues, and district testing requirements.

Juan found an opportunity with the Los Alamos Energy and Environmental Science Academy, a National Science Foundation (NSF) program, for a research experience in industry at an environmental testing laboratory near his town. Through the program he created an ongoing collaboration between his classroom and the scientists working to monitor water quality in the local streams and lakes. His students enjoyed the relationship with the working scientist. Juan was able to keep his laboratory skills sharp with the latest testing instruments, and his teaching took on a whole new dimension as the students began to take the lead in determining what data to collect and what tests to run, and in following the long-term changes in their local water supply. The class was given responsibility for monitoring a section of the river that was close to the school, and many students developed science fair projects from the data they collected. One student later interned for the testing laboratory during a summer break and planned to continue this work as a career. Several students became environmental activists for a river conservation group helping to bring native fish species back to the river.

Juan's immersion PD brought to his classroom practice new energy, a higher confidence level in content, new ideas, cutting-edge technology, and a sense of collaboration with real-world science. Students became engaged in the environment of their community and knowledgeable about changes in their local river. Rather than reading about how scientists monitor the environment, looking at pictures, or watching a movie, these students and their teacher were actively participating in it. Juan continued to maintain a long-term record of the data each year's class collected so that students could see changes over time in their community. He took on a new leadership role as well, providing other teachers with the skills they needed to begin a long-term environmental project in their classrooms. Juan prompted a network of like-minded teachers to share in the development of instructional methods, classroom activities, and content skills, and to work alongside the network of scientists that continued to partner with his students on data collection.

Keeping pace with new developments in one's content field is always a challenge and is often set aside as the demands of the classroom take priority. Having an immersion PD experience in research or industry is an exceptional way for teachers to keep their skills, content knowledge, and worldview fresh. Though not always a paid job, and certainly not a career change (but a career-changer!), keeping one foot in the adult world of one's content area will help teachers keep on top of developments in their field and provide a fresh perspective to share with their students.

EXPANDING YOUR HORIZONS

The desire to further your professional expertise in this manner must originate with you. The next step comes in locating a program within your teaching field. Considerations should include time available to participate in a program, location of the program (on a college campus, in a scientific or industrial setting, or in the field), and the type of experience you feel would be most beneficial in engaging your students. Some examples of current programs that provide teachers with high-quality immersion PD experiences with science groups, industry, and research projects are provided here. Opportunities vary. Some select early-career teachers; others select teachers with 5 or more years' experience. Some are state-specific or regional-specific; others look nationally for applicants. Some provide a stipend; others do not. Details may change, but this listing below will give you a start in finding an experience that works best for you.

Science Teacher and Researcher (STAR)

STAR is a 9-week summer research program for pre-service science and mathematics teachers. Participants work alongside a research mentor on a group or independent research project and attend workshops that focus on transforming the teaching of science in the classroom. The program is coordinated by the California State University system in partnership with the Department of Education (DOE), the National Aeronautics and Space Administration (NASA), the National Oceanic and Atmospheric Administration (NOAA), the National Science Foundation (NSF), and the National Optical Astronomy Observatory (NOAO). STAR offers applicants positions with national research laboratories around the country. Visit http://starteacherresearcher.org/ for details.

NOAA Teacher at Sea

The NOAA Teacher at Sea program provides a unique learning opportunity for K–16 teachers by sending them out to sea aboard a NOAA research vessel. Participants work closely with scientists and crew to increase their oceanic and atmospheric literacy. The program provides an interdisciplinary research experience and a clearer insight into how our oceans work for teachers to take back to their classrooms. By participating in this program, teachers increase their content knowledge and gain insight into how science works. Details can be found at http://teacheratsea.noaa.gov/#/home/.

Earthwatch Educator Fellowships

Earthwatch looks for passionate K–12 teachers who want to learn more about environmental issues, who desire to make a difference with their teaching, and who are committed to engaging their communities by bringing about awareness and action. Teachers collaborate with a team of educators for a 10–14-day expedition to conduct scientific research. Fellowship recipients develop their skills and environmental understanding as they work alongside field researchers. Visit http://www.earthwatch.org/education/teacher-fellowships for details.

PolarTREC

The PolarTREC program brings K–12 educators and polar researchers together to advance polar science education and understanding through hands-on field experience. Participants improve their content knowledge and instructional practices by bringing inquiry-based polar science learning to their classrooms. By translating their field experience, teachers increase students' engagement in the polar regions and interest in STEM careers. Learn more at https://www.polartrec.com/.

School of Rock (SOR)

The SOR Expedition for Earth and Ocean Science Educators is the adventure of a lifetime for educators interested in experiencing scientific ocean drilling research with the experts. During this multi-day workshop on board the JOIDES Resolution or in the IODP Gulf Coast Core Repository, educators from across the world work with real core material and lab technology to learn how the science reveals clues about the Earth's history. Find details at http://joidesresolution.org/node/3298.

Los Alamos Opportunities for Teachers

For teachers in New Mexico only, Los Alamos Research Labs offer professional development experiences designed for teachers working at the high school and middle school levels. Participants selected to join the program will work directly with scientists from the Los Alamos National Laboratory for summer programs in order to master content knowledge and scientific methodologies in either energy or environmental science. Details on these opportunities can be viewed at http://www.lanl.gov/community/education/teacher-resources/index.php. To locate similar programs in your region, check out http://www.science.gov.

Lawrence Livermore National Laboratory (LLNL)

The Teacher Research Academy (TRA) at LLNL offers unique professional development experiences to middle school, high school, and community college faculty. Participants increase their understanding of science and technology and improve their ability to provide students with a context in which to understand how science is applied. Teachers work through several levels of PD starting with a few days and increasing to several weeks as they grow in their knowledge and skills. Visit https://www.llnl.gov/ for details.

Toyota International Teacher Program

The Toyota International Teacher Program is an international PD opportunity focusing on environmental stewardship and global connectedness for U.S. secondary school teachers. Selected teachers travel on a short-term (2–3 week) study tour to a country that is at the forefront of innovative solutions to environmental challenges. The teachers explore environmental issues through hands-on activities and incorporate what they learn into interdisciplinary and solution-focused lesson plans to share with their students and communities in the United States. Learn more by visiting http://www.iie.org/Programs/Toyota-International-Teacher-Program.

National Endowment for the Humanities (NEH) Summer Grants Workshops

The NEH summer programs offer a variety of opportunities for teachers lasting 1 or more weeks, both here in the United States and abroad. Many of the programs offered are STEM related, but these offerings may not be available every summer. Teachers get reimbursed for much or all of the expense and get a great training experience to boot. Find out if there's a program to suit your interests by visiting http://www.neh.gov/divisions/education/summer-programs.

CONCLUSION

Whether you have been teaching for 3 years or 30 years, have been dissatisfied with the professional development provided for you, or just want to have an extraordinary experience, it is time to participate in an RET or immersion PD opportunity that will change the way you think about science education. Each of the teachers mentioned in the case studies in this chapter began with a desire to do something outside the classroom that would enhance what they did inside the classroom. All sought an experience that would bring energy, content knowledge, and new ideas to their careers. I encourage you to "kick-start" your science content knowledge, rejuvenate your teaching practice, and expand your horizons on STEM career pathways through this type of experience. Share the results with your students and watch what happens.

   Conduct your search for an immersion PD experience and find a program that resonates for you. These opportunities will change the way you think, feel, and approach your teaching forever. What are you waiting for?

APPENDIX

Visit Pathways to Science for links to national offerings of immersion PD and Research Experiences for Teachers (http://www.pathwaystoscience.org/programs.aspx?descriptorhub=SummerResearch_Summer%20Research%20 Opportunity).

Additional Sources

Osborne, M. (1998). Teacher as knower and learner: Reflections on situated knowledge in science teaching. *Journal of Research in Science Teaching, 35,* 427–435.

Penuel, W. F. (2007). What makes professional development effective? Strategies that foster curriculum implementation. *American Education Research Journal, 44*(4), 921–958.

Supovitz, J. A. (2002) The effects of professional development on science teaching practices and classroom culture. *Journal of Research in Science Teaching, 37*(9), 963–980.

WORKS CITED

Davis, K. (2003). "Change is hard": What science teachers are telling us about reform and teacher learning of innovative practices. *Science Education, 87,* 3–30.

Dixon, P., & Wilke, R. A. (2007). The influence of a teacher research experience on elementary teachers' thinking and instruction. *Journal of Elementary Science Education, 19*(1), 25–43.

Dresner, M. W. (2006). Teacher research experiences, partnerships with scientists, and teacher networks sustaining factors from professional development. *Journal of Science Teacher Education, 7*(1), 1–14.

Garet, M. A. (2001). What makes professional development effective? *American Education Research Journal, 38*(4), 915–945.

Hanuscin, D. L. (2007). Schools's IN for summer: An alternative field experience for elementary science methods students. *Journal of Elementary Science Education, 19*(1), 57–67.

Hashweh, M. Z. (2003). Teacher accommodative change. *Teacher and Teacher Education, 19*(4), 421–434.

Kopkowski, C. (2008, April). *Why they leave.* Retrieved from http://www.nea.org/home/12630.htm

Loucks-Horsley, S., Love, N., Stiles, K., Mundry, S., & Hewson, P. (2003). *Designing professional development for teachers of science and mathematics* (2nd ed.). Thousand Oaks, CA: Crown Press.

National Research Council. (1999). *How people learn.* Washington, DC: National Academies Press.

Pop, M. M. (2010). Research experiences for teachers (RET): Motivation, expectations, and changes to teaching practices due to professional program involvement. *Journal of Science Teacher Education, 21,* 127–147.

Seymour, E., Hunter, A., Laursen, S., & Deantoni, T. (2004, April 26). *Establishing benefits of research experiences for undergraduates in the sciences: First findings from a three-year study.* Retrieved from http://www.interscience.wiley.com

Wei, R. C., Darling-Hammond, L., Andree, A., Richardson, N., & Orphanos, S. (2009). *Professional learning in the learning profession: A status report on teacher development in the United States and abroad.* Dallas, TX: National Staff Development Council.

Woolfolk, A. (2007). *Educational psychology* (10th ed.). Boston, MA: Pearson Education.

Yoon, K. D. (2007). *Reviewing the evidence on how teacher professional development affects student achievement.* Issues & Answers Report, REL 2007-No. 033. Washington, DC: U.S. Department of Education, Institute of Education Science, National Center for Education Evaluation and Regional Assistance, Regional Educational Laboratory Southwest.

CHAPTER NINE

Research Experiences FOR Teachers Can Enhance THE Teaching OF Science

SUE WHITSETT

INTRODUCTION

As an undergraduate, I was a biology major and chemistry minor and earned a degree in secondary science education. I did not complete any research as part of my degree, and at the time, the university I attended was not a research university. I engaged in many "cookbook" labs during college but did not have the chance to go beyond what was asked. Many times we teach as we were taught. For many years, my students completed the labs in the textbook lab manuals and never went beyond what was asked, which usually involved a finite answer to the question proposed in the lab activity. I was a teacher of science, but not a scientist. Yet I wanted my students to practice science and to think like scientists. This is a difficult task for individuals if they have never had the chance to do so themselves. It would be analogous to a surgeon that has never completed an entire surgical procedure. I personally would not want that surgeon performing surgery on me. My opportunity to continue my own professional development by doing science research was about to begin.

A headline—"Research Experiences for Teachers"—and a small, one-paragraph advertisement in *NSTA Reports* (National Science Teachers Association) stated: "Do research for the summer and earn money in Rochester." The money was much more than I could make working a minimum-wage (or thereabouts) job for the summer. The advertisement continued: "Please e-mail for an application and more information" (National Science Teachers Association, 2004).

I was excited. A chance to do research at a university was an experience I had never had. A few weeks later I found out I was accepted into my first RET (Research Experiences for Teachers) program at the University of Rochester. I had many questions. What type of research would I do? What if I did not know what the scientists were talking about? What if I did not know how to use the equipment in the lab? How would this help me in my teaching? Five days after school let out I landed in Rochester, New York, to begin my summer experience. Little did I know that I would become an RET junkie for the next five summers! That was the summer of 2004, and it changed how I would approach the teaching of science in my biology classes for the rest of my career.

Before we get started on the core of RET, I want to paint a picture of my classroom in action.

Fall Semester 2008, After 5 Summers of Research Experiences

Day 1

Good afternoon and welcome to BIO 1. Today is the first day of this class, and beginning today and throughout this course you will learn science by doing science. At your desk you have a small pipette filled with water, a ruler, a note card, and a petal from either a daisy or a rose. Please add one drop of water to the petal and make as many observations as you possibly can about that petal and the drop of water on the note card provided. You do have one tool to use at your desk to help you make some of the observations. (Wait about 2 minutes as I walk around.) Please share your information with the people next to you, in front of you, and behind you. (Wait about 5 minutes while students share.) Can anyone volunteer an observation (write on board)? Another observation? (Continue until there are no other observations. If students do not mention shape of the drop [perfect sphere] or the size of the drop, encourage students to look again and describe.) Please draw a line on the note card.

Now, take your petal with the drop of water on it and turn the petal upside down. Please write any other observations you can make below the line. Next, try shaking the petal and write down any other observations. Please share with the person next to you. Would anyone like to share the new observations? (Write on board all new observations. Around this time the students start asking why the drop of water did not fall off.) Would anyone share a question they may have? (Write these on the board also.)

Today you're beginning your journey to do science the way scientists in a scientific research lab do it. This past summer, in a lab at the University of Wisconsin-Madison, I performed the same research activity we've just completed. I got the idea from an article I saw last spring during the last week of school. I happened to have a rose on my desk, and being a teacher of science, I tried the same thing you just did. Knowing that I would be at UW-Madison for at least 6 weeks during the summer, I decided I wanted to do more research on this phenomenon known as the "Rose Petal Effect," a phenomenon caused by the nanostructure of the surface of the petal. This is a new field of science, and the article I saw last spring had been published only 6 months before I saw it. Tomorrow we will begin our year of biology by doing more science, learning about safety as we do science, and doing what some research scientists are currently doing in labs around the world. So far, scientists know of very few flowers that exhibit this effect. So we will help scientists by trying to find out what other flowers exhibit this effect and what benefit to society this effect could have. I know this is the first day of school, but I am going to ask that you do some homework tonight (many groans). Please find a flower that is not a rose or a daisy, and bring one flower to class with you tomorrow. Please do not buy a flower. Find one in your garden, your neighbor's garden (please ask if you may have one flower), or on one of the nature paths around school. Please bring it to the room before school starts and put it into the vase marked with your hour. Thanks, and have a great first day of school.

Little did my students know they were being introduced to scientific inquiry, biomimicry, and nanotechnology on the first full day of general biology.

BACKGROUND

In 2001, the National Science Foundation's (NSF) Directorate for Engineering began the RET program. The program was developed based on a similar program for undergraduate students, Research Experiences for Undergraduates (REU). The purpose of the program was to "bring knowledge of engineering and technological innovation to the pre-college classroom." NSF also wanted to "build long-term collaborative partnerships between K–14 science, technology, engineering, and mathematics (STEM) teachers and the NSF research community by involving the teachers in engineering-funded research and then helping them translate their research experiences into classroom activities" (Russell & Hancock, 2007, p. ES-1). At the time of this writing, the program still exists with slight modifications in the synopsis of the program.

> [The] program supports the active involvement of K–12 science, technology, engineering, computer and information science, and mathematics (STEM) teachers and community college faculty in engineering and computer science research in order to bring knowledge of engineering, computer science, and technological innovation into their classrooms. The goal is to help build long-term collaborative partnerships between K–12 STEM teachers, community college faculty, and the NSF university research community by involving the teachers and community college faculty in engineering and computer science research and helping them translate their research experiences and new knowledge into classroom activities. (National Science Foundation, 2013, p. 3)

With the release of the *Next Generation Science Standards* (NGSS), which emphasizes integrating scientific practices and engineering design processes rather than simply learning content, teachers need to be exposed to real scientific research to understand what scientists do and how they do it before they can be expected to have their students "do science." The following statement is from the pre-publication NGSS release document of the *Next Generation Science Standards*: "Coupling practice with content gives the learning context, whereas practices alone are activities and content alone is memorization. It is through integration that science begins to make sense and allows students to apply the material" (Achieve, Inc., 2013, p. 2). In their article "Summer Scientific Research for Teachers: The Experience and Its Effect," Westerlund and colleagues state: "To teach science by inquiry methods requires that teachers have familiarity and/or training with scientific inquiry" (2002, p. 64). They go on to say:

> Participation in research in scientific laboratories immerses teachers in the culture of science. Asking questions, writing proposals based on library research, learning techniques, designing new protocols, using new scientific vocabulary, analyzing data, presenting results, and just being with research scientists, graduate students and technicians on a daily basis are all part of the culture of science. (Westerlund, Garcia, Koke, Taylor, & Mason, 2002, p. 64)

They further state:

> In order to emulate the research experiences of working scientists, teachers need total immersion in the scientific research experience without the distraction of teaching responsibilities. Generally, this can most readily be accomplished during extended breaks from school such as during the summer. (Westerlund, et al., 2002, p. 64)

Further support for teachers participating in summer research experiences is found in a report written by Harold Wenglinsky (2000). He analyzed 7,776 eighth-graders' science scores from the National Assessment of Educational Progress (NAEP) science assessment in 1996. He found that students outperformed their peers by 40% of a grade level if their teachers were involved in professional development in laboratory skills.

An evaluation by NSF (Russell & Hancock, 2007) of the RET programs found many benefits to teachers involved in these programs. The evaluation was a series of surveys completed by over 800 teachers who had participated in RET from 2001 to 2006. Positive project outcomes were identified by 80% or more of the respondents in the following ways:

- increase in teacher's general knowledge base in the areas of science, technology, engineering and mathematics (STEM)
- increased motivation by the teachers to find ways to improve their students' learning

- increased confidence as a STEM teacher
- increased awareness of current STEM research issues
- teachers included examples or applications from RET activities in their teaching
- teachers were able to communicate more effectively about new technologies in their teaching, and increased use of technology in the classroom. (Russell & Hancock, 2007)

In their article "Teachers' Participation in Research Programs Improves their Students' Achievement in Science" (2009), Silverstein and fellow researchers measured the impact of teacher participation in a summer research program at Columbia University by using the test results of students taking the New York Regents Exams. Scores were compared between students who had a teacher in the program and students who did not have a teacher who had participated in the Columbia University program. The students were in the same course being taught the same curriculum (i.e., all students may have been in a general biology course but had different teachers who were required to teach the same curriculum). After teachers had participated in the summer research for three or four summers, their students' scores on the New York Regents exam were 10.1% higher ($P = 0.049$) than the scores of students whose teachers did not participate in a research program during the summer. Two other benefits noted by the study were a decrease in teacher attrition rates and a savings in cost to the school of $1.14 per every $1.00 invested by the program. The savings came from students not taking the Regents exam a second time and through savings from teacher retention rates, since the districts did not have to recruit for new teachers.

Columbia University's Summer Research Program's premise is "that experience in the practice of science improves the quality and authenticity of science teaching and thereby increases student interest and achievement in science" (Silverstein, Dubner, Miller, Glied, & Loike, 2009, p. 440). This program is slightly different from other RET programs in that participants at Columbia University have more time invested in the program: 16 weeks over two summers compared to a more typical 6-week program for the RET program.

RET programs sponsored by the National Science Foundation also have different formats. There are two types of RET programs: site or supplement. The site programs are typically set up to have multiple teachers participate at a school or college. The supplement is a single professor who has a National Science Foundation-funded project, bringing in a single teacher or teachers (possibly two or three) to his or her lab. Both programs involve a stipend for a teacher and usually occur during the summer for 6 to 10 weeks. I participated in both types of RET programs and found the site program to be more beneficial. Working with other teachers to brainstorm ideas and problems was very helpful, since many scientists do not understand the needs of teachers or the needs of students in a typical K–12 classroom. An article based on an RET site program in Virginia paired their participating teachers in labs for the following reason.

> This paired placement allows participants to problem-solve with a fellow K–12 teacher throughout the development of research investigations and classroom extension materials. It also helps to prevent feelings of isolation, and seeks to empower both participants to ask questions and make progress while working in a busy university research laboratory. (Trail, 2011, p. 83)

The National Science Foundation sponsors RET programs in the following directorates: Engineering, Geosciences, Biological Sciences, and Materials Research Science and Engineering Centers.

Bram Duchovnay and Caroline Joyce state: "'Doing science'—conducting actual research side by side with researchers—is perhaps the best way to achieve scientific literacy" (Duchovnay & Joyce, 2000, p. 1597). From my RET experiences, I learned so much that I was able to incorporate into my classes. My philosophy of teaching science was that students needed to "do science" in order to understand science. I was able to bring content and activities into my classes that were not there prior to my summer research experiences. I was also able to bring into my classes new technologies, or information about new technologies, that I did not know about prior to the RET programs. Duchovnay and Joyce state:

> Introducing current research into the science curriculum transforms the traditional role of educator. The ongoing support of a global, intellectual community encourages teachers to learn new pedagogies and try new curricula. Teachers cannot rely on textbooks to instruct their students about cutting-edge research, but must constantly learn in their classrooms. (Duchovnay & Joyce, 2000, p. 1597)

They further stated:

> Continued professional development for teachers in both science education and using educational technology is an international need. We believe that technology taught in a research context has a far greater chance of being incorporated into a teacher's repertoire of strategies than technology training alone. (Duchovnay & Joyce, 2000, p. 1597)

Summer RET programs typically last from 6 to 10 weeks, or from 240 to 400 hours. These hours are professional development for the teachers who participate. Some of these programs require teachers to engage in other professional development activities besides the actual research. In all the programs I was required to take a full-day lab safety-training program before I could enter the lab. Hearing about lab safety from experts in the field was beneficial when I went back to my own classroom and had to teach lab safety to my own students.

A different type of professional development occurred during my first year. I had at least 10 hours of training in how to prepare for a scientific presentation using a scientific poster. I was able to bring some of this training back to my students in helping them prepare for events that required public presentations that were timed or required a scientific poster. The programs I participated in required attendance at "brown bag lunches." Brown bag presentations are typically held during the lunch hour and given by professors on campus to share their research. What a fantastic way to learn about many different types of cutting-edge research throughout the summer!

There were many other professional development opportunities I participated in during my RET experiences. All of these opportunities eventually came back to my classroom. In a report by Yoon and colleagues entitled *Reviewing the Evidence on How Teacher Professional Development Affects Student Achievement*, the authors state: "This report finds that teachers who receive substantial professional development—an average of 49 hours in the nine studies—can boost their students' achievement by about 21 percentile points" (Yoon, Duncan, Lee, Scarloss, & Shapley, 2007). The summer programs provide many more than the 49 hours in the report. I do believe my students' achievement benefitted from my involvement in scientific professional development programs.

RET programs not only benefit the teachers who are involved. Researchers have also stated that they have benefitted from having teachers in their lab. In an article written by a K–12 teacher, a GK–12 graduate student, and a Louisiana Tech faculty member, all the authors agree: "Collaboration aids all parties involved" (Bearden, Mainardi, & Culligan, 2009, p. 278). They continue by stating: "The partnership gave the members a better appreciation for the opportunities for collaboration available at the university and K–12 levels of education that provide direct channels for research to be integrated into the K–12 classrooms" (Bearden et al., 2009, p. 278). This campus and many others have met the goals set out by NSF to develop the partnerships stated in the solicitation mentioned earlier in this section.

BEST PRACTICES

My first RET program was part of an RET site at the University of Rochester in New York. There were five RET participants during the summer, and they were placed in different labs across campus. We met on a weekly basis and participated in various "brown bag" presentations during the summer. These presentations were not part of the research taking place in the labs we worked in, but they exposed us to other cutting-edge research occurring on campus. I found these sessions to be incredibly informative about other "new" research. I could use these lectures to broaden my content knowledge, and that led to communicating that information to my students.

My first RET program was spent doing research in Professor Thomas Eickbush's biology lab next to scientists researching the location of the insertion of retrotransposons in the DNA of *Drosophila melanogaster* (fruit flies). I had to learn about retrotransposons, which I had only remotely heard of, before I could begin to help with the research in the lab I was assigned to. In my first 2 weeks I spent long days at the library learning all I could about the research that not only the lab I was assigned to was involved with, but the research engaged in by other labs around the world. I learned that a retrotransposon is a segment of DNA that makes a copy of RNA, and then reverses the process and makes a new copy of DNA, which then inserts into a new location on the DNA strand. With the help of scientists in the lab, and the mentor who was assigned to me, I became part of a scientific research team for the summer. This meant that I attended weekly lab meetings and eventually had to present my findings to my lab team. At the end of the summer I had to participate in a scientific poster session, which was another activity that I had never been part of in my career. A very unique experience was the chance to work with a college student who was participating in the REU (Research Experiences for Undergraduates) program. We were able to work together on various parts of the research project that was assigned to us. I learned a lot about students pursuing a science degree, information that I was able to share with my students.

My research topic for the summer was "Cloning and Mapping the Missing rRNA Genes of *Drosophila melanogaster*." Once I knew more about the research that was being conducted in my lab, I began to do the same work as some of the lab team. I learned the necessary skills to construct a library of Bacterial Artificial Chromosome (BAC) clones, followed by an analysis of individual clones and their assembly into larger groupings known as contigs. To analyze the clones I used various enzymes to determine the size of the clones and the distance between retrotransposons. I used a specific set of enzymes to determine if a clone came from the X or the Y chromosome. I used Polymerase Chain Reaction (PCR) and gel electrophoresis to determine if there were any retrotransposons present in the DNA. I used a polyacrylamide gel electrophoresis to help determine the size of fragments and a pulse field gel electrophoresis to obtain a high resolution of large fragments. Many of these procedures were not present when I attended college as an undergraduate, nor were they present during some of the coursework I took after I graduated. I had never attended a "research university." Based on my work, I was able to put together aligned contigs showing the possible locations of inserted retrotransposons. If some of these terms are unfamiliar to you, imagine how I felt stepping into a lab with a research focus on retrotransposons, a term I barely understood until I was notified that I was selected. One of many benefits to a teacher in an RET is the learning of new content and new technologies that we never encountered while we were in school. This experience definitely helped me in teaching AP Biology, especially the areas of cellular and molecular biology. It is always easier to teach a topic when you have experience with the process and content.

The following summer I became involved in a different RET at the University of Wisconsin-Madison doing environmental field research. This RET program was a supplemental grant to an ongoing NSF grant by Professor Stanley Dodson in the Limnology Department. My title for the research was "Limnology Study on the Black Earth Creek Watershed." I was involved in studying water quality after a heavy rain event and its effect on the chemical composition of a river. I collected water samples at seven different sites along a 3-mile stretch of river—above, below, and between wastewater treatment plants. The research was an attempt to help discover the reason for a large fish kill after heavy rain events in the area. I worked with a doctoral candidate in Dr. Dodson's lab and helped him with his research for his thesis.

My project was focused on collecting water samples and chemically analyzing the samples. I used professional-grade lab sampling equipment to collect some of the data, along with the reliable Hach chemical water quality tests. I was trained on how to use new lab equipment that analyzes Dissolved Organic Carbon/Nitrogen (DOC/N). I also obtained sediments from the river and performed a protocol to determine the rate of nitrification occurring in the sediment.

I had monthly meetings with Dr. Dodson, who had set up the RET supplement for his lab. There was one other RET teacher who worked directly with him on his research project; we did not do the same research. The other RET participant and I had very different experiences and outcomes from our 6-week project. RET programs can be structured differently, and as this was the first year that this professor had an RET supplement, the program was not as structured as the site program that I had participated in the previous summer. At the same time I was in this program, my husband was in another RET site program across campus. As we compared notes nightly, I realized how different programs could be, not only between campuses but also on the same campus.

My third RET (which continued through the next 3 years) took place at an RET site in the Materials Research Science and Engineering Center (MRSEC) at the University of Wisconsin-Madison (MRSEC, University of Wisconsin Madison, 2013b). For each of these four summers I was involved in research in the area of nanotechnology. The first year involved taking an online course in nanotechnology to understand the concepts of this emerging field of science while working with a professor on my project. During these four summers I was required to develop a curricular project that involved an inquiry experience for my students. Some of these activities can be found at the website Exploring the Nanoworld (MRSEC, University of Wisconsin-Madison, 2013c). Each summer

I had a mentor on campus who helped me develop my project, and I had access to a lab that I was able to work in to do my research for a given topic. Additionally, every summer we were required to participate in brown bag lunches on campus with another RET group from the Nanoscale Science and Engineering Center (NSEC) (NSEC, University of Wisconsin Madison, 2013), along with a large REU group working with professors in the field of nanoscience. These presentations were focused on cutting-edge nanotechnology research that was occurring on the University of Wisconsin campus. And in the last RET we began our experience by attending a "boot camp" on lab techniques before we began our research on campus. Based on feedback we had given the previous summer regarding some of the lab techniques we needed for our research, our RET program collaborated with the local technical college to give us some training in routine lab procedures done in a research lab. These routine procedures for a university are not necessarily what we do every day in our own classroom.

My first research project in the field of nanotechnology was titled "The 'Lotus Effect' (Nanotechnology of the Lotus Plant)." My mentor was a friend and professor on campus better known for his work with Wisconsin Fast Plants™, Dr. Paul Williams. Together we developed an inquiry activity using the concept of "structure affects function." We used two varieties of Wisconsin Fast Plants™, one that showed the "Lotus Effect" (a sphere of water rolls off a surface taking any dirt particles with it) (MRSEC, University of Wisconsin Madison, 2013a) and the wild-type plant that does not show this phenomenon. Along with this inquiry-based activity, I developed a model of surface structure and size of particles down to the nanoscale level using various grits of sandpaper to demonstrate how water acts differently on various surfaces. The wealth of knowledge that Dr. Williams contributed about Wisconsin Fast Plants™, combined with the use of a scanning electron microscope (SEM) to observe the surface structure of leaves, led to an unforgettable summer. It also led to a curriculum project that was used on multiple occasions during several school terms in all levels of my biology classes. This experience, along with the course on nanotechnology, sparked my interest in the field of nanoscience. I wanted to come back for more the following summer and, more important, I wanted my students to know about this emerging scientific field.

After 3 years in RET, I was ready to change how I was teaching biology to my students: I wanted them to do some form of inquiry. The activities I developed in my third RET were used with all of my biology classes from remedial biology to advanced biology. I decided to begin slowly with the idea of inquiry and developed a guided inquiry activity for my students using Wisconsin Fast Plants™. Students grew wild-type plants and a plant known to show the "Lotus Effect," but they were not told which seeds belonged to which type. Students had to compare plant structure and the function of water drops on the leaves and pods of the plants to

SEM photographs of pods and leaves to determine which plant showed the characteristic, and then describe why it occurred.

Students also had to bring in leaves from home or outdoors and test those for the "Lotus Effect." This did not go as planned because by the time the students were going to bring in leaves, it was December, and in Wisconsin there are not many leaves left on the trees. I suggested a Plan B: bring in leaves from houseplants. This strategy led to a few surprises. It appeared that a poinsettia may exhibit the "Lotus Effect." This was unexpected, as we did not find any evidence in the literature at that time that a poinsettia had a leaf structure to exhibit this phenomenon. Students were disappointed that we did not have a scanning electron microscope at our disposal to look at the surface structure of the leaves to verify their observations. The students realized the value of having more detailed microscopic photos to help them with their analysis of determining which plants truly exhibited the "Lotus Effect."

The following summer, I chose to work in a lab studying diatom structure. The title of my research was "The Nanofabrication of Silica in Diatoms." I worked with a graduate student in Dr. Michael Sussman's lab to set up an inquiry into population dynamics of diatoms based on the amount of silica in their environment. Diatoms are unicellular algae that encase themselves in glass-like shells made of silica. Because they are unicellular, they are quite small, most often under 10 micrometers. But because of this size range, many students do not think of these organisms as having any dimension when viewed through a microscope. This activity was combined with developing a scale model of the cell wall of a diatom, magnified 10,000 times through the use of 1-mm beads representing the silica nanospheres of the diatom. Scientists discovered that silica was deposited in structures that were in the shape of spheres in the size range of approximately 100 nanometers that make up the cell wall. The intent was that students would be able to perform the activity and also learn about the size of silica nanospheres by looking under a microscope at a diatom, measuring the size of the diatom, and then making a model 10,000 times larger than what they saw under the microscope.

The concept of objects being in the range of nanoscale (under 100 nanometers) is very difficult for students to grasp. I hoped that by building a scale model, students might better understand this concept. Additionally, I did some research on the fabrication of the silica nanospheres inside the diatom and tried to come up with an animation of how silica is extracted from the ocean, taken into the cell, and expelled as silica nanospheres. I found that I was a little ahead of my time, as the exact cell pathway process at the cellular level was not known at that time. It was good for my students to know that scientists do not yet know all the mechanisms of cellular processes, and that there is still much work to be done to understand how cell organelles work together to produce different materials.

The third summer I was more on my own to develop an activity and again decided to work with plants, this time doing research on the "Rose Petal Effect." I again had the opportunity to collaborate with Dr. Williams to find out if Wisconsin Fast Plants™ exhibited this effect. The title of my research for the summer was "The Petal Effect (Nanotechnology of the Rose Petal)."

You may be wondering how one comes up with ideas for a summer research project. The director of the RET site sent out a list of possible topics in May, based on recently released research articles in the field of nanotechnology. One article that caught my eye concerned the "Rose Petal Effect." Having a rose on my desk that day, of course I had to try the phenomenon described. I was hooked and did more research on the nanostructure of the rose petal that causes a drop of water to form spheres and adhere to the petal. In the research article, some scientists had developed a method to make an imprint of the surface structure of the petal by using two different polymers. The polymer, when hardened, had the same nanostructure as the petal itself. My immediate thought was to incorporate this idea into my curricular project for the summer. The only problem was that the materials needed for the procedure included chloroform. Today most schools, mine included, prohibit the use of chloroform with students. My research for the summer included trying to find other flower petals that exhibited the Petal Effect and then coming up with a suitable material that could make an imprint and also be handled safely by students. The curricular activity developed was testing various flower petals for the Petal Effect and then making an imprint of the petal based on the polymer mixture that I developed. Again, having access to a scanning electron microscope for looking at both the surface of the petal and the surface of the polymer was a plus in working at the nanoscale level. An added bonus to the summer was applying for a patent to the procedure I developed. Since the procedure was very similar to the procedure in the article (minus the chloroform), I did not receive the patent, but the process of applying was worth the effort.

My fourth summer at the University of Wisconsin-Madison, and my last as an RET participant, was again in MRSEC. The title "Producing Silver Nanoplates to Show the Superhydrophobicity of the Petal and Lotus Effects" was a culminating activity that combined my work in summers one and three at the University of Wisconsin-Madison. Much of my nanotechnology research dealt with the surface structure of materials, specifically the hydrophobicity of the surface based on the nanostructure of the surface of the materials I had been working with. An activity that had been used at MRSEC—the silver mirror reaction—was modified to allow different amounts of silver, in the form of nanoplates, to adhere to a surface. This activity was based on the concept that varying the amount of time allowed to form nanoplates would result in different surface structures. When water was added to the surface, different properties were observed, ranging from hydrophobic to hydrophilic. Additionally, the hydrophobic surfaces also exhibited both the Lotus Effect and the Petal Effect. One more visit to the scanning electron microscope allowed photos to be taken of the different surface structures. The final activity developed for the classroom was the procedure for students to use—to plate the silver nanoplates onto aluminum foil—and then match the SEM photos to the actual samples produced.

ADAPTATIONS OF THE BEST PRACTICES

Following my first RET experience during the summer of 2004 at the University of Rochester in New York, I was able to bring some of my experiences back into the classroom in a couple of ways. Because of the lack of equipment in my school and the lack of funds to perform the procedures used in the research lab, I brought back other experiences to my students. After the summer RET program, I put my poster from my scientific presentation in the back of my room. I could not replicate the experiments I had performed because of the lack of equipment (thermocycler), chemicals (to perform PCR and electrophoresis), and money (to order various primers, buffers, and photo materials). My school did not have a biotechnology class for students to learn the techniques necessary to perform this level of research. I was able to do basic gel electrophoresis with students using methylene blue to stain their gels. I compared their results to those photos in the poster of gels that used ethidium bromide staining. Students were able to see the difference in resolution of the gels from the research lab. This led to a discussion of the various stains used and the benefits of both techniques.

I was able to describe lab procedures to the students, along with showing them how to use a data lab notebook. It was during this year of teaching that all my students began using lab notebooks any time they did a lab. Students discovered how to enter data and learned the reasons for not erasing data even if they were very different from the other data collected. I was able to share information with my students about a career in the pure sciences because of the discussions I had with the REU student I worked with. My seniors were especially interested in this information.

One unique activity that occurred during the school year was made possible when the University of Rochester shipped a first-generation model atomic force microscope (AFM) to my school. Part of my RET program during the summer at Rochester was spent learning about an AFM, what it is used for, and how it is used, along with a basic training session. For participating in the training session, the University of Rochester agreed to send its traveling model to my school. This generated much interest not only from my students but also from other students in other classes in the science department, including students in chemistry and physics. Students in my school had only seen and used compound microscopes that most often were used to magnify objects 400 times. Seeing what an atomic force microscope looked like, and what it could do, gave students an experience that is usually not associated with high school.

My second year in the classroom following a summer of RET at the University of Wisconsin-Madison was very similar, though there was not a new large scientific poster. I was able to use my field experience to talk about teamwork, the importance of collecting multiple samples, and how science is not always done in just an indoor lab. My lab for the second RET summer was a river, complete with leeches (the students really liked the pictures of me with leeches on my feet after one outing!) and high water above my waist at times. Because I had to collect water after major rain events, the research required some strange scheduling during the day and week. When I asked students if they would come in on weekends to collect data for an experiment they had set up, they would laugh and say no way. I could drive home the idea that sometimes science does not happen on your preferred timetable. I could not control the weather, but I could determine the best time to collect the data needed for the research, which included some weekends and evenings.

The inquiry-based projects I developed during the next three summers researching and developing activities in the area of nanotechnology were implemented in my classroom the following school year. Students had a chance to learn science by setting up their own "experiments" based on research that I did during the summer. In biology, most experiments with living materials require a much longer time frame than 1 or 2 days, and that allowed for teaching of some of the major concepts that went along with the inquiry activity. As part of the RET, we did have to develop a curriculum that fit into our school's required curriculum. This was difficult in itself: how can you add more into a full curriculum and still be sure students are learning the district's curriculum? Part of the development of a curriculum during the summer RET program required that we fit our activity into our current curriculum, possibly replacing activities we had previously used with "new" activities. The RET activities were not meant to be stand-alone activities that took more time to teach, but rather were to be integrated into what we were already teaching.

One of the activities I developed in my second year at the University of Wisconsin-Madison MRSEC for my students required the purchase of some equipment that we did not have in our school. (Some of the RET programs include a stipend for the teacher to purchase supplies to implement the activity that is developed.) I purchased some aquarium equipment and gathered many plastic water bottles to set up individual "environments" for my students to design and collect data on their experimental designs. This was the first time my students and I tried to raise diatoms, and, in all honesty, our results were not what we had hoped for (many of the environments we created killed the diatoms), but it actually was very beneficial for my students to see that not everything turns out as planned.

This in itself was incredibly helpful in getting students to see that often in science, the first time you run an experiment will require many adjustments to procedures. Also, just because the original procedure did not work, they were not downgraded, and it was not considered a failure, since they learned useful information about their procedure.

The school year after participating in MRSEC for the third summer is described in the vignette at the beginning of this chapter and continued at the end of the chapter in more detail than the above activities. The students and I had a working science lab. We did not know at the beginning of the activity what the outcomes would be for the materials the students brought in. We learned science by doing science, just as it is done in scientific labs around the world.

The RET experience had a profound effect on how I taught science. I saw scientists in action—not doing the "scientific method," but doing scientific practices. I truly learned that scientists do not work alone, which meant that my students began to work in "teams" for research projects and not individually. I learned the importance of designing an experiment based on previous research, communicating data to others, having replicable data, and the concept that not all results are the "right" answer. I truly experienced the practices of science and learned more about the nature of science during my RET programs than anything I had done in college to prepare me for teaching science. With the release of the National Research Council's *A Framework for K–12 Science Education* in 2011 (Board on Science Education, National Research Council of the National Academies, 2012), these scientific practices have become a foundation for the *Next Generation Science Standards* and are critical in helping students understand the nature of science. In an ideal world, all pre-service teachers would have an opportunity to spend one semester in a research lab prior to doing any teaching, and then integrate their research into activities as they begin the practice of teaching. Research experiences are not only good for the pre-service teacher, but should be part of a regular teacher's ongoing professional development.

My experiences are very similar to those described in the above-mentioned research articles about RET. Personally, the RET programs had many benefits beyond the knowledge they provided in the different fields of science. One of the strengths of the last four summers as an RET teacher was the opportunity to work with a group of four other teachers. We met formally on a weekly basis to share our ideas and to problem solve with one another. We critiqued each other's projects, bounced ideas off each other, and did formal presentations throughout the summer for one another. Scientists in a research lab do not work in isolation, and this concept was followed in our work as nanotechnology RET participants. We mirrored how research is conducted.

Another benefit was the confidence I gained in myself as a teacher of science, even in areas I was not certified in. That confidence was in both technology and content. The confidence in technology was in the knowledge of different types of technology used in research, the new cutting-edge technology being researched, and actually using more technology in my classes. The confidence in scientific content was in the areas of physics, a stronger content in application of math, and in nanotechnology. I became a stronger STEM teacher because of my experiences.

There are various programs that give opportunities for teachers to do research, some during the summer and some during the school year. The programs will vary depending on who sponsors the program. Some programs will be set up by an individual scientist or engineer in a corporation and be run like an internship. Some will be funded by other government agencies—for example, the Teacher at Sea program that the National Oceanic and Atmospheric Administration (NOAA) manages. The National Science Foundation funded the program I am most familiar with and the one this chapter is based on. RET programs offered by NSF for K–14 teachers can be accessed at http://www.pathwaystoscience.org/programs.aspx?descriptorhub=RET_NSF. If you are interested in participating in an RET or similar program, begin looking for programs in the winter, as many are filled by the spring. If you are acquainted with a research organization, you may want to inquire if it offers programs allowing teachers to work in its lab during the summer. The National Science Teacher Association website has a calendar of events (http://www.nsta.org/publications/calendar/) that also lists other opportunities.

CONCLUSION

At the end of the first year of teaching after having done an RET, I felt that students had more respect for me as a teacher of science than in my previous years of teaching. Students realized that I actually worked with "real" scientists in a "real" science lab. My enthusiasm, the knowledge I gained from working with a college student pursuing a science career, and the knowledge gained in the science itself allowed me to share information with my students that I never had been able to share previously. I saw more of an appreciation for learning and doing science from my students. I overheard comments from other students in different classes: "I wish I could be in your class; you *do* science." I had other teachers in my department send their students into my room after school to seek information about becoming a scientist or doing research.

Students, for the most part, enjoyed doing activities that students in the other biology classes were not doing. I noticed that as I became more comfortable with the practices of science, my students became more comfortable, too. When they performed the activities that I developed, the atmosphere in the classroom seemed to be more "scientific." No longer were my students looking for the "right" answer.

It became acceptable if something did not turn out as predicted—it was not wrong! They were given the freedom to think more about what was happening and how they could manipulate variables if necessary. In addition, students were exposed to more technology (for example, the scanning electron microscope photos of their petal imprints, or the 1-year period when they were able to use the atomic force microscope) than I had previously used in teaching. And an interesting benefit was bestowed on my classes: I was no longer hearing from my students the question, "Why do I have to learn this?"

My experiences in an engineering building for four summers was not only a major learning experience for me, but provided much insight into why my students needed to be well rounded in all areas of science during high school. Because of this, my students were no longer just biology students. We all became STEM students. I found that I was teaching more physics and chemistry concepts in my biology classes than in previous years and using more math with my students than I had before. My students and I were using the practices of science, and I had learned the engineering design processes during my research. The content knowledge and curriculum that I brought back for my students was very different from my biology background. Learning about cutting-edge science and being able to communicate that knowledge and the skills needed to be successful in these newly emerging fields kept my students' interest. If you don't know about a topic, you typically do not teach about that topic. My students left biology learning about nanotechnology and could communicate with others in such a way that convinced me they had learned the material and understood it.

Scientists do not always understand teachers and their classrooms. Many have not been back in a classroom since they were students themselves. A research lab is not the same as a science lab in a K–12 school. Many times we had to explain why we had to modify the projects we were attempting. As an example, the original paper I read about the "Rose Petal Effect" used a procedure that called for chloroform as one of the materials to make an imprint of the petal. I needed to find a safe, suitable replacement for this material, and that search became part of my research for the summer.

What scientists take for granted may be something we as teachers of science have never heard of, especially if science was not our major in college. A bridge was built each year I was an RET participant between K–12 education and the scientists I worked with on my projects. There was a much better understanding by both of us about each other's field of expertise and what is required to be successful in both of our fields.

The best benefit of being an RET teacher was the way in which students responded to my efforts to teach biology in a way that is more like a research lab atmosphere. I enjoyed seeing and hearing the excitement of new discoveries ("You mean no other scientist in the world knows that white phlox has the petal effect?").

I felt that I had developed a community of learners rather than a classroom of students. These benefits were worth the six summers of research.

To end this chapter, I will finish the vignette I began at the beginning of this chapter. This narration continues with days two through six of the school year in my classroom after completing my fifth summer of research experiences. The vignette is based on the curriculum I developed from my research during that summer.

Day 2 of the Semester

As we begin our study of science safety and the "scientific method," we will become scientists. This past summer I developed a procedure that can be used safely by students in a high school classroom to study the "Rose Petal Effect." Please find your flower and bring it to the back of the classroom. You will see some materials on the lab stations. Before we move to a station let's find out what the materials are and how they should be used. Who can tell me what these are (holding a pair of goggles)? What are they used for? Where should they be worn? (Students answer until someone gives the correct answer.) Please put on the goggles when you get to the lab bench and leave them on until we leave the lab area.

Please move to a lab station, two people per side. In front of you is a more precise micropipette than the Beral pipette you used yesterday, a small labeled glass beaker of water, and a blue safety mat that serves two purposes—absorbs spills and helps prevent the glass from breaking if the glass tips over. Please be sure the glass is always on the mat. There is also a container that is labeled with 15% PVA. PVA is a chemical that will be used to make a negative imprint of the surface of your petal. PVA stands for Polyvinyl Alcohol. Please do not handle. You will also find a small plastic weigh dish, a metric ruler, a note card and writing utensil for each student along with one Sharpie per lab station.

At this time I'd like to demonstrate the correct technique for using the micropipette. You will be adding a 5-microliter drop of water to your petal. (Demonstrate use and then have each student handle the micropipette before adding water to the petal.) Handling the petal by its edges, remove one petal from your flower. Proceed as you did yesterday and make observations on the note card concerning what happens to the water when applied to the petal, and the size of the drop of water. (Walk around observing students testing their petals.) When finished set the petal on the blue mat.

(When all the students have their flowers on the table, I continue.) The next step is to label your weigh dish with your initials and hour. (Wait while everyone labels.) Next, you will place your petal at the bottom of the weigh dish. Once you have your petal in the weigh dish, I will pour the PVA over the top of the petal. Please do not touch the PVA once I have poured it. Then carefully carry the weigh dish to the fume hood, come back to the station, remove your goggles, and gather the note card and return to your desk.

RESEARCH EXPERIENCES FOR TEACHERS CAN ENHANCE | 199

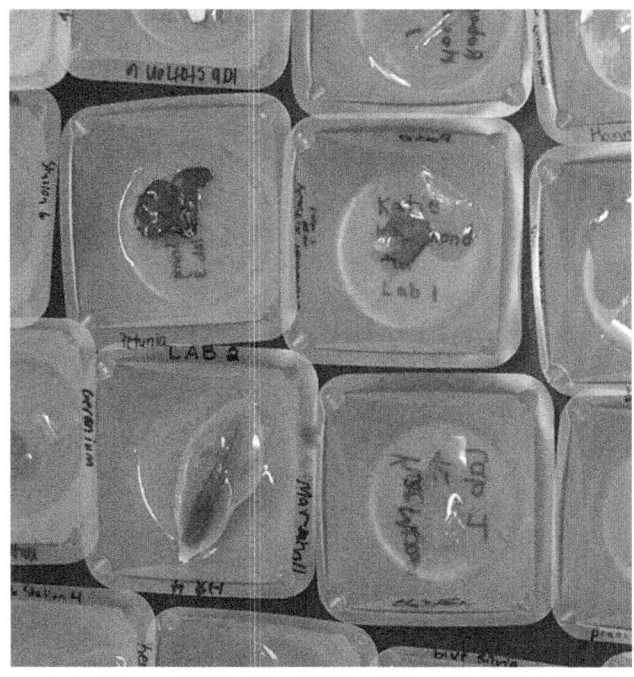

Photo 9.1. Petals That Students Removed Covered with PVA.

(Photo by Sue Whitsett, 2008.)

Day 3

Today we will remove the negative imprint from the petal and pour a different chemical over the negative imprint to make a positive imprint, which will have the exact surface structure as your petal. The chemical today is PDMS, abbreviation for polydimethlysiloxane. I will need each of you to wear a pair of goggles and a pair of gloves while working. I have at the lab station a forceps. Carefully remove the petal from the PVA. Place the PVA with the imprint upright in the bottom of the cup. When you have the PVA negative imprint in the cup, raise your hand and I will pour the PDMS over the imprint, and then you will return the paper cup to the fume hood. We will look at these next week to give the PDMS time to harden.

Days 4 and 5

While we waited for the PDMS to harden, I spent the first 20 minutes of each day giving students background by using my final RET presentation. Students learned about nanotechnology through the use of an activity on dilution, they learned about surface structure, and they learned about lab safety.

Day 6

Today you will be removing the PDMS from the PVA. We will test the PDMS the same way we tested the original petal. We will have a contest to see which petal imprint that has a drop of water on it can hold the most weight. We will collect class data and learn how to write a full lab report. Finally, we have access to a scanning electron microscope. I will be sending some of your imprints to a local college; photos will be taken of the surface and returned in a few weeks along with the original imprint. There will also be images taken of some of the flower petals if the professor can find the same flowers in his location. You will need to write up a lab report following the guidelines provided. (Students receive a sheet on how to do a lab write-up.) Your conclusion will include the following points: Which flowers in class exhibited the Petal Effect? Which petal imprint was able to hold the most weight via a drop of water? In the future, what could this property be used for? Additionally, you will need to write one paragraph summarizing in your own words what you learned from this activity.

WORKS CITED

Achieve, Inc. (2013, May 1). *Next generation science standards*. Retrieved from http://www.nextgenscience.org/sites/ngss/files/Final%20Release%20NGSS%20Front%20Matter%20.pdf

Bearden, K. K., Mainardi, D. S., & Culligan, T. (2009). In search of the active site of PMMO enzyme: Partnership between a K–12 teacher, a graduate K–12 teaching fellow, and a research mentor. *Chemical Engineering Education, 43*(4), 273–278.

Board on Science Education, National Research Council of the National Academies. (2012). *A framework for K–12 science education: Practices, crosscutting concepts, and core ideas*. Washington, DC: National Academies Press.

Duchovnay, B., & Joyce, C. (2000, March 3). Essays on science and society: The spirit of discovery. *Science, 287*(5458), 1595–1597.

MRSEC, University of Wisconsin-Madison. (2013a, May 2). *Exploring the nanoworld MRSEC*. Retrieved from http://education.mrsec.wisc.edu/Edetc/cineplex/lotus/index.html

MRSEC, University of Wisconsin-Madison. (2013b, May 2). *Materials Research Science and Engineering Center*. Retrieved from http://mrsec.wisc.edu/

MRSEC, University of Wisconsin-Madison. (2013c, May 2). *Exploring the nanoworld*. Retrieved from http://education.mrsec.wisc.edu/

National Science Foundation. (2013, May 1). *NSF*. Retrieved from http://www.nsf.gov/pubs/2011/nsf11509/nsf11509.htm

National Science Teachers Association. (2004, March). Summer opportunities. *NSTA Reports*. Arlington, VA: NSTA Press.

NSEC, University of Wisconsin-Madison. (2013, May 2). *Nanoscale Science and Engineering Center*. Retrieved from http://www.nsec.wisc.edu/

Russell, S. H., & Hancock, M. P. (2007). *Evaluation of the research experiences for teachers (RET) program: 2001–2006*. SRI International. Retrieved October 11, 2013, from http://csted.sri.com/sites/default/files/reports/RET2_FINAL_REPORT_June_30_06.pdf

Silverstein, S. C., Dubner, J., Miller, J., Glied, S., & Loike, J. D. (2009). Teachers' participation in research programs improves their students' achievement in science. *Science, 326*(5951), 440–442.

Trail, J. (2011). University of Virginia's Research Experience for Teachers (RET): Program overview and outcomes. *Journal of Virginia Science Education*, v4, n2. 81.

Wenglinsky, H. (2000). *How teaching matters: Bringing the classroom back into discussions of teacher quality*. Princeton, NJ: Milken Family Foundation and Educational Testing Service.

Westerlund, J. F., Garcia, D. M., Koke, J. R., Taylor, T. A., & Mason, D. S. (2002). Summer scientific research for teachers: The experience and its effect. *Journal of Science Teacher Education, 13*(1), 63–83.

Yoon, K. S., Duncan, T., Lee, S. W.-Y., Scarloss, B., & Shapley, K. (2007). *Reviewing the evidence on how teacher professional development affects student achievement.* Issues & Answers Report, REL 2007–No. 033. Washington, DC: U.S. Department of Education, Institute of Education Sciences, National Center for Education Evaluation and Regional Assistance, Regional Educational Laboratory Southwest.

CHAPTER TEN

Modeling Sustainability Through STEM Service-Learning

LEIGH JENKINS

INTRODUCTION

Service-learning embedded into the curriculum in K–12 schools provides students with meaningful, integrated, hands-on connections to the broader world of STEM careers. Through science, technology, engineering, and math (STEM), students are encouraged to embark on educational pathways that potentially lead to meaningful employment in STEM careers. Modeling sustainability in our schools and communities through STEM service-learning can empower students to become environmental and economic stewards and to serve as life-long agents for community change.

Service-learning is defined as a teaching and learning strategy that integrates meaningful community service with instruction and reflection to enrich the learning experience, teach civic responsibility, and strengthen communities (Eyler, Giles, Stenson, & Gray, 2001). Service-learning is distinct from volunteerism by its use of reflection and its integration into the curriculum in formal education (Strage, 2000).

This chapter will provide examples of how service-learning projects can be integrated into any school curriculum and how STEM sustainability projects can support service-learning. At the same time, the chapter will speak to the research that demonstrates student engagement through service-learning and how a range of positive outcomes for students can be tied directly to these experiences (Meyer, Hofshire, & Billig, 2004). The project featured in this chapter relies heavily on reflection as an evaluation tool and uses the building of community partnerships to support the sustainability of the service-learning project.

High-quality service-learning is often described as having positive impacts on students' personal and social development, academic achievement, citizenship, and career awareness (Billig, 2002). The reasons for this vary, but it is believed that successful service-learning can be attributed to students feeling empowered as they take on leadership and adult-like roles, gain competence in the practical application of what they are learning, and reflect on their experiences. High-quality service-learning as a part of school curriculum is also dependent upon having a supportive administration, sufficient resources to support professional development, strong ties to the community, and districts or counties with key educational strategies for increasing student achievement (Strage, 2000). In formal education, the responsibility of the educator is to work toward student achievement of defined learning standards, and although we scoff at directing all of our teaching "to the test," we are held responsible for student achievement to some degree. Thus schools and school districts that have a goal to increase student achievement will welcome and incorporate successful service-learning into the school curriculum.

The methodology for integrating service-learning into one's curriculum can take many forms. In the context of science, service-learning has the unique potential of touching on each traditional STEM content area, as well as showing great promise in its cross-curricular potential. STEM, however, is not limited to science. So much of what students need to know involves all aspects of STEM, including technology, engineering, and math. For example, this type of learning could potentially benefit communities that are looking at ways to evolve from oil-based energy systems to greener technologies in transportation, business, and government. Through STEM service-learning projects, students might design a new way to fuel a municipal fleet of vehicles with natural gas. Institutions, including schools, are interested in incorporating energy-saving engineering, as well as modeling greener technologies for the community to see. For example, students could design ways to incorporate solar panels or wind turbines to mitigate excess energy consumption in the county schools. Funding for alternative energy enhancement is often available for schools and non-profit organizations, especially if it is used for educational purposes.

Integrating lessons to enhance student awareness of the environment and sustainable living are pregnant with possibility. Sustainability is often defined broadly as "development that meets the needs of the present without compromising the ability of future generations to meet their own needs" (*Our Common Future*, 1987, p. 40). Issues concerning the environment, health, and sustainable living are community issues and therefore provide promise for service-learning projects.

Support for incorporating service-learning into K–12 curricula is timely in our attempt to engage students in STEM learning. STEM learning is directly related to workforce development (U.S. Department of Labor, 2007). If students are engaged in STEM learning early in their education, there is a far greater likelihood that they will choose STEM majors in college and eventually pursue STEM careers (Fairweather, 2008). So much of what we attempt in the traditional classroom is met, understandably, with protest from students over the relevance of what they are learning. Basic concepts must be taught, but how do we turn those basics into tools students can use to become independent learners?

Service-learning can provide that transition. The idea of providing students with high-quality STEM service-learning opportunities in their community has had positive learning outcomes, especially in helping students apply what they have learned to the real world (Eyler et al., 2001). It is our charge as educators to make learning meaningful and relevant to our students and to assist them in a successful matriculation to college and career.

BACKGROUND

Traditionally, the ideas within *sustainability education* are addressed through formal and informal environmental education. Environmental education in the classroom is not a new idea. Since the latter part of the nineteenth century, naturalists have gifted us with the image of our oneness with nature. Wilbur Jackman was one of the first educators to promote environmental education in elementary schools with his *Nature-Study for the Common Schools*, published in 1891. Not long after, in 1911, Anna Botsford Comstock published her *Handbook of Nature-Study for Teachers and Parents*, providing elementary school teachers with a comprehensive compendium for the study of natural history.

As the twentieth century progressed, natural disasters were highlighted in the media and witnessed firsthand. The economic and environmental tragedy of the Dust Bowl in the 1930s heightened our awareness of the importance of soil conservation and the need for new, more conservative farming techniques. The Soil Conservation Service was established in 1935 as a means of protecting our soil and water resources (Helms, Pavelis, Argabright, Cronshey, & Sinclair, 1996). The post-World War II period brought the mentality of "man over nature" with the production of pesticides, herbicides, and fertilizers, all of which were overused in the attempt to make our world more comfortable, profitable, and secure for humans. When we didn't know what to do with what was left of these products, we buried them or dumped them into our waterways. It was not until the 1960s, when Rachel Carson published *Silent Spring*, that the world began to understand the consequences of irresponsible and ignorant actions, and to imagine the impact of a world without the sound of birds (Carson, 1962).

The environmental movement began to take hold with the hope that environmental legislation would return our world to its nineteenth-century purity.

Worldwide Earth Day celebrations "were a landmark expression of public support for a realignment of values and a new respect for the environment" and "set the stage for the transition of education *about* the environment and *in* the environment, to education *for* the environment" (Pandey, 2006, p. 193). International organizations joined the movement with the United Nations Conference on the Human Environment, which called for the establishment of environmental education programs. The United Nations Educational, Scientific, and Cultural Organization (UNESCO) sponsored a series of international environmental education workshops and conferences. As a result of the Belgrade Charter of 1975, environmental education could formulate a goal:

> The goal of environmental education is to develop a world population that is aware of, and concerned about, the total environment and its associated problems, and which has the knowledge, attitudes, skills, motivation, and commitment to work individually and collectively toward solutions of current problems and the prevention of new ones. ("Belgrade Charter," 1976, p. 2)

In 1970, reflecting the national commitment to environmental education, the United States passed the National Environmental Education Act. The act was designed to promote environmental awareness among citizens, encouraging them to take the responsible actions necessary to assure our survival and improve the quality of life. The act was reauthorized in 1990, establishing the National Environmental Education Foundation (NEEF), whose purpose was to leverage private support for disseminating environmental knowledge. In 1996, the North American Association of Environmental Education (NAAEE) published guidelines for the initial preparation of environmental educators and environmental materials: guidelines for excellence. Again, in 1999, the NAAEE published *Excellence in Environmental Education: Guidelines for Learning (K–12)*, establishing standards for quality environmental education.

The National Environmental Education Act was then reauthorized in 2010 with, among other components, the intention to "clarify the definition of environmental education to focus on formal and non-formal education, conveying knowledge and skills to equip the general public to work collectively and individually toward solutions to current environmental problems and prevention of new ones" (*National Environment Education Reauthorization Act*, 2010, p. 1). Unfortunately, not much progress has been made. The latest statistics on this topic reveal that

> eighteen of fifty states have a formal environmental education learning objective, eight out of fifty states have assessments that include environmental education, three states, Arkansas, Pennsylvania, and Wisconsin, require environmental education teacher training prior to certification or licensing, and twelve of fifty states have a bylaw or requirement for K–12 environmental education in instruction. (*National Overview*, 2007)

These numbers are shockingly low given the need for an environmentally literate populace. In 2011, Maryland became the first state to require environmental science as a graduation requirement.

Environmental education is fundamentally different from education for sustainable development. Although it is extremely important, one might argue that environmental education, as a model for the future, is no longer enough. Education for Sustainable Development (ESD) was first endorsed at the United Nations General Assembly in 1987. Sustainability education and education for sustainable development are often used interchangeably. However, one level speaks to awareness while the other speaks to using education as a tool for achieving sustainability. ESD "calls for giving people knowledge and skills for lifelong learning to help them find new solutions to their environmental, economic, and social issues" (McKeown, Hopkins, Rizzi, & Chrystalbridge, 2006, p. 9). The word "development" often implies a foreign underdeveloped nation, but, in a sense, we are all developing when it comes to issues of sustainability. All nations are working on becoming sustainable, as are all communities and the individuals in those communities. This is where we find an opening for discourse on the topic of sustainability in our classrooms. It is here that we engage students in determining the kind of future they envision.

Whereas environmental education is traditionally funneled into the science curriculum, sustainability education is, by its very nature, cross-curricular. A project-based curriculum can quite naturally be developed to meet the learning objectives of math, science, social studies, and language arts. Including a related art project and/or connecting your curriculum to a foreign culture can help you meet the learning goals of all core subjects as well as some of the required electives your students need. One of the reasons sustainability education is so attractive to the educator is its cross-curricular nature, which brings into focus governmental policy, economics, human geography, engineering, and scientific concepts. Current issues such as climate change, extreme weather events, energy consumption, water quality, and water scarcity can all be combined to complete the perfect recipe for learning in a real-world context.

Sustainability education has developed a set of standards, some targeted at students in grades K–12. The *National Education for Sustainability K–12 Student Learning Standards* "defines what K–12 students should know and do to be sustainability literate" (U.S. Partnership, 2009). The three essential learning standards include the following:

- Students understand and are able to apply the basic concepts and principles of sustainability.
- Students recognize the concept of sustainability as a dynamic condition characterized by the interdependency among ecological, economic, and

social systems and how these interconnected systems affect individual and societal well being. They develop an understanding of the human connection to and interdependence with the natural world.
- Students develop a multidisciplinary approach to learning the knowledge, skills, and attitudes necessary to continuously improve the health and well being of present and future generations, via both personal and collective decisions and actions. They are able to envision a world that is sustainable, along with the primary changes that would need to be made by individuals, local communities, and countries in order to achieve this. (U.S. Partnership, 2009)

These standards are designed to provide educators with guidelines for how to incorporate sustainability into existing cross-subject curricula. Sustainability standards also help create and guide new ways of thinking about systems. By presenting environmental issues through a real-world lens of systems thinking, students are able to synthesize the broad impact of these issues on communities and their economies. Researchers argue that "both the economy and society must operate within the finite constraints of the ecosphere" (Pitt, 2009, p. 39). Unfortunately, students are rarely presented with this kind of systems approach to learning. Information is still presented in silos rather than in interdisciplinary, real-world terms that help students understand the relevance of what they are being taught. In order for students to become environmentally literate citizens of the world, they must be given practical, hands-on approaches to real issues that face their communities.

We can no longer think of the planet's resources as infinite. In our consumption-driven world, quality of life, fairness, and human relationships take a backseat to wealth and prosperity. However, if you create the model of prosperity to look like a world filled with clean air, clean water, and strong communities, then you can begin to create a new way of thinking about the system. Educators have the ability to initiate "learning as a tool to facilitate choice between alternative futures" (Pitt, 2009, p. 39). As educators, we are responsible for providing all our students this "new prosperity" by adhering to the following sustainability precepts:

- living within environmental limits
- ensuring a strong, healthy, and just society
- achieving a sustainable economy
- using sound science responsibly
- promoting good governance. (Pitt, 2009, p. 40)

Sustainability, as a brand new paradigm, is impossible to teach in isolation. More effectively, it should be modeled as a "way of life," as if we are already living the reality of a world with diminished resources, overpopulation, and food scarcity. The difficulty is the overarching mindset most citizens of our planet manifest:

that our resources are unlimited and that comfort and wealth are, and always will be, the status quo. Educators must pose the following questions to their students: Is Earth running out of natural resources? Will the world population eventually reach Earth's carrying capacity? Is Earth's atmosphere being irreversibly altered by humans? To not posit these questions about the truth of these realities is a huge disservice to our students.

One of the most difficult tasks in integrating sustainability education, STEM education, and service-learning into our curriculum reflects the proverbial attempt of trying to fit a square object into a round hole. Most schools are still operating inside the antiquated factory model of education, in which students rotate in age groups between classes that are 50 minutes long and are subject-area specific. Teachers are rarely given time for collaboration with colleagues to develop project-based curricula to provide students a cross-curricular way of seeing the world. Life simply does not exist in a silo called chemistry or one called geometry. If you are lucky enough to be teaching at a STEM or project-based institution, then you are likely ahead of the game with integrated learning models, flexible schedules, and co-teaching strategies.

However one defines STEM, all students should become STEM literate. According to Rodger Bybee (2010), STEM literacy refers to the following:

- Acquiring scientific, technological, engineering, and mathematical knowledge and using that knowledge to identify issues, acquire new knowledge, and apply the knowledge to STEM-related issues.
- Understanding the characteristic features of STEM disciplines as forms of human endeavor that include the processes of inquiry, design, and analysis.
- Recognizing how STEM disciplines shape our material, intellectual, and cultural world.
- Engaging in STEM-related issues and with the ideas of science, technology, engineering, and mathematics as concerned, affective, and constructive citizens.

Students, regardless of their career interests, must have exposure to traditional studies in science, math, engineering, and technology. Fortunately, most schools at the secondary level still require 4 years of science and math as graduation requirements. This knowledge, or—if you will—*literacy*, is necessary for living sustainably in the world. Bybee further points out that "STEM-related issues such as energy efficiency, climate change, and hazard mitigation" (2010, p. 31) lend themselves to problem-solving instructional strategies that are age and grade-level appropriate. Somehow we must learn to fit the basics into these real-world issues.

Educators are tasked with making STEM subjects such as science and math exciting and stimulating to all of their students. This is, at times, a challenge when covering esoteric topics such as protein synthesis or the structural bonding angles of molecules. Basic facts are important to the overall understanding of science and math, but how do we then set students on a course of exploration toward gaining these basic facts on their own and fitting these details into the broader picture?

According to Blust and Pinnell (2007), "innovative, hands-on, inquiry-based activities that augment science and math experiences make these subjects more appealing to students and active discovery has been found to promote interest in mathematics in girls," (p. 3), one more important justification for becoming STEM literate. Blust and Pinnell further state that "service-learning and experiential learning methods, when integrated into STEM curriculum, result in positive attitudes towards these fields and increase the likelihood of students pursuing further education in this area" (2007, p. 3). The importance of making sure that the STEM curriculum is accessible to all, regardless of gender or ethnicity, is that when these subpopulations are excluded, inadvertently or intentionally, we pass up a whole population of potential ideas and innovative ways of thinking and viewing the world.

So our challenge is to make these connections between science literacies and to motivate students to use that knowledge to live sustainably in the world. The argument for this type of learning is evident, but making it a reality in the classroom is often a daunting task. A determination to transform our students' worldview from that of wanton consumerism to a conservation worldview is what has driven the following program of study. The project incorporates service-learning strategies at its core. In addition, it includes efforts to incorporate sustainability, as well as STEM, into a traditional science curriculum in the context of a rural West Virginia high school.

BEST PRACTICE

The roots of service-learning projects often can be located in the identification of a problem or a need in the community. Our high school had a small greenhouse on our school campus that was not being used to its full potential. The greenhouse had been constructed by students enrolled in a former agricultural program around the year 2000. It had been put together by students to support horticulture and agriculture education at a time when these two areas of study were more highly valued by our school system and our community. However, with the retirement of the agricultural education teacher and the movement from a traditional agriculture curriculum to an agriculture mechanics program of study, the greenhouse was not being used for agriculture or any growing program.

Fortunately, by 2009, the community within which our school is nestled was beginning to move in a "buy local, buy fresh" direction. Farmers markets were being developed, and local growers were interested in using community resources to grow food. After all, schools are community resources, and the growers in this community wanted to see this community resource on our school campus being used. My goal in developing this project was to reinstate the greenhouse to its year-round growing potential and incorporate the renovation into my environmental science curriculum.

On the first day of school, I handed my students clipboards, pencils, and paper as I announced their first project. The assignment was to renovate the greenhouse. The structure itself was in fine condition, but the task at hand had more to do with sustainability. Could the greenhouse be used year-round to grow food—food for the community, food for the school, or food for local businesses? The limits of my students' imaginations were the boundaries for this project.

The parameters of the assignment included working in seven groups of three, the inclusion of a presentation of their group project, the development of a funding source and sustainability plan for their project, and the requirement that they had to incorporate a form of renewable energy into the project plan. The students in this class were a combination of seniors, juniors, and one sophomore. Their research skills were mixed, but all had some basic skills in documenting sources and how to conduct searches. Given the scope of the assignment, I decided to present the idea to my principal.

Photo 10.1. 2009/2010 Advanced Placement Environmental Science Students Pose for a Group Picture After Taking Measurements and Making Plans for the Greenhouse Renovation.
(Photo by Leigh Jenkins, 2009.)

The principal of the high school was incredibly supportive of our ability as teachers to come up with creative ideas. He had supported a pilot project at our school in which four teachers, one from each of the core subject areas, came together to teach a 21st Century Symposium class, another project-based, student-led, research-oriented elective course offered to the student body. The principal had personal experience as a vocational director in another rural county in West Virginia. While there, he had developed a sustainable program where students found initial investment funds and, with these funds, purchased the land and materials to build a house. Students were trained in construction skills, and when the house was finished, it was sold and the proceeds were used to fund the next year's building project. Having developed his own sustainable program of study, he thought the greenhouse project was an excellent idea and believed that it offered great potential not only for the agriculture program, but for the science program as well. In addition to the support of the principal, our project had the support of the superintendent of schools, as he was interested in incorporating alternative energy models as educational tools throughout Morgan County.

Berkeley Springs High School is a relatively small school located in the eastern panhandle of West Virginia, nestled in eastern Appalachia. Except for a sand mining facility, the major industry in Morgan County is tourism. Family farms still exist, but many have been sold off to make way for vacation homes and retirement developments. Agriculture in Morgan County has changed in the last decade. Traditional farming has given way to the local growing of specialty crops, with niche markets being created all the time.

My hope in pursuing this project was to allow students to see the importance of growing food in their community, the importance of food security, and the importance of sustainable practices in their everyday living. The fact that agricultural pursuits had become scarce had caused some concern in our community. Our county is isolated, and the question of how we could ensure food security for our county was often a topic of great concern. A tour of local farms provided my students with a big picture of the variety of agricultural operations located within the county.

I organized a county field trip for my class to visit some of the growing businesses in the area. I wanted my students to see the importance of sustainable agriculture practices, as well as the viability of careers in entrepreneurship through agriculture. Our first stop was a small group of organic growers who used a solar-powered pump to provide water for their house and crops. Their operation was small, but they supported themselves through Community Supported Agriculture (CSA), where a box of seasonally produced food is sold for a monthly fee. This operation, Wind Dance Farm, was further supported by their environmental school for local students in grades K–8, in which many local home-schooled children participated. This farm gave students a taste of what it is like to be a successful small grower in the community.

We then visited another local grower who had established a successful regional business through hydroponic growing. This facility provided to the students an example of a successful business that was able to take advantage of a regional market. Paul Mock's greenhouses were designed to specialize in the growing of bib lettuce, watercress, and hothouse tomatoes. He had established a large-scale growing method specifically for these three products, successfully selling in the Baltimore/Washington regional market. This grower demonstrated economical sustainability through his ability to market his product regionally to well-known companies such as Whole Foods.

Finally, we visited a grower who used a method known as biodynamic growing. Biodynamic farming, originally developed by Rudolph Steiner, is an esoteric method of growing that honors the relationship among the soil, atmosphere, plants, and microbes in a holistic, sustainable system. In addition, biodynamic farmers adhere to strict organic techniques of natural composting and fertilizing. Biodynamic farmers follow the astronomical calendar as a guide for planting and sowing. Many biodynamic farmers broadcast homeopathic preparations designed to strengthen the growing relationship energetically. Research shows that biodynamic methods result in reduced yields. However, soil quality, energy efficiency of production, improved organic carbon yield, and greater soil biodiversity, including increased populations of microbes and earthworms, are realized (*What Is Biodynamics?*, 2012). In other words, this biodynamic farmer focused on quality, as opposed to quantity. This farmer also incorporated the use of a pond within a greenhouse to serve as a heat sink for internal temperature control. This unique way of farming demonstrated to students that growing can be accomplished in historically successful and recently renewed ways.

Students were able to take newly acquired information back to school to incorporate into their research projects. They participated in the research needed to calculate construction costs for the greenhouse renovation. They located vendors and executed comparisons to find the most cost-effective supplies. They used their creative initiative to design alternative energy applications and worked with a local builder to determine strategies for including designs in the final renovation plan. Students then developed a detailed project proposal to submit to funding agencies for review. Two additional weeks were spent on their research assignment, and presentations were then made to their fellow classmates and our principal. Our principal was invited to the classroom presentations not only to observe the demonstrations of student projects, but also to be a part of the planning of our project.

Peer review was used to constructively critique projects, and from these comments revisions were made. From the revised projects, information on all aspects of the renovation was compiled. It was the compilation of these ideas that enabled us to receive the State Farm Youth Advisory Grant in January 2010, which allowed us to renovate the greenhouse and showcase some of the local applications of alternative energy. It is necessary, at this juncture, to stress that funding is not needed to initiate a service-learning project. In fact, funding is typically replaced by the work of volunteers and the support of community stakeholders. In our situation, funding was not needed to put the greenhouse back into production, but was needed instead to incorporate the use of an alternative form of energy.

The State Farm Youth Advisory Board "supports school-based service-learning projects because it promotes excellence in public schools" and believes that service-learning is a "method that has been proven effective to improve student engagement and [positively] impact student achievement" (State Farm, 2012, Service-Learning tab). The Berkeley Springs Greenhouse Renovation Project was designed to take an underutilized existing structure and restore it to its original use. The project was funded under the environmental service category at $41,763. The grant monies covered the installation of eight 230W photovoltaic solar arrays to provide electrical support for fans and lights, and a three-panel, 120-gallon Velux solar water heating system for supporting the root-bed heating to provide for increased germination and plant growth. Movable propagation tables were installed to maximize the number of students who could work in the greenhouse at any given time as a part of the materials component of the grant, and a computerized temperature system was installed to provide more stable temperature control during the growing season. The greenhouse already contained a back-up propane heater, and the grant covered the installation of a larger propane tank. Revenue from the growing program was designed to cover the cost of fuel needed to keep the greenhouse actively growing during the colder months of the year. Along with the physical components of the greenhouse, the grant covered the cost of hiring four interns during the summer of 2010 to manage the greenhouse and to promote growing programs in the community.

The project was designed to engage students in both the renovation of the greenhouse and the planning of the subsequent growing program. Community stakeholders played an important role in the success of the project. Mountain View Solar, a local solar contracting company, successfully bid for the solar installation. The company was willing to involve students in the actual installation of the solar panels and equipment and has since advised on all aspects of the design and operation of the greenhouse. Its interest in educating the students and the community in the benefits of solar has made it a successful contractor, both locally and regionally. Other community stakeholders include the Morgan County Farmers Market, which was willing to allow students to operate a booth at a reduced rate for selling produce and promoting the school's farm-to-school program. (Farm-to-school programs promote the purchase of local, fresh produce to be served in local schools.) Mock's Greenhouses graciously offered to provide funding for propane should the program not be able to fund fuel in the first year.

The intention of our stakeholders was to ensure that the greenhouse would operate year-round. Community support was extremely generous.

The students became active participants in creating a climate of environmental responsibility and sustainability through the promotion of food security, healthy eating habits, sound economic programs, and the use of alternative energy. Students used creativity and innovation to grow, market, and manage the products grown in the greenhouse. Students began by developing a business plan to determine what they would need to begin their growing program. They determined how much in materials would be initially required to germinate, propagate, harvest, and sell their produce or plants at the local farmers market. Students were assessed throughout the project development and were frequently asked to reflect on various aspects of the service-learning project as it evolved.

Specialists were brought in to teach the interns about the management and care of the greenhouse, including how to keep the greenhouse clean and the plants free of disease. Students also learned the fine art of vermiculture, which uses worms to compost waste. Although I anticipated the female interns cringing at the thought of handling worms, they were fascinated and jumped right into the project without hesitation. The worm castings and worm tea were used to fertilize plants closing the loop in the plant-growth cycle. Students also had written into their business plan the incorporation of food grown in the greenhouse to be used in the high school cafeteria, with unused produce going to the local soup kitchen. As part of their service to the community, students agreed to work with younger students at some of the local schools to teach them the basics of growing. Students worked with the Natural Resource and Conservation Service to provide third graders with spruce saplings to take home to their parents to plant. In the hopes of making the growing program financially sustainable, students took the produce and plants they had grown to the Berkeley Springs Farmers Market to sell.

The overall program was not without stumbling blocks. For one thing, the installation of solar panels did not meet all the energy needs of the greenhouse. The temperature in the greenhouse proved unpredictable and therefore had to be carefully monitored throughout the growing season. Many features of the greenhouse, such as the movable tables, were only installed during the last winter and only had one spring to prove their viability. As the growing tables include a pressurized system that requires hot water to be transported a relatively great distance, the ability to maintain healthy and consistent root-bed heat is still being determined.

Overall, the project was successful in demonstrating what service-learning can provide for schools and communities. This service-learning project could have been as simple as high school students teaching younger students how to grow vegetables and flowers in their own school garden, or teaching students that vegetables come from the ground and not the local supermarket. The fact that the project had the added dimension of establishing a sustainable greenhouse operation truly stretched the students' understanding of entrepreneurship and maintenance of a successful business operation.

Our state is finally supporting the farm-to-school movement, serving the produce of local farmers, such as tomatoes and cucumbers from Mock's hydroponic greenhouses, in all of our county school cafeterias. Our greenhouse now provides specialty lettuce and a variety of sprouts for our cafeteria salad bar. Students are able to choose what sprouts they prefer and are given healthy choices from a variety of vegetables. Students are also provided with the nutritional value they are gaining when they eat local produce. It is hoped that the added information students receive from the program will begin to educate them on healthy eating habits and show great gains in the overall well-being of our school community.

MODELING SUSTAINABILITY IN YOUR SCHOOL AND COMMUNITY

How can you begin to model sustainability through STEM service-learning in your own community? The answer to this question depends on the type of community in which you live. If your community is small, your ability to secure a financial partner may be limited, and resources may be scarce. However, your project can be quite successful without a huge financial investment. Larger communities may have the luxury of securing a financial partner (for example, Walmart or Lowe's) through educational grants or promotion. Reliably, most communities possess an inherent need for service of some type. Many service-learning projects are initiated in the local schools as a way to introduce real-world, practical, hands-on learning. Many schools require a certain number of service hours as part of the high school graduation requirement. In many instances, students may begin to accrue these service hours in the eighth grade. One certainty is that students are able, willing, and often eager to serve. They crave practical learning experiences in meaningful contexts.

One place to begin a service-learning project is in your own school, whether you are a student at a local college or employed at an educational institution. Many communities, depending on their size, have active recycling programs. As landfills reach their capacity in many regions of the country, numerous larger metropolitan areas have adopted mandatory recycling. Other medium-sized communities make recycling more attractive by offering curbside pickup. In rural communities, recycling is still voluntary and requires transporting recyclables to a fixed location. Recycling programs are relatively easy to initiate in many institutions, including hospitals and schools. This is also where these programs are greatly needed, given the populations they serve.

The massive amount of paper used in educational institutions makes recycling an economic and ethical necessity. In addition, vending machines often dispense plastic water bottles, aluminum soda cans, or glass bottles, all of which can and should be recycled. For relatively little money, the containers needed to collect these items can be purchased and placed in convenient locations. Students can begin by calculating the amount of paper their school uses in one year. Once they have this figure in their heads, they can make visual comparisons as to the amount of space this might occupy in the local landfill. Then they might look at how many trees would have been consumed to supply their school with this amount of paper. In addition, having students develop an advertising campaign to educate the student body on the merits of recycling is a great project for students of any age and can be easily incorporated into all areas of the curriculum.

Another opportunity for schools and communities to model sustainable living is through the state or local farm-to-school movement. Buying locally grown foods and locally made products is a fabulous way to demonstrate community sustainability from the perspectives of economics, ecology, and good health. The high cost of transportation is one of the cogs that turn this wheel, but an awareness of the health benefits of eating fresh foods as opposed to foods that have travelled halfway around the world is another important driver. Fresh food tastes better than week-old produce. In addition, it just makes more sense to support local businesses, including farmers, growers, and craftspeople. The economic good sense of putting money back into one's own community is a no-brainer. Farmers markets are becoming quite common in most communities. Creating a connection between local farmers and food service administrators is the first step to bringing locally grown produce into your cafeteria.

Many states now support the farm-to-school movement that brings fresh, healthy foods into our schools. Part of this movement is a response to the nation's obesity problem, reflected in the fact that more than one-third of children and adolescents are overweight or obese (Centers for Disease Control and Prevention, 2012). Healthy food in cafeterias has been shown to produce marked gains in student achievement and behavior. Developing a service-learning project around growing food is relatively easy for most educators. School gardens are already common as educational tools for teaching students about native plants, invasive species, and wildlife attractants. Why not develop gardens to teach students how to grow their own food, an important skill in a world that will soon be challenged with feeding 9 billion people by 2050?

Students thrive on these skills, and the fruits of their labors can be used in many parts of the curriculum at all grade levels. Students can grow sprouts for their salad bar. Some schools have designed edible landscapes, as opposed to ornamental plantings, to contribute to the fresh foods offered in the cafeteria. Raised beds can be established to demonstrate how seasons are extended in colder regions of the country. Herbs and vegetables can become part of the family and consumer science program. Teaching students how to dry herbs and preserve foods is a lost skill that can be revived in schools and through agricultural extension agencies.

Food delivery systems are vital to food security and can be demonstrated through service-learning projects. Students can also become involved in preparing and distributing food at their local soup kitchen. Another great service-learning opportunity presents itself through one's community farmers market. Many local farmers markets have co-operative booths that allow small or seasonal growers to share market space, an ideal situation for student growers who may not be able to pay the full cost of a booth. Growing programs enhance STEM learning in the biological and agricultural sciences, but also require math, engineering, and technology skills for the execution of a business plan, the development of marketing strategies, and entrepreneurial skills that are vital to making any small business a success. Schools can demonstrate sustainability by making these programs return enough in revenues to start the same successful program in subsequent years with new groups of students.

The availability and cost of energy is another area of concern for most communities. As community facilities look for ways to cut energy costs, new, greener technologies offer a variety of solutions. Schools are perfect institutions for modeling how new energy technologies can save money and serve as important learning tools for incorporating math, science, and engineering into the curriculum. Students can witness firsthand the economic gains achieved by gathering energy directly from the sun or indirectly from the sun through wind power. Students can conduct energy audits of their school and look for ways to conserve energy and incorporate new technologies to address energy costs. Students can then calculate the savings in both kilowatt-hours and dollars if certain energy conservation measures are taken. Significant savings could be generated district-wide by simply educating the faculty to turn lights off in their classrooms when they leave. Savings can be demonstrated by calculating what can be saved simply through the replacement of fluorescent and incandescent lighting. Students can organize competitions between schools to see which school is best able to lower its relative energy consumption. Considering that some of these projects may require some initial investment, students could argue for the long-term advantage of incorporating new alternative energy technology into systems that are only going to grow larger. As traditional resources are depleted, these alternative sources of energy become even more essential to addressing issues of sustainable systems. The possibilities for incorporating STEM learning through the development of energy projects are endless.

Each of these areas of concern can be used to model sustainability in our STEM classrooms. Students become active participants in greening their school campuses. Real-world issues become fuel for real-world solutions right where we live. Students connect with their peers to create solutions to the unique issues faced by their very own school. This, in turn, leads to innovative solutions to real-world problems that might exist in each student's home. Students educate their parents on living more sustainably. Sustainable living then becomes a way of life for individual families and eventually for communities as a whole. The mindset of a community is key to modeling sustainability through STEM service-learning.

Changes take place slowly; therefore, one must be patient and willing to participate in the evolution of creating sustainable communities. Your students are your community's future citizens. The more you can involve students in their communities as youth, the greater the chance they will put their youthful energy to productive use and begin to engineer creative responses to the challenge of developing sustainable communities in which to live.

CONCLUSION

Service-learning through STEM education and with a focus on sustainability can be successfully implemented in the school curriculum at any grade level and across all disciplines. Service-learning through STEM education allows an educator to incorporate meaningful curriculum into science, technology, engineering, and math programs. Service-learning provides a means for students to accumulate mandatory community service hours toward graduation or to simply introduce students to the satisfaction rendered through service projects. Adding a STEM focus to service-learning projects provides teachers with tools to empower youth to become interested and skilled in areas leading to educational pathways toward STEM degrees and professions. In addition, these skills and added knowledge can provide students with new ways of looking at their world, thus stimulating more sustainable ways of approaching problems and issues that face them in the future. STEM service-learning projects can be incorporated into the K–12 curriculum at any grade level and across all subject areas. Through STEM service-learning, students are empowered to become environmentally and economically sound citizens in their communities. Students involved in hands-on, integrated, real-world learning, learn to use critical thinking and problem solving abilities—twenty-first-century skills so urgently needed in developing a productive workforce.

In addition, incorporating service-learning projects into the K–12 curriculum requires few resources. Most projects call for little or no funding and are established through the development of community support from the many agencies and district offices involved in educational outreach. A beginning point for any service-learning project is to survey the needs of your community. What environmental or economic issues does your community face? What stakeholders would be interested in seeing this issue resolved? How can students play a role in providing ideas and solutions to these problems? With very little effort, service-learning can play an important role in the education of all children.

APPENDIX

Resources

Energy

- Kilowatt Ours, http://www.kilowattours.org/about.php, http://www.kilowattours.org/curriculum/KilowattOurs-Curriculum.pdf
- National Wildlife Federation—Cool School Challenge, http://www.nwf.org/Eco-Schools-USA/Become-an-Eco-School/Cool-School-Challenge.aspx

Farm to school

- Fair Food Network, http://www.fairfoodnetwork.org/
- National Farm to School Network, http://www.farmtoschool.org/

Federal stakeholders

- Department of Environmental Protection (DEP); check your state
- Environmental Protection Agency (EPA), https://www3.epa.gov/
- USDA, United States Department of Agriculture, Natural Resource Conservation Service (NRCS), http://www.nrcs.usda.gov/wps/portal/nrcs/site/national/home

Funding

- Lowe's Toolbox for Education Grants, http://www.toolboxforeducation.com
- State Farm Youth Advisory Board, http://www.statefarmyab.com/
- Whole Kids Foundation, https://www.wholekidsfoundation.org

Recycling

- American Forest & Paper Association School Recycling Award, http://www.paperrecyclingawards.com
- National Youth Leadership Initiative, Los Angeles Unified School District, Recycling Service Learning Project, http://www.chavezfoundation.org/uploads/14._model_Recycling.pdf

STEM workforce development

- Change the Equation, http://changetheequation.org/

Vermiculture

- Planet Natural Research Center, https://planetnatural.com/worm-composting
- The Worm Guide, http://www.calrecycle.ca.gov/Education/Curriculum/Worms/default.htm

WORKS CITED

The Belgrade Charter: A global framework for environmental education. (1976, January). *Connect: UNESCO-UNEP Environmental Education Newsletter, 1*(1), 1–10.

Billig, S. H. (2002). Support for K–12 service-learning practice: A brief review of the research. *Educational Horizons, 80,* 184–189.

Blust, R., & Pinnell, M. (2007). *Using service learning to develop a K–12 STEM service learning and experiential learning site.* Retrieved October 16, 2013, from http://icee.usm.edu/icee/conferences/asee2007/papers/2384_USING_SERVICE_LEARNING_TO_DEVELOP_A_K_12.pdf.

Bybee, R. W. (2010). Advancing STEM education: A 2020 vision. *Technology and Engineering Teacher, 70*(1), 30–35.

Carson, R. (1962). *Silent spring.* New York: Houghton Mifflin.

Centers for Disease Control (CDC). (2012, June 7). *Childhood obesity facts.* Retrieved November 15, 2012, from http://www.cdc.gov/healthyyouth/obesity/facts.htm

Comstock, A. B. (1911). *Handbook of nature-study for teachers and parents.* Ithaca, NY: Comstock Publishing Company.

Eyler, J. S., Giles, D. E., Jr., Stenson, C. M., & Gray, C. J. (2001, August 31). *At a glance: What we know about the effects of service learning on college students, faculty, institutions, and community, 1993–2000* (3rd ed.). Manuscript submitted for publication, Vanderbilt University, Nashville, TN.

Fairweather, J. (2008). *Linking evidence and promising practices in science, technology, engineering, and mathematics (STEM) undergraduate education.* Paper presented at the National Research Council's Workshop Linking Evidence to Promising Practices in STEM Undergraduate Education, October, Washington, DC. Retrieved from http://www7.nationalacademies.org/bose/Fairweather_CommissionedPaper.pdf

Helms, J. D., Pavelis, G. A., Argabright, S., Cronshey, R. G., & Sinclair, H. R., Jr. (1996). National soil conservation policies: A historical case study of the driftless area. *Agricultural History, 70*(2), 377–394.

Jackman, W. S. (1891). *Nature-study for the common schools* New York: Henry Holt and Company.

Johnson, J., Thompson, A., & Naugle, K. (2009). Place-conscious capacity-building: A systemic model for the revitalisation and renewal of rural schools and communities through university-based regional stewardship. *Rural Society, 19*(3), 178–188.

McKeown, R., Hopkins, C. A., Rizzi, R., & Chrystalbridge, M. (2006, October). *Education for sustainable development toolkit.* Paper presented at United Nations Decade of Education for Sustainable Development (2005–2014).

Meyer, S. J., Hofshire, L., & Billig, S. H. (2004). *The impact of service-learning on MEAP: A large-scale study of Michigan Learn & Serve Grantees: Year two evaluation report.* Denver, CO: RMC Research Corporation.

The National Environmental Education Reauthorization Act of 2010. (2010). H. R. Misc. Doc. No. 110th-P.L. 101–619. Retrieved October 17, 2012, from http://beta.congress.gov/bill/111th/senate-bill/3833/text

National overview: State level EE legislation/policy [Fact sheet]. (2007). Retrieved October 28, 2012, from http://www.fundee.org/campaigns/nclb/brief5b.htm

North American Association of Environmental Education. (1999). *Excellence in environmental education: Guidelines for learning (K–12)* (Publication No. EPA-NT902897-01-1). Retrieved October 16, 2013, from http://www.fs.fed.us/outdoors/naturewatch/implementation/Curricula/Guidelines-for-Learning.PDF

Our common future: The World Commission on Environment and Development (1987). G. H. Brundtland & M. Khalid (Comps.). Oxford, England: Oxford University Press.

Pandey, V. C. (2006). Environmental education programmes for attitude changes among children. In V. C. Pandey, *Environmental education* (pp. 191–216). Adarsh Nagar, Delhi, India: Isha Books.

Pitt, J. (2009). Blurring the boundaries: STEM education and education for sustainable development. *Design and Technology Education: An International Journal, 14*(1), 37–48.

State Farm Service Learning. (2012). Retrieved October 30, 2012, from http://www.statefarmyab.com/service-learning/

Strage, A. A. (2000). Service learning: Enhancing student learning outcomes in a college level lecture course. *Michigan Journal of Community Service Learning, 7,* 5–13.

U.S. Department of Labor Employment and Training Administration. (2007, April). *The STEM workforce challenge: The role of the public workforce system in a national solution for a competitive science, technology, engineering, and mathematics (STEM) workforce.* Washington, DC: U.S. Department of Labor.

U.S. Partnership for Education for Sustainable Development. (2009). *National education for sustainability K–12 student learning standards* [Fact sheet]. Retrieved October 16, 2013, from http://s3.amazonaws.com/usp_site_uploads/resources/123/USP_EFS_standards_V3_10_09.pdf

What is biodynamics? (2012). Retrieved October 30, 2012, from https://www.biodynamics.com/

CHAPTER ELEVEN

Outdoor Ecological Inquiry Brings Students AND Nature Together

DAVE OBERBILLIG

INTRODUCTION

Many of us grew up in a time when outdoor experiences were as regular as a day in school. Time outside was either the natural next step once the school day finished, or was mandated by our parents as a way to get us out of the house. While network television provided an indoor distraction to the exploration and fun to be had outside, it offered limited options: three commercial network stations and public television. The result was ample independent wandering, discovering, imagining, and creative play in the wonderland of the natural world. Building tree forts, wading creeks, catching frogs and snakes, and rounding up neighborhood kids for outdoor games played well into the evening was routine. That was then. The experience for today's youth, especially those living in urban and suburban landscapes, is markedly different. Children's lives are much more structured, with afterschool programs to provide supervision for kids with working parents, school and club sports for exercise and socialization, and less unstructured outdoor time—all to satisfy the minds of parents fearing crimes against children. On top of it all, children lack easy access to undeveloped landscapes that allow exploration through the lens of their own imaginations.

The typical school day—and thus the STEM (science, technology, engineering, and math) education experience—provides students with a structured environment within the four walls of a school building. Students learn the routine of standard school practice, moving from room to room by following a daily schedule, sitting in organized rows for instruction and in their STEM classrooms participating in laboratory investigations. For most schools, the world beyond the walls of the school building is not a part of the regular school day. A strict division exists between school (inside) and the rest of the world (outside). Due to the overall lack of outdoor time and a lack of outdoor emphasis in formal education, students are unfamiliar with their natural surroundings. When high school students are asked to name their local play areas and parks, native plants, common birds, or migratory species, most of them struggle to come up with answers. Very few have ever looked at insects under a magnifying loop or birds through binoculars. Even in the most humble school environments, plantings exist around the foundations of buildings, ants find cracks in the pavement, local birds search for seeds and bugs, and life is interacting in a variety of ways that await discovery by curious minds.

While knowledge of local natural history and fundamental ecology are critical to whole child development, an equally important concern is for students to understand that ecological principles also inform us about access to healthy food, consideration of healthy food choices, and food security. Except for claiming "from the store," most students have no idea where their food comes from, much less how food production is tied to nature's cycles and principles.

As we progress into the twenty-first century with a global population approaching nine billion and uncertainty about crop production because of climate change, access to food for many in the world will become a critical issue. The combination of ecological and economic constraints will require future generations to seek ecological solutions to the food problem. Development of local agriculture through the use of locally adapted food crops, developing innovative irrigation, harvest, and food storage systems adapted to use in a variety of ecosystems around the world, and developing efficient, low-energy food transport systems will be necessary to distribute food equitably to even the poorest people on the planet (Godfray et al., 2010). For future generations, development of creative technologies and markets to address food security issues requires a keen understanding of the ecological principles that lie at the foundation of producing abundant, healthy food.

As the population of the United States and the world continues to grow and shifts toward urban living, fewer students will directly experience farm life or unstructured play in local natural areas. What do our youth lose by substituting plentiful and unencumbered outdoor experiences with time spent accessing the tremendous amount of information and entertainment available through digital media? They lose outdoor experiences that guide their future understanding of the required connection between people and the land, both physiologically and spiritually. They lose perspective about how connected we all are to a world that is biologically diverse and interesting, sometimes dangerous, and often awe-inspiring. Without even knowing it, their worlds are diminished. In fact, for many of our youth, the natural world has been pushed into a realm that engenders suspicion and fear, further alienating them from a true understanding and appreciation of it.

Trends in the way children spend their time do not suggest a quick fix to this situation, but schools have the opportunity to provide substantive outdoor experiences while promoting academic achievement and positive attitudes about school, community, and nature. This holistic form of education can happen with the patient instruction of teachers who understand that a whole child is one who not only understands the content and concepts of academic disciplines, but does so by experiencing the connection of those concepts to the world of nature beyond the walls of the classroom.

Given that today's school buildings are increasingly networked, providing students with incredible access to information about the world, what possible advantage could be gained by asking students to prepare for an outdoor experience and venture onto the school grounds or to a local park to learn science (and more) through direct experience? The advantages range from expanding students' awareness of their school and neighborhood to gaining experience with scientific inquiry processes of observation and experimentation, to learning fundamental concepts of ecology, biology, and earth science. The prospect of fresh air and novel stimuli brought by seasonal change provides a bonus to the experience. In spite of the apparent limitations to childhood participation in outdoor activities and school-related outdoor investigations, the possibilities for standards-based investigations in outdoor settings are as limitless as those in any laboratory. With a combination of creativity, adaptation to surroundings, and careful planning, student engagement in the natural world can work in almost every setting. The benefits are worth the effort to establish regular outdoor ecological investigation; the research demonstrates gains in student achievement, reduced absenteeism, and improved student behavior (Lieberman & Hoody, 1998).

Ultimately, ecological education is critical for every student and the concern of every citizen. Whether we like it or not, or understand it or not, we are inextricably tied to the natural world. Therefore, our health and well-being is also tied to the natural world. When we provide students with relevant, outdoor science-based experiences, they begin to understand the important connections between themselves and a healthy environment, and develop the capacity to make educated environmental decisions, no matter what their future may hold. The burgeoning global population will continue to place a strain on natural resources and the health of the ecological systems that sustain us. The issues that develop locally, regionally, and globally around environmental health will always require bright, educated minds to come up with ecologically sound solutions. A presentation of the myriad problems that human-caused environmental degradation have created is not within the scope of this work, except to say that every problem, to be solved well, requires an informed citizenry—and a well-solved problem is one that conforms to the rules imposed by nature and frees future generations from having to invest time, energy, and money correcting the mistakes of those ignorant of fundamental ecological principles that guide environmental and economic sustainability.

BACKGROUND

In *Last Child in the Woods: Saving Our Children from Nature Deficit Disorder*, Richard Louv (2008, p. 226) states, "An environment-based education movement—at all levels of education—will help students realize that school isn't supposed to be a polite form of incarceration, but a portal to the wider world." David Orr echoed that sentiment in *Earth in Mind* (1991) when he wrote that all education should be environmental education. Many K–12 students have a limited understanding of how nature works because of limited exposure to the natural world and a lack of guidance about how components of nature fit together. Our current educational system has simply not made a commitment to engaging students with nature. The summary of research that follows is not all about the educational attributes of outdoor experiences for youth, but rather points toward cultural phenomena that devalue time spent in nature and hence contribute to a lack of concern and commitment for developing academic activities in natural settings within the bounds of the school day. Nonetheless, additional research here documents achievement and affective gains for students provided with academic opportunities in outdoor settings.

Downward trends in time spent engaged in outdoor experiences for children and in the landscape available for outdoor experiences have been well documented (Clements, 2004; Hofferth, 2009). According to the *Connecting America's Youth to Nature Report* (The Nature Conservancy, 2011), almost 90% of American teens spend time online every day, and 69% of the same group play video games. A 2010 Kaiser Family foundation survey (Rideout, Foehr, & Roberts, 2010) found that children age 8–18 spend, on average, 7 hours and 38 minutes each day engaged in entertainment media (more than 53 hours per week). Researchers were shocked that the amount of time "plugged in" had increased by almost an hour from 2006. While it may be pointed out that listening to digital music can be done while outdoors, the Nature Conservancy report (2011) also pointed out that fewer than 40% of teens engaged in typical outdoor activities *every week*, including outdoor activities as commonplace as visiting a local park. Clearly there has been a radical shift away from time spent in nature over the last 50 years. The instant gratification of digital entertainment might completely overwhelm the more subtle attractions found outdoors, but research suggests that when given more opportunities to enjoy the outdoors, teens will be more likely to value nature and engage in it into their future (*Connecting America's Youth*, 2011).

This is not to say that children have completely turned away from all outdoor experiences. During the last 20 years, participation in organized youth sports has jumped dramatically. Outdoor sports such as baseball, softball, soccer, tennis, track, and football engage our youth, but the focus is decidedly different than spending time in nature. Why has this shift occurred? It is no stretch to reason that changes in socioeconomic structure related to parenting in America have reduced the time available for parents to spend with their children in unstructured outdoor time. Between 1975 and 2009, the presence in the labor force for mothers with children under age 18 increased from 47.4% to 71.6% (Bianchi, 2011). Time that might have been spent outdoors with the kids has turned into work time, and kids have fallen into the ready supervision of organized play. A final concern about the move toward organized athletics is that the demand for playing fields has gone up, eliminating open spaces that support environments important for unstructured play and exploration.

A structured practice or competition aimed at developing the nuances of a game is a decidedly different cognitive activity than either unstructured exploration of a hidden corner of a park or a detailed observation of insect assemblies during springtime hatches. Undoubtedly, children have the potential to realize great benefits from the socialization aspects of team sports: learning about collaboration and cooperation, risk taking, and how to take control of emotions through the ups and downs of winning and losing. The added benefit of developing lifelong skills that can promote an attitude of fitness provides further support for team sports. Experiences in nature add another important dimension to time outside. As Kellert (2005, p. 83) states, "Play in nature, particularly during the critical period of middle childhood, appears to be an especially important time for developing the capacities for creativity, problem-solving, and emotional and intellectual development." The capacity for creativity and problem solving fit with current ideas about important skills necessary for survival in the twenty-first-century workforce. Since nature constantly presents novel environments and situations, intellectual development through outdoor experience makes sense. Solving the problems of reading landscapes, crossing streams, and understanding the language of animal tracks requires awareness and decision making. Working independently or with others at these tasks provides a measure of self-confidence for a child that continues into adulthood.

Adults support nature experiences in general, but adults are concerned about children's safety in wild places, especially areas described as woods, creeks and ponds, or open fields and hills (Fraser, Heimlich, & Yocco, 2010). Not surprisingly, areas such as these constitute the majority of what might be called natural areas and point to an underlying fear of the bogeyman that has been accentuated with heightened media promotion of crimes against children. These concerns limit adults' willingness to allow children to participate in unsupervised play or free exploration, even if the adults had these experiences when they were children.

Several other factors work against opportunities for school children to learn in outdoor laboratories. Pressure about student performance on state and national standardized tests shifts the focus to seat time dedicated to math and reading practice. Claims that students lose much of what they learned the previous year over long summer vacations have led to the suggestion of increasing both the length of the school year and the school day. These efforts are related to student performance on standardized tests, especially international comparisons. However, the idea that more hours in school will help U.S. students catch up with the rest of the world is questionable, given that Finland is consistently at the top of international mathematics and science scores while having the lowest number of hours spent at school among the developed nations tested (Helm, 2011). Funding for field experiences has been reduced, as has funding for all science subjects, to focus reduced budgets on the tested subjects of English language arts and mathematics. The reduction in funding impacts the ability of teachers to take students off campus on field trips. Ellerson (2010), in a report for the American Association of School Administrators, noted that by the 2010–2011 school year, half of schools nationwide had eliminated all field trips because of funding cuts. That dismal statistic was predicted to worsen. In some situations, even recess time has been curtailed to provide more time for students to practice for testing (Barth, 2008). While recess remains intact in most elementary schools, the trend over the last 5 years has been a reduction in minutes. A Robert Wood Johnson Gallup Poll survey (*The state of play*, 2010) found that principals are supportive of recess time and find that recess improves student achievement and reduces behavioral interventions. In a study of 8- and 9-year-old students, researchers found that teachers rated those students with daily recess as having better behavior than those with either intermittent or no daily recess (Barros, Silver, & Stein, 2009).

While the research above points to a general decline in youth contact with the outdoors and the world of nature, it does not inform us about why outdoor ecological education should be a dedicated part of the formal education experience. It might be argued that schools are the hub of every community and both reflect and guide the values of those communities. It is also clear that parents are supportive of their children's interaction with natural landscapes but harbor concerns over their children's safety. After more than a dozen years of guiding students on field trips in Yellowstone National Park, I can confirm parental concerns about safety, even on a trip into a National Park that caters to tourists and has a reputation for doing a tremendous job protecting its guests. Even with mandatory parent informational meetings, many parents need reassurance that their children will come home safely. School grounds and nearby parks typically have much less potential for incidents and accidents than does a place like Yellowstone, and the research demonstrates that with careful planning, the benefits to student learning and well-being should lead us to incorporate outdoor investigations as part of the school program.

One of the greatest challenges for teachers is to effectively and regularly stimulate the intellect and emotions of their students. Regular outdoor experiences interrupt the routine of school by engaging children in novel environments. Even the same outdoor area appears unique when visited at different times of the year. While good teachers rearrange their classes to guide students to new ways of thinking, nature's seasons do this regularly and with no effort required from educators. Gaining more access to nature has positive impacts on student performance. Wells and Evans (2003) revealed that the cognitive functioning of children in poor urban environments improved when they moved closer to urban areas with greater access to natural environments. Their research provides more support for the positive impacts of nature on mental function and demonstrates that, especially in urban environments, a connection with nature has important positive physical and mental outcomes.

Dedicated outdoor experiences solidify ecological concepts by providing a local framework upon which to test new ideas. With enough exposure to their local environs, students begin to map their bioregion and understand the ecological and socioeconomic benefits of the natural world around them. They become more intimately connected to their local landscape. The added benefit of a dedicated outdoor program is increased student achievement in all academic subjects, better attendance, more enthusiasm for learning, and fewer behavioral problems (Lieberman & Hoody, 1998; Barros et al., 2009).

Accommodating dedicated outdoor education to the typical delivery of K–12 formal education is no easy task, especially in this age of high-stakes testing. This is not due to outdoor education just arriving on the formal education scene as a novel experiment. The progressive movement of John Dewey in the late 1800s and early 1900s, best exemplified by the laboratory schools at the University of Chicago, experimented with experiential learning in the laboratory and exported the ideas to farm schools. A few schools have made this transition (Lieberman & Hoody, 1998) and demonstrate the holistic educational benefits for their students, but there is still much resistance to taking students beyond the confines of the classroom. It may ultimately require significant federal and statewide leadership to convince educators and the public of the benefits of taking students outside. The evidence provides strong support for doing so, and educators and scientists who understand the benefits of regular contact with nature must continue to promote this valuable idea.

A Framework for K–12 Science Education (National Research Council, 2012), the basis for the *Next Generation Science Standards (2013)*, presents significant changes to our approach to teaching K–12 science over the original *National Science Standards* developed in the mid-1990s. Organized around science and engineering practices, cross-cutting concepts, and disciplinary core ideas, the framework helps students and teachers organize science and engineering knowledge in a more systematic fashion. Since the current best explanations about life science, physical science, and earth science are integrated around the cross-cutting concepts (patterns; cause and effect; scale, proportion, and quantity; systems and systems models; energy and matter: flows, cycles, and conservation; structure and function; and stability and change), students are better able to organize the myriad facts and concepts of science around relatively few central ideas. Outdoor ecological investigations lend themselves well to helping students better understand core concepts from all three science disciplines, because the cross-cutting concepts are easily observable in nature. Observations of plant and animal adaptations to demonstrate the cross-cutting concept of structure and function are one example. These observations also lend themselves to integrating engineering design by co-opting nature's time-tested designs for human systems. The concept of applying nature's solutions to our design problems has been well developed through biomimicry (Benyus, 1997). Energy flow in ecosystems can be used to demonstrate the cross-cutting concept of energy and matter: flows, cycles, and conservation. Observations of natural phenomena outside the classroom easily incorporate all seven cross-cutting concepts. More important, by revealing the cross-cutting concepts to students in the context of the world they see outside the classroom, they will invariably connect more directly with their local landscape. What is revealed to them about the world they live in will stick with them in ways that classroom experiences can rarely approach.

The long tradition of classroom-based education lives deep in the psyche of many educators and administrators. Time outdoors represents recess or a break from learning, and the tools of learning—computing, laboratory experiments, Smart Boards, and PowerPoint lectures are found inside. The reality is that sophisticated technologies have been developed to explore the natural world, and students can engage in these technologies in many aspects of outdoor exploration. Through simple thermometers to more complex data loggers, handheld Global Positioning System (GPS) technology, Geographic Information System (GIS) systems, telescopes, and laptop applications, students can explore technology and use it to further their understanding of ecological relationships. Data gathered outdoors can be further processed indoors for students to learn about programs designed for statistical analysis and modeling. Connecting with online data sources such as monitoring stations, web cams, and satellites adds to the potential technologies that students can use to research environmental parameters.

In March 2009, members of the White House grounds crew and Sam Kass, the White House chef, broke ground for the White House vegetable garden, a project inspired by First Lady Michelle Obama. Local school children were enlisted to do the planting along with the First Lady, and by mid-June over 200 pounds of healthy, organic vegetables had already been harvested. The First Lady established the garden as a way to promote healthy food for youth and gardening as an active and productive activity as part of her "Let's Move!" (2010) initiative, to focus attention on the problem of childhood obesity. A well-planned school garden represents a tremendous opportunity for entry into the outdoors. While school gardens create a space for students to actively engage in learning about food and nutrition, they also present opportunities to consider weather, water cycles, insect pollinators, life cycles of plants and plant biology, genetics, and reproduction, as well as a host of other possibilities in other content areas.

A garden is a great idea when you have the White House grounds crew to weed and water, but how can this work in an urban environment where open space is at a premium? Stephen Ritz developed a gardening program known as the Bronx Green Machine. This program completely dispels the myth that urban school gardens are not possible. Beginning in 2011, Ritz expanded his idea of helping urban students become academically and community engaged by moving to the Renaissance Charter High School for Innovation in Harlem. Here he promoted a can-do attitude among his students by encouraging them to build garden walls at a variety of locations around the Bronx, where they could grow and harvest vegetables to be sold at local markets and given to the needy (Trangle, 2012). While producing healthy food, his students learn science and math, marketing, and a host of twenty-first-century skills that cannot be delivered through classroom experience alone. Outdoor spaces in the Bronx are mostly concrete, brick, and asphalt, but the gardens connect students to nature through their attention to the natural cycles involved in growing and producing healthful food.

Stephen Ritz saw an opportunity to academically engage his students beyond the four walls of the classroom and, in doing so, has provided a unique, relevant, and challenging educational experience for them. There is little doubt that he had to work hard to overcome challenges along the way to make this engaging program a reality, but the results speak for themselves. Student engagement in areas of academic growth is what we all seek. If we do not take advantage of possibilities for doing so in the outdoors and in our community, we are selling our students short. The Bronx Green Machine that Stephen Ritz developed may be somewhat unique to his situation in New York City, but every school is surrounded by relevant nature, and we must be clever and creative enough to recognize it and make use of it. There are myriad examples of teachers doing so across the country. My situation is unique. I live in Montana, one of the most unspoiled and beautiful states in the nation. For me, even in downtown Missoula, there are natural areas near enough to access on a short walk from school. However, to provide the best possible experience for my students, one that they are not likely to forget, I have crafted, over several years, a unique immersion into one of our nation's most incredible natural areas, Yellowstone National Park. I share that experience with you below.

BEST PRACTICE

May in Montana is a splendid time of year. While the weather can still be unpredictable and catch us off guard with a day of cold and snow or spin up a windy and wet afternoon thunderstorm, there are also true spring days with warmth and bright sunshine. It also brings green-up: grasses on the hillsides spotted with wildflowers, and the raw green of newly emerging leaves from hillside shrubs and cottonwoods along the rivers. May is also the time when my advanced biology class travels to Yellowstone National Park, my favorite place on the planet. This rite of spring is a highlight of the year, one that the students look forward to and that I feel fortunate to get to lead.

Our planning begins in early January when we agree on the dates of the trip. Typically, we go for a weekend in mid-May—as late in the school year as possible to avoid the week prior to finals. The mid-May date guarantees that most campgrounds in the park are open, that we have a better chance of avoiding snow (never guaranteed!), and that the elk and bison are down in the valleys for calving. The presence of herbivores in the valleys means that wolf and bear viewing are at their peak. Not only that, but the park is abuzz with predator groupies—those folks who follow the big carnivores around the park. This knowledgeable community of experts can provide my students with intimate knowledge of the animals, something that I don't always have. I never tire of this experience. I get to share with my students a place that is referred to as "the Serengeti of North America," providing for many of them a first-ever camping and hiking experience, as well as a wide-ranging lesson in ecology and geology.

Two weeks before the trip I hold a mandatory parents' meeting. At this meeting the parents are required to sign all field trip forms and an additional form guaranteeing that they understand the inherent dangers of traveling, hiking, and camping in Yellowstone country. Once they understand the rules of the game, we carefully go over the essential checklist of personal camping gear, the trip itinerary (including meals), the parent drivers and chaperones, and emergency precautions. Typically 12–15 students and 2–4 parents make the trip. This allows us to use parent vehicles for travel and to fit students and all necessary equipment easily into four or five vehicles.

The trip lasts 4 days and 3 nights from Friday through Monday. Part of the fun is shopping for food on the Thursday night before we leave. I take three or four students, and we make the rounds to stock up on 4 days' worth of food for hungry teens. Since it requires a formidable kitchen to cook for 20 people, we also have to gather kitchen gear. I've accumulated quite a collection of utensils, pots and pans, and even a couple of stoves over the years, but the parents always add essential extras like lanterns, additional stoves, water carboys, and more. One of the best items I've purchased is a screen tent that can completely cover a picnic table. On blustery, rainy, or snowy days it keeps the cooks happy and the kitchen dry.

On the Friday morning of the trip, we all converge at the school parking lot to consolidate gear, load the vehicles, and hit the road for a 5-hour drive to Gardiner, Montana, and the original entrance to Yellowstone National Park. We always stop at the entrance arch for pictures and to read the sign above the arch: "For the Benefit and Enjoyment of the People." Thus begins our adventure into the world's oldest, and arguably best, national park.

Once in, the learning begins. Actually, it begins even on the road to the park. Every student is given a simple, spiral-bound journal to begin the trip. Before we leave the high school parking lot, I ask them to modify their perspective on the world by noticing the animals around them in every possible environment and to either list or describe everything they see. Before we leave the parking lot, robins, crows, fox squirrels, and maybe even something a little more exotic are added to the list. I also ask them to unplug—no cell phones, iPods, or game stations. By doing so, I hope they will notice a little more wildlife and pay closer attention to the landscape around them. Eliminating digital connections is easier said than done, given that I impose no strict control over the adult chaperones in each vehicle (though I do explain to them the intent), and I do not take their phones. The idea is to plant the seed that nature offers a richness and originality that is different from electronic media but in many ways more entertaining. It only takes a different focus on the world. As the weekend progresses, most of the students forget about connecting digitally and immerse themselves in the beauty of Yellowstone and with their travel companions.

Our trip focuses on the northern and western part of the park, from Mammoth Hot Springs east to the Lamar Valley in the north, and from Mammoth south to Old Faithful in the west. As we enter the park, I ask the students to observe how people interact with the wildlife. Sadly, many Yellowstone visitors consider the park to be zoo-like, rather than a sanctuary for the protection of wildlife. In the quest for the perfect photo, visitors will push animals to a flight-or-fight response. After we set up camp and settle in for dinner, we talk about wildlife viewing ethics and viewing practices that are safe for people and less stressful for animals. We address critical energy budgets for animals that remain in harsh winter environments, including hibernation, food storage, torpor, as well as hyperphagia and fat accumulation among the strategies for energy conservation for winter animals. This discussion addresses concepts of organismal biology and adaptation, but also gets students to consider that animals are not here for our sole benefit, and that stress-free interactions help them conserve energy that is necessary for survival and reproduction.

On Saturday morning we rise with the sun (5 a.m. in May), grab a quick hot drink and morning snack, and drive directly to the Lamar Valley. Starting early gives us the best chance to see two of the park's top predators, grizzly bears and wolves. The northern route to the Lamar Valley is the most likely place to see them. Once a bear or wolf is spotted in Yellowstone, a "jam" forms fairly quickly. Often, park wolf or bear specialists, or aficionados who follow the animals, will be among those viewing. I try to gather my students with these people to learn more details about the individual animals or packs we are viewing. During these discussions we consider the trophic effect that wolves have on the Yellowstone ecosystem. While wolves and bears compete with each other, it is wolves that provide "meat-on-the-ground" in the form of elk (and sometimes bison) carcasses that bears take over, and that many other species share in, until the carcass is picked clean. These include eagles, ravens, coyotes, magpies, several beetle species, and even bluebirds, which consume maggots that emerge from older carcasses. This helps the students begin to put together Yellowstone food webs and consider the complex interactions between the diverse organisms of this largely intact ecosystem. For many students, this is their first view of wolves or bears, and many express their awe with child-like excitement.

Following a tailgate lunch, we begin a field excursion with a hike up and along Specimen Ridge (between the Lamar Valley and Tower, just to the south of the northern road). During lunch, the students look over maps to determine the route, elevation gain, and distance to travel. This helps them to learn a little about reading topographical maps and route finding on such maps, and to determine their location within the park. The hike is a challenging physical experience for many of my students. It is just over six miles, with 3,000 feet of elevation change—1,500 feet steeply up at the beginning of the hike and then a long descent along the ridge until we meet the northern road again. Over the years we have seen pronghorn antelope, badger, Uinta ground squirrel, bighorn sheep, black bear, grizzly bear, moose, elk, bison, and even a boreal toad on one occasion, plus a host of bird and plant life. While we've never seen them all on one hike, we usually see four or five of these species.

On the way up to the ridge the students are thrilled to find elk antlers that have been dropped by the previous year's bull elk. These cast-off antlers are called sheds. To come across a dozen sheds on one hike is not out of the ordinary. These discoveries provide the backdrop for our discussion of wolves as a keystone species in Yellowstone. Keystone species are the organisms that are a key link to the survival of many other species in an ecosystem and provide a foundation for increased ecosystem biodiversity. For this conversation we stop in an aspen grove. I ask the students to carefully observe the aspens they see before them and then share their observations. Many good observations emerge: the spacing of the trees, newly emerging spring leaves, homes for cavity nesting birds, and so on. Typically, one student will notice that the aspens in the grove are either very old and large, or are about knee height, stunted, and browsed. There are no aspens of intermediate size

Photo 11.1. Hellgate High School Students Begin the Hike up Specimen Ridge, Yellowstone National Park. (Photo by Dave Oberbillig.)

and age. The students quickly surmise that an unchecked elk population keeps the emerging aspen browsed to a height that matches late spring snow depth. They also realize that in time all the mature aspen will succumb to age, taking with them feeding and nesting opportunities for a host of other members of this complex ecosystem.

When I started this trip 15 years ago, wolves had been introduced only a few years before, so we hypothesized the impact that wolves might have on elk and aspen over time. Now, my students can examine data about elk population dynamics in the park and see that elk populations are declining and that elk behavior is changing so that elk are less apt to concentrate their feeding in one area. Now elk tend to spend time in smaller herds and stay nearer good cover or escape routes, which reduces the pressure on aspen and willow stands. As we continue to visit the aspen grove, patches of different age-class aspen are beginning to appear.

As we continue along the trail, we stop to make observations of animals, identify tracks and scat, and take in the grand vistas of the Yellowstone backcountry. Toward the end of the hike we choose a comfortable place to break for water and snacks. While the students relax, I read aloud Aldo Leopold's essay "Thinking Like a Mountain" from his classic conservation work, *A Sand County Almanac* (Leopold, 1973, p. 137). Before beginning, I ask the students to listen carefully to the form of the essay. In it, Leopold asks us to consider how an inanimate but natural feature, the mountain, might think about the extirpation of wolves from the surrounding landscape. This essay follows nicely from a day of predator viewing and discussions about the impact of wolves as a keystone species as the students hear Leopold's view of predator control. After the reading, the students take time to write their own essay that must begin with, "Thinking like a. ..." The students allow their own personal experiences and observations to flow onto the page as they think like clouds or rivers or broad valleys, and place themselves in the perspective of nature. This exercise adds an interesting piece to their journals and helps them see a relationship between the observations and conclusions made by scientists with the power of creative writing.

By now the students are tired and ready to get back to camp for a hot meal and rest. Nonetheless, I ask them for one more assignment before bed. As the chaperones and I prepare dinner, the students dig into my Yellowstone "library," a small storage box with a collection of books about Yellowstone history, lore, ecology, biology, geology, and even general works on astronomy and the night sky. The students are allowed to choose any topic that interests them and prepare a short presentation (under 5 minutes) to educate us about another aspect of Yellowstone. They can give a short lecture, write a poem, or even perform a skit with others. This is an informative and entertaining end to a long and adventurous day. The students need little prodding to turn lights out in preparation for the next day.

On Sunday we break camp and head south toward Old Faithful, stopping at a couple of geothermal attractions along the way—the Fountain Paint Pots and the Firehole River. At the Fountain Paint Pots, every geothermal feature that the park has to offer is present on a short walk: mud pots, fumeroles, hot pools, and geysers. There is even a brief introduction to the variety of microbial life forms that live in and near the geothermal features. The Firehole River walk goes from Biscuit Basin to Old Faithful, where a range of geothermal features can be observed over the course of a few miles. Along the way, the students learn about the unique microorganisms that inhabit Yellowstone's thermal features. The color gradients around the edges of the pools represent organisms with different temperature tolerances. It was here that *Thermus aquaticus*—the thermophilic bacteria with the enzyme taq polymerase—was discovered. This discovery made possible polymerase chain reaction techniques, which have furthered myriad discoveries in genetics research.

After viewing Old Faithful, we leave the park to head northwest through the Madison River valley and on to Lewis and Clark Caverns State Park. On Monday morning we tour the caverns before packing up one last time to head back to Missoula and home. Before the drive home, we gather for a final closing about the trip. This is an informal and fun time to hear from all participants about their trip highlight. Some statements are profound, others funny, and some even inspiring. We hear about new interest in camping, the wonders of geothermal features, the beauty of the landscape, and amazing animal sightings, but also how some students and adults have changed their attitudes about conservation as a result of the trip. I am rarely surprised—that is the combined power of outdoor experience and Yellowstone National Park.

Since this trip is a school-related academic activity, I do assess student work. The journal is scored for the detail of the wildlife list, notes about predator-prey relationships in Yellowstone, wolves as a keystone species, organism adaptations to winter survival and energy budgets, notes on preparation of students' individual presentations and any reflective notes they might include in their journal. Students who choose not to go on the trip and stay behind at school are given a packet of Yellowstone ecosystem-related materials to research and review. In this way, all students are able to respond to test items about ecology and population dynamics that are related to Yellowstone and might appear on the final exam.

Invariably, I have left out of this description of the trip significant portions of planning, logistics, funding, and communication. To take on an excursion like this requires much advance planning and ample communication between school officials and parents. It also takes time to develop. I have learned much about what works from my experiences with students, parents, and camping logistics over the years. However, this trip is the highlight of my year and worth every extra moment. While most of you will not have Yellowstone as a potential destination, every place has a local nature preserve or wild landscape within a few hours' drive. Given that I have had students tell me—even years after they have left school—that it was their all-time best school experience, it is more than worth the time.

APPLYING BEST PRACTICES IN THE CLASSROOM

For many teachers, the idea of organizing an outdoor learning experience is daunting. The following not only describes an outdoor ecological inquiry but also describes the necessary preparation for a safe and effective experience. The idea that students will emerge from the classroom and immediately embrace the local wilds might be a little naïve. The time spent preparing the experience is the difference between a chaotic disaster and an engaging adventure for both you and your students. Remember, the research supports both cognitive and behavioral gains, so this effort is not a break from the routine but rather an effort to become part of an established routine.

Prior to departing from your classroom with students eager to explore the wonders that await them outdoors, a few essential tasks must be completed. Be sure that administrators and parents are aware of the activities you are planning with your students by demonstrating the educational benefits and connections with standards. If you are a trailblazer in your school, it's imperative that advance notice of your outdoor plans be given to all administrators. In some instances you may have to present the research (see the prior section for appropriate articles of support for outdoor ecological education) to your principal, superintendent, and school board. However, do not be deterred by having to invest this initial energy, and be prepared for some level of pushback, especially in places where heavy emphasis is placed on test results. Having administrative support provides many benefits: administrators will see that you are making an effort to be innovative and creative in order to increase academic achievement; administrators can support you when parents or colleagues question changes to the routine; administrators can help coordinate with local parks personnel and knowingly communicate with community members who see students outside the classroom during the school day, and may even be encouraged to attend so they can observe the investigations in action. Your presentation to administrators must include how your plan connects with district and school goals for student achievement, the important affective gains documented in the research, your attention to the organizational and safety details for your chosen activities (including notification to parents and school medical personnel), and your specific lesson plans. Once you have taken the proper preparation steps, you can be assured that your outdoor investigation is fully supported, and you can be confident about the success of your adventure.

Another vital preliminary step is to adequately map your outdoor resources. The investigation presented below requires a space that can accommodate 4–6 groups of students (class of around 25), providing each group with enough space to move freely in a 2 × 10-meter area. The investigation can be modified to use a smaller space quite easily, so be sure to consider the modifications before giving up on outdoor activities. As stated above, even the sparsest environments may harbor hidden havens, however small, for ecological investigation. If you decide the resources are just not there, or are too limited to engage several classes of students, you might require the more involved preparation step of gathering community support and resources for developing a habitat garden or vegetable garden on your school's grounds. Don't forget to carefully explore the neighborhood for local parks. The wild, weedy nature of abandoned lots cannot be ignored in evaluating potential study sites. For the comparative study presented below, having two (or more) different kinds of landscapes is important to the success of the investigation.

Ideally, the mapping step becomes part of the investigation for the students. The development of neighborhood maps that incorporate the ecological components of their neighborhood, town, and region aligns them with their local landscape in ways that most of them will not have considered. It not only sparks their curiosity about what might be out there, but also activates their spatial awareness through the design of scaled maps. The students can document the local landscape through sketches, an exercise that develops the observational skills necessary for scientific inquiry, or use photographs to document their chosen landscape over time. Accounting for seasonal changes further connects students with their local ecosystem and the processes that govern it.

Once you've found a couple of sites that are large enough to support the activity—they should be sites with areas of different vegetation characteristics—you are ready to proceed with an investigation that will engage your students in ecological inquiry. An expanse of mowed lawn works just fine for this activity. (Your students will be surprised at how much biodiversity is found there!)

Briefly, the study you will undertake requires students to make a comparison of the biodiversity of two (or more) distinctly different sites. Biodiversity is a representation of the different kinds of organisms in a given habitat or ecosystem at a given time. There are two components to biodiversity: (1) species richness, or the number of different kinds of organisms present, and (2) species number, which represents the number of each species present. Both can be measured in this study, but the focus is on a comparison of species richness.

During this investigation, students are asked to consider the process of science by making observations, asking ecological questions, developing hypotheses and making predictions, controlling for variables while collecting data, and organizing the data in a way that allows them to come up with a logical conclusion to their study. This investigation works well with a class of around 25 students and can be adapted for use by middle and high school students. Few knowledge prerequisites are necessary for this investigation. Since this is a comparative study, students can examine differences in biodiversity rather than making more strict taxonomic identifications. More advanced students will benefit from having a deeper understanding of taxonomy, because it will help them to explore subtle differences in ecological communities that a more simplistic comparison might not catch. Students should possess basic observational skills (which should be further developed through this investigation), be able to use metric measurements accurately, be able to record data accurately, have a fundamental understanding of taxonomy, and be able to follow directions. Following are the instructions for students in conducting a comparative investigation of the biodiversity of two areas near their school.

The equipment needs are minimal for this investigation. They include a clipboard, a 10-meter measuring tape, a meter stick, containers for plant collection, data sheets, and a stopwatch. If you have prior knowledge of your investigation site, and if you have limited access to equipment, you may choose to modify this list. The measuring tape and meter stick are used to line out a transect (a randomly chosen line or area where observations are made) in the study area so that students can make consistent observations between study areas. The transect can be any pre-determined size, so a variety of "frames" can be used. A meter square of PVC pipe or a "square" made of meter sticks can be used. Hula hoops also work well (but beware the distraction for your students!). A predetermined length of cord can substitute for the measuring tape. Many students have a stopwatch function or timer on their phones, or you can simply have each timed period start at the same time and use a central timer. Ultimately, the important thing is to sample consistently for all trials. As long as you use the same size transect and sample for the same amount of time, you will be fine.

By demonstrating the techniques and asking the students to practice prior to going outside, you will save a great deal of time and trouble. Since this is both a lesson in the nature of science and biodiversity, emphasizing attentiveness to the procedures and good observational skills is important. Demonstrate how to set up a transect within your classroom. You can discuss with students why it's important to randomize the transect site within your study area (so the students don't pick "more interesting" sites that might produce data that's not representative of the site) and why having different groups all using the same procedure produces more reliable data. Since the students are asked to literally crawl over the surface they are observing, they should practice exploring on hands and knees and be made aware of dressing appropriately for days in the field. Practicing the amount of time spent for each trial is also important so that students get a feel for the pace necessary to cover the transect thoroughly. Provide the data sheet to familiarize students with tallies they need to make for each trial. Be sure to warn students of potential hazards to avoid as they explore the transect. Examples are animal waste, trash, and broken glass. Making initial observations from outside the study transect and moving slowly and carefully will help to avoid these issues. Finally, help the students anticipate what organisms to look for by providing examples of the types of fungi, plants, and animals they are likely to come across in their investigation. Most students will never have observed a patch of grass, an abandoned lot, or a hedgerow with nearly this kind of attention before. They will be surprised.

The data sheet can help guide the students toward observations they might not otherwise consider. The presence of organisms will likely depend on the temperature, time of day, recent weather (for example, recent heavy rain may saturate soils, causing earthworms to emerge at the surface), time of year, and other site-specific factors that may influence the presence of organisms. Having spaces on the data sheet for environmental and temporal conditions can help guide student awareness. Spaces should also be provided to record organisms viewed, evidenced or collected for each "pass" over the transect. Examples of organism evidence include tracks, burrows, feathers, droppings, and castings. Since students will not likely know how to identify many of the organisms, especially invertebrates, sketches and quick measurements can help students to better categorize organisms on return to the classroom. If students are equipped with iPads, cell phones, or cameras, digital pictures offer an even better "description" to be used for future identification. On the other hand, simply recognizing the number of different kinds of organisms is all that is necessary when making comparisons of species richness.

CONCLUSION

Our youth face significant environmental challenges moving forward. The monumental prospect of mitigating and adapting to climate change, the difficulty of providing plentiful food and clean water for a burgeoning human population, the challenge of maintaining native landscapes and biodiversity in the face of continuing habitat destruction, and the task of protecting people from emerging zoonotic diseases as changing landscapes, climate, and human mobility influence the scope of epidemics, necessitates a well-educated and solutions-oriented STEM workforce. To develop sustainable solutions to these impending issues, an intimate understanding of ecological concepts and systems-level thinking is required. While theoretical knowledge of ecological and systems-science concepts can be gained through the traditional progressions of K–12 and higher education, the reasons to care about the necessity of working within the limits of ecosystems, valuing our place in them, and developing sustainable solutions requires that all citizens engage with their environment and develop a sense of place throughout their education.

At every level of community we clearly promote the concept that education is vitally important to our democracy and our economy. Every new administration promotes educational reform to improve the model, and yet a closer examination suggests our caring is explicit about particular aspects of education at the expense of others. Students are asked to be proficient at English language arts and mathematics by emphasizing their importance through testing. States and schools demand that teachers develop this knowledge through teacher evaluation and testing. Test-based proficiencies apparently equate to future community and economic success for those who participate in our public K–12 system. Unfortunately, we fail to emphasize the foundations that are the basis for sound community function and economic success. To maintain our collective prosperity, sustainable environmental systems are essential. By explicitly and specifically engaging students in all educational content areas through outdoor experiences, and through a lens of ecological and environmental education, students begin to understand how all aspects of community and economy are tied to the natural world around them. What's more, they begin to realize that sustainable economies and communities require diligent environmental stewardship. However, moving the institution of public education to embrace this ideal is another matter.

STEM education reform has had many incarnations in the 50 years since the launch of Sputnik—from the Woods Hole conference of 1959 and the introduction of the "New Math," to the emergence of "back-to-basics math" in 1980, to the publication of *A Nation at Risk* (Gardner, Larsen, & Baker, 1983), which described the ineffectiveness of math and science education in our country. STEM reform efforts continued into the mid-1980s with the development of Project 2061 through the American Association for the Advancement of Science, the National Science Education Standards in the mid-1990s and promotion of science as inquiry, the advent of No Child Left Behind in 2001, with math and science assessment, and into the present with the emergence of the *Next Generation Science Standards* and STEM integration. Reform is the norm, and with every new presidential administration, a new twist on delivery of the elements of STEM emerges. What doesn't seem to change much is the structure of the school day and how time in school is organized for students. By continuing to follow traditional school schedules and practices, the STEM reforms will change little about the outcome of student attitudes toward nature and how we interact with it. K–12 education systems, at their base, continue to promote the kind of education that got us into an unsustainable relationship with the environment to begin with. The key concept that we fail to embrace is that sustainable communities and economies can only exist with dedicated environmental stewardship that takes into account ecological rules. Translating those rules into STEM reform (and education reform in general), including the dedicated outdoor and community time that students need to develop a sense of place, is the change necessary to help future generations live well in a healthful and productive environment.

What does this mean for K–12 educators? It means providing a variety of interesting and challenging lessons through dedicated, locally relevant, outdoor exploration. If you do not have access to pristine nature or local parks within easy walking distance of your school, even those school grounds that are seemingly devoid of nature might surprise on close inspection. The key feature of student outdoor experiences is getting kids out of the building to make observations of natural phenomena. Observations can be made in spaces as small as a sidewalk tree box or as big as a regional park. Subjects for observation can range from patches of grass to a school garden to passing migratory bird populations.

While considering science process and ecological principles through direct observation is the positive academic outcome, a potentially more important experience for many students is providing a new lens on time spent outside. Even in urban environments, students can begin to see elements of their world that, though having previously gone unnoticed, can spur awareness and interest in the natural world at large. As teachers and students gain more practice in engaging nature, the enjoyment and personal growth they experience adds a new dimension to the typical education experience provided inside the school building.

APPENDIX

Measuring Biodiversity Lesson Plan

Introduction

Our biodiversity study at and near our school will be ongoing. We will attempt to document biodiversity at the school and at study sites chosen for their level of disturbance and variety of organisms. This is a group activity. Our goal is to use a systematic study to make a valid comparison of the biodiversity of local habitats that represent different levels of human disturbance. By following a systematic and consistent procedure between the study sites, we can draw meaningful conclusions about the biodiversity differences, plus make predictions about similar habitats we may encounter.

Materials

- Clipboards for each team member
- One 10-meter tape for each team
- One meter stick for each team
- Container for plant samples
- Data sheets for each team
- Stopwatch

Part I

Within your group, come up with a **definition for diversity**. Brainstorm, and then write a final definition that you all agree upon in the space below. Be sure to provide an example of diversity based on your definition.

Now, **write a definition for biodiversity**. Again, brainstorm this definition. Think about this on a planetary scale as well as for your local region, your neighborhood, and the area we are sampling. Yes, provide examples!

You will be comparing two different habitats to determine their level of biodiversity. One is a **disturbed habitat**—it has been significantly altered from the native habitat that once existed there. The other is a **restored habitat**—it was disturbed but has been replanted and reseeded with vegetation more typical of native or natural landscapes.

What is the ecological question that we should be attempting to answer for this research?

Now that you have a research question, what is your hypothesis about biodiversity relative to disturbed versus restored habitats?

Part II

Observe your instructor's directions for establishing a transect. Write down what you observe **in your notebook**.

Part III

1. Measure your transect. Be precise. This should be 1m x 10m. BE SURE YOU DO NOT DISTURB THE AREA OF YOUR TRANSECT IN ANY WAY PRIOR TO MAKING OBSERVATIONS.
2. Observe any and all **animals** that might cross an imaginary line that runs down the center of your transect. These may fly, crawl, walk, or move in any way (this might be birds, butterflies, crawling insects, etc). Do this **for 2 minutes**. Record your data!
3. The **two animal surveyors** in your group **crawl** shoulder-to-shoulder along the 10-meter line and search out every animal or animal sign they possibly can (tracks, scat, feathers, burrows, everything). Use the containers for the live stuff to help you identify it. Record all the data in your field notebook, both types *and* number. Do this **for 5 minutes**. Pace your crawl so you observe the entire transect with the same level of effort.
4. After the animal survey, **plant samplers** will go over the transect and take ONE complete plant sample for every different kind of plant observed. Collect enough of the plant to show how it is different from other plants in your collection. **Record** the number of different types of plants and the approximate number of each type within the transect.
5. Record the number of individual types and the total number of **fungi** observed in your transect.

We will repeat steps 1–5 for two different sites, taking care to be as consistent as possible about how we make observations at both sites.

Part IV

Chart the data for each site. List the number of **different kinds** of animals and plants as well as the **total number of each**.

Record the data of other student groups for EACH site, and then compare the DIFFERENT sites. This will be done in class as we summarize data.

For your conclusion, note and discuss any trends that you see in the data. Write the conclusion in your notebook. **Your conclusion must refer to your hypothesis.**

Your performance on this lab will be evaluated by:

- notebook entries, including a chart of the data
- participation in all group and fieldwork activities
- the final report of the project

Biodiversity data

Names of all group members:

_____

_____

Weather conditions:
Temperature (degrees Celsius): _____
Cloud cover (% and type): _____%_____
Wind (speed and direction): _____

Name of study site:

_____

Description of study site:
Overall area:

_____

1 × 10-meter transect (make specific notes in the box below):

_____

One meter

_____

Ten meters

_____

Data Collection:
Use sketches or descriptions when listing each individual species observed
  Observation Data (3 minutes)—do your best to leave your transect UNDISTURBED!
  List every animal that you observe moving across your transect:

The Animal Crawl (10 minutes—about 1 minute per meter):

Animal species diversity list (richness)	Approximate number (abundance)

Use additional pages if necessary to add to your list

Plant species diversity (richness)	Approximate number (abundance)

Use additional pages if necessary to add to your list
Class data collection and processing will occur in the classroom.

WORKS CITED

Barros, R., Silver, E., & Stein, R. (2009). School recess and group classroom behavior. *Pediatrics, 123*(2), 431–436.

Barth, P. (2008). Time out: Is recess in danger? The Center for Public Education. Retrieved from http://www.centerforpubliceducation.org/Main-Menu/Organizing-a-school/Time-out-Is-recess-in-danger

Benyus, J. (1997). *Biomimicry: Innovation inspired by nature.* New York: HarperCollins.

Bianchi, S. (2011). Changing families, changing workplaces. *The Future of Children, 21*(2). Retrieved October 17, 2013, from http://futureofchildren.org/futureofchildren/publications/docs/21_02_02.pdf

Clements R. (2004). An investigation of the status of outdoor play. *Contemporary Issues in Early Childhood, 5*(1), 68–80.

Ellerson, N. (2010). *A cliff hanger: How America's public schools continue to feel the impact of the economic downturn.* Arlington, VA: American Association of School Administrators.

Fraser, J., Heimlich, J., & Yocco, V. (2010). *Project grow outside: American beliefs associated with encouraging children's nature experience opportunities; Development and application of the EC-NES scale.* Prepared for Children and Nature Network. Edgewater, MD: Institute for Learning and Innovation.

Gardner, D. P., Larsen, Y. W., & Baker, W. (1983). *A nation at risk: The imperative for educational reform*. Washington, DC: US Government Printing Office.

Godfray, H., Beddington, J., Crute, I., Haddad, L., Lawrence, D., Muir, J., Pretty, J., Robinson, S., Thomas, S., & Toulmin, C. (2010). Food security: The challenge of feeding 9 billion people. *Science, 327*(5967), 812–818.

Helm, B. (2011). Instructional time versus "seat time." *College Summit, Launchpad Blog*. Retrieved October 17, 2013, from http://www.collegesummit.org/blog/content/instructional-time-versus-seat-time

Hofferth, S. (2009). Changes in American children's time—1997 to 2003. *International Journal of Time Use Research, 6*(1), 26–47.

Kellert, S. (2005). Nature and childhood development. In S. R. Kellert (Ed.), *Building for life: Designing and understanding the human-nature connection* (Vol. 1). Washington, DC: Island Press.

Leopold, A. (1973). *A Sand County almanac: With essays on conservation from Round River*. New York: Ballantine.

Let's move! (2010). *School garden checklist*. Retrieved October 17, 2013, from http://www.letsmove.gov/school-garden-checklist

Lieberman, G., & Hoody, L. (1998). *Closing the achievement gap: Using the environment as an integrating context for learning*. San Diego, CA: State Education and Environment Roundtable.

Louv, R. (2008). *Last child in the woods: Saving our children from nature deficit disorder*. Chapel Hill, NC: Algonquin Books.

National Research Council. (2012). *A framework for K–12 science education: Practices, crosscutting concepts, and core ideas*. Washington, DC: National Academies Press.

Orr, D. (1991). *Earth in mind*. Washington, DC: Island Press.

The Nature Conservancy (2011). *Connecting America's youth to nature report*. Retrieved October 17, 2013, from http://www.nature.org/newsfeatures/kids-in-nature/youth-and-nature-poll-results.pdf

Rideout, V., Foehr, U., & Roberts, D. (2010). *Generation M^2: Media in the lives of 8 to 18 year-olds*. Menlo Park, CA: Henry J. Kaiser Family Foundation Study.

The state of play: Gallup survey of principals on school recess. (2010). Robert Wood Johnson Foundation. Retrieved October 17, 2013, from http://www.playworks.org/sites/default/files/d6/StateOfPlayFeb2010.pdf

Trangle, S. (2012, December 5). Green Machine given more room to grow. *The Riverdale Press*. Retrieved October 17, 2013, from http://riverdalepress.com/stories/Green-Machine-given-more-room-to-grow,51480

Wells, N., & Evans, G. (2003). Nearby nature: A buffer of life stress among rural children. *Environment and Behavior, 35*(3), 311–330.

CHAPTER TWELVE

Twenty First Century Skills Inspired Through Global STEM Projects

DAN CARPENTER, FLORENTIA SPIRES AND JOSEPH ISAAC

INTRODUCTION

Increasing the science, technology, engineering, and mathematics (STEM) pipeline is essential for the U.S. to remain globally competitive. Global collaboration in STEM education serves as a useful vehicle for teachers and students to develop, implement, and hone their STEM skills in an international arena (see Figure 12.1). Global STEM education is pertinent for providing students with the implementation of twenty first century skills that characterize global digital citizens. Global citizenship entails an awareness of the interdependence of individuals and their social, scientific, political, systems and a sense of responsibility for cultivation of principled decision-making (Green, 2012). Global science collaboration is developed via ongoing communication between individuals through the increasing use of free and low-cost Web 2.0 platforms. Global science can be implemented by novice and experienced educators nationally and internationally. Global projects emphasizing STEM content subsets, such as bioscience, environmental and space science, engineering, and robotics can be used to support global STEM citizenship through projects and develop students as global leaders.

The Next Generation Science Standards (NGSS) represent a vision to reform science education (Bybee, 2014; Hayes, Lee, DiStefano, O'Connor, & Seitz, 2016), and to increase student proficiency and competitiveness in STEM disciplines (National Research Council (NRC), 2013; National Science Teachers Association (NSTA), 2016). New and innovative science standards promoted by the NGSS

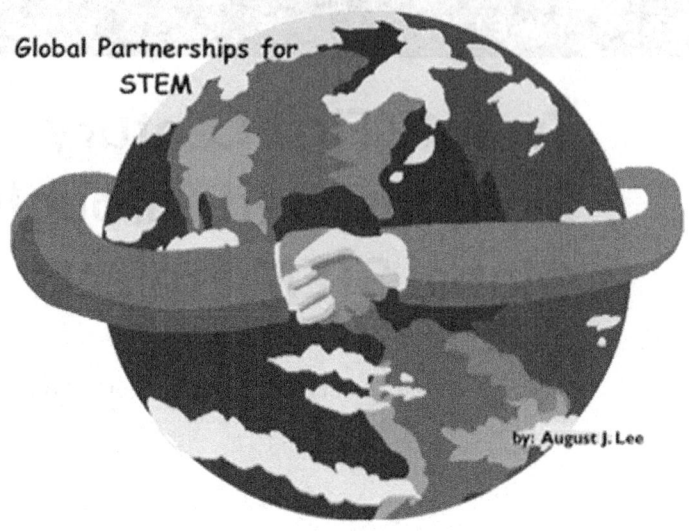

Figure 12.1. Global Collaboration Through Partnerships.
(Source: Lee, 2016.)

incorporate concepts that focus on critical thinking and communication as fundamental to student success in today's global economy. As a result, the standards and requisite skills in them require a complete shift in how teachers approach STEM instruction to ensure that science and engineering practices are intertwined within the standards to promote critical thinking and communication in students.

Ultimately, the educational changes introduced by the NGSS imply a major paradigm shift in teacher practices. The NGSS and state standards provide opportunities for educators to interact with their students like scientists and engineers. Educators and students learn behaviors that scientists and engineers engage in as they investigate the natural world (NRC, 2013). They focus on problem solving, the nature of science, and engineering as associated with twenty first century skills.

Effective implementation of the standards and development of twenty first century skills require educators to shift their instructional approach from teaching isolated topics, concepts and processes to providing authentic experiences that develop critical thinking, problem solving, collaboration and creativity. Effective implementation of the standards requires teachers to further develop twenty first century skills with students. Thus, teacher focus in the curriculum and their approach to teaching and learning will need to shift from science as a body of knowledge to presenting science as a way of knowing, and from treating science as an isolated subject to connecting science with other disciplines (Bybee, 2014).

However, a lack of concrete and coherent implementation strategies such as global STEM education, have left educators feeling unprepared for the reform movement (Cunningham & Carlsen, 2014). In this chapter, you will be provided with best practices for a global STEM collaboration model designed to develop global awareness and problem solving through collaboration with teachers and students separated by geography (Nugent, Smith, Cook, & Bell, 2015). What is most notable in this model is that teachers and students from schools in the United States collaborate on real world, authentic global problems with other teachers and students from around the world using Web 2.0 tools to communicate like researchers. This model is about creating teacher and student citizen scientists who can solve real-world problems. Citizen science presents opportunities for problem solving, alignment with rigorous teaching standards, and the development of twenty first century skills that move teachers and students from global awareness to problem solving on a global scale (Bonney et al., 2009).

BACKGROUND

Global education began in the 1960s and 1970s with educational initiatives in the "thinking versus doing" movement, where students were asked to solve contextually-based problems that we now call inquiry. In the 1980s and 1990s, computing and digital technologies grew the amount and types of information exponentially, making inquiry and the problem solving process a bit easier for students and teachers. However, the interaction and collaboration needed to solve problems beyond the classroom remained a struggle (Schwartz, Lederman, & Crawford, 2004).

The global education reform movement began with collaborative work between education, non-governmental organizations (NGOs) and intergovernmental agencies (Standish, 2014). What came from this reform movement has led to further globalization of education where teachers and students share data, collaborate on global problems and take steps to solve problems (Zahabioun, Yousefy, Yarmohammadian, & Keshtiaray, 2013). Projects, such as the eMammal citizen science project for example, are a collaborative effort between several organizations, including the Smithsonian and the Student Discover Project. Participants use infrared cameras to capture and monitor mammalian biodiversity and conservation. In projects like these, students from all over the world interact via the Internet using Web 2.0 technologies to identify organisms captured by a wildlife camera. Essentially, teachers and students from any country in the world can get online and identify the organisms that appear on an image that may originate in China, Kenya, Germany or the United States. This type of global collaboration project is authentic even though limited in terms of how teachers and students engage in inquiry. Other projects involve more inquiry-based practices, using classroom websites, blogs, Wikis, Google Hangout, and similar tools to get classrooms anywhere in the United States to solve STEM-related problems alongside a classroom not so nearby.

There are many levels of collaboration (from limited communication to engaged collaboration) and different twenty first century skills (e.g., creativity, critical thinking (problem solving), collaboration, communication) addressed through global projects that the concept of global education has taken on variety of asymmetrical meanings (Parker, 2011; P21 Framework for 21st Century Learning, 2016). In particular, creativity, critical thinking, collaboration and communication are referred to as the 4Cs. For example, the term "global education" has often been used interchangeably with "international education," "global learning," "global dimension," and "global citizenship," all capturing a global approach to learning. Here we define global education as an approach that engages students from different nations using STEM-related skillsets while they encircle dimensions of citizenship, developmental education, sustainability, intercultural education, and education for peace (Scalzo, 2012). In exercising these attributes, students hone twenty first century skills to solve real world problems, which require not only the 4Cs but also innovation (P21 Framework for 21st Century Learning, 2016). Students become global citizens and advocates as they connect what they learn in subject areas and engage with students across the globe to solve real world problems (Nugent et al., 2015). Many of the existing definitions of global education are grounded in themes associated with Hanvey's (1979) vision of five dimensions of a global perspective: (a) perspective consciousness, (b) "state of the planet" awareness, (c) cross-cultural awareness, (d) knowledge of global dynamics, and (e) awareness of human choices (Kirkwood, 2001; Landorf, 2009). As educators integrate the combined standards and twenty first century skills and principles into their curriculum, four of Hanvey's dimensions (i.e., "state of the planet" awareness, knowledge of global dynamics, awareness of human choices, and cross-cultural awareness) may serve to inform instructional methods and provide a contextual framework for student learning.

Global STEM collaboration is relatively new as a philosophical and pedagogical application in classrooms. In recent years, it has grown steadily with the advent of "social media" technologies and their applications to the citizen science movement in communities. A simple Google search for citizen science projects yields an increasing number of projects to engage.

The global STEM education continuum is a conceptualization of social engagement through collaboration that promotes constructivist teaching and learning experiences from a global perspective (Nugent et al., 2015). Nugent et al. (2015) describe instructional examples along the continuum moving from global awareness to contribution in order to make a difference in the world through project-based and problem-based learning methodologies. Smith (2013) describes the continuum ladder from no collaboration to parallel activity, to shared data, then to limited collaboration, and finally to engaged collaboration. According to Coughlin, (as cited in Lindsay & Davis, 2012) the key to designing meaningful integrated global collaborations is to fully understand expected social interactions, task depth and complexity, and foster complex reasoning for critical thinking application. Regardless of the place on the global STEM education continuum, all types and forms of global collaborative practices can be transformational through the power of sharing (Ertmer et al., 2011; Lindsay & Davis, 2012; Neal, Mullins, Reynolds, & Angle, 2013; Stegmann, Wecker, Weinberger, & Fischer, 2012).

FIVE CONCEPTIONS

The five conceptions that define and demonstrate the engagement levels for conducting global education include global awareness, parallel activity, shared data, limited communication, and engaged collaboration (Cook, Bell, Nugent, & Smith, 2016). Each level of global collaboration and engagement intensifies, requiring a greater skillset and organizational plan that lends itself to increased connectivity, communication, and ultimately full engagement. Teacher leaders must implement an advanced skillset for effective execution as they guide students to be successful in global collaboration (Vance, Kulturel-Konak, & Konak, 2014).

Global Awareness

Global awareness is the initial phase of incorporating global education and does not include collaboration between global teachers, students, or classrooms. In this phase, students often become conscious outside of their immediate community and country using digital learning platforms (Nugent et al., 2015). Platforms that engage students beyond their community and country allow them the unique experience of thinking globally through the use of software. Some examples are: (a) Google Earth's mapping challenge (to learn about historical, geographical, natural history, or just the proximity of a place) ("The Great Global Puzzle Challenge with Google Earth," 2011), (b) Discover Education's Virtual Field Trips, (c) National Geographic Map Projections Project, or (d) Journey North's migration of birds and butterflies to Canada, South America and beyond ("From the NSTA Calendar: Journey North Begins," 2010). While students learn through these platforms, they become globally aware of other land, air, and water spaces beyond their own place of residence. For a list of links to some of these projects, please see Appendix A, Table 12.1.

Parallel Activity

Students working on the same project using the same timeline in different nations: this is the essence of parallel activity (Nugent et al., 2015). This form of engagement does not require student use of technology, per se, for implementation. While students are not interacting with global partners in this phase of engagement, global teachers interact at this level and guide student engagement using a common timeline. In parallel activity, students know they are conducting an activity simultaneously with another class in a different part of the world. The Global Citizen Science Moth Week is an example of such an activity (Moskowitz & Haramaty, 2013; Appendix A, Table 12.1).

The Global Citizen Science Moth Week is a global project that invites people from around the world to learn about moths and contribute to the body of knowledge about moths. The online project is typically held the last two weekends of July and focuses on the study and data collection of moths around the globe. The online system allows groups or individuals to add observations and images to an online database while interacting through social media with others doing the same (iNaturalist, 2017). Data entry is a social interaction, but also scientific in that it requires like observations through a matched database from other collaborators around the world. These parallel observations provide students and teachers with collaborative interactions that go beyond their classroom or community.

Shared Data

The implementation of work that is accomplished simultaneously between at least two global partners along with the collection of shared data constitutes the phase of global collaboration. An example of a software program that immerses students in this type of engagement is National Aeronautics and Space Administration's (NASA) Signals of Spring, where students are able to track marine animals around the world (Signals of Spring—*ACES*, 2004). Students collect data and share it synchronously or asynchronously with global partners in different nations using a blog format (Hillis, da Silva, & Raguseo, 2008). Collaboration through blogging has increased student connectivity and collaboration beyond the classroom. This type of communication has been shown to increase student motivation to want to solve problems in a collaborative way, while also promoting an awareness for other people's ideas in how to solve problems. It has also been shown to increase student interest in solving authentic problems (Meinecke, Smith, & Lehmann-Willenbrock, 2013; Schuenemann & Wagner, 2014). Other formats available for convergence of data sharing include wikis, Google classroom, and video chat platforms (Meyer, 2008). With synchronous or asynchronous methods of communication, students begin to understand the collaborative accessibility with global partners.

Limited Collaboration

Global collaboration that engages students in activities on a limited basis may require a heightened skillset of digital citizenship and aptitude for technology use (Zahabioun et al., 2013). For example, the social skills for collaboration and networking in this phase require students to communicate via blogs, but also suggest a synchronous communication format using multimedia platforms like Skype, Oovoo, Yahoo messenger and Google Hangout video chats (Brown, 1984; Harshman & Augustine, 2013). The teacher leader provides guidance for the students to account for the time difference between their own nation and that of their global partner's nation (Tichnor-Wagner, Parkhouse, Glazier, & Cain, 2016). Student opportunities to interact in real time provide insight that every person on the Earth is interconnected and closer in real time than ever possible due to Internet access.

Engaged Collaboration

Complete immersion of real time communication of global education include experiences of students developing interactive projects with their global partners that fully engage all in problem solving (House, 2011). Students may participate in brainstorming ideas on how to approach their project and designing the work while remaining in constant communication synchronously and asynchronously, addressing project development with their global partners. An example of a collaborative project of full engagement is the World Smarts STEM Challenge whereby high school students from two nations vote on the greatest need in their community, brainstorm solutions, and engage human potential to ensure that fellow citizens thrive based on student contributions through a global educational and collaborative experience (IREX World Smarts STEM Challenge, 2016).

There are many examples of IREX World Smarts STEM Challenge success. One finalist project was a collaboration between a school in Ghana and a school in Maryland, U.S. In this project, students worked collaboratively to create electricity using a soil-based generator. The idea was that by creating electricity through more organic means, the amount of fossil fuels consumed to produce electricity would decrease, thereby decreasing environmental impact (IREX World Smarts STEM Challenge, 2016). This example of full engagement maintains a sense of seamless communication and collaboration that indicates there are no longer borders between two nations.

BEST PRACTICES FOR GLOBAL COLLABORATIVE PROJECTS

Authentic Global Project Reflections

Implementing global projects should be done with the goal of aligning learning to relevant STEM curricular components and standards as well as twenty first century skills. In order to ensure success, the decision-making and planning process should rest in the hands of the participating teachers. Global teachers may include teachers from different classrooms in the same or different countries, but the conceptualization of global collaboration is to increase awareness of different cultures in different classrooms. Organizations exist whose purpose is linking global classrooms to one another (see Appendix B, Table 12.2). However, it is important to determine global partners that are a "best fit" to yield the greatest success for the collaborative project outcome. For that reason, it is advised to use a checklist that weigh the propensity for success as you compare one potential global partner to another. Nonetheless, global teachers in both nations ideally and ultimately take on a shared responsibility for implementation. The proposed project learning should also relate to real world situations and issues important to all stakeholders. Real world problems extend the student beyond the classroom into global issues on the spectrum of inquiry as a learning process, thereby exposing students to project-based learning and other inquiry based teaching and learning methods (Crawford, 2007; Thomas, 2000; Wenning, 2011).

Authentic Projects

The authors of this chapter have led and conducted several Global Science Projects. Florentia, Joe and Dan discovered through their experiences the common practices recommended for a successful transnational collaboration. Moreover, the authors present specific teaching, learning and assessment methods that came about through the planning and implementation stages of development. All of these projects were problem solving based activities that developed components of the 4Cs as well as global awareness. All three projects served as an engaged collaboration.

Florentia's project example

Florentia's global science project created an enthused reaction from participants engaging in the Global Precipitation Measurement Mission platform led by National Aeronautics & Space Administration (NASA) where she served as a NASA Master Teacher. This project required middle school students from two nations to learn about the quantity of precipitation in their region, the use of water, and the endless ways that the local water supply is spent in their nation. These students then engaged in the collection of precipitation data over time. They were required to analyze data for the time of year measured and ultimately conduct comparative analysis of the data collected for each nation. Each nation learned about the weather that existed for their global counterpart at the time the project was conducted. They extended the shared learning to unveil the variations of the climate and seasons. In some cases, the global partners extended the length of the project in order to unveil the comparisons for the variations of seasons more authentically. Students exchanged information surrounding their daily uses of water, highlighting the importance of precipitation. A student participant from Nigeria spoke about the importance of agriculture in the lives of the people in his community, who primarily rely on rainfall for irrigation. As a result of project engagement, teachers and students became more aware of community use and value of water as a natural resource in each nation. Students discovered that as citizen scientists they have a choice and a responsibility to use their voice to make a difference in their decisions related to how water is used, wasted, and potentially conserved in their community. Students ultimately generated a list of suggestions on how they would begin newfound practices of conserving water.

Joe's project example

One of Joe's projects allowed students from different locations to collaboratively explore the details of the cell cycle and its checkpoints with the purpose of developing the skills and knowledge to comprehensively address a driving research question: "How can we stop the spread of cancer cells?" Working in teams, students from high schools in Washington, D.C. and St. Thomas, U.S. Virgin Islands researched a type of cancer, visited and interviewed a cancer researcher or oncologist in their respective region, and developed a short, multimedia presentation describing the pathology of their particular cancer. The students also researched cancers that are prevalent in their communities and summarized efforts and methods currently being used to treat cancers. The results were not only to be shared with their classmates, but their counterparts in the partner school. The ultimate goal was to share their findings with the American Cancer Society. Students from both locations were tasked to research the prevalence of cancer clusters in their communities, while also finding local sources of carcinogenic reservoirs in those areas. After compiling initial data and results, the two groups of students compared and contrasted those results from their two respective locations by way of video-conferencing. The project allowed the students in both locations to collect real data from local hospitals, health departments or university research labs to find out the most prevalent types of cancer found in their respective communities.

The hands-on and interactive portions of the project took place over six 90-minute class periods. Day one started off with students watching an engaging video on homeostasis and given their Project Guidebooks. The remainder of the class was devoted to student understanding that homeostatic controls prohibit cells from growing indefinitely. The homeostasis project uses hard-boiled eggs and dye. The students put a hard-boiled egg into a glass beaker of water and then add a couple drops of dye, then predicted how cell size related to cell efficiency. The class ended with the students completing an opening questionnaire. Day two had the students participating in an interactive game as an overview of the cell cycle. Students examined live onion cells and came up with a classification system for the cell cycle and the mitotic stages they observed. They then compared the predicted stages with the official stages of the cell cycle and mitosis by analyzing pre-treated onion root tip slides. Observations were then recorded in students' Project Guidebooks. Day three continued with students identifying stages of the cell cycle by looking at the activity of chromosomes at each stage of the cell cycle and noticing distinguishing checkpoints. In so doing, students were able to determine if tissue was cancerous or non cancerous by accounting for variations between the two. Day four began with students separating into groups of three to four and choosing a specific type of cancer to research. Each group of students in each location researched unique types of cancer. Student groups were then given several days to conduct research outside of class for their specific cancer type. Students were encouraged to visit hospitals, health clinics, doctor offices, and local health departments to collect as much information as they could. Day five continued with a comparison of student projects as groups compared, contrasted and critiqued each other's research. Groups created a presentation of a specific tissue, the cell cycle abnormalities that cause it to become cancerous and the cancer's associated treatment methods. The last day of the project was a presentation of student findings between Washington D.C. students and U.S. Virgin Islands students in a videoconference. The sharing of research provided opportunity for students to discuss similarities and differences in cancer rates, mortality and treatment options in their respective communities. The students then ended with open dialogue, exchange of ideas and the completion of a closing questionnaire.

Dan's project example

One of Dan's many projects includes collaborations, strategic problem solving and interactions between students in a biology class with environmental organizations in the Midwest U.S. In one such environmental project, students investigated human impact on a surrounding nature area. Students collaborated with several environmental organizations to determine the natural history and current-day biodiversity of the area, then ultimately determined human impact on the area. Students began the project by taking several nature hikes into the surrounding nature area and making observations about the types of plants, animals and microorganisms in the nature area. This biodiversity study then led to student observations and discussions on succession. The succession discussion and observations then led to the natural history of the surrounding nature area. The natural history investigation then led students to develop research questions, which further led to groupings of students that then set out to design methodology to measure biodiversity and human impact on the area. Research questions and methodology then led to student problem solving with a local University and environmental organizations on how to decrease human impact on the nature area. Student findings over several years led to the area being classified as a nature preserve. Student projects and inquiry also led to sustainable biodiversity classification of the preserve as a protected area, as well as nature trails and outdoor classrooms for community use. While this example is considered international, it is one that can easily be expanded beyond our nation's border for global collaboration. Dan did this project with different students over different years. Five teachers and their students got involved in the project. Some students reached out to organizations to engineer deer paths into walkways for humans to reduce off trail contamination of the ecosystem. One year a teacher and several students worked with local engineers. several organizations, businesses and the University to create a nature trail with mulch to reduce further human impact, while also encouraging further student interaction in the nature area. Another year a teacher worked with students, engineers, a construction company and local businesses to build an outdoor classroom structure and several garden boxes for vegetable production. The mathematics and engineering required to complete the projects involved teachers from the math department, the woodshop and technology educators as teachers had to interact to create a shared learning environment to complete these projects. Moreover, the cross disciplinary approach these projects took provided authentic learning experiences that required complex communication, collaboration, critical thinking and creativity with respect to city code and the outdoor nature area. The engaged collaboration provided an opportunity for high level problem solving with teachers, community organizations, businesses and even political parties to ensure the projects were long-standing integrated STEM-oriented for student learning as well as community needs.

The Global Precipitation Measurement project partnership between U.S. and Nigeria, the Cancer Cluster Collaborative between U.S. and Caribbean students, and the collaboration between students in the Midwest U.S. and environmental organizations are just three examples of how project-based learning methodologies might contribute to globally aware and competent students. Projects of this magnitude stretch student understanding of the nature of science across the interconnected global spectrum. Projects like this align to the standards, satisfy cross cutting concepts, integrate disciplinary core ideas, but also promote the construction of knowledge through authentic experiences.

One of the most difficult, yet rewarding aspects of projects like these is the ability to work productively with students, without regard to time zones, culture, and socio-economic differences. Differences contribute to students and teachers global awareness and provide the strides in fostering a more just and verdant global community. International and international collaborative projects foster greater understandings and empathy while building leadership skills and environmental stewardship found in twenty first century skills, but often times difficult to implement in normal classrooms. The ability to navigate away from long-standing silos of subject specific instruction will go a long way in providing a rich and refreshing educational experience for the educators involved as well. As long as planning is succinct and comprehensive in the simultaneous implementation, the benefits reaped by student and educators alike will far outweigh any challenges associated during the execution and enactment stages.

ADAPTING THE GLOBAL STEM EDUCATION PRACTICE

Developing an Outline for the Global Project

The first step for initiating a global project is identifying a project of focus. Thereafter, the process for declaring a targeted population and location most conducive for meeting the project goal must be determined. There are a number of organizations that can help provide potential global partners. For example, one must reflect on the following: Does the global project require a specific continental location? Does the project location require proximity to a natural body of water? Is farmland required for the project or is there a different consideration needed to be taken into account for the success of the global project? In order to conduct a global science project, the global teacher leader must have a starting place. Identifying global partners can be time consuming and intimidating. Organizations that provide opportunities to connect global collaborative partners can decrease the time needed to find partnerships and decrease intimidation. Other organizations provide strategies that help identify global partners by extending a teacher's collaborative network through shared learning goals and social interactions that promote meaningful authentic projects. (see Appendix B, Table 12.2)

Importance of a project outline

A global project must be meaningful with intended outcomes. Meaningful projects are authentic projects in which students should be provided with opportunities to solve real problems, directly connected to their real lives. Intended outcomes for students should be aligned to the standards but also authentic in terms of student products that serve their learning as well as their collaborative partners. As a product of teacher planning of authentic experiences, it is important to produce and share a scope and sequence that outlines the authentic partnership collaboration in a timeline for the project steps (see Appendix C, Table 12.3 and Appendix D, Table 12.4). The timeline required for implementation should be shared with and between all global participants several weeks in advance. In order to ensure the project takes the direction for achieving the desired outcomes, the outline is meant to be developed for guidance throughout the project. In a meaningful way, the questions of: (1) who and what nation the global partnership includes; (2) the goal of the content gained, (3) how the goal will be achieved, and (4) why the project is important for global learners must be considered. The answer to these questions provides a clear vision of the importance of the global project.

While this timeline is not a hard deadline, it is important that the global partners have adequate time to digest the information. It is also imperative to consider the timeline for engagement as compared to their school calendar events in each country, while ensuring that the impacts for engagement are doable for their full-fledged participation. Implementing a global project does require some degree of flexibility to avoid developing emotional stress surrounding elements that may be out of the global participants' control. The outline provides a glimpse for each component that stakeholders are expected to engage. The specifics outlined help the global partners make an informed decision about their choice to participate, while cerebrally gearing up for the upcoming project. Partners that work collaboratively on a global project with fidelity benefit from being informed of touch points of the global process upfront.

The outline with the timeline should explicitly include the actions of the students. This approach allows the global teacher partners to begin familiarizing themselves with the contextual information that students will experience. A pre-assessment is recommended for use prior to the global STEM project to determine student entry-level knowledge and skills. The instrument is meant to effectively measure students' knowledge base prior to the onset of the project launch in order to assess each student's growth in their contextual knowledge of the global project as well as its real world application at the conclusion of the project. Formative assessment should be used to measure student progress through the global student collaboration. Finally, the global leaders will provide optional approaches for sharing the global project results learned by students to the public. During the phases of the entire project, pictures are encouraged to capture evidence of student engagement. However, it is imperative to ensure that authorization is sought prior to capturing and publishing images of participants, especially students. Advocacy with the face of students, lends itself to bold and progressive initiatives when shared with the public. To disseminate information, a brochure is one possible way to distribute the results to community members. Students may also endeavor to present their learning at a Town Hall meeting to educate the public, or write a letter to a government agency. Students may also consider contacting local politicians who can impact decision-making. Alternatively, students may simply choose to enact different strategies on their school campus to improve the welfare and conditions for students.

Identification of Global Leaders

The global leader from the U.S., as the origin country in all shared projects, takes on the responsibility to lead and facilitate the global project. The U.S. global leader has the vision on the development of the project. In order to ensure that all parties are involved and endorse the global collaboration experience for students, all stakeholders should be introduced to the authentic experience, content and potential process prior to launching the global project. Contact administrators in the upper stratosphere of the organizations for approval.

Depending on the flow of authority and level of autonomy on the school level, district leaders, in addition to the school leaders, may need to be contacted. In some instances of identifying the point person for a global project, it is recommended to start at the school level and request information regarding the hierarchal structure that may suggest other pertinent stakeholders to contact at the start of the project. In contacting the administrators, the global leader should state which staff persons need to be on board for a successful project. Important administrators might include the district school leader, principal, content specific leader, and technology resource leader.

Upon establishing that all project stakeholders are involved, it is recommended that you, as the origin country lead teacher, serve as the senior global leader of the project. Additionally, clarity will need to be established concerning, "Who will be designated global lead teacher to monitor the progression of the project?" This will establish clear lines of information dissemination to the global partners to allow relevant information to filter through that point person. The point person, in turn, has the endorsed authority and responsibility to ensure the same information trickles down to all other stakeholders. This will also eliminate redundancy and missteps that might occur when multiple contacts are being used for the same project. The point person also ensures a clear line of acceptable communication is followed to ensure that protocol is adhered to for members of each nation.

The global leader, in this case, the teacher in the United States, should initially share information and train their colleagues involved. This will help formulate allies early on in the process of implementation. Training should be succinct and help teachers conceptualize how the project will conducted in the classroom with respect to how the teacher manages the learning environment. Moreover, the United States global leader should also lead the training of the global partner teachers in similar fashion, keeping in mind that classroom management may differ between nations. With this in mind, it is recommended to discuss the differences that may exist prior to training teachers.

It is recommended that each global partner leader brief their teaching staff of the upcoming collaborative project and virtual meeting schedule for the sake of ease in having all members onboard from the beginning stages of introduction for the upcoming tasks. By doing so, global teachers are able to carefully explore the curriculum standards and make connections to solidify knowing and understanding the significance in their participation for the project.

Job-Embedded Professional Learning

Professional development for global projects should be built into the normal work day for teachers. Job-embedded professional learning on how to plan, implement and evaluate the impact of global collaborative projects can and should be built into professional learning communities of teachers working together to plan out the project, align it to the standards and curriculum and assess its impact on student learning. It is necessary for novice global teachers to gain clarity of the entire process for effective implementation, but this process does not have to be any more difficult than implementing a new teaching and learning strategy.

Professional learning is a job-embedded process that is meant to equip the global teachers with the skillset needed to promote their self-efficacy for leading global components of instruction. As the global leader of the project and facilitator, professional development and learning is needed to ensure that all curricular materials required for project implementation are pre-planned and ready for a teacher dry run of the project through the planning stage of professional learning. The planning and dry run can be conducted with all teachers either face-to-face in person or by way of a jointly held collaborative group in a virtual meeting. If a synchronous professional learning of planning and dry run of the project are not possible, then a video may be recorded of United States teachers carrying out the process of implementation for the global teachers to view. As teachers experience the project through the dry run, synchronous or asynchronous sharing should help all teachers learn how to do the project with their students to ensure the collaboration fit within the curriculum of all classrooms involved. In all professional learning sessions, teachers must record questions that need to be addressed and openly discuss questions and or concerns of the execution process. It is imperative that all issues concerning the classroom implementation are resolved between global teachers prior to the start of student collaboration. The questions asked from the global teacher leaders are meant to ensure clarity in knowledge and conducting procedures with their students is gained.

As a result of the professional learning, all teachers should gain some degree of confidence to lead their students with the tasks of the global project. All teachers engaged in the global STEM project will be equipped to confirm the exact requirements needed to implement the project with their students as it relates to materials and time for the project tasks, the availability and supply of items needed, the collection of materials required for a product of a successful global project, and plan how to lead the unpacking of the project components with their students. Being able to identify and collect the materials ahead of time help ensure that project components are in place-ready to utilize on day one when launching the global project.

Cultural Considerations

For teachers and students, it is important to be aware of what it means to be a global citizen. In order for teachers and students to interact with global partners, digital technology must be used. As such, teachers must educate students about responsible digital citizenship and cross cultural norms to ensure cultural sensitivity. Teachers must find ways to ensure that students avoid offensive behavior or mannerisms, due to the lack of understanding of other cultures. Global collaborative teachers should find ways to develop a professional, collegial, and collaborative rapport with partner school participants. Teachers and students benefit from the sharing and exchange of cultural norms to ensure that each global stakeholder is educated to effectively and collegially work with people from different cultures. Teachers and students as global participants also benefit by proactively gaining a head start via public resources, such as books and websites that reference customs and cultural norms. The success to promote harmony is based in knowing how to interact with peoples from specific nations. Some of the big ticket items for initiating a successful global partnership include understanding: (1) how to greet nationals; (2) non-verbal communications; and (3) meeting etiquette, to name a few (see Appendix E, Table 12.5). Additionally, sensitivities to dress codes, group dynamics, and forms of direct and non-direct eye contact may also impact how the host global partners may be perceived based on how their communication approach is first received; this may determine the tone for a positive and successful global experience.

Florentia worked with global partners from China, Botswana, Nigeria, and Pakistan. Joe worked with global partners in the British Virgin Islands, Antigua, and Jamaica. In many parts of Asia, negligence in understanding cultural norms can decrease the ability to be granted an opportunity to begin a global project. For example, Florentia noticed differences in meeting formats; her Asian global counterparts were almost always in a group setting, where she was most often the sole U.S. individual during the global meetings. The Asian global partner collaboration was collective as they bounced ideas off of one another during virtual discussions. The Asian global partners were less flexible about the times they were willing to meet. The meetings usually occurred during their morning hours between 9 a.m. to 11 a.m. while the corresponding time for Florentia was most often between the hours of 11 p.m. and 1:00 a.m. Additionally, the Pakistani students were specifically isolated by gender groupings; the U.S. students were thus required to enter into isolated gender groupings in order to interact with their Pakistani global partners.

On the other hand, Florentia's Nigeria global meetings conducted with the Nigerian global partners were more flexible in meeting as a group or with an individual global teacher partner. The time set for meetings with the Nigerian global partners was more flexible with the U.S. partners in that the time allotment occurred in the mornings, afternoon, or evening hours. Provisions were met to have students remain after school in order to meet with global partner students, in some cases. The Nigerian global students were in mixed gender groupings with fewer restrictions. For example, a male gender group could be near the female gender group in the same room. Male and female genders interacted with both genders of their U.S. global partners.

Time zone changes for projects with Caribbean schools were of little consequence since the schools worked with are only one hour ahead of Eastern Standard Time. The main note of difference was simply getting timely responses back from the island-based teacher partners. Sometimes the "easy pace" of the island life interrupted the flow of urgency beholden to air-tight schedules and deadlines surrounding project pacing. As either the partner or coordinator of the project stateside, ongoing communication will prove to be crucial for effective implementation.

While taking precautions to avoid interruptions in the timeline for a global project, it is important to be mindful of the need to embrace one's ability to be flexible. This state of being is important to eliminate unnecessary stress when series of uncontrollable events occur and program interruption is not in the control of either global partner. For example, Florentia's GPM project in Nigeria experienced the partner school shutting its doors due to an ebola outbreak in the country. In another case, Florentia experienced a long silence with her Pakistani partners when there was a terrorist attack that reportedly killed 72 women and children in a park located in the same small town where the global partners lived and attended school. The global leader must be willing to sit quiet in some instances and wait for the flames to simmer before he or she can reignite the project. In the event the project continues, it may require that the schedule and timeline be amended or abbreviated in order to achieve the intended goal as all of the limitations are considered.

The exchange of school calendars is an essential part of the planning process for global project collaboration. While the exchange of calendars allows the partners to plan the global project to correlate with schedules for both nations, it also allows the global teacher leaders to hone in on the differences in cultural celebrations that could be highlighted as teachable moments within the confines of the project. One of the most rewarding experiences for the author team is the acknowledgement and awe our students have experienced as a result of learning cultural differences about their global partners. Exchanging school calendars also ensures that instructional time, important testing periods, calendar holidays or closings are taken into account to reduce the number of glitches in the completion of the global project. It is also highly recommended that both global partners request information about internal school events and programming that may not be indicated on the calendar to ensure that all dates are taken into consideration.

As a result of using both school calendars and the acknowledgement of internal programming, the global project will intentionally be aligned to both school calendars to ensure that there is no interruption in the execution of the global project.

Global Communication for Project Implementation

Establish the digital platform for use

Global projects may have a plethora of options to choose from for communication. Teachers can opt to interact globally via blogging, Google Classroom, Skype, Zoom, Oovoo, and other wireless chat formats for virtual communication. What is interesting is that not all countries use the same communication platform, which creates a unique challenge. Beyond Facebook, the most widely used communication platform is Google+ and Twitter. Zoom is a relatively new platform for communication and can be used in some countries as well. On the other hand, Skype has a limited market and can only reach 24 countries around the world, some only with payment. With variability in how communication tools are used and what is available, it is important to research what might be available with global partners in their country.

The computer technology used requires a test run prior to the start of the project to confirm effective usability. This will allow the global partners from each nation to determine the efficiency of the technology. The range of frequencies for transmitting signal can be slow, medium, or fast. Testing for Internet speed and bandwidth capacity is important to establish prior to the official launch of the global project. These logistical pre-checks will decrease the probability of extenuated delays and issues involving communication between the partner schools. These types of logistical confirmations are best crosschecked by involving a technology expert to perform a test run for using the platform of choice. For example, the technology expert may reconsider the Internet Service Provider that determines the bandwidth or reevaluate the connection to the router. Specifically, the bandwidth is important to ensure that notes during the email transactions, chat messaging synchronously or asynchronously, downloading large files, and virtual face to face meetings that occur in real time can be successful. Also, it is vital that the technology expert confirms that computer networks process tasks rapidly in order to avoid wasting valuable planning or instructional time for the global teacher leaders or their students. Checks and balances for technology efficiency is a priority task to confirm for successful communication globally.

Email communication

Email communication should occur with the point person identified by the most senior person of the global stakeholders. The global senior leader is the person expected to delegate the local school global leader from the host school or district. Communication should occur a minimum of three times per week to ensure that fluency regarding project development occurs. This amount of communication, however brief, seems to be optimal in order to keep partner teachers engaged, involved and accountable. This communication could entail information regarding updates, additional information for project implementation, feedback from lessons, or the sharing of pictures of the host students engaged in their global work.

Video chat communication

Virtual meetings are a means of communication that provide global partners an opportunity to ask clarifying questions face to face. The more personal approach is one of the strengths to help stakeholders develop a rapport during global collaborations. The meeting leader should include an agenda for each meeting to develop a culture of learning that provides clarity about project development and implementation. The merging of collaborative work implemented will require communication via email and virtual platforms agreed upon based on technology capability and access.

Before any work begins, a virtual handshake is regarded as one of the most significant components for getting the collaboration off to a great start with other nations on a global project. It is recommended that the virtual handshake occur soon after email communication has been established. In our experiences, our initial virtual communication ranged from one week to two weeks after teachers and students first direct contact with global partners. The virtual handshake can take on different formats, such as an exchange between the global leaders via e-mail, blog or even personal social media accounts. The purpose is meant to have the global partners establish communication on a more personal level in hopes that contact will be maintained on a regular basis during and potentially after the project. During the virtual handshake, the global leaders benefit by sharing background knowledge of oneself, the school history, the demographics of the student participants, and how the work fits into the school curriculum and possibly the community.

Following the virtual handshake, a project-specific virtual professional learning meeting for teachers from each nation will be provided as described earlier. The initial meeting and conversation is recommended for developing the collaborative partnership between teachers. The open collaboration helps global teacher partners develop a rapport that ultimately trickles down for the success of the project to the global student for effective engagement.

Project launch meeting

It is recommended that a video chat occur between teachers at least two weeks prior to officially launching the global project. This event is very important as such, all stakeholders should be included. An agenda should be provided to organize the ordered steps needed to accomplish the intended goals and establish cultural norms for the meeting. Stakeholders in attendance should be introduced and might include administrators, teachers, students and technology experts for each nation. All stakeholders should be free to ask pending questions to gain clarity of their role and the implementation of the global project.

Now that all stakeholders have met, asked questions and conducted troubleshooting strategies for greater capacity, the final outline should be reviewed by the global leaders. They will then make final adjustments, and review the final draft for release to all other stakeholders. It is important to note that stakeholders might only include teachers, one from each classroom participating in the Global collaboration experience. The launch does not have to be complex; the experience should be simple and to the point to begin the project respectively.

Additional Considerations

Unique materials and shipment

Resources vary from nation to nation. As a result, it is important to identify the resources needed for the execution of a project well in advance. It is important to communicate in advance with the global partners to discover what materials they can obtain locally versus what they may need to provide support for the project. At least four weeks prior to the launch of the project, the global leaders should solidify a list of materials required, funding available, accessibility, and materials needed. It is essential to verify if materials for the global project require shipping. The needs assessments may be a deciding factor for selecting global partners. A research assessment of the timeline for departure from the U.S. to the arrival destination at the postal location is important. For example, if you learn that mail often vanishes in a specific country, then you may determine that it is not a risk that you would want to take for the sake of being successful. The global facilitator may decide that this country is not an option to begin a partnership due to the possible risks (see Appendix F).

The global leader initiating the project may be required to support the project by sending a materials package to the global partners well in advance to ensure they are ready to begin the project by the start date. By finalizing this information with a minimum of four weeks in advance, all materials can be purchased, gathered or sent from the host global partners. This strategy will ensure that all pieces of the project are in place for project launch. In this final stage of preparation, the global project is provided a final date and is ready to be launched.

For example, Florentia used a professional rain gauge for their global project. In order to ensure there was a fair strategy used to collect data, she sent the same type of rain gauge to the global partners to ensure their calibrations were aligned due to variations from the manufacturer. Additionally, it is important to compare which shipping carrier to use for more prompt service and delivery of materials to global partners, as costs and delivery methods vary. If required, the facilitator or other global leaders can fundraise to ask for donations for purchasing the materials required for the global partner.

Reflection and recognition

Teachers are sometimes motivated to get involved in a global project by knowing that they will receive acknowledgement that verifies their participation in a project of global magnitude. In cases where global teachers ask "How does this work help me?", the author team has learned that a small token of appreciation goes a long way. For example, the teacher earning a Professional Development Certificate that states their completion of a global project, thus certifying the teacher as an instructional global leader, moves mountains. More specifically, the recognition aids their professional development continuing education component as a professional and helps build the teacher's portfolio. As for the lead teacher in the U.S. facilitating the project, affiliations with organizations (i.e. a school district, a university, a science or a STEM organization) that support the line of work for the project may be generous when approached prior to the project implementation. The lead teacher will promote the work and seek supporting agencies to endorse the project. For example, Florentia was able to gain endorsement from Texas Tech University and the project leadership to gain a certificate that acknowledged global teacher participants for their contributions to global science work (see Appendix G, Figure 12.2). Finally, participation in a global project of this magnitude is significant to include in one's resume. However, it is important to have a contact person, the facilitator, verify that the global project was, in fact, implemented for the sake of transparency.

Global STEM Collaboration

Global STEM collaboration is no longer a new component of teaching and learning. Global STEM collaboration has been shown to provide inquiry-based teaching and learning opportunities while also promoting twenty first century skill development. The conceptualization of Global STEM collaboration as a teaching and learning opportunity has been shown to provide social engagement in and out of classrooms that further promote constructivist teaching and learning experiences. The spectrum of collaboration demonstrates that engagement in global projects promotes global awareness, parallel problem solving activities, shared data in collaboration and problem solving, and engaged collaboration to solve authentic STEM problems.

The collaboration continuum provides a foundation for twenty first century skill learning and development. The Partnership for 21st Century Learning framework can be implemented in global STEM projects to promote the development of needed and requisite knowledge, expertise, skills and supports to better prepare students for the next generation workforce. We have shown in this chapter that effective global collaborative projects promote the development of specific twenty first century learning skills, also known as the "4 C's"; (1) creativity (and innovation); (2) critical thinking (and problem solving); (3) communication; and (4) collaboration.

Global STEM collaboration problems like those discussed in this chapter can be seen as examples of how twenty first century skills can be developed through delivery of authentic teaching and learning experiences in current and modern STEM classrooms. What is perhaps most profound is that STEM teaching and learning leveraging only single STEM disciplines is not realistic. Rather, true authentic problem solving requires a cross disciplinary approach in order to truly be authentic as seen by projects discussed in this chapter.

An increase in cross-disciplinary connections that compliment STEM- and STEAM-centered projects and curricula rose out of the increased criticism of rigid, silo-based subject-matter curricula. With inevitability, researchers discovered that the approach led to gaps between student mastery of subject content and their relatability to real-world situations. For example, a group of students could be learning calculus at mastery levels of proficiency. However, those same students may have extreme difficulty incorporating the same calculus concepts in a molecular biology class while attempting to successfully make correct solution dilutions. Another example might include students in an Anatomy and Physiology or Biology class that can be taught the genetics, pathology and etiology of cancer. But until it's placed in a context of community health, health care and environmental health issues, cancer could still easily be looked at as a malady reserved for the unfortunate. This was an important lesson to learn for the students in the Cancer Cluster Project as they were also concurrently being exposed to related lessons in their social studies and health classes that looked at the societal and economic impacts of cancer.

Teacher implementation of global STEM collaborative teaching and learning requires training. Training must include mechanisms that teach teachers how to create and solve authentic problems with students. Teacher training must include what, when and how to align authentic problems in STEM education to solving the problems aligned to the NGSS and state standards. In the STEM disciplines, the standards and the curricular scope and sequence are inclusive components to constructivist pedagogies such as the spectrum of inquiry-based teaching and learning discussed in this chapter. Project- and problem-based teaching and learning has proven to provide the foundational constructivist inquiry-based approach needed to align the scope and sequence of the combined standards. These components provide what is needed to include in global STEM collaboration as a critical piece of current day classrooms.

Teacher training in constructivist inquiry based pedagogies such as project-based teaching and learning must include when, where, and how to incorporate global STEM collaboration projects within the scope and sequence of the standards. Moreover, constructivist pedagogies require teacher training on how to capture student learning aligned to the combined standards. Student capture of their learning that leverages progressive assessment mechanisms is critical to the implementation of global STEM collaboration. Progressive assessment methods include a balanced assessment approach for multiple data capture methods. Selected response assessment, rubrics that guide students to critically think and problem solving types of assessment that capture student products, and even teacher-student communication that assess formative assessment of student progress at mastering the combined standards are needed in teacher training. Planning and aligning global STEM projects to the standards along with progressive assessment systems to ensure student mastery of the standards is critical to the proper execution of global projects. The planning and alignment of global projects require teachers and administrators to collaborate with all participating teachers, further extending the idea of collaboration as a mechanism to instructional improvement. Essentially, teachers will teach teachers how to plan, develop, implement and reflect on global projects. The training is job-embedded as the teacher facilitators educate teachers and collaborate by way of the curricular scope and sequence. Global collaboration also reflects on teaching effectiveness by using student data in a teacher collaborative environment for instructional improvement.

CONCLUSION

We hope that by sharing the collaboration spectrum, readers can find ways to implement global STEM projects in ways that increase teacher and student efficacy at doing real science, engineering, mathematics and technology based applications to teaching and learning. It is our hope that by sharing our experiences, other educators around the world may implement projects like these, find opportunities for students to interact in cross cutting concepts and ideas, while also having fun in the learning process.

We truly believe that global STEM projects are one of many examples of inquiry-based teaching and learning and that it can serve as a vehicle for educational change in and between classrooms, in particular in STEM classrooms around the world. As schools prepare students to be productive global citizens, cross-curricular collaboration will remain a viable vehicle for education reform in pedagogical context and student learning.

APPENDIX A

Table 12.1. Platforms That Engage Students Beyond Their Community.

Name of program	URLs
Google Earth's Mapping Challenge	1. https://geoguessr.com/ 2. https://www.amazon.com/dp/0753467216?tag=amz-mkt-chr-us-20&ascsub-tag=1ba00-01000-a0049-mac00-other-nomod-us000-pcomp-feature-scomp-wm-4
Discover Education's Virtual Field Trips	http://www.discoveryeducation.com/Events/virtual-field-trips/explore/
National Geographic Map Projections Project	http://www.nationalgeographic.org/activity/investigating-map-projections/
Journey North Citizen Science	https://www.learner.org/jnorth/
Global Citizen Science Moth Week	http://www.inaturalist.org/projects/national-moth-week

APPENDIX B

Table 12.2. Identifying a Global Partner.

Global Partnership Programs	Audience	Web Address
1. Global Partnerships	International for all grade levels	http://www.globalpartnerships.org/
2. Global Classrooms	International for all grade levels	http://www.globalclassroom.org/collaboration/globalpartners.html
3. Global School Net	International for all grade levels	http://www.globalschoolnet.org/gsnpr/
4. Level Up Village	International for all grade levels	http://www.levelupvillage.com/
5. SEED	International for all grade levels	http://media.corporate-ir.net/media_files/IROL/97/97513/global_stewardship/seed.html
6. Digital Human Library	International for all grade levels	https://www.digitalhumanlibrary.org/teachers/global-connections-for-teachers-and-students/
7. e-Teacher Challenges	International for all grade levels	https://teacherchallenge.edublogs.org/step-10-connect-other-classes/
8. British Council	International for all grade levels.	https://schoolsonline.britishcouncil.org/partner-school
9. E-pals	International for all grade levels	https://www.epals.com/#/connections
10. Peace Corps World Wise Schools	International for all grade levels	https://www.peacecorps.gov/educators/
11. Edmodo – Global Chats	International for all grade levels	https://spotlight.edmodo.com/product/global-chats-using-google-hangouts-skype-pen-pals-blogs,326017/
12. Global Studies High Schools	Washington, D.C. high school	http://theodorerooseveltdc.org/apps/pages/index.jsp?uREC_ID=321800&type=d

APPENDIX C

Table 12.3. Part I: Timeline for Global Collaborative Partnership.

Start Date	Activity	Audience	Materials
6 months–1 year	Identify the global project	Global leader	Original project or adopt a project from an organization collecting ongoing global data
4–6 months	Contact multiple prospective global partners	See Global Partnership Platforms	Internet connection and email contact for ongoing communication
4–6 months	Begin to introduce the global project to be launched; Identify the leadership for the global organization	Responding recipients from the email communication	Letters, description of the project, and email access
12 weeks prior	Make relevant contacts with organizational leadership; request a technology check for capability	Hone in on the Global Partners of Interest that are intrigued by the proposed project	Emails, letter of introduction with elaborations (this may use an informal back and forth approach to answer queries
11 weeks prior	Follow up with updates on acceptance or denial of participation	Leadership, Lead Teacher	Emails, syllabus, establish inventory checklist via email for global partner
10 weeks prior	Global Leader initiates the Virtual Handshake/Meeting; Leadership share the project with grassroots global partners	Leadership, Administrators and Teacher Leaders	Power point to introduce the global project; discuss mailing norms in the global country
9 weeks prior	Identify group that you choose to conduct a global partnership	A variety of potential global partners	Internet, email; Send Materials/inventory to global partners via USPS, FedEx, UPS, DHL, or the relevant Embassy
8 weeks prior	Follow-up meeting – reaction to project by grassroots global partners; relationship building	Leadership	Internet, email, Virtual platform
7 weeks prior	Virtual meeting; virtual handshake	Grassroots global partners	Virtual platform
6 weeks prior	Provide a Pre-Project questionnaire to learn about strengths and fears	Global teacher partners at the grassroots level	Google Forms

APPENDIX D

Table 12.4. Part II: Timeline for Global Collaborative Partnership.

5 weeks prior	Present global project in the form of a virtual PD; synchronous vs. non-synchronous; Questions and Answer	Global teacher partners at the grassroots level	Technology; Power point presentation
4 weeks prior	Email syllabus with activities; global teachers become familiar with content and timeline; carry out activities (how many?) Have materials/inventory arrived?	Administration and teacher leaders	Technology; syllabus; timeline
3 weeks prior	Virtual meeting/PD; answer questions for global teacher partners; address any fears; Have materials/inventory arrived?	Global teacher partners	Technology
2 weeks prior	Have materials/inventory arrived? Set up Google Classrooms for teachers & students; Set up Global student groups	Administration/leadership; Technology expert	Email; Google Docs
1 week prior	Set up materials for all of the activities	Global teacher partners, Administrators, Leadership	Email
Week 1	Student Pre-Assessment	Students	Internet and Google classroom access
Weekly	Email communication; Project activities implementation; share and analyze data; Global teacher partner communication or virtual meeting; Global student partner communication	Leadership; teachers; students	Internet access; materials, virtual meeting, Google classroom
End of Project	Community Presentation; Global teacher partner communication or virtual meeting	Students	Resources; community members
Final Week	Students Post-Assessment	Students and community members	Resources; community members
Post Project – 1 week later	Teacher post questionnaire; will provide information on professional development and growth	Teachers	Google Form

APPENDIX E

Table 12.5. Samples of Cross Cultural Competencies to Consider.

Cultural Norm	American	Latin American	African	Asian
Handshake	Solid	Gentle handshake, a hug or kiss on the cheek	Bow with supporting elbow	Bow with
Eye contact	Direct	Indirect with elders	Indirect with elders; direct professionally	Indirect with elders
Communication	Direct – mean what you say and say what you mean	Indirect – does not necessarily mean what is stated	Indirect – does not necessarily mean what is stated	Indirect – conflict avoidance
Group behavior	More individualized	More focused on core groups and looking out for others	Comfortable as part of a group	Comfortable as part of a group
Time consciousness	Time wasted is never returned and is a valuable resource	Time is relative to what is happening in life	Time is relaxed without urgency in general	Eastern Asia – time is valuable; in Southeastern Asia – time is not important
Dress code	Casual but neat clothing; Christians dress more conservative	Town vs. Village dress code varies; Christians dress more conservative	Casual clothing; Christians and Muslims dress more conservative	Don't show too much skin; traditional, Christian, and Muslim dress is more conservative

APPENDIX F

Table 12.6. Use of a Checklist for Potential Collaborative Partner.

Guidelines Required for Global Partners	Sufficient	Insufficient
Is there interest in the global science project?		
Does project location have resources needed to carry out the global project?		
Is there a language barrier that hinders effective collaboration?		
Are global leader(s) responsive in the beginning stages of communication?		
Is the student population sufficient for a transnational global collaboration?		
Does the time zone difference allow for face-to-face virtual collaboration?		
Is the amount of technology available sufficient for global communication?		
Is the bandwidth sufficient for communication? Have technology checks been performed?		
Are resources/materials available for full engagement?		
Does the postal system have a reputation for delivering mail in that country?		
How long does it take mail to arrive from the U.S. to the global country?		
Are global teacher leader participants interested and willing according to their questionnaire responses?		
Total		

APPENDIX G

Engaging Students Globally in Collaborative STEM Educational Activities

GLOBAL PRiSE
Pragmatic Researchers in STEM Education

In Honor of

Corey Anderson

and students from

Prince George's Schools

for their Participation with Students from Pakistan
in a Global Science STEM Project in 2015

TEXAS TECH UNIVERSITY
College *of* Education

Walter S. Smith
Founder, Global PRiSE
PhD Program
Texas Tech University

Figure 12.2. Certificate of Recognition Sample.
(Source: Smith, 2015.)

WORKS CITED

Bonney, R., Cooper, C. B., Dickinson, J., Kelling, S., Phillips, T., Rosenberg, K. V., & Shirk, J. (2009). Citizen science: A developing tool for expanding science knowledge and scientific literacy. *BioScience, 59*(11), 977–984.

Brown, J. L. (1984). *Global learning teacher education manual: A model inservice or graduate course for elementary and secondary teachers.* Montclair, NJ: Global Learning, Inc.

Bybee, R. W. (2014). NGSS and the next generation of science teachers. *Journal of Science Teacher Education, 25*(2), 211–221.

Cook, L. A., Bell, M. L., Nugent, J., & Smith, W. S. (2016). Global Collaboration Enhances Technology Literacy. Technology and Engineering Teacher, 75(5), 20–25.

Crawford, L. (2007). Global body of project management knowledge and standards. *The Wiley Guide to Project Organization & Project Management Competencies (Vol 8)*, 206–252.

Cunningham, C. M., & Carlsen, W. S. (2014). Teaching engineering practices. *Journal of Science Teacher Education, 25*(2), 197–210.

Ertmer, P. A., Newby, T. J., Liu, W., Tomory, A., Yu, J. H., & Lee, Y. M. (2011). Students' confidence and perceived value for participating in cross-cultural wiki-based collaborations. *Educational Technology Research and Development, 59*(2), 213–228.

From the NSTA Calendar: Journey North Begins. (2010). *NSTA Express*, 2–3.
Green, M. F. (2012). Global citizenship: What are we talking about and why does it matter?. *International Educator*, 21(3), 124.
Hanvey, R. 1979. "Cross-Cultural Awareness." Toward Internationalism: Readings in Cross-Cultural Communication. Rowley, MA: Newbury
HouseHayes, K. N., Lee, C. S., DiStefano, R., O'Connor, D., & Seitz, J. C. (2016). Measuring science instructional practice: A survey tool for the age of NGSS. *Journal of Science Teacher Education*, 27(2), 137–164.
Harshman, J. R., & Augustine, T. A. (2013, October). Fostering global citizenship education for teachers through online research. *The Educational Forum*, 77(4), 450–463.
Hillis, M., da Silva, J. A., Raguseo, C., & Center, A. B. (2008). Cartoon festival: An international digital storytelling project. *TESL-EJ*, 12(2).
House, J. D. (2011). Effects of computer activities and classroom instructional strategies on science achievement of eighth-grade Students in the United States and Korea: Results from the TIMSS 2007 assessment. *International Journal of Instructional Media*, 38(2), 197–208.
iNaturalist (2017). iNaturalist.org home. Retrieved March 30, 2017, from https://www.inaturalist.org/home
IREX World Smarts STEM Challenge. (2016). Retrieved from https://www.irex.org/project/world-smarts-stem-challenge
Kirkwood, T. F. (2001). Our global age requires global education: Clarifying definitional ambiguities. *The Social Studies*, 92(1), 10–15.
Landorf, H. (2009). Toward a philosophy of global education. In *Visions in Global Education: The Globalization of Curriculum and Pedagogy in Teacher Education and Schools: Perspectives from Canada, Russia, and the United States*, Vol. 29, 47–71. Peter Lang. Peter Lang Publishing, Inc. New York, NY.
Lee, A. J. (2016). *Global partnerships for STEM* [Digital Art]. Washington, DC: Private Collection.
Lindsay, J., & Davis, V. (2012). *Flattening classrooms, engaging minds: Move to global collaboration one step at a time.* Chicago: Pearson Higher Ed.
Meinecke, A. L., Smith, K. K., & Lehmann-Willenbrock, N. (2013). Developing students as global learners: "Groups in our world". *Small Group Research*, 44(4), 428–445.
Meyer, N. (2008). Technology to accomplish student achievement: Blogging, second life, wikis, collaboration, multimedia, creativity in the classroom. *Medium*, 32(3), 10–11.
Moskowitz, D., & Haramaty, L. (2013). National moth week-a new global citizen science project focused on moths. *Terrestrial Arthropod Reviews*, 6(3), 185–200.
National Research Council. (2013). *Education for life and work: Developing transferable knowledge and skills in the 21st century.* Washington, DC: National Academies Press.
National Science Teachers Association. (2016). *National Science Teachers Association, about the next generation science standards.* Retrieved March 30, 2017, from: http://ngss.nsta.org/About.aspx
Neal, G., Mullins, T., Reynolds, A., & Angle, M. (2013). Global collaboration in teacher education: A case study. *Creative Education*, 4(09), 533.
Nugent, J., Smith, W., Cook, L., & Bell, M. (2015). 21st-century citizen science. *The Science Teacher*, 82(8), 34.
Parker, W. C. (2011). International education in US public schools. *Globalisation, Societies and Education*, 9(3–4), 487–501.
P21 Framework for 21st Century Learning. (2016). National Framework for 21st Century Learning. Retrieved March 30, 2017, from: http://www.p21.org/our-work/p21-framework
Scalzo, J. (2012). What is global education? *Connections (07010400)*, 33(1), 6–7.

Schuenemann, K., & Wagner, R. (2014). Using student-generated blogs to create a global perspective on climate change. *Journal of Geoscience Education*, *62*(3), 364–373.

Schwartz, R. S., Lederman, N. G., & Crawford, B. A. (2004). Developing views of nature of science in an authentic context: An explicit approach to bridging the gap between nature of science and scientific inquiry. *Science Education*, *88*(4), 610–645.

Signals of Spring—*ACES*. (2004). Retrieved from http://www.signalsofspring.net

Smith, A. (2013). *Nations and nationalism in a global era*. Hoboken, NJ: John Wiley & Sons.

Smith, W. S. (2015). *Certificate for "Engaging Students Globally in Collaborative STEM Educational Activities", 20 December 2015*. Lubbock, TX: Texas Tech University.

Standish, A. (2014). What is global education and where is it taking us? *Curriculum Journal*, *25*(2), 166–186.

Stegmann, K., Wecker, C., Weinberger, A., & Fischer, F. (2012). Collaborative argumentation and cognitive elaboration in a computer-supported collaborative learning environment. *Instructional Science*, *40*(2), 297–323.

The Great Global Puzzle Challenge with Google Earth. (2011). *Publishers Weekly*, *258*(34), 60–66.

Thomas, J. W. (2000). *A review of research on project-based learning*. The Autodesk Foundation Report, San Rafael, CA.

Tichnor-Wagner, A., Parkhouse, H., Glazier, J., & Cain, J. M. (2016). Expanding approaches to teaching for diversity and justice in K–12 education: Fostering global citizenship across the content areas. *Education Policy Analysis Archives*, *24*(59).

Vance, K., Kulturel-Konak, S., & Konak, A. (2014, March). Assessing teamwork skills and knowledge. In *Integrated STEM Education Conference (ISEC), 2014 IEEE* (pp. 1–6). IEEE.

Wenning, C. J. (2011). The levels of inquiry model of science teaching. *Journal of Physics Teacher Education Online*, *6*(2), 2–9.

Zahabioun, S., Yousefy, A., Yarmohammadian, M. H., & Keshtiaray, N. (2013). Global citizenship education and its implications for curriculum goals at the age of globalization. *International Education Studies*, *6*(1), 195.

CHAPTER THIRTEEN

Alternative Reality

Gamifying Your Classroom

REMY DOU

INTRODUCTION

You all know them: the Gamers. The XBOXers. The Playstationers. The PCers. The students in your classroom who are as likely to be discussing the most recent video game release as they are gossiping about the latest celebrity rumor. Times have changed. While once students grouped themselves by what kind of music they listened to (i.e., rappers, rockers, punks, etc.), these days children often group themselves by the kinds of games they play or the type of gaming system they use to play those games (Comulada, Rotheram-Borus, Carey, Poris, Lord, & Arnold, 2011).

The "gaming" phenomenon is hard to ignore. For classroom educators, a typical day includes "catching" students playing games on their laptops. If students do not own laptops, you might catch them playing on their mobile gaming device or on their cell phones. It is a ubiquitous occurrence. And contrary to popular belief, girls are nearly as likely to play video games as boys (Entertainment Software Association, 2012).

Games themselves have changed. I remember growing up and looking forward to the next Super Nintendo game release. Once the cartridge sat snugly in the console, my siblings and I would spend hours playing. Depending on the game, the whole thing could be over in about 10 to 20 hours of (sometimes continuous!) play. Nowadays, players can spend months or years on a game, always exploring new facets (e.g., *The Elder Scrolls* and *World of Warcraft*).

And siblings are no longer required. Gamers can play with others around the globe using an Internet connection. Game makers today focus increasingly on the social connectivity of games. This is evidenced by the fact that most top-selling games have strong social components, often facilitated by a PC or console wirelessly connected to the Internet (Entertainment Software Association, 2012).

How can educators compete with the likes of *Angry Birds*, *Candy Crush*, or whatever the newest craze is these days? Games engage students on many levels: their fast-moving colorful images, their instant feedback, the sense of accomplishment they provide, and so on. While some educators are disdainful or apprehensive of games and/or Gamers, this should not lead to discouragement. Rather, games present opportunities for learning. What is it about games that can capture, in class, the full attention of the child who typically cannot sit still for a minute?

The principles of gaming, which we will explore, can teach us how to keep our students engaged. By using these principles to redesign the classroom, we can reach students in ways the traditional classroom structure does not (Gee, 2003). In order to understand why this is so, let us take a closer look at the gaming world.

BACKGROUND

In 2012, the Entertainment Software Association released its annual "Essential Facts" report, providing some statistics about the computer and video game industry. These will help put gaming's ubiquity in perspective:

- The average American household owns at least one gaming console.
- 47% of Gamers are female.
- 62% of Gamers play with other Gamers in the same room.
- In 2011, consumers spent nearly $25 billion on the games industry.
- The average game player's age is 30.

Much can be gleaned from this data. It tells us that a high percentage of people play games, that males and females are almost equally involved, that Gamers come from all age brackets, and that gaming is largely a social event. Video games have rooted themselves in our culture.

Cognitive research has greatly increased our knowledge about the way people learn, and, surprisingly, much of it lines up with the way good games are designed (Gee, 2003). For example, we now know that self-efficacy and learning significantly improve when students are provided with projects or assignments that are neither too easy nor too hard, but rather are challenging to the point where growth occurs without frustration. In other words, students learn best via activities through which they can expect outcomes that accurately reflect their effort

(Sherer, Maddux, Mercandante, Prentice-Dunn, Jacobs, & Rogers, 1982). This is not unlike Vygotsky's zone of proximal development (National Research Council, 2000b). Similarly, good games tailor a user's experience to match his or her skill level. Some games can make difficulty-level adjustments as a player's skills increase or decrease (Holland, Jenkins, & Squire, 2003). Games provide the "fruitful order" of problem complexity that cognitive science tells us leads to mastery (Salen, Torres, Wolozin, Rufo-Tepper, & Shapiro, 2011).

The use of obstacles is a familiar tool used in making games interesting. The idea sounds counterintuitive, but, in fact, working through complex challenges in a game world can be inherently satisfying (Holland et al., 2003). Many are familiar with *Pac Man*, a game where a player's main goal is to run around collecting tiny pellets. In and of itself, this concept does not sound like much fun. To change that, game makers added a maze through which players must carefully navigate in order to collect these pellets. Still, that was not fun enough. What makes this game really engaging is that while players maneuver through mazes collecting pellets, they are also being chased and thwarted by enemies seeking to terminate them. Each time a player gets past a particular level, having collected every pellet, the enemies in the following level become more relentless in their pursuit. Clearly, challenging but surmountable obstacles make games fun (McGonigal, 2011).

Open, inquiry-based learning strategies greatly benefit diverse audiences partly because students participate in producing knowledge rather than consuming it—active as opposed to passive learning (National Research Council, 2000b). Good games often give students opportunities to design their own worlds and/or make their own decisions in order to learn from the effects of those decisions. This kind of experimental encounter, which is common to scientific practices, provides fertile ground for higher-order thinking (Salen et al., 2011).

Behavioral research shows that a person's self-identification can affect the outcome of his or her career choice (Bhattacharyya, 2008; Brickhouse, Lowery, & Schultz, 2000). STEM (science, technology, engineer, math) educators who hope that some students pursue STEM-related careers must not only prepare them content-wise, but also help reshape their perspective of what a "STEMist" looks like. Games provide flexibility for players to refine their perception of themselves and others by allowing users to create new identities or redefine their own. By personifying a scientist in a game, students may be more likely to feel comfortable transferring that identity to other areas of their lives (Gee, 2003).

Collaboration is also a component of video game design. Collaboration is crucial for increased learning among a diverse group of students and is an inherent part of many—if not most—games. Studies show that minority students who develop habits of studying and working in groups achieve greater academic success (Gillman, 1990). This is not just true of minority students. Gaming challenges provide common ground on which students can come together to collaborate and learn.

James Paul Gee, Kurt Squire, and others have written extensively on these and other gaming principles that promote learning and engage players (Gee, 2003; Squire, 2003; Shaffer, 2006). Most, if not all, can be applied to the classroom. The following list narrows these principles down to a few of the most basic themes for successfully "gamifying" your STEM classroom.

- Games promote active learning rather than passive experiences.
- Games allow players/students to build and reflect on various identities and choices.
- Games promote success through collaboration.
- Games offer players/students multiple pathways for success.

Quest to Learn

Quest to Learn (Q2L) is a public school in Manhattan that serves grades 6–9 and whose overarching curriculum was developed in large part by game designers. Similar to other schools, Q2L expects students to acquire knowledge directed by a standards-based curriculum, but the way in which content is delivered differs. Rather than teaching "units," teachers lead their classrooms on "missions." Missions can last about 10 weeks and vary in scope. Instead of activities and homework, student teams participate in "quests" directed at fulfilling missions. At the end of every mission, students in the school are confronted with an epic task called the "Boss Level" (Quest to Learn, n.d.).

This may sound surreal to many—like an imaginary world—but real and concrete learning takes place. Students on quests can often be found "gathering resources," "creating models," "designing tools," performing "scientific experiments," or "analyzing text," all of which are scientific habits of mind. Q2L is designed so that its graduates can solve complex problems and make strategic decisions based on analysis of evidence. According to the Q2L website, "Quest is not a school whose curriculum is made up of the play of commercial videogames, but rather a school that uses the underlying principles of games to create highly immersive, game-like learning experiences" (Quest to Learn, n.d.; Salen & Zimmerman, 2003).

So, what have its outcomes been? Q2L is already ranked in its district's 97th percentile in terms of student engagement. The question many are awaiting the answer to is whether or not schools like Q2L increase students' standardized test scores. So far, students at Q2L have shown the same level of testing success as those at traditional schools, although full curriculum implementation has yet to occur. Q2L will continue adding one grade level each year so that by 2015 it encompasses grades 6–12. While it will be interesting to analyze students' learning gains at Q2L in the coming years, examining the strategy by which the school effectively engages students is of equal consequence.

Game designers are well aware that properly designed games can keep Gamers motivated, even late into the night. This level of engagement could greatly benefit student learning. While engagement is not the only necessary piece for academic success, it is a large component of behavior transformation (Lent, Brown, & Hackett, 1994). The principles of game design can alter the course of a classroom and the way students think. Gamifying the classroom is more than just a STEM thing; it is a learning thing.

Interview with a Gamer: Dr. Shannon Mortimore-Smith

Inspired by James Paul Gee's 36 principles of learning found in games, Dr. Shannon Mortimore-Smith asked herself a question: "What if I actually designed a course that ran much more like a game?" Dr. Mortimore-Smith, a professor at Shippensburg University in Pennsylvania, took the risk of converting her classroom into a game. The results intrigued her, to say the least. Sitting outside the conference building at the National Council of Teachers of English 2012 Annual Conference, she expounded on what took place in one of her undergraduate courses (Mortimore-Smith, 2012).

Having taught high school for 7 years, she understood the importance of providing engaging learning experiences to her students, many of whom came from impoverished backgrounds. "In order to learn, they want someone to make it meaningful to them," she said (Mortimore-Smith, 2012). While working on her doctorate, she found that teaching literacy by incorporating pop culture, particularly graphic novels, added meaning to students' learning. Though she felt somewhat skeptical at first, it turned out her method served as an effective tool to harness student interest while giving birth to complex content discussion surrounding classroom concepts. She took this teaching strategy a step further into the world of gaming, where the goal of good games is also to provide meaningful experiences (Salen & Zimmerman, 2003).

The *Zelda* and *Final Fantasy* series had offered Dr. Mortimore-Smith rich entertainment when she was in her 20s; she had played these games avidly. The communities she encountered and built in gaming worlds became significant to her, something non-Gamers may find odd. These relationships provided emotional rapport and opportunities to learn from others and to work together. She reasoned that if games have the capacity to evoke memories, friendship, and collaboration, then their design principles should be able to nourish the same in a school setting.

She approached her superiors at Shippensburg and suggested she teach an elective course implementing this creative concept: English 336—Theories and Approaches to Teaching Language Arts and Literacy. Somewhat surprised by the lack of resistance from her department chair, Dr. Mortimore-Smith went on to structure her course like a game. She "gamified" her classroom.

Her course semantics changed, as did many other things. Students no longer worked on tasks or assignments; they collaborated on quests. There were smaller quests within larger quests, which were part of three major learning pathways students could choose from. With each completed quest, students earned experience points. Anywhere from 1,800 to 2,000 experience points garnered them an "A" in the course, and there were many ways to accomplish this.

The planning process itself was like a game. Over a period of about a month and a half, Dr. Mortimore-Smith created quests and respective rewards for her students. She drew primarily from the learning principles outlined by Gee, a learning and linguistics researcher whom she admired. Three principles in particular stood out: Active Critical Learning, Multiple Routes, and the Psychosocial Moratorium Principle.

Successful games allow students to learn by doing. Dr. Mortimore-Smith understood that Gee's Active Critical Learning Principle states just that. Students learn by being involved in content practice. Unsurprisingly, this is similar to the literature on open inquiry that permeates every fiber of the *Next Generation Science Standards*—the practice of science. Science practices lend themselves to a second concept—the Multiple Routes Principle—in that it encourages a variety of avenues for students to progress through an activity, not all of which are required for success. But more than that, the Multiple Routes Principle gives students opportunities to structure their learning trajectory. For Dr. Mortimore-Smith, this was a way of "designing pathways for students" (Mortimore-Smith, 2012). Not knowing exactly how students would choose to present task-achievement created a sense of adventure and novelty.

If a student failed to pass a quest, he or she had the opportunity to retry. This is a component of most games (the ability to "try again" or "continue") and is part of the Psychosocial Moratorium Principle. Students feel encouraged to take risks because of reduced consequences of failed learning tasks. Rather than forcing students to prove content mastery by performing on a single assessment, they have opportunities to assess themselves, reconstitute, and retry, thereby acquiring mastery.

By the end of the course, students had grown socially connected and had developed skills to think critically—not only about classroom content, but also about their own learning. Students responded well, saying things such as "this was the first class where I was allowed to have an opinion" (Mortimore-Smith, 2012). They learned the value of collaboration as it pertains to success attainment, a value shared by the STEM community at large. Yet the wheels of change turn slowly, especially at the university level. Despite sharing her success with colleagues, Dr. Mortimore-Smith found it difficult to help others in academia to view these strategies as more than juvenile.

Not all students displayed eagerness or interest at the beginning of the course. Dr. Mortimore-Smith encountered some resistance and anxiety, primarily from students uncomfortable with asking questions or directing their own learning. They wanted to be told what to do. Yet this kind of prescriptive teaching method, wherein students are expected to follow a list of synthetically processed procedures with preordained outcomes, takes away from the development of the critical thinking skills they will require for success in a rapidly changing technological world (National Research Council, 2012).

"Learning is exploration; it is discovery," says Dr. Mortimore-Smith (Mortimore-Smith, 2012). The basic design of good games, such as the ability to probe and reflect on actions, can transform the way students engage with subject content. Introducing new approaches like this one can be challenging, particularly in light of the teacher performance pressures of high-stakes testing. "But if [teachers] are willing to play with the curriculum in more meaningful ways, they could meet those objectives" (Mortimore-Smith, 2012).

BEST PRACTICES

Introduction to Gaming Principles: Adam Mortague

Early on in my career, I implemented a simple, game-like activity as part of my unit on genetics. This was the premise:

> Adam Mortague, an incredibly gifted young researcher-turned-entrepreneur, is an expert in the field of genetic engineering and nuclear cloning. Thanks to him and his team's research efforts, drought-resistant strains of rice, corn, and wheat have been brought to various stages of development, bringing relief to ailing countries where soil is arid. Moreover, species once thought to be on the brink of extinction are now finding hope in the methods developed by Adam and the experts at the Cloning Advancement Institute of Nevada (CAIN), which he founded. Adam also killed his clone. It is the year 2057.

I played this weeks-long game every year with my biology students. In the game, students took on one of the following roles: defense attorney, prosecutor, or judge. Their job helped convict, defend, or decide whether or not Adam Mortague committed murder when he terminated his clone, Abel, in order to harvest organs for his own survival. To complicate matters, Abel was in a state of induced paralysis that began prior to his birth. Adam argued that the clone was nothing more than a set of organs developed in a metaphorical petri dish.

While the concept of the game radically introduced children to the world of bioethics, it also served as a platform on which to pursue learning on biotechnologies, cloning, cell replication, genetics, protein synthesis, and much more. Although the game sounds more like a movie than reality, it was plausible enough for students to accept the illusion and the premises of the case. But there was more to it.

Over the years I developed a series of official documents used to lure students to investigate. The documents ranged from Adam Mortague's criminal and medical records to presidential memos and the contract signed between CAIN and Abel's surrogate. For weeks, students engaged in acquisition of content knowledge, presentations, evidence analysis, and discussions about the case. A lot of this took place without teacher prompting.

At the end of the unit, we held a trial. On that day, I was no longer Mr. Remy. I sat in the defendant's seat, playing the role of Adam Mortague. The judges were in charge of the classroom and its management. With an agenda in front of them, they commenced the trial, which lasted about 45 minutes, with each side taking turns presenting evidence and asking questions. When it was all over, the judges walked into a separate room to deliberate, something they took very seriously, and returned with a verdict. Some students cheered in celebration of their victory; others scowled, knowing they put their hearts into their work but did not get the result they expected. In the following class, we debriefed and discussed the case.

The Case of The People v. Adam Mortague was a game I used to engage my students in ways that sitting in class filling out worksheets or organizing presentation slides did not. Like all good games, it had characters, a setting, challenges, a system of value, and imaginary boundaries. The experience seared itself into students' minds. Years later, after graduation, when they come to visit or see me out in public, they mention Adam Mortague; they mention the game we played, and the trial. They always remember what role they played and whether or not they won their case. They also seem to remember some of the content-related concepts I hoped would stick. As an educator, it fills my heart to see glints of excitement in my students' eyes as they reminisce years after our interactions. They learned and, most important, were interested in learning.

Using Rewards Systems

Another example of "gamifying" my classroom involved a rewards system similar to those used in video games.

At the beginning of the school year, students in my high school biology courses read a sign on the wall that stated: "One SmartyBuck = Candy, Two SmartyBucks = Homework Pass, Three SmartyBucks = Extra Credit." Part of my first-day-of-class spiel included a description of what these were and how to earn them. I specified that classroom participation, show of effort, and/or outstanding academic success would earn each of them a SmartyBuck. I made a conscious effort to award them fairly, without prejudice toward any student regardless of academic standing, gender, disability, or race. Students quickly realized that earning SmartyBucks resided within their reach and were worth the effort. SmartyBucks motivated even those without strong track records of classroom participation.

Elementary teachers know exactly why SmartyBucks worked. They use these kinds of rewards systems with their young students all the time. We assume that at some point children grow up, and those systems no longer have the capacity to motivate. Yet in actuality they continue to be effective through adulthood, though the systems' manifestation may differ in maturity and complexity.

Developing a classroom environment based on the principles of gaming, such as the use of rewards, is an effective way to engage students, but there is more to it than giving out candy in return for participation. Creating the ambience that generates interest and investment is what we as educators should be after. That can be achieved through a deeper understanding of the learning principles of game design.

STRUCTURING CLASSES LIKE A GAME

The time has come to convert your classroom environment into a game-like oasis—an alternative reality. Your students will find themselves interested in achievement, and you will enjoy exercising your creativity and expertise. From the second students walk into your classroom, they will be part of the "game": you are the rule-keeper and they are the players. You are also the game designer. As students navigate their way through your course, they will face obstacles (e.g., projects, tests, assignments, tasks, etc.). They will overcome these obstacles, score as many points as possible, and win the game.

For successful classroom transformation, consider the following more like a set of guidelines than a rigid blueprint.

The Avatar

All games have a main character that Gamers control. Be it a simple paddle, as in the case of *Pong*, to a complex persona, as in the case of Commander Shephard from *Mass Effect*, the avatar is the digital ambassador through which players interact with game features. For the classroom, avatars do not need to be digital, but that is an option. Avatars can be represented by the roles students take on. The benefit of having students create avatars lies in the interaction between students' identities and their game/avatar identities, which can elicit reflection on and development of STEM identities. That is to say, students who do not normally identify with STEM professionals and STEM careers can use their roles in a game as a way of exploring those self-schemas. This can be particularly helpful for students from diverse cultural backgrounds (Foster, 2008).

Students may be given the opportunity to select their roles from a list of possibilities. The options should include STEM-based identities. For example, students in my botany classes had the option of being engineers, designers, supervisors, or public relations specialists. Role terminology can be adapted according to grade level. In addition, you may want to include non-STEM roles that appeal to students who struggle to connect with STEM. These can include artists, audiovisual experts, writers, and so on. Introducing students to STEM through roles with which they already identify may help ease early anxiety. As you define and select student roles, keep in mind that their effectiveness is dependent on their meaningful contribution to team-based learning tasks.

The avatar, or character, that students develop in your classroom will serve different purposes. These include a conceptual leveling of the playing field. All students should begin at the same point and have the same potential for growth. This can be particularly helpful for students coming in with poor academic records, since it gives them a chance to start over (National Research Council, 2011). Affording a low-ranking student the opportunity to associate with a STEM-based role may be the first time he or she considers that a realistic possibility.

In some cases, it may benefit students to visualize their avatar. Doing so can allow for options to customize avatars' physical appearance. Not only can this create stronger student associations with the avatar identity; the process can also be used to help keep track of skills students have learned. For example, certain avatar accessories (e.g., clothing, jewelry, and markings) can demarcate particular skills learned. Students may gain access to certain accessories only after they have demonstrated concept mastery in one or more areas. Additionally, by measuring growth in terms of skills learned (as opposed to grades received), students focus more on their strengths than their weaknesses. Dozens of free online tools exist for developing characters and avatars; one I recommend is Face Your Manga (http://www.faceyourmanga.com).

Challenges

Challenges are the heart of an alternative gaming environment. Whatever they are called, be they quests, missions, adventures, treks, or levels, they are the obstacles students have to overcome in order to progress through a game. In their most basic form, challenges include all the assignments presented throughout the course (i.e., take-home work, activity sheets, group work, exams, projects, etc.). Challenges come in all shapes and sizes.

With the advent of the *Next Generation Science Standards*, foundational changes to the way science is taught will continue to take place, as will changes to the assignments we fashion for students. Students should be "doing" science, an exercise in which learning is more about generating knowledge than absorbing it through lectures or textbooks. This goes beyond hands-on labs and taps into the heart of science: identifying a problem, structuring investigation, gathering evidence, analyzing evidence, developing conclusions, and sharing knowledge

(National Research Council, 2000a; National Research Council, 2012). These are the activities students should be mastering. They are also inherent to the structure of gameplay, where players take active roles and analyze game feedback to reflect on progress (Squire, 2003).

There are free resources to help facilitate the "practice" of science. Government data sets, be they federal or state, present an opportunity for students to perform genuine data analysis. These include Global Information System (GIS) records, satellite imagery, and all sorts of other measurements. Federal agencies such as the National Aeronautics and Space Administration and National Oceanic and Atmospheric Administration provide accessible data sets that are sometimes accompanied with educator guides and/or lesson plans. A quick online search should be all it takes to get you started on these. You can also try starting at Data.gov (http://www.data.gov). The goal is to facilitate student use of evidence in argument formation and sharing of knowledge.

In addition to data sets, myriad citizen science projects engage students in work that impacts the scientific world. Endeavors like the *Monarch Larva Monitoring Project* and *Zooniverse* provide a place for students to contribute to growing bodies of research. For a website with a collection of citizen science projects, check out *SciStarter*. These can be used to provide motivation for mastery. The more students believe their contributions have meaningful impact, the more likely they are to put effort into understanding content.

As a final suggestion, competitions can also be used to engage students in meaningful scientific practices. The *Siemens We Can Change the World Challenge*, *eCybermission*, the *National STEM Video Game Challenge*, and *FIRST Robotics* are examples of engaging, student-centered STEM competitions. They motivate students to think critically about their participation, build on prior knowledge, collaborate, design, build, measure, analyze data, and share knowledge.

Additional details and resources on data sets, citizen science projects, and competitions are located in the appendix to this chapter.

Designing student challenges and trajectory should occur with forethought. In games, overcoming obstacles not only presents players with immediate rewards, but also provides a sense of approaching a specific target or reaching a particular goal. Recreating this ambience will require creativity and—even more—reflection on the purpose of the assignments solicited from students. Often, class work and homework are assigned without much thought as to how one piece relates to another in a given chapter or unit. Sometimes assignments are simply given to meet grade quotas. While external pressures may make it difficult to plan out assignments ahead of time, designing them so that they scaffold student knowledge while leading to content mastery will better facilitate a gaming environment, not to mention learning.

A final word on this topic: the activities that emerge from gaming environments should include independent activities as well as group tasks. Individual assignments that build student understanding can be used as gateways for more challenging group tasks. The goal should be to combine individual work with robust group activities in such a way that students perceive an explicit learning direction: concept mastery.

Choice and Teamwork

In an ideal setting, teachers can create an environment where students can choose from a variety of missions in order to personalize their learning trajectory. Tasks can be structured in an open format in which multiple activity pathways exist or can be developed, each of which will earn students a particular number of points. Think of it like a restaurant menu: students with a budget (i.e., the number of points required to pass) choose from a list (i.e., menu) of activities of differing values that will meet their budget. This process allows for more personalized learning and task ownership. Some tasks may require individual, independent work, while others may be designed for collaboration and sharing. Students should be able to decide on the path they will take to ensure that they earn enough points by the time a particular chapter or unit has ended.

Collaboration should be encouraged, since it is one of the best outcomes of social gaming. Teamwork does more than teach children how to work together; it is a proven learning method (Gillman, 1990). STEM research is very much like that as well. Rarely do people work in isolation. Research often requires a team of experts and students in different fields to make significant discoveries. Classroom activities should be designed to replicate the same process.

Bosses

In a world where the classroom is a game, capstone assessment activities such as exams and end-of-unit projects would be granted the status of "boss level." In any game, the boss level presents a significantly taxing challenge to test the skills players have accumulated. In the classroom, boss levels are the culmination of activities and assignments related to a particular unit or chapter in a course. They could be anything from a classroom presentation to a portfolio-like submission, depending on how your course is designed. By the end of each section, students should get the sense that all activities and assignments (i.e., missions) have been leading to the boss level.

Point System

The point system is a key aspect of any game, and it should play an important role in any course designed to feel like a game. The point system creates a major source of feedback for players. By keeping track of how many points they have earned, players can trace their growth or determine whether playing strategies should be altered. To meet goals, players are often required to accumulate a certain number of points. The number of points required to achieve success is often obvious. Classroom grading systems are not designed to be so intuitive.

The point system does not assert any judgments. It is simply a testament of a fact: you earned this many points. Ultimately, it can be argued, points and grades are the same: they are both ways of making quantitative judgments on the quality of various activities. The difference lies in the behavioral implications of each system. Point systems tend to promote positive outlooks, since the number of points students earn is always increasing, thus motivating them to focus on a very specific goal (i.e., acquiring enough points to pass a course). In traditional grading, by contrast, letter grades can fluctuate and make it difficult for students to visualize and/or mathematically calculate what is required of them to successfully complete a unit or course.

To be true to the principles of gaming, points should be assigned to a variety of activities according to their level of difficulty, and students may earn them only if they completely meet the requirements of each activity. In other words, points can only be earned when activities are done correctly. This kind of pass-fail system can be intimidating and discouraging unless students are given many opportunities to resubmit assignments in the same way that games allow players to retry certain levels or obstacles. Although lowering the consequences of failure may sound counterintuitive to our thinking about student motivation, in fact it encourages students to take more positive risks with regard to their learning (Gee, 2003). While the burden on the educator assessing activities more than once may sound overwhelming, efficiently laid-out assignments, clear grading rubrics, and the effort exerted by students wanting to improve will provide relief. Additionally, the concept of multiple submissions can be a motivator for designing activities that solicit unique results from students, thus minimizing levels of plagiarism.

For the sake of remaining compatible with a school's existing grading system, the number of points earned on assignments can correlate with relative letter grades. For example, students successfully completing 70% of quests or missions would receive 70% of the maximum number of points; according to traditional grading systems, this would be the equivalent of a "C" (wherein A = 90–100%, B = 80–89%, C = 70–79%, D = 60–69%, and F = 0–59%). Units or chapters taught in class can be designated as "levels" or "stages" of a game, respectively. Students may be required to earn a particular number of points in each stage within a predetermined period of time in order to move on to the next, and the same could be true of levels. With forethought, teachers can effectively designate the total number of points needed to pass a course.

Points make it possible to motivate students with more than just the allure of successfully completing an assignment. Points can be used to grant special game-related privileges. Student avatars may be allowed to change in appearance or skill level based solely, or in part, on the number of points he or she has earned. Also, accruing a certain number of points could earn students the right to participate in special missions or activities, be designated as classroom leads, or receive other rewards.

Atmosphere of Gaming: Nomenclature

Breaking students out of the school paradigm and into a gaming atmosphere will also require creative nomenclature. It is hard to imagine being part of a fun game if tasks are called homework, worksheets, and exams. Just hearing those words might disanimate students. The trick is to apply words that stimulate the imagination. The terminology does not require an expert wordsmith, nor must it be perfectly synced in theme and context, but it should be purposeful.

Tasks like classroom assignments or in-school activities can be called "missions," "explorations," "expeditions," "achievements," "treks," or many other terms that provide a sense of doing something exciting. While these can take the place of school terms like class work, homework, assignments, tests, quizzes, and projects, keep in mind that activities should aim to engage students with the practice of science rather than perpetuate information regurgitation (see "Challenges," above). Students or student teams can be given titles descriptive of the work of STEM professionals, such as "Engineers," "Designers," "Data Gatherers," and "Content Experts." Themes could also play a role in naming components of the classroom environment. In a physics course, terms like "quantum," "vector," "Einstein," and "particle accelerator" can inspire creative team names or activities. Biology teachers can use words like "folding," "fieldwork," and "viruses." The use of content-related terminology will not only be a continual source of terminological exposure to students; it can also serve as a mnemonic device.

Proper game terminology does not need to be a perfect science, but it will help shape the atmosphere for entering students. Although students are smart enough to know that a rose by any other name is still a rose, it is the sense of illusion, the suspension of disbelief provided by honoring new nomenclature, that will allow them to engage in learning quests.

Atmosphere of Gaming: Boundaries

Even in simple games, players have to accept and understand the boundaries of a game. Take, for example, "freeze tag," where the concept is to run away from whoever has been assigned the role of "it." Those touched by the player who is "it" must stand as still as possible until a teammate comes to their rescue. In reality, unless players are zapped with yet-to-be-created freeze-ray guns, no one actually freezes. For the duration of the game, players suspend their disbelief. When frozen players move or cross the boundaries of this game (in other words, cheat), others express their disdain. If cheating continues to occur, then player cohesion and acceptance of imaginary boundaries disintegrate. The game is no longer fun.

To keep games fun, their pretenses must be kept up. In the case of game-like learning environments, the same applies, but the imaginary rules are different. Many of these rules are based on players' roles. In the classroom, players will keep up the roles of captain, adventurer, scientist, social media manager, and so forth, and will be accountable to each other and the teacher for the responsibilities of their roles. When someone "cheats" or is not fulfilling his or her duties, steps should be taken to ensure that other students do not become discouraged. One way of resolving this tension could include the use of promotion, demotion, or probationary status as player designations. For example, the "designer" who failed to do his or her part of a project may be asked to reconsider the role he or she plays, switch teams, and/or participate on a different assignment.

Fairness is a big part of game play. Rules can be presented or developed in conjunction with students in order to maximize acceptance. Once they are established, it is important for the teacher or facilitator to enforce those rules. This includes ensuring that all students "play" correctly, but it also means that educators must abide by the rules established. Lending credence to the imaginary world and encouragement for suspension of disbelief must be sustained.

Are You Ready? A Basic Outline

By now you should be feeling excited and have some great gaming ideas to implement in your classroom. You may be wondering, "Okay, are there step-by-step procedures I can follow?" If you have been keeping up, you will not be surprised by my hesitation to provide them. My belief is that the very act of thinking purposely and creatively about implementing these concepts in your classroom generates ownership of the project. Few other things are as critical to the success of this type of endeavor.

Nevertheless, guidance can help. Below are a few questions you may want to consider as you develop your alternative reality:

- Have students been assigned appropriate roles with which they can identify?
- Are student roles clear and meaningful? Do they encourage collaboration and leadership?
- Do students have a way of visualizing their roles and tracking progress?

- Have tasks been designed to encourage critical thinking skills such as evidence analysis, formulation of conclusions, and data communication?
- Are challenges purposeful and clear? Do they help scaffold learning? Is concept mastery their ultimate goal?
- Do students get to collaborate often?
- Are students given grade level-appropriate options from which they can choose performance expectations and a personalized learning trajectory?
- Are points commensurate with challenge difficulty?
- Have students been given rubrics that make clear what successful task completion looks like?
- Do students have multiple opportunities to depict mastery?
- Has classroom nomenclature changed? Do terms pique the imagination?
- Are teacher- and student-designed boundaries that reinforce proper comportment in place within the alternative reality?

Not all of these questions need to be addressed for successful implementation. It probably goes without saying that different teachers in different schools have different needs. Classroom management and policies vary across states, districts, schools, and classrooms. A one-size-fits-all model would fail to take that diversity into account. It would also reduce the level of emotional and creative investment that educators place in the game. For this concept to work, an educator should be invested in its success. Understanding the goals and basic components of game-based classroom design allows anyone to structure courses around existing resources, interests, and needs.

ADDITIONAL THOUGHTS: MMOGs

Other gaming phenomena such as massively multiplayer online games (MMOGs), which capture the attention of millions of individuals every day, are worth taking a closer look at in the context of STEM education. MMOGs are not just simple single-player games like *Bejeweled* or *Angry Birds*; these games immerse users in immense, complex, alternate worlds. Over 12 million people play *World of Warcraft* (*WoW*), one of the most popular MMOGs (Entertainment Software Association, 2012). In this game, players are represented by avatars belonging to either the Orc race, called the Horde, or the human race, called the Alliance. Players interact with objects in this world, battle enemies, join guilds, buy and sell items, text or live chat with other players, design clothing, and engage in a host of other behaviors. Located on the virtual planet Azeroth, cities and territories exist that would require thousands of hours of gameplay to fully explore. It is not uncommon to run into players who spend months' worth of hours in this environment. With even its own economy, *WoW* practically stands as an alternate universe.

Azeroth and other virtual worlds found in MMOGs are not just play spaces. Scientific conferences and even grant review panels have been held in these settings. In 2008, sociologist William Sims Bainbridge, who spent years studying and writing about these worlds, put together an international science conference in the *WoW* game space. For three days, 120 scholars from Australia, Europe, North America, Russia, and other parts of the world entered Azeroth to attend plenary sessions, participate in expeditions, and share information (Bainbridge, 2010). Additionally, Bainbridge, who also works at the National Science Foundation (NSF) as a program director, uses *Second Life* to create a secure space in which to bring together expert panelists to review grant proposals. Reviewers follow traditional protocol with the distinction that everything occurs through a virtual world, including a discussion, which saves NSF thousands in travel costs (Bohannon, 2011).

These worlds further the development of scientific thinking habits. Players occasionally create online forums to support game play in which they develop "systems-based reasoning" and "knowledge construction" (Steinkuehler & Duncan, 2008; Squire & Steinkuehler, 2005; Bainbridge, 2007). Collaboration, which is often crucial to the success of scientific endeavors, is also a strong component of player mastery in these worlds. With all the potential benefits for STEM learning that can take place in these virtual environments, further analysis of what works and does not work inside of MMOGs is worth pursuing (Steinkuehler, 2004).

CONCLUSION

While reading the questions in the section "Are You Ready? A Basic Outline," you may have gotten the impression that those questions had more to do with proven, successful classroom structures than gaming principles. Indeed, the purpose of this chapter has been to depict how both align very well. The principles woven into the fabric of a successful game mirror those fashioning an engaging classroom environment, and vice versa. Participant engagement and learning experiences engross both.

Much remains to be learned about "gamifying" STEM classrooms. Current studies tend to over-generalize results, but the time is ripe for more investigation. Basic research on its effectiveness with particular groups of students would greatly advance its implementation. Grasping how students transition from doing activities because they are fun, to doing them for inherent worth, will also contribute to our understanding of how games help learners engage and increase their STEM content knowledge. Motivation and learning, in this case, are two sides of the same research coin (Foster, 2008). Incorporating meaningful play is a way to keep students engaged and therefore maintaining high levels of cognitive development. If a student is not engaged, that student is not learning.

If "gamifying" an entire course seems like an overwhelming endeavor, consider "gamifying" a portion, such as a unit or chapter. I anticipate that you will notice greater student engagement and be motivated to expand this pedagogy to other lessons and activities. While the road ahead may seem uphill, consider the fun you will have employing your creativity and professionalism—no more business as usual. The instinct to play games resides in all of us, and gaming is as much about learning as it is about having fun.

APPENDIX

Sample resources for data sets:

- NOAA Education Resources (http://www.education.noaa.gov)
- My NASA Data (http://mynasadata.larc.nasa.gov)
- Quantitative Environmental Learning Project (Math and Environmental Sci.) (http:// www.seattlecentral.edu/qelp)
- Basic Local Alignment Search Tool (BLAST) (http://blast.ncbi.nlm.nih.gov)

Sample resources for citizen science activities:

- See Appendix A in Chapter 13 of this book, "Integrating Informal STEM Learning into Your Classroom," for a list of citizen science activities.

Sample resources for STEM competitions:

- National STEM Video Game Challenge (http:// stemchallenge.org)
- Siemens We Can Change the World Challenge (http://www.siemens-foundation.org/en/programs/the-siemens-competition-in-math-science-technology/)
- eCybermission (http://www.ecybermission.com)
- Plus an existing variety of STEM Bowls, Olympiads, and Fairs

WORKS CITED

Bainbridge, W. S. (2007). The science research potential of virtual worlds. *Science, 317*(5837), 472–476.
Bainbridge, W. S. (2010). Virtual worlds as cultural models. *Transactions on Intelligent Systems and Technology, 1*(1), 3:1–3:21.
Bhattacharyya, G. (2008). Who am I? What am I doing here? Professional identity and the epistemic development of organic chemists. *Chemistry Education Research and Practice, 9*, 84–92.
Bohannon, J. (2011). Meeting for peer review at a resort that's virtually free. *Science, 331*(6013), 27.

Brickhouse, N., Lowery, P., & Schultz, K. (2000). What kind of girl does science? The construction of school science identities. *Journal of Research in Science Teaching, 37*, 441–458.

Comulada, W., Rotheram-Borus, M., Carey, G., Poris, M., Lord, L., & Arnold, E. M. (2011). Adjustment of trendy, gaming and less assimilated tweens in the United States. *Vulnerable Children and Youth Studies, 6*(3), 263–275.

Entertainment Software Association. (2012). *Essential facts about the computer and video game industry.* Retrieved October 11, 2013, from http://www.theesa.com/facts/pdfs/esa_ef_2012.pdf

Foster, A. (2008). Games and motivation to learn science: Personal identity, applicability, relevance and meaningfulness. *Journal of Interactive Learning Research, 19*(4), 597–614.

Gee, J. P. (2003). *What video games have to teach us about learning and literacy.* New York: Palgrave/MacMillan.

Gillman, L. (1990). Teaching programs that work. *Focus: The Newsletter of the Mathematical Association of America, 10*(1), 7–10.

Holland, W., Jenkins, H., & Squire, K. (2003). Theory by design. In B.A. Perron (Ed.), *Video game theory.* London: Routledge.

Lent, R. W., Brown, S. D., & Hackett, G. (1994). Toward a unifying social cognitive theory of career and academic interest, choice, and performance. *Journal of Vocational Behaviour, 45*, 79–122.

McGonigal, J. (2011). *Reality is broken: Why games make us better and how they can change the world.* New York: Penguin Books.

Mortimore-Smith, S. (2012, November 17). Interview. (R. Dou, Interviewer).

National Research Council. (2000a). *How people learn: Brain, mind, experience, and school* (expanded ed.). Washington, DC: National Academies Press.

National Research Council. (2000b). *Inquiry and the National Science Education Standards: A guide for teaching and learning.* Washington, DC: National Academies Press.

National Research Council. (2011). *Learning science through computer games and simulations.* Washington, DC: National Academies Press.

National Research Council. (2012). *A framework for K–12 science education: Practices, crosscutting concepts, and core ideas.* Washington, DC: National Academies Press.

Quest to Learn. (n.d.). Curriculum. Retrieved November 16, 2012, from http://q2l.org/curriculum

Salen, K., Torres, R., Wolozin, L., Rufo-Tepper, R., & Shapiro, A. (2011). *Quest to Learn: Developing the school for digital kids.* Cambridge, MA: The MIT Press.

Salen, K., & Zimmerman, E. (2003). *Rules of play: Game design fundamentals.* Cambridge: The MIT Press.

Shaffer, D. W. (2006). *How computer games help children learn.* New York: Palgrave Macmillan.

Sherer, M., Maddux, J., Mercandante, B., Prentice-Dunn, S., Jacobs, B., & Rogers, R. (1982). The self-efficacy scale: Construction and validation. *Psychological Reports,* 51(2), 663–671.

Squire, K. (2003). Video games in education. *International Journal of Intelligent Simulations and Gaming, 2*(1), 49–62.

Squire, K., & Steinkuehler, C. (2005). Meet the gamers: They reseach, teach, learn, and collaborate. So far, without libraries. *Library Journal, 130*(7), 38.

Steinkuehler, C. A. (2004). Learning in massively multiplayer online games. In *ICLS '04: Proceedings of the 6th International Conference on Learning Sciences* (pp. 521–528). Pittsburgh, PA: International Society of the Learning Sciences.

Steinkuehler, C., & Duncan, S. (2008). Scientific habits of mind in virtual worlds. *Journal of Science Education and Technology, 17*, 530–543.

University of Minnesota. (2010). *Home.* Retrieved December 21, 2012, from http://www.mlmp.org

Zooniverse. (2012). Retrieved December 21, 2012, from http://www.zooniverse.org

CHAPTER FOURTEEN

Using Whiteboards to Create a Student-Centered, Collaborative Classroom

BUFFY CUSHMAN-PATZ

INTRODUCTION

A Whiteboarding Classroom

It is 12:50. The bell for C-block is still ringing as the last students walk through the door. They rush to the board at the front of the room to see which problems are left or what their partners have chosen. They then quickly head to their tables, drop their backpacks, and check in with their group-mates: "What are we working on?" "Which one did you pick?" or "How did you get that?" they ask. They grab markers, quickly pull out their homework, and jump into further conversation with their partners.

It is algebra class, and these students are whiteboarding.

Within the next 10 minutes, they will work in their groups of two or three to present their solution to one of last night's homework problems. Meanwhile, I will walk around the room, looking quietly over each group member's shoulders to check his or her homework for completion. Sometimes the group members stop me to ask a quick clarifying question; sometimes they are too engrossed in their work to notice me. (I leave the answer key open at the front of the room so they can check for themselves the "Is this the right answer?" queries. The answers are not what whiteboarding is about. The *process* is, so letting them have access to them does not spoil anything.)

They take their roles seriously. They know that by choosing the problem they have chosen, they are taking responsibility for helping their classmates solve this problem and others like it. They know that if there is widespread confusion, I will step in to help, but otherwise the task is up to them. They know their classmates will have real, probing questions for them, so it is in everyone's best interest for them to be clear and thorough in their explanations, including organizing their work neatly on the whiteboard.

After I have made the rounds, I call out, "Finish up your whiteboards in one minute." As they complete their whiteboards, they bring the boards to the back of the room and set them down, move to their tables, and pull out the supplies we have agreed they need for whiteboarding presentation sessions: completed homework, colored pen for editing, and their original source material (e.g., textbook, worksheet) from which they are all working, open to the relevant page.

By the time I reach my seat at the back of the room, the first group is usually up at the front of the room with their whiteboards set in the tray of the wall-mounted chalkboard, and the three presenters are standing next to it, waiting for the silent cue from me. The presenters read the original question and then explain to their classmates how they solved it and what their group consensus was about the answer. The audience members compare their own homework to the solutions presented and ask the presenters questions about differences they see. My role is to sit, watch, and listen. This is their process, for their learning, and I try to be just a "fly on the wall."

Once the group has finished its presentation and the other class members have had a chance to compare their work and ask questions, the audience applauds the presenters, and the next group heads up to the front to repeat the process.

Whiteboarding as a Practice

I was originally exposed to the concept of whiteboarding at a Modeling Physics workshop I attended at Arizona State University in Tempe. As in all the best professional development, we learned by doing. We spent the 3-week workshop learning the content in the same way we would later help our students learn it, so that we could experience it through the eyes of a learner. During that workshop I was able to feel what it was like to be a student in a whiteboarding classroom, thanks to an expert teacher-facilitator committed to developing us as a community of learners. I felt the way that whiteboarding (as well as the modeling approach) democratized the classroom and empowered me as a learner—even in a room full of people who, I was sure, knew more about the content than I did.

I came back from that summer's modeling workshop so committed to creating a whiteboarding culture in my classroom that I decided to try to implement it in all of my classes at the time: seventh-grade Math and Chemistry (in addition to Physics). Later, when I began teaching Algebra, I incorporated it from Day One. I also spread the gospel to my colleagues in other disciplines—English and history included—and within a year 2–3 whiteboards could be found in several classrooms throughout the school. The practice of whiteboarding changed my classroom, changed my relationship with my students, and enabled student learning and depth of understanding at a level I did not know was possible until I experienced it myself.

This chapter documents what I learned along the way and the system I developed through years of refinement, so that you can implement it in your classrooms, too.

BACKGROUND

Decades of education research have spoken volumes about how people learn. Our role as educators is to create learning environments for our students that take into account what we know about the conditions that enhance and enable learning. A successful whiteboarding classroom puts into practice many of the principles we know are true about student learning.

Designing the Learning Environment

First and foremost, in a whiteboarding classroom, students have the opportunity to talk about the content they are learning and teach each other. They have the chance to collaborate with other students, test out their ideas, and then present them to an authentic audience. Teachers, in turn, have the opportunity to use these formative assessments to learn what their students have learned and, if necessary, adjust instruction. Whiteboarding plays a critical role in the series of steps necessary for students to retain information and to truly learn it.

Learning through application and repetition

"The more elaborately we encode information at the moment of learning, the stronger the memory," reports Medina (2008). The practice of whiteboarding can play an important role in creating meaningful and contextual experiences in the early stages of the learning process.

Wiggins and McTighe, the authors of *Understanding by Design*, describe the cyclical nature of well-designed learning experiences: "part to whole, whole to part—that is how we come to understand and use our knowledge" (2005, p. 251). They describe the cycle as one familiar to any coach or performer—the movement back and forth, from content to performance and back again, from discrete skill to strategy and back again (Wiggins & McTighe, 2005). Whiteboarding as a classroom practice plays a critical role in this cycle by providing students an opportunity to "perform" skills. Repeated whiteboarding sessions within a unit of content allow students to encode information in meaningful ways consistent with using this cycle.

Students talking

In *Ready, Set, Science!* the authors describe the importance of student talk:

> In order to process, make sense of, and learn from their ideas, observations, and experiences, students must talk about them. Talk, in general, is an important and integral part of learning, and students should have regular opportunities to talk through their ideas, collectively, in all subject areas. Talk forces students to think about and articulate their ideas. Talk can also provide an impetus for students to reflect on what they do—and do not—understand. (Michaels, Shouse, & Schweingruber, 2008, p. 88)

Whiteboarding provides multiple opportunities for talking, first in small groups as students prepare their whiteboards, and then as a larger class community during whiteboarding presentations.

Students teaching each other

A decade's worth of data about peer instruction (PI) shows the positive impact of students learning from and teaching each other. Peer instruction, as described by Crouch and Mazur (2001), who implemented PI in physics courses at Harvard University, is a process of peer teaching that "engages students during class through activities that require each student to apply the core concepts being presented, and then to explain those concepts to their fellow students" (p. 970). They found that through peer instruction "students' grasp of the course material improves according to a number of different measures ... both during class, and when tested for retention at the end of the semester" (p. 970).

Whiteboarding provides opportunities for students to collaborate and teach each other—not once, but twice within a class session. Students peer-teach during the stage of the whiteboarding process when they create their whiteboards; they then re-teach the material, even if they have only just learned it from a peer, during the presentation of the whiteboards to the whole class.

Learning in a safe classroom climate

In *How People Learn*, the authors describe the attributes of learning environments that must be cultivated, noting that "learning is influenced in fundamental ways by the context in which it takes place" (Bransford, Brown, & Cocking, 2000, p. 25).

> The norms established in the classroom have strong effects on students' achievement. In some schools, the norms could be expressed as "don't get caught not knowing something."

Others encourage academic risk-taking and opportunities to make mistakes, obtain feedback, and revise. Clearly, if students are able to reveal their preconceptions about a subject matter, their questions, and their progress towards understanding, the norms of the school must support their doing so.

Teachers must attend to designing classroom activities and helping students organize their work in ways that promote the kind of intellectual camaraderie and the attitudes towards learning that build a sense of community. In such a community, students might help one another solve problems by building on each other's knowledge, asking questions to clarify explanations, and suggesting avenues that would move the group towards its goals. (Bransford, Brown, & Cocking, 2000, p. 25)

Formative assessment to inform instruction

The *Understanding by Design* process emphasizes the regular use of formal and informal assessment, noting the ability of formative assessment or "assessment-in-progress" to "ferret out the apparent from the genuine understandings" (Wiggins & McTighe, 2005, p. 247).

Teachers in a whiteboarding classroom are able to use whiteboarding sessions as formative assessment tools to learn what their students have learned and to adjust instruction.

Meeting next generation standards

Common Core State Standards (CCSS), *Next Generation Science Standards* (NGSS), and 21st Century Skills Standards all recognize that student learning is not only about content, but also about students' ability to deeply understand content, transfer it into new settings, and communicate it to others. These new standards are based on much of the same research referenced here. Whiteboarding is a strategy that can help implement some of these best practices while helping students become proficient at the performance expectations associated with new standards.

The *Common Core State Standards for Mathematical Practice* describe "varieties of expertise that mathematics educators at all levels should seek to develop in their students. These practices rest on important 'processes and proficiencies' with long-standing importance in mathematics education." These include:

1. Make sense of problems and persevere in solving them.
2. Reason abstractly and quantitatively.
3. Construct viable arguments and critique the reasoning of others.
4. Model with mathematics.
5. Use appropriate tools strategically.
6. Attend to precision.

7. Look for and make use of structure.
8. Look for and express regularity in repeated reasoning. (*Common core state standards*, 2010)

Whiteboarding, as a classroom practice, overtly provides students with opportunities to construct viable arguments and critique others' reasoning. Further, it creates rich opportunities to attend to virtually *all* of the other mathematical practices, especially with well-thought-out questioning practices from the audience of the students' peers and teacher.

The eight practices of science and engineering, identified by the *Framework for K–12 Science Education* (National Research Council, 2012) and incorporated into the NGSS, are similar in scope to the CCSS Math Standards. The practices of science and engineering are:

1. Asking questions (for science) and defining problems (for engineering)
2. Developing and using models
3. Planning and carrying out investigations
4. Analyzing and interpreting data
5. Using mathematics and computational thinking
6. Constructing explanations (for science) and designing solutions (for engineering)
7. Engaging in argument from evidence
8. Obtaining, evaluating, and communicating information. (National Research Council, 2012)

Again, whiteboarding provides rich opportunities for practices 6, 7, and 8, and has the potential to play a role in the other practices, as well.

BEST PRACTICES: HOW TO ADAPT WHITEBOARDING TO YOUR CLASSROOM

The goal of whiteboarding is to enable students to take ownership of their own work and also to play a role in helping their fellow students learn. This requires an intellectually safe classroom culture where collaboration is valued over competition, and where the success of individuals depends on the success of the group, and vice versa. For many students (and many teachers!), this is a shift that takes time to develop. Effective and efficient whiteboarding in a smoothly running classroom, like the one I described in the introduction, requires clear roles and expectations for students (and teachers), introduced early and referred to often.

What I describe below are the details of the practices I developed over a 5-year period as I implemented whiteboarding in my mathematics classes with an 80-minute block schedule. Even though I use—and love—whiteboarding in other classes, I wrote this chapter looking through my lens as a mathematics teacher, because the transformation in my teaching practice and in the classroom environment was significant as a result of whiteboarding.

Whiteboarding can be used in any classroom, in any discipline; the processes and practices described can be modified as needed. Specific suggestions for modifications are included in the next section of this chapter.

[**Note.** Throughout the rest of this chapter I will use the word "whiteboard" as both a noun and a verb. *Whiteboard* (*noun*): A two-foot by three-foot (by 1/8-inch) sheet of melamine or similar substance with a surface that can be marked on with dry-erase markers or crayons. *Whiteboard* (*verb*): The process of writing or drawing on whiteboards, and/or presenting the written-on whiteboards.]

Building the Culture

In a smoothly running whiteboarding classroom, after the systems have been developed and the collaborative culture has been fortified, the students, in essence, run the class by themselves. The teacher's role becomes the "guide on the side" in a dynamic learning community.

It is important to establish norms and guidelines for the various steps of the process: creating the whiteboards, presenting the whiteboards, and being an audience member for whiteboarding presentations. Each of these steps will take practice, and it is important to evaluate the class's success as you go along and, if necessary, modify the guidelines appropriately or discuss specific strategies to adhere to the expectations.

Embedded in the following sections are sample sets of guidelines about the different steps of whiteboarding that my students and I developed over time. I share these with new groups of students at the beginning of a course, and they provide a clear set of expectations for how we will develop a whiteboarding culture. At first, students (and teachers) will need regular reminders to stay within the established guidelines; over time, the guidelines will develop into standards of practice for the classroom as a community of learners. Persistence is key. Consistency is essential. It will take patience—with your students and with yourself—as you collectively develop the habits of a whiteboarding classroom.

Creating Productive Student Groups

Developing student groupings before class increases the efficiency of the whiteboarding experience. The number of groups ideally ranges from four to seven, with two to four students per group, but depending upon class size, teachers may need to create more or larger groups. For the sake of time and audience patience, it is best to have no more than six whiteboard presentations per class period; the exact number your class can tolerate will depend on your students and the length of your classes.

Whiteboarding provides excellent opportunities to create mixed groups so that there are both teaching and learning opportunities within each group. I make sure that each group has a student who has demonstrated mastery of related topics and who will likely be able to successfully teach some of his/her peers, and that each group includes a student who is challenged by the topics and may benefit from working with peers with a more thorough understanding of the ideas. The whiteboard creation phase provides excellent opportunities for student-to-student teaching in small groups. Students are especially good at explaining concepts using language that is accessible to their fellow students.

Regular rotation of students within groupings allows them to work with a variety of their classmates over time. It also provides relief for less successful groupings. I prefer to rotate student groups weekly. Students know when they walk in on Monday that they will be getting new group-mates, and they look first thing to the board to find out who their group-mates are. Classroom seating arrangements are not permanently fixed; groups sit together within the classroom for ease of collaboration.

Selecting Problems for Students to Whiteboard

What students whiteboard about is actually the least important part of the whole whiteboarding process and is entirely dependent on the subject material and instructional goals. One method of selecting whiteboarding topics is described here; another is described at the end of this chapter in the "Modifications" section. There are infinitely more variations, which I encourage you and your students to explore. In my mathematics class, I had students whiteboard homework or classwork questions, and I describe how to replicate that process here.

Allow students to whiteboard questions or problems they have already had time to think about individually or in groups, either as homework or as classwork. To facilitate students getting to work as soon as they enter the classroom, write the problem numbers on the board before class begins and let the student groups select problems they want to whiteboard as soon as they enter the classroom.

One significant consideration in selecting problems is to aim for each student group to work for relatively equal amounts of time to prepare their boards. Depending on the problem sets, this may manifest in any of a number of ways:

- If all problems are fairly similar, each group can prepare a whiteboard about one problem.
- If there are multiple problems that are solved similarly, some groups may be responsible for all (or some) similar problems.
- If there is a particularly long investigation, multiple groups may present different parts of the investigation (e.g., Group 1 works on #5a while Group 2 presents #5b and #5c), ideally after checking with each other.
- If the homework or classwork set included more problems than there are groups to prepare and present (e.g., the homework had 10 problems but there are only 6 groups), include the problems that are most critical for students to understand or review. As an alternative, list more problems than there are groups to solve them. Then you can either encourage student groups who finish their preparations early to take on additional problems, or create whiteboards yourself for those problems. (As long as this is reserved for rare occasions, it can be fun for the teacher to pop into the lineup of presentations right along with the students.)

In light of all these options, it is critical to know the mathematical challenges that may crop up so that you can anticipate relative lengths of time it will take groups to prepare and present the problems.

Choosing Their Own Problems

Teachers who thoughtfully select a variety of problems from which students can choose empower students to then choose those problems that they feel most comfortable explaining and presenting.

Whenever possible, I let student groups choose without any interference. However, there are times when I circumvent this. If there is a particularly challenging problem, and a group who I think would be best suited to handle it, I let them know I think they are up for a challenge and ask if they will take on the tough one. Sometimes this gets negotiated with a promise of extra help from me during preparation, but this works out just fine in the end. Even if I help them through a problem they were completely stuck on themselves, *they* are the ones who end up teaching it to the rest of their classmates. This way they learn the concept even better during the process of teaching it, even if they had not worked it out on their own to start with.

If there are more groups than problems, allow more than one group to sign up for the more challenging questions/problems, so that you can either have both groups present and compare solution methodology, or draw straws right before presentations to choose which group will present (being sure the group that does not present today gets a chance to present tomorrow).

Preparing Whiteboards

Once class begins, student groups have approximately 10 minutes to prepare their whiteboards for presentations. The details on how their group wants to approach the preparation are up to them. Every group has a slightly different approach. Only in extreme situations do I intervene; it is important that students take ownership of their learning and understanding during this phase. The parameters are clear: they have a limited amount of time, and they have resources available to them (their completed homework, their textbooks, their notes, and their group mates). They have a clear "product" that they are working to produce: a presentation to their peers that will enable them to understand the problem.

Some groups have a clear leader who delegates roles to the group members. Many groups pull out their homework and compare answers first, check disparate answers with the answer key available to them, then decide how they want to prepare their board. Some groups begin writing immediately, then discuss, using their board for visual cues. All of these are effective approaches, and it is revealing to watch students as they learn to work with new (as well as familiar) partners and better develop a sense of their own learning needs and approaches. As the year goes on, students become more willing to assert their own needs and desires for effective learning. If necessary, I coach students on effective collaboration strategies and language.

Guidelines for Whiteboard Creation

- Think about what information is essential and what is not. Include only essential information!
- Show all relevant and non-trivial steps of your work.
- Use board space efficiently and organize your work logically.
- Write neatly, clearly, and large enough so that someone in the back of the room can see your board.
- Include appropriate figures and drawings relevant to the problem.
- Use color effectively: write words in easy-to-read colors. Consider color-coding your work to distinguish different parts.
- All group members are responsible for understanding every part of what is on your whiteboard, and any of you may be asked questions about the material, so make sure you understand and approve of everything on your board before you present.

Sidenote: Checking Homework

In this 10-minute period during which students are working together to prepare whiteboards, I circulate the room to check students' homework for completion.

Students reference their homework as they are preparing their whiteboards, so it is already lying on the table or desk. A quick glance for completion and effort takes only a few seconds; it is possible to record each student's completion rating (e.g., on a clipboard) and move on to the next group, often without disturbing the students' collaborative work session. Circulating while students are working also allows for listening to students' work processes, as well as making you available in case students have pressing questions they can't work out within their groups.

Preparing for Presentations

Getting students to stop preparing their whiteboards and transition to becoming audience members for presentations can be challenging, especially when student groups are productively engaged in the work. It is helpful to give warnings (e.g., 2 minutes left; 1 minute left) before asking students to stop preparing their boards. Using an automated timer, especially one that students can see as they work, may also be helpful.

When whiteboard preparation time is over, use a set of procedures to signal this shift. Students must put markers down, move their marker bins back to their storage place in the room, and move their completed whiteboards to the designated place in the room (ideally the back of the room, where they remain out of sight until they are presented). Students then move to their seats and pull out the supplies they need for actively listening to and engaging with the presenters: their completed homework, the source material for the questions/problems so that they can reference what the question asked and what information was given, and a pen to make notes or corrections to their homework as they learn from the presentations. (I require students to use a color pen so that these changes can be clearly distinguished—by them, most importantly, but also by me—from their original work.)

Presentations begin once the audience is seated and prepared. The teacher, acting as an audience member along with the students, takes a seat near the back or side of the classroom. Transitioning from preparing whiteboards to preparing for presentations should use no more than 2 minutes of precious class time, so students need plenty of warning and clear systems to guide them.

Giving Presentations

Presenting groups stand at the front of the room with their whiteboards (ideally set on an easel or in the tray of a wall-mounted board) and describe the question or problem they are presenting and the method and thought process they used to approach it. Presenting students should reference their boards as they speak, but speak directly to their audience (their peers, not the teacher), making eye contact and using good presentation skills such as appropriate volume, posture, and so on. All presenters in a group should have a role in the presentation. Initial descriptions of the boards and the process they used to solve the problem should take no more than 2 minutes. The remaining time is used for questions and answers.

Guidelines for Whiteboard Presentations

- Explain how you thought about the problem and why you decided to solve it the way you did. (This is the most important part of your presentation!)
- Briefly explain the process and/or steps you used. Explain any possible variations that other students may have (or that you found within your group).
- Expect questions. Be willing to listen carefully to other ideas or methods for solving the problem, and compare them to your own.
- If, while presenting, you determine that revisions or corrections are in order, note them by drawing a line through the original work and writing new/revised information above or beside it (ideally with a different-color pen). Mistakes are instructive; please leave them on the board! Also, it may confuse your audience if the answer changes, and they happened to be looking at their own papers while you made the change.
- Stand up straight, speak loudly and clearly, and be succinct.
- Be sure that all group members have a significant role in the presentation.
- Be conscious of where you are standing—make sure all your classmates have a clear view of the board.

Asking Questions

The audience members' objective during presentations is to understand the presenters' method of solving the problem at hand in relation to their own method for solving it. In some cases it will be the same; in many cases it will be different. The primary role of audience members during presentations is to ask questions that elucidate the various ways to approach a problem, and for the class to come to a consensus about the correct answer(s) to the problem at hand, despite the various approaches for finding it. The secondary role of audience members is to check their own homework answers against this group consensus once it is achieved, and to notate their work accordingly.

Developing students' questioning skills is a key component of the presentation stage of the whiteboarding process. The intent of this whiteboarding process is to create a culture of collaboration, with students learning from each other rather than competing against each other.

A basic rule for audience members during presentations is that they only ask questions; they may not make declarative statements. For some students (and teachers!), it is a challenge (a necessary one) just to turn statements into questions. For example, instead of saying "That is not what the question asked for," a student might ask, "Are you sure that is what the question was asking for?" A next-level challenge, once students have made a habit of asking questions, is to make sure those questions are open-ended, with no implication of right or wrong. One example of this would be when a student might ask, "Can you clarify what the question was asking for?" Another example of a progression, as a student develops his/her technique, might be, "You did not do the second step correctly," to "Are you sure your second step is right?" to "Can you explain again how you did the second step?"

When students are able to consistently engage in non-judgmental questioning of one another, they are able to drop their defenses and open their minds to new ideas and new ways of doing things. They begin to stop worrying so much about who is right and focus more on understanding and helping each other understand.

Guidelines for Whiteboard Audience

- You are responsible for checking the work on the board and comparing it to your own finished work. If your work disagrees with the presenters' work, ask questions of the presenters to understand their reasoning and methods. (There is no guarantee that the presenters' answers are right, so it is to everyone's benefit for you to ask them clarifying questions!)
- You may only ask questions; you may not make declarative statements.

Instead of:	*Ask:*
You didn't put a unit on your answer.	What are the units for your answer?
You got 15. I got 8. Is my answer okay?	Can you explain how you got 15?
You wrote the question wrong.	Can you clarify the question asked?

- You are responsible for making any necessary revisions/additions to your own work/homework in colored pen.
- Applaud (clap for) all presentations once everyone has had a chance to check the work and ask questions.
- You may not make changes to your own whiteboards once presentations have started.

Guiding Q&A Sessions

The role of the teacher during presentations is to listen as much as possible and to intervene as little as possible. The presenters, and not the teacher, should call on audience members who have questions. The students should self-correct their questioning approaches (to the extent possible) if they slip into making statements.

Students are responsible for paying attention to the presenters during their presentations, as well as other questioners, such that if an audience member asks a question that has already been answered while (s)he was not paying attention, the other students (rather than the teacher) will point that out to him or her. Again, the overarching goal is to give the *students* the primary responsibility for their learning experience.

In some circumstances it is necessary for the teacher to intervene. If the discussion has become overly circular or chaotic, and the presentation is confusing a majority of audience members rather than helping them, the teacher should intervene with a directed question that clarifies. If the presenters have given an incorrect solution or methodology, and the audience's questions have not helped the presenters to successfully revise their boards, then the teacher must ask a question that draws attention to the source of the error. At any point that the teacher needs to intervene, (s)he should use questioning techniques consistent with the expectations for student questioning.

The teacher may also need to guide the conversation in one of several ways. If only one presenter is responding to audience questions, the teacher may follow up the next question by asking a particular presenter to respond. If students are mispronouncing content vocabulary, the teacher may need to provide the correct pronunciation. (Usually this is necessary only once or twice before students take on the role of helping each other to correctly pronounce words.) If an audience member is not adhering to the guidelines set forth, or is distracting classmates or the presenters, the teacher may need to address an audience member directly. Teachers should limit themselves to one intervention per presentation.

Finally, the teacher serves as the ultimate timekeeper. Although (s)he may delegate the formal timekeeping role to a student helper, the teacher may need to extend a presentation's Q&A segment when it is especially beneficial to all students, or may need to speed up a presentation when time is running short. The teacher has the overarching pacing goals in mind, and (s)he must budget the precious class time accordingly.

Finishing Presentations

After each presentation, the designated timekeeper or teacher cues the audience by saying, "Let's thank the presenters." Audience members applaud in recognition of presenters' energy and efforts. Presenters return their whiteboards to the back of the room and set them down in their designated spot. Whiteboards can be returned without being erased. This is both for efficiency and for reference in case any further questions come up during the class period about the solutions presented. Presenters then sit down with their classmates and become audience members for the next groups.

Assessing

The purpose of whiteboarding is, of course, learning: learning content, learning to collaborate, learning to explain one's thinking, and learning to understand others' thinking. In a successful whiteboarding session, the teacher is able to hear all students in the class use content-specific language in context and get a sense of their thinking processes. This is formative assessment at its purest, as it in*form*s the teacher about student learning and gives him or her plenty of in*form*ation about where students—individually or collectively—are having difficulty or are showing improvement or understanding.

Grading the whiteboarding process itself may be counterproductive to its goals. We want to encourage *learning*, not a grade, as the primary motivation for whiteboarding. Teachers and students both learn during productive whiteboarding sessions, and this learning can be transferred to other potential assessment situations. I prefer to use whiteboarding as a primarily informative assessment situation and let students experience the learning potential in the low-stakes, collaborative culture that the whiteboarding classroom creates.

If it is necessary to "grade" students during whiteboarding, here are a few ideas for how to do so while still remaining consistent with whiteboarding's philosophy.

1. Develop a rubric, consistent with the guidelines that you lay out, by which students may self-assess their presentations. Alternately, let peers assess each other's presentations using the rubric, and/or based on their understanding of the concepts after the presentation.
2. At the end of the week during which they worked together, develop a rubric by which students may assess each other's collaborative work skills and contributions to the group.
3. Allow students to improve their homework scores by turning in corrections or clarifications they made to their own work (clearly noted in colored pen as different from their original work!) by listening to the presentations.
4. Allow students who attempted all problems and showed all work (as confirmed by the teacher while students were preparing their whiteboards at the beginning of the class session) to earn a homework score of "exceeds proficiency" if the student determined, by listening to the presentations, that (s)he answered all homework questions correctly.

Improving the Culture

The first segment of the Best Practice section was titled "Building the Culture," and this final segment is about culture as well. An intellectually safe, collaborative culture is the foundation of any successful whiteboarding classroom, and creating one is an iterative, ongoing process. It will take constant work—and constant learning and improvement along the way.

As you are transitioning to a student-centered classroom from a teacher-centered one, hold yourself and your students to the guidelines laid out throughout this chapter, even if it feels hard or unnatural at first. Be consistent with your students and yourself. Pay heed to the pitfalls described below. You will find that you will change as a result of developing these practices into a habit. And your students will change. So will your relationship with your students and their relationship with the content.

Whiteboarding works when the students feel safe and comfortable enough to share their ideas and their thought processes with their classmates and with the teacher. This safety develops from being asked (and asking) open-ended, non-judgmental questions. It develops when negative language ("No, that's not right" and "Don't do it that way") is replaced by positive suggestions ("Let's see how you got that" and "Let's try this a different way"). And it develops from being a part of a learning community where students know that it is not only okay to be wrong, but that significant learning can come from mistakes.

How will you know if your learning community is succeeding? How will you know how your students feel about the whiteboarding culture? Ask them! After a few weeks (or months) of implementing whiteboarding, spend 10 minutes at the end of a class instituting a "Plus/Delta" session, in which you encourage students to give (out loud, in written form, or digitally) feedback in the form of Plusses—things that are working well and that they like—and Deltas—things that they would like to see improved upon or changed, and suggestions for how. Plus/Deltas are an excellent opportunity to give students an even greater voice in the classroom community, but it is essential that the teacher be willing to respond to the Deltas and implement changes and to take ownership of his or her own role in shaping the culture.

Modifications

What I described above are the step-by-step details of how I used whiteboarding to transform my mathematics classes from dull, practice-and-repeat, teacher-centered lessons (even I was bored!) to student-centered, *fun*, investigative approaches to learning and teaching mathematics. But whiteboarding can also be used in a variety of science, technology, engineering, and math (STEM) classes, and in more ways than have been described so far.

Two other particularly effective ways to use whiteboarding are Board Meetings and Gallery Walks.

Board meetings

A variation on one-group-at-a-time whiteboarding presentations is the "board meeting" or "meeting in the round" concept. In this instance, all groups stand in a circle and present their findings or solutions, usually to the same problem or challenge, at the same time. This allows students to compare notes with each other and discuss similarities and differences in their data, their responses, and their approaches to solving and/or presenting. Board meetings are an especially good way to build consensus and to develop the concept of a community of learners. In chemistry class, we used board meetings for students to compare how they classified a variety of substances as Matter or Not Matter, and then used the collective responses as fodder for discussion about the properties of matter, allowing students to develop their own working definition of matter. In physics class, we used board meetings to compare data collected from experiments—for example, position versus time data for a cart on a ramp—and students were able to compare the shapes of their graphs even if they had not used the same slope or time intervals for their data collection.

Gallery walks

Gallery walks can be used in a variety of ways, but are particularly useful when students are examining similar or related content. In this case, student groups prepare whiteboards, then place them in a part of the room where they can be easily accessed by groups of people standing nearby. All groups set up their whiteboards simultaneously in different parts of the room, and class members circulate ("walk") the room ("gallery") examining the various boards. Depending on the task, one group member may stay stationed at the group's board to explain the board to viewers.

Gallery walks are useful after students have brainstormed on a topic in small groups. Groups can then circulate to learn what other groups brainstormed and to push their own thinking on a topic. Gallery walks can also be a useful extension of the jigsaw cooperative learning technique, in which a large chunk of material is subdivided and each group takes ownership of sharing a section of the material with classmates. Whiteboards can facilitate presentations of what individuals or groups learned, and that they now need to share with their peers.

Pitfalls to Avoid

Developing a smoothly functioning whiteboarding classroom takes time, and the process continuously evolves as the students and teachers do. The practices I describe here are those that I developed over years of working with students and developing my whiteboarding classroom. The more I work with students, the more I learn and the more ideas I have for improvement. The same will be true in any classroom.

There are many potential stumbling blocks in the process of developing a highly functioning whiteboarding classroom. Here are some of the ones I have encountered or have seen other whiteboarding teachers encounter, along with my suggestions for avoiding or adjusting them.

Students splitting the work without collaborating

Like all of us, students look for shortcuts and time savers when possible. For some student groups, this amounts to splitting the work of preparing a whiteboard into sections, and each student taking responsibility for a part of it. If the students within a group do not collaborate on their parts, this can lead to presenters making statements like "I don't know, that wasn't my part," or discovering when their whiteboard is being presented that they disagree with the work that one of their classmates has done.

One way to avoid the "divide and conquer" mentality among students is to allow only one dry-erase marker per group. This encourages student groups to discuss their ideas collectively before deciding what will be written on the whiteboard. Once students have established this as a practice, which may take a few weeks or months, it may then be possible to allow more markers per group for efficiency, to encourage students to discuss before writing.

Another way to ensure that students take collective responsibility for the work on their whiteboards is to make sure that all presenting students respond to questions from the audience. Ideally, the presenting students learn to self-regulate this process and take turns responding to questions. However, if one or more presenting students dominates the questioning session, then the teacher may need to gently redirect audience questions to the silent group member(s). As a general practice, each presenting student should make at least one substantial contribution to each presentation. Even group members who were previously absent can contribute by reading/describing the problem at hand. Having been absent does *not* disqualify students from being able to contribute to their group during discussions or presentations!

Reverting to a teacher-centered classroom

There are two common traps that ensnare teachers who are still transitioning from a teacher-centered classroom to a student-centered one: invading students' whiteboard space, and allowing students to direct presentations to the teacher rather than to their fellow students.

Invading Students' Whiteboard Space. The goal of whiteboarding is to give *students* ownership of their work, their knowledge, and their understanding. To fully give them this control, teachers must treat the whiteboards almost as a part of their bodies. (We would not straighten a student's shirt for them or fix their belt—we would, instead, tell them about a problem we notice and have them fix it themselves.) We should do the same for whiteboards and consider the board as students' property. The most disempowering thing a teacher can do is walk up to a whiteboard that students are presenting and mark on it or take control of the conversation by taking over their physical space.

A teacher should intervene on a whiteboard presentation only if all student questioning has been exhausted or the presenters have become confused themselves to the point of being entirely unhelpful to the class's understanding of a concept. In such a case the teacher should *ask* the presenters if he or she may come up and help. Likely they will say yes, at which point the teacher can then invite the students back to their seats and lead a brief clarifying discussion. If at all possible, this discussion should use a different chalkboard or whiteboard, enabling the students' whiteboard to remain intact as a point of reference for the teacher-led conversation.

Students Presenting to the Teacher Rather Than Their Classmates. During presentations, presenters should speak to the audience of their peers, not the teacher. Students should make eye contact and interact with their fellow peers during presentations. If necessary, remind students who their audience is. I have often told my students that I already know the content; it is their peers with whom they are in the learning process. It is critical, of course, that you actually mean what you say and let the students genuinely *be* the audience, not you.

Being taken by surprise by the answers (and processes) students present

Not every whiteboard presentation that makes it to the front of the room needs to have perfectly correct answers on it; a lot of learning can come from discussion of incorrect answers. However, it is to your benefit, as the facilitator, and to your students' benefit, as learners, if you have a sense ahead of time of which presentations *must* have more prolific discussions in order to ensure that all students understand the processes.

While students are preparing their whiteboards, use the time to walk around the room and get a sense of the board contents. In some cases it is helpful to talk a group through their challenge and let them correct their board before they formally present to the rest of the class. In other cases, it is useful to clarify with them verbally, but ask them to leave the mistake on the board so that the audience finds the error and asks questions about it. This ensures a fruitful, efficient, student-centered discussion during the presentation. In some cases, it is instructive to let whiteboards be presented with mistakes or errors on them without letting students know that you have noticed them. I would allow errors to be deliberately presented only when the group members were strong, confident students who I knew could handle an on-the-fly reaction to being challenged by their classmates.

Knowing your students' strengths, weaknesses, and sensitivities comes into play in this decision-making process. In groups with a highly sensitive or shy student—especially during the early weeks of whiteboarding, while you are still building the class culture—I prefer to (at least) alert them of (and usually allow them to edit) something that other students may challenge or have questions about, so that they are not caught completely off-guard in front of the audience. In cases of confident students, or students that fellow class members tend to revere as "always right," I am more apt to let them present with a mistake on their board. The culture of the classroom can benefit from strategically allowing "wrong answers" to make it to the front of the room.

Spending too much (or not enough) time whiteboarding

Time is the most precious commodity we have in schools and in our classrooms, and how we spend our time should reflect what we value most. Whiteboarding, an opportunity for students to be collaboratively engaged in talking and thinking about content, is a highly valuable activity and deserves a large chunk of class time, even if that means re-organizing other activities that we had formerly spent class time on. That said, it is important to develop systems and routines that encourage students to be efficient.

Whiteboarding works very well with block schedules. Working with an 80-minute block, our time breakdown (for math class) looked like this:

- 10 minutes—creating whiteboards, teacher checking homework;
- 35 minutes—whiteboard presentations, Q&A, and transitions;
- 35 minutes—teacher-led discussion of new concepts, students practicing those concepts;
- 0 minutes—students choosing problems or writing down homework questions for that night. The problems are posted on the board as students enter the classroom, and can be selected even before the bell rings. Homework is posted on a small section of the board at the front of the room and remains there until the next day; homework is also posted online.

Additional pitfalls

In order to be completely honest about pitfalls, I must admit that there are several challenges that I am still working on and still figuring out the best way to handle. Some of the struggles that continue to challenge me include the following.

Ensuring Students Make the Most of Their Group Time. While I think it is important that students have autonomy in determining how their group interacts during the collaboration/preparation phase, I need to do a better job of introducing them to strategies, practices, protocols, and language that would enhance their ability to interact with their peers in constructive ways.

Enabling Students to Focus on Understanding the Process. Since most students have grown up in a school system in which getting the "right" answer is the end goal, it is hard (and understandably so) to break them of the habit of trying to make sure they have the right answer above all else. Throughout the year, as our classroom becomes more and more of a whiteboarding classroom, the students begin to value more and more the processes and understanding they are experiencing. However, I have noticed that this seems to translate more to how they give presentations and less to how well they listen to them. There is room for improvement in students' listening for understanding during whiteboarding presentations. I am still trying to figure out the best way to dis-incentivize the focus on the "right answer" above all else.

Managing Time Wisely. Ah, time. We never have enough of it, do we? It is a constant struggle to allow the necessary discussions and interactions that whiteboarding brings to a classroom *and* to introduce students to new concepts in the investigative and student-centered way I want to, all in the same block of time. I admit that the 80-minute block schedule I sketched out earlier in this chapter is my ideal; we do not always stick to that schedule as well as I would like to, and we end up needing to make adjustments to fit everything in. I am not sure this will ever get resolved, but it is a good problem to have when students have so much to say about their thinking that I do not want to stop them.

CONCLUSION

Students love whiteboarding. On days when our schedule was shortened, or when we needed to adjust our schedule to review for an assessment (or take one), students would walk into the classroom and see that there were no problems posted on the board and say with sadness and some indignation, "What? No whiteboarding today?" If two or more days passed without whiteboarding, they would whine, "When are we going to whiteboard again?" Students often remarked, as the bell rang at the end of an *80-minute math block,* "Really? Class is over? Time always passes so quickly in here!"

What is it they love so much? I daresay they love learning. They love being in control of their own learning. They love working with their peers and having a sense of control over how they spend their time. They love being able to stand up while they make their whiteboards if they want to, and move around the room a bit when it is their turn to present. They also love drawing with markers!

In our whiteboarding classroom, we were able to create an environment in which learning was fun, even though the work was challenging. As educators, we all know this is possible. We love learning. I share this best practice with you in the hope that it will help you create an environment in which students become as passionate about learning as you are about teaching. As you implement whiteboarding and internalize these practices, *they* will change *you*. And the culture of your classroom will change as a result.

APPENDIX

Table 14.1. Recommended Supplies.

Item	Details	Quantity	Notes
Whiteboards	2' × 3' pieces cut from ⅛" melamine	10–15	• Ideal number of whiteboards is at least twice as many as the number of groups in your largest class. • Large ⅛" melamine sheets are available at Home Depot for under $20; ask them to cut them into six 2' × 3' sheets for you.
Markers	Four-color packs are useful	1 set of 4 colors per group	• It is useful to store the markers in small plastic bins, in sets of four with erasers.
Erasers		1 per group	• Store in bins with markers.
Whiteboard cleaner solution		1 bottle	• Cleaning whiteboards a few times a month is a good task for student helpers, and it helps maintain the boards' lifespans.

(Source: Author).

WORKS CITED

Bransford, J., Brown, A. L., & Cocking, R. R. (Eds.). (2000). *How people learn: Brain, mind, experience, and school*. Washington, DC: National Academy Press.

Common core state standards for mathematical practice. (2010). Washington, DC: National Governors Association Center for Best Practices (NGA Center).

Crouch, C. H., & Mazur, E. (2001). Peer instruction: Ten years of experience and results. *American Journal of Physics, 69*(9), 970.

Medina, J. (2008). *Brain rules: 12 principles for surviving and thriving at work, home, and school*. Seattle, WA: Pear Press.

Michaels, S., Shouse, A. W., & Schweingruber, H. A. (2008). *Ready, set, science! Putting research to work in K–8 science classrooms*. Washington, DC: National Academies Press.

National Research Council. (2012). *A framework for K–12 science education: Practices, crosscutting concepts, and core ideas.* Washington, DC: National Academies Press.

Next generation science standards. (2013). Achieve, Inc. Retrieved May 4, 2013, from http://www.nextgenscience.org/next-generation-science-standards

Wiggins, G. P., & McTighe, J. (2005). *Understanding by design.* Alexandria, VA: Association for Supervision and Curriculum Development.

CHAPTER FIFTEEN

Communicating Science TO Public Audiences Through Media IN High School

Improving Students' Attitudes and Motivations in Science

BERNADINE OKORO

History has recorded the United States as a dominant force in science, technology engineering, and math (STEM) since World War II. However, the last two decades have foreshadowed and witnessed the decline of U.S. students pursuing STEM careers.

As the U.S. math, reading and science test scores have fallen (as evidenced by the Programme for the International Student Assessment [PISA] study), the need to identify, cultivate and nurture the next generation of scientists and engineers who will tackle the nation's grand challenges has risen.

In the May 2010 report, *Preparing the Next Generation of STEM Innovators: Identifying and Developing Our Nation's Human Capital*, the National Science Board defined science, technology, engineering and mathematics innovators as:

> individuals who have developed the expertise to become leading STEM professionals and perhaps the creators of significant breakthroughs or advances in scientific and technological understanding. Their capabilities include mathematical and spatial abilities either alone or in combination with verbal aptitude along with other factors such as creativity, leadership, self-motivation, and a diligent work ethic. (2010, p. 1)

Some of the key recommendations the National Science Board made at the K–12 level were to provide opportunities for excellence, identify and develop all populations of students in all different demographics, and foster a supportive ecosystem.

In light of these recommendations, this chapter will try to infuse the idea that in order to provide opportunities for excellence and foster a supportive ecosystem for science, students must discover an affinity for the natural world around them and document that process. Students, whether they attend schools that are public, parochial, urban, rural, alternative or are homeschooled, all share untapped curiosity about science and how they learn about the natural and engineered world. This chapter invites readers to look at science and its learning through the eyes of students who learned to use their artistic expressions to make commentary about the science concepts they were acquiring. This chapter also attempts to lay the framework for examining the following questions:

- Can art and science converge in high school with a traditional science curriculum and achieve good results?
- How can science educators expand the kinds of opportunities that broaden science understanding to populations of students?
- How can educators increase student motivation and interest in STEM fields?

Communicating scientific concepts to public audiences through media is a strategy used to foster student interest in STEM fields. This strategy could also help students obtain the communication skills desirable for many STEM professions as well as in the workforce.

INTRODUCTION

The Case for Media Literacy in Today's Schools

Participants at the 1992 Aspen Media Literacy Leadership Institute defined media literacy as "the ability to access, analyze, evaluate and create media in a variety of forms" (Aufderheide, 1993). The Center for Media Literacy now cites media literacy as a framework "to access, analyze, evaluate, create and participate with messages in a variety of forms—from print to video to the Internet" (What Is Media Literacy? A Definition ... and More, 2013, para. 1). Media literacy uses the knowledge of the role media plays in society as well as necessary inquiry skills and self-expression needed for democratic citizens.

Formal media education and research in the United States is not presented as a dominant issue in education. In this respect we lag behind other English-speaking countries in the world. One speculation about the reason for this involves the arrival of U.S. media production into other countries. Other nationals, including Australians, Scots, and Canadians have regretted what they perceive to be "American" values filtering into their media systems. They hope that media literacy will make their young citizens more critical and also more appreciative of indigenous programs.

However, such concerns, which provide a driving force for media education, are somewhat absent in the United States (Andersen, 1992).

David Considine, author and professor at the Reich College of Education at Appalachian State University, cited many benefits to implementing media literacy in schools:

- Media literacy is a competency, not a course and therefore interdisciplinary.
- Media literacy is experiential, hands-on and consistent with learning styles.
- Media literacy fosters responsible citizenship in a democratic society.
- Media literacy nurtures group work, cooperative learning and partnership.
- Media literacy is sensitive to stereotyping, bias and multicultural education.
- Media literacy is consistent with the critical thinking skills movement.
- Media literacy has been successful with at-risk students and retention rates.
- Media literacy connects the curriculum of the classroom with the curriculum of the living room. (Considine & Haley, 1999)

Making a Case for Media Literacy in the Science Classroom

The need for media literacy, or methods of media literacy education, is absent or unclear in classrooms outside of English, journalism, and the social sciences. Science teachers, among others, continue to struggle with high student illiteracy rates, new educational technologies, and other conditions that make media literacy seem more like an extracurricular activity. If media literacy had a presence in the curriculum, much could be achieved. Currently there is limited research exploring various classroom uses of mass media by secondary science teachers to address scientific phenomena. The education field is ripe for an increase in the use of mass media to address scientific issues.

Klosterman (2012) cited a possible approach described in a naturalistic study where science teachers' use of mass media can be directed to address socioscientific and sustainability issues. The idea of bringing mass media and science together is not entirely new. Sadler (2009) proposed that socioscientific issues (SSI) are situated learning contexts that engage students in authentic practices, such as hands-on activities and stimulate their moral thinking and judgment about issues that affect society.

Zeidler, Sadler, Simmons, and Howes (2005) argue that SSI education aims to stimulate and promote individual intellectual development in morality and ethics, as well as awareness of the interdependence between science and society. SSIs encompass general ethical dilemmas or personal moral concerns that require students to exploit arguments based on content knowledge, informal reasoning, explicit reflections on relevant epistemological aspects, and personal connections at the micro- (familiar), meso- (state citizenship), and macro- (human race-perspective) levels. In the SSI setting, environmental, economical, political, moral and ethical considerations are required to offer students learning opportunities to prepare them to act as contributors to the life of the society in which they live, or as future active citizens (Bencze, Sperling & Carter, 2011; Mueller & Zeidler, 2010).

Proponents of SSI argue the potential benefits of the use of SSI in the science classroom:

- SSI can nurture students into scientifically literate citizens who can apply evidence-based scientific content knowledge to real world socioscientific scenarios.
- Foster a socially conscious classroom whereby students consistently reflect on their reasoning and its implications.
- Encourage and build argumentation skills needed for reasoning and thinking processes.
- Help facilitate the types of discourses used in the real-world scientific deliberations.
- Promote critical thinking skills such as inference, explanation, evaluation, interpretation, and self-regulation.

Previous studies have shown that SSI approaches may be powerful contexts to enhance key skills as argumentation (Dawson & Venville, 2010; Evagorou, 2011), reflective judgment (Zeidler, Saddler, Applebaum, & Callahan, 2009), informal reasoning (Wu & Tsai, 2007; 2010) and decision-making (Grace, 2009; Gresch, Hasselhorn & Bogeholz, 2011). Despite these results, the extent to which students actually learn science contents through SSI approaches remains unclear (Von Aufschnaiter, 2008). To this concern, although positive outcomes of SSI approaches on students' understanding of science concepts relevant for the addressed issue have been reported (Sadler, Barab, & Scott, 2007; Castano, 2008; Dawson & Venville, 2010), few studies have investigated in depth the relationships between students' science knowledge and the quality of their informal reasoning about a given SSI from the scientific viewpoint (Zohar & Nemet, 2002; Sadler & Zeidler, 2005).

The spectrum of communication technologies that we encounter in our everyday lives is a mismatch to what is encountered in schools. Though schools have modernized and begun to increase the use of educational technology to supplement the curriculum, today's society demands a more expansive, innovative curricula to prepare students to become more productive and efficient citizens who understand and can solve local, regional, and global problems. Communities, parents, and teachers each have a role to play in assisting students in understanding and using new forms of technology. As more educational technologies infiltrate the classroom, such as video gaming and flipped classrooms, the need for teacher training to assist students in using the technology continues to expand. However, strategic planning in education must link the technology in the classroom with student engagement of science outside the classroom. Just as a novelist or playwright raises issues in a play or novel, schools have the potential through media literacy to address the important role that issues play in newscasts, paintings, feature articles, poems, sitcoms, or magazine ads. These aesthetics could be used to show students how to compare and contrast scientific messages and how they are constructed in the media. For some students, there may be a correlation between the amount of media output they consume and their functional illiteracy.

Electronic mass media communicates not through print, but through sound and images. Newspapers are no longer the primary source of information. Additionally, as students engage in electronic mass media, educators have another opportunity to facilitate students as they learn to vet media sources for accuracy in science content.

Society has recognized that media is not just a source of entertainment and information, but an experience that defines lives. There is a connection between students' attitudes and their ability to learn—a positive attitude facilitates learning; a negative one obstructs it. Educational theorists support the notion that students are more attentive to ideas they recognize as directly relevant to their lives; thus they can make meaning of the data (Hynd, Holschuh, & Nist, 2000). If they perceive a topic to be personally relevant to their world, their attention span is enormously increased.

With collaboration from other organizations, educators can find methods to employ the media's connection to other, more pressing societal problems. Images are very powerful. Images, good and bad, have the potential to shape and influence people's lives. By understanding this power and influence, educators—particularly science educators—have an opportunity to help students become more scientifically literate and exhibit more citizenship skills by empowering them to interact positively with their society through media literacy. This empowerment can occur when students realize the possibilities of their interaction and develop the tools needed to interact with their environment.

Achieving scientific literacy is one goal while achieving media literacy is another goal. Educators cannot develop proficiency in teaching reading, mathematics or oral expression in year or two. They learn to convey each subject by studying it for years, teaching it, reflecting upon in and discussing it. Media literacy could be achieved in the same manner. The challenge stems from a perceived lack of media expertise, a lack of media instructional models, and a lack of time needed to create relevant science curricula. However, the hours spent listening to the radio, cruising the Internet, and watching television can provide educators with relevant media expertise. The other challenges that stem from science curricula can be met using experimental instructional methods.

Furthermore, students need to understand different forms of media, because doing so helps them learn. Those media forms convey the content of their culture. We daily consume movies, radio, television, newspapers and magazines well into adulthood. However, there is a need to assimilate the cultural and scientific content, to deal with media-created problems, and to pass along the understanding to all students. With some strategic planning and instructional support, science educators can help students acquire reading skills, functional literacy, and scientific literacy skills that would promote greater self-esteem and achievement in many students. Science educators can also use media to help underserved students get into the mainstream culture. Science educators have a mutual interest to facilitate students' ability to understand and confront society's challenges on a local, regional and global level. Strategies using STEM literacy and media literacy can help students make meaning of the science around them and solve many of their problems.

BACKGROUND

Research from the PISA study suggests a correlation between a student's scientific literacy and his or her motivation and attitude in science learning. The term "scientific literacy" can be placed within a framework that uses personal, social, and global contexts, scientific competencies, scientific knowledge and attitudes toward science (Bybee, McCrae, & Laurie, 2009). Personal, social and global contexts encompass life situations that utilize science and technology. Scientific competencies include identifying scientific issues, explaining phenomena scientifically and using scientific evidence. Scientific knowledge refers to a students' body of knowledge about science, the nature of science, and the natural world. Attitudes toward science refers to the student's interest in science, support for scientific inquiry, and responsibility toward resources and environments. Attitudes about science play an important role in scientific literacy by influencing an individual's interest in, attention to, and response to science and technology. Attitudes in science are included as an important outcome of science education in many education systems around the world; however very few education systems evaluate such outcomes (Bybee et al., 2009).

The National Science Education Standards define scientific literacy as the following: "scientific literacy is the knowledge and understanding of scientific concepts and processes required for personal decision making, participation in civic and cultural affairs, and economic productivity" (Scientific Literacy, 1996).

A scientifically literate person is defined as one who has the capacity to:

- ask, find, or determine answers to questions derived from curiosity about everyday experiences

- describe, explain, and predict natural phenomena
- read with understanding articles about science in the popular press and to engage in social conversation about the validity of the conclusions
- identify scientific issues underlying national and local decisions and express positions that are scientifically and technologically informed
- evaluate the quality of scientific information on the basis of its source and the methods used to generate it
- pose and evaluate arguments based on evidence and to apply conclusions from such arguments appropriately. (Scientific Literacy, 1996)

In the 2006 PISA study (Bybee et al., 2009), scientific literacy is referred to as four interrelated features that involve the following:

- an individual's scientific knowledge and use of that knowledge to identify questions, to acquire new knowledge to explain scientific phenomena, to draw evidence-based conclusions about science-related issues.
- an individual's understanding of the characteristic features of science as a form of human knowledge and inquiry.
- an individual's awareness of how science and technology shape our material, intellectual, and cultural environments.
- an individual's willingness to engage in science related issues and with the ideas of science as a constructive, concerned and reflective citizen.

Norris and Phillips (2003) made an important distinction between the fundamental sense of literacy—which is the ability to read and construct meaning from scientific texts—and the derived sense that is the ability to use such knowledge in new contexts.

Science is a powerful enterprise that can improve people's lives in fundamental ways. Science can also provide a foundation for continued science learning, as well as for the study of other academic subjects. Students who learn to talk with peers in scientific ways, tracing logical connections among ideas and evidence and criticizing ideas constructively, may employ those skills in other areas (Michaels, Shouse, & Schweingruber, 2007).

The goals of science education can be communicated to students with vigor such that students develop interest in and support for scientific literacy. Students can learn to apply their scientific and technological knowledge for personal, social, and global benefit. A goal of science education could also include finding new ways to reinforce to students the role of scientific literacy in a person's attitudes, beliefs, and motivations that influence personal actions. Communicating scientific data to public audiences through media through creation of a scientific magazine for high school students is a project-based strategy that can be used to reinforce for students the role of scientific literacy as well as improve their attitudes and motivations in science.

Research Suggests Scientific Literacy Can Be Beneficial for STEM and Non-STEM Fields

Massachusetts Institute of Technology Engineering professor Richard Larson made the case for widespread STEM literacy, and explained why it is as relevant to our twenty-first century "information driven" economy as basic reading/writing literacy has been to the industrial economy of the past two centuries.

> STEM literacy is a way of thinking and doing: A person has STEM literacy if she can understand the world around her in a logical way guided by the principals of scientific thought. A STEM-literate person can think for herself. She asks critical questions. She can form hypotheses and seek data to confirm or deny them. She sees the beauty and complexity in nature and seeks to understand. She sees the modern world that mankind has created and hopes to use her STEM-related skills and knowledge to improve it. (Berger, 2012)

Numerous studies have observed that despite the continuous high unemployment and underemployment rates in the United States, employers are having trouble filling many positions that require technical skills. The United States continues to experience a wide skills gap, primarily due to low numbers of students who are majoring and graduating in STEM fields. Students are avoiding STEM courses in school if they are not required—whether in high school, post-secondary education, or college. As a result, their skills do not match those needed for the jobs that are most in demand.

Additionally, there exists a skills gap between the knowledge students use to generate data through social media networks versus the knowledge needed to create websites, media presentations, designing T-shirts and creating computer applications.

Larson suggests the engineering mentality and approach are needed in virtually all aspects of society. Although many students may be inclined to think that they do not plan to be a scientist or engineer, they therefore conclude they don't need to know STEM. He purports the bigger misperception is that the engineering mentality can apply to just about everyone in the job market, not just those pursuing STEM careers. Larson cites that prior to the Industrial Revolution, only a small percentage of the world's population was literate. Literacy rates increased throughout the nineteenth century, as people started migrating from the countryside to towns and cities for job opportunities in the newly industrialized societies (Berger, 2012). Many of the new jobs, particularly those offering higher salaries, required the ability to read and write. With universal education rising globally in almost all countries around the world, literacy rates have soared over the past century. Larson further notes a significant difference between being proficient at reading and writing and being a playwright, literary critic, book editor or journalist (Berger, 2012).

However, when it comes to STEM learning, students may not see the difference between achieving STEM proficiency and pursuing a STEM profession. Consequently, the general public, if polled, might agree that basic literacy is critical for the majority of jobs today. However, the idea that STEM literacy is an increasingly important qualification for a wide variety of jobs is something they might question.

U.S. science communication methods and programs may further contribute to this misconception. Scientists and other STEM professionals have made the case for STEM literacy, have defined what being "scientific literate" means, but have not adequately made the case why quantitative reasoning, familiarity with sophisticated machines, and dealing with complex systems, problems and decisions are important job skills in this rapidly changing, technologically advanced world. Every day, humans are consumed with numbers and statistics: utility bills, gas prices, inflation, financial aid, mortgages, and taxes. The list goes on and on. Quantitative reasoning skills, Larson suggests, are important for many jobs, so that one can understand what is going on and be able to adequately explain it to colleagues and customers (Berger, 2012). Quantitative reasoning skills enable one to use his or her math abilities and to perform due diligence. For instance, understanding the interest rates consumers pay for credit cards, car loans or mortgage requires a certain degree of quantitative reasoning. If consumers do not apply such reasoning by themselves, they fall prey to those who may not always have the consumer's best interest in mind.

Students are surrounded by smart machines: Kindles, iPods, iPhones, Androids, tablets, digital Blu-rays, LCD TVs, video recorders, Macs, PCs, and more. Some of these accessories have become invaluable tools at school and work. Though students may use these smart machines, the ability to deal with their sophisticated nature and use them effectively and efficiently to help address complex problems is another very important STEM skill. When people use smart machines effectively, they become more productive.

Larson also states that "dealing with complexity is another very important STEM skill" (Berger, 2012, p. 2). Working with complex systems, such as traffic grids, air traffic control, and satellite operations, particularly those involving a large number of steps or components, requires having a good overall understanding of the system, so that one can efficiently plan how to get the work done. In engineered systems, unanticipated situations often arise that have not been encountered before and thus require good, logical problem solving skills. Similar skills are necessary to make decisions about the environment, propose business solutions and evaluate the various options involved in making complex decisions. In general, people who are comfortable dealing with complexity are better able to handle more demanding, higher paying jobs. The "information driven," technology-intensive society will drive the need for the general public to become more STEM literate.

What Are Some of the Skillsets Needed to Become Media Literate?

Media education is, perhaps, the right of every citizen, in every country in the world. The U.S. Bill of Rights mandates freedom of expression. The right to information is key in building and sustaining democracy; thus media education should be introduced wherever possible within national curricula. In addition, non-formal and lifelong education enables people to gain an understanding of the way in which media operate in their society, and it helps them to acquire proficiency in using these media to communicate with others (Goteborg, UNESCO, & Nordicom, 2001).

A media–literate person is one who has the ability "to access, analyze, evaluate, and produce both print and electronic media in various forms" (Anderson, 1992, p. 1). Critical thinking skills are predominantly taught through complex text and solving problems. Critical thinking skills could be achieved by linking these skills to the media world.

For students to access media, they will need instruction in how to assess the value and validity of websites in all areas.

For students to analyze media, they will need instruction on how to detect advertising and publicity. Students will need to understand that people construct media messages. Nearly all media messages are designed to inform, entertain, or persuade the masses. Scientists use media messaging for controversial scientific issues such as climate change, evolution and stem cell research.

For students to evaluate media, they must learn to judge the value of media products for themselves. Students will have to develop the ability to determine the value of any given film, television program, magazine or newspaper article. In addition, media literacy would allow students to develop their own modes of criticism, interpretation and evaluation to assess media.

For students to produce media, they must learn the process of creating their own media message. It can be empowering to witness students seeing for the first time that they, too, can participate in making art, print, or news, a film, a television program, or a website that affects other people. By producing media, students learn in a more personal way that media messages are "constructed." They learn to construct scientific messages that are factual and entertaining.

Science educators, with some training, can show students how simple edits in film or television can dramatically change the meaning and emotional impact of a scene or of an entire story. Students can learn how to include or edit out a particular shot, how music choices influence how the audience experiences a character, or how to create story moments. These skills are essential in making students media literate. Students become producers of their own works, make editorial decisions of their own, and decide for themselves how to present clear and concise material. They can incorporate media into their science learning and, in the process, develop a new way to communicate their understanding of a scientific concept.

The Assessment & Teaching of 21st Century Skills (ACTS21s) categorizes abilities internationally into four groups:

- Ways of thinking deals with creativity, critical thinking, problem-solving and decision-making.
- Ways of working deals with communication and collaboration.
- Tools of working deals with information and communications technology (ICT) and information literacy.
- Skills for living in the world deals with citizenship, life, and career which includes personal and social responsibility. (Assessment & Teaching of 21st Century Skills, 2013, p. 2)

Other skills that exceed content area and simultaneously prepare students for post-secondary education, referred to as 21st century skills, include the following: communication, teamwork, ethics, leadership, flexibility, problem-solving and critical thinking.

Communicating science to public audiences in urban settings through art, film and video or through magazine creation is a multi-step process. First, students must engage with the content. Whether the content is physics, chemistry, anatomy and physiology, biology or environmental science, students should have access to the content and should be engaged with the science. Second, educators must ensure that they provide students multiple opportunities to understand the foundations of the science behind a concept. Students will have to conceptualize its meaning and ultimately express the different representations of the meaning using various art forms. It will require students to become scientifically literate. They must engage the content in a variety of ways in order to achieve scientific literacy.

CASE STUDY

Communicating Scientific Concepts in Chemistry to Public Audiences Through Student-Created Magazines

The following SY 2009–2010 vignette presents some real classroom experience with a science teacher using print media as a strategy to increase student motivation, student interest, and student confidence in science in an urban setting.

The vignette also supports how implementation of the *Next Generation Science Standards* (NGSS) can be used to address diverse student groups. Some considerations should be factored. First, the vignette is focused on a limited number of performance expectations. It should not be viewed as showing all instruction necessary to prepare students to fully understand the performance expectations. Neither does it indicate that the performance expectations should be taught one at a time. Second, consider that student understanding builds over time. Some topics or ideas will be revisited extensively throughout the course of the year. Therefore, science instruction should take student understanding into account. Performance expectations will be achieved by using coherent connections among disciplinary core ideas, scientific and engineering practices, and crosscutting concepts within the NGSS. Finally, the vignette is intended to illustrate specific contexts and practical strategies to engage all students in the NGSS.

The goal of the unit was to explain the kinetic molecular theory and use it to explain changes in gas volumes, pressure, and temperature. Students were asked to apply their understanding of Boyle's Law: the relationship between pressure and volume at constant temperature (pV = constant at constant temperature and number of moles), Charles' Law (the relationship between volume and temperature, V/T = constant at constant pressure and number of moles), and Gay-Lussac's Law (the relationship between pressure and temperature P/T = constant at constant volume and number of moles) to solve gas stoichiometry problems. Students were also asked to solve problems using the Ideal Gas Law, $pV = nRT$, and the Combined Gas Law, $P_1 V_1 / T_1 = P_2 V_2 / T_2$.

District of Columbia Public Schools encompasses 111 of the 238 public elementary and secondary schools and learning centers located in Washington, DC. During the SY 2009–2010, the total student enrollment consisted of 43,366 students. Approximately 18 high schools serviced approximately 10,000 students in the DC Public school system. With a population of more than 1,500 students, Woodrow Wilson Senior High has a very diverse student body (Table 15.1).

Table 15.1. Woodrow Wilson High School Demographics.

Student Population During 2009–10 School Year	1512
% African American Students	49
% Caucasian Students	20
% Latino Students	19
% Asian Students	8
% Mixed Races	4
% English Language Learners	9

Student Population During 2009–10 School Year	1512
% Students with Special Needs	11
% Students with Free and Reduced Lunch	37
% Students who Live in the Neighborhood	52
No. of Advanced Placement Courses Offered	27
No. of Small Learning Communities Present	6
No. of Varsity Sports Programs	26

(Source: District of Columbia Public Schools.)

Introduction to the Core Idea: Gases and Their Properties

The 10th–11th-grade afternoon chemistry class at Woodrow Wilson Senior High had an average attendance of 27 students. The students varied in age from 15 to 18 years old. Ms. B. had a diverse classroom community with approximately 5% of the class being diverse learners English Language Learners (ELLs) and 6% of the class receiving special education services. Ms. B. spent the first semester building science concepts, a supportive classroom and maintaining her high expectations.

- *Tip for implementation: Science educators will need to create a learning environment that builds on previous science concepts, provides for a supportive classroom, and sets high expectations so that they can introduce media literacy concepts to their student population.*

Ms. B. had challenges making science accessible for all the learners in her classroom. Teaching was further complicated due to the fact that many students had uneven or disrupted school careers and thus arrived with gaps in their understanding of basic science concepts. Class sessions were 90-minute blocks. The classroom was outfitted with laboratory tables throughout the room with a small space for desk chairs to be dispersed throughout the room.

The classroom was shared with the biology teacher. Ms. B. rolled her mobile cart with her projector, papers, and lab materials into the classroom in the afternoons for the fourth-period class. Students brought their textbooks to class. The class had use of interactive science notebooks. These notebooks were graded portfolios consisting of months' worth of Cornell notes, assessments, classwork, journal reflections and lab data.

- *Tip for implementation: Science educators that share classrooms can schedule the use of technology with the other classroom teacher to help facilitate learning.*

Reese, a very conscientious student, would wheel the cart into the classroom upon Ms. B.'s arrival and set up the projector. Ms. B. passed out the warm-up for the day. Elena and Frank passed out graded papers to the students after they completed their warm-up.

- *Tip for implementation:* Incorporating students early into the classroom routine will help students transition into thinking and assigning student tasks associated with developing scientific and media literacy.

Prior to the day's lesson on properties of gases, Ms. B. had the students do a warm-up activity to grasp the concept of gases. Students were allowed to pick their lab partners. They reviewed their roles as reporter, facilitator, experimenter, time keeper, and note taker, as these roles were established and practiced in several labs throughout the first half of the school year.

Students blew up a latex balloon and released the air. Students then took the balloon and placed the balloon's opening on top of a polyethylene bottle. The balloon was secured onto the bottle with a rubber band. Students were asked to make observations and hypotheses about the device they created. Different student groups tried to blow the balloon up with the bottle. They noticed that the balloon could only expand after a certain point. Several red-cheeked students tried to blow enough air to get the balloon to expand inside the bottle. Next, student groups had to think about what they observed and ask the following questions:

- What caused the balloon to expand?
- What happened to the air that is trapped in the bottle when you blow into the balloon?
- How could someone design a modification and include a sketch in the design that would allow the balloon to expand? [The NGSS scientific and engineering practice of asking questions in grades 9–12 arises from careful observation of phenomena or unexpected results to clarify and / or seek additional information. Asking questions as a scientific and engineering practice in grades 9–12 can also involve defining a design problem that involves the development of a process or system with interacting components and criteria and constraints that may include social, technical, and /or environmental considerations (NGSS Release, 2013).]

All of the students brainstormed ideas. More students in the groups began to get involved in designing a modification. Students decided on various levels to cut holes or slits within the bottles to see if it would improve airflow. Ms. B. walked around the room as students began to show their design. After Ms. B. gave the groups design approval, they formulated a new hypothesis about the airflow in their modified design and began to blow into the balloon to test the design. Students reported out with 2- to 3-minute oral presentations. Using whiteboards, they drew their design modifications onto the board and presented on their hypotheses, their data, their design modifications, and their results. Students then spent the remainder of the class period writing their laboratory report while helping their peers express their own ideas about the observation made about gases. [Another practice, constructing explanations and designing solutions in grades 9–12 builds on K–8 experiences and progresses to constructing and revising explanations based on valid evidence obtained from student's own investigations, models, theories, simulations and peer review. The assumption is that the theories and laws that describe the natural world operate today as they did in the past and will continue to do so in the future (NGSS Release, 2013).]

Incorporating English Language Arts into Chemistry Through Poetry

The next lesson on matter and its interactions continued the subject of properties of gases. Ms. B. did a twenty-minute read-aloud on characteristics of gases. In this lesson, Ms. B. wanted to engage the students further into the concept of gases. Using her equity sticks (popsicle sticks with student names) Ms. B. called on students at random to answer questions. She read the first section on the properties of gases. She called on students to read different sections of the text. Ms. B. used Cornell notes as her primary note taking strategy. Ms. B. took the objectives out of the text and demonstrated for students how to turn the objectives into questions that students could jot down on the left hand side of their Cornell and then answer after they reviewed the text. Students were instructed how to convert the main idea titles of each section into questions. For example, a section entitled "properties of gases" would become "what are some properties of gases?" Students understood that other subheadings gave the details of the paragraph, thus providing details for the answers. Ms. B. allowed 25–30 minutes for the close reading activity to give students time to engage with the text.

Ms. B. then introduced a new activity. She gave out strips of paper. She explained that students would write haikus about the properties of gases. Ms. B. asked students if they knew what a haiku was. A few students raised their hands. Ms. B. then wrote the definition of a haiku on the board. (Haiku is a form of Japanese poetry having unrhymed verses of three lines containing usually five, seven, and five syllables, respectively.)

- *Tip for implementation:* Science teachers could use this lesson as an opportunity to collaborate with English teachers to review Haiku poetry with students a day prior to the lesson.

Ms. B. modeled a haiku about chemical reactions. She explained that the haiku on chemical reactions should present ideas that help students understand what chemical reactions were. Students used the remaining fifteen minutes of class working in pairs to write their own haikus. Ms. B. placed previous concepts on the board like a word wall. Students chose previous concepts they knew: matter, atoms, periodic table, bonding, moles or chemical reactions. Ms. B. asked a couple of students to share their haikus to the class so that students could hear how haikus were created by their peers. For homework, students were given another opportunity to create a haiku. Students were also given ten reading comprehension questions about the properties of gases.

Developing Gas Law Explanations: Boyle's Law and Charles' Law

The next lesson gave students multiple opportunities to engage with the matter and its interactions through the concept of gases and gas laws by studying the relationships between the gas properties through hands-on activities. From these activities, students can develop the scientific and engineering practice of constructing explanations for the causes of the phenomena they observe. The students spent time studying the relationships between pressure, temperature, volume, and number of moles through a series of laboratory activities. Ms. B. introduced a task which would allow students to do experiments and apply their new knowledge of gas laws. Students looked at the concept of Boyle's Law. For the warm-up, students read an article from the CHEMatters magazine and wrote their responses in their reading log.

- *Tip for implementation:* If a teacher does not have CHEMatters, he or she can Google the topic and find a related article.

The question the class had to investigate was: What is the relationship between the pressure exerted by a gas and its volume? Students were tasked with graphing data on pressures and volumes to determine the relationship between pressure and volume. Student groups had several sets of data to plot. Students had to pull from their prior knowledge of independent and dependent variables, range and scale to determine whether the relationship between pressure and volume was either a direct or indirect relationship. Students discovered through graphing the data that pressure-volume relationship showed an inverse relationship. Graphing data on pressure and volume, reinforced students' mathematical computation and graphing skills. [The practice of developing and using models, in this case, a mathematical or computational model is used to generate data to support explanations, predict phenomena, analyze systems and/or solve problems (NGSS Release, 2013).]

Students then moved to the next concept: Charles' Law. Each laboratory table was set up with a hot plate, two Erlenmeyer flasks, two balloons, and a 1000-ml beaker filled with ice water. Students were tasked to investigate the following questions:

- What is the relationship between the temperature and volume of a gas?
- What is the correlation between change in temperature, change in shape, and change in pressure of the balloon?
- What is the mathematical relationship one might use to explain the hypothesis?
- How might someone go about finding such a relationship?

Students put on their safety gear before beginning the investigation. A student would place about 10–20 ml of water from a graduated cylinder into the Erlenmeyer flask. They then took one of the balloons and blew it up.

- The student let the air out of the balloon and put the opening of the balloon over the Erlenmeyer flask.
- The flask was set over the hot plate to boil for fifteen to twenty minutes.
- Students recorded their observations of the flask over a period of time.

Students watched in awe and delight as the balloon started to jerk and move. A few students took out their phones and began video recording. Students observed the water in the flask bubbling quickly and the balloon beginning to expand. Afraid that the balloon would burst, students yelled for Ms. B. to come to the table as other students started to crowd around one table as the balloon seemed to be expanding almost out of control.

Ms. B. inquired, "How long have you been observing the balloon on the hot plate?"

Students exclaimed excitedly, "almost 18 minutes!"

Ms. B. questioned, "Do you know if you have enough information to make a correlation between volume, temperature and the pressure of a gas? You have formed out your hypothesis. Start looking at your data. What information can you draw from your observations? Continue with your investigation."

Another student used some hand warmers to grasp the hot Erlenmeyer flask and put it on the desk for several minutes. Afterwards, one student put the same Erlenmeyer flask on the 1,000 ml beaker of ice cold water. Students looked on to see the balloon slowly deflate into the flask. The students then recorded their results. Students repeated the experiment with a different variable to see if they would achieve the same results. Some groups had trouble getting the balloon opening over the Erlenmeyer flask. Some groups had trouble figuring out whether more water or less water would make the balloon expand faster or slower than their previous experiment. Several groups figured out a new variable but wanted to keep using balloons to repeat the experiment. Ms. B. reminded the class that they still had write up their lab report and do their oral presentations with their whiteboard illustrations.

Through this laboratory experiment, students begin to process gas properties and explore what makes a gas expand and contract. They also explored the properties of temperature, pressure, and volume and the correlation between a gas and its ability to expand and contract within open and closed systems.

Developing Explanations for Gas Laws While Capturing the Science Visually: Gay-Lussac's Law

The next afternoon, Ms. B. rolled her cart into the room. For this class, Ms. B. wanted to take advantage of the excitement created by the students about the reactions they saw during the Charles' Law activity. Ms. B. saw a media opportunity and needed to figure how to incorporate it in the lesson. She wanted to determine how she could use her film and video skills in the lesson to have the students take simple video or pictures while they were learning the science. She had no camera equipment and knew she would, at some point down the road (time permitting), teach something about science, media and messaging. The class explored Gay-Lussac's Law.

"Today, we are going to look at Gay-Lussac's Law," Ms. B. announced. "Before we do that, I want everyone to gather around the demonstration table. I am going to fill this tub with water." Ms. B. added soap to the water. She used her hand to spread the soap suds around. "Can I have one volunteer with goggles to spread the soap around?"

"Ok, I am going to turn on the gas, and stick the tube into the tub. Anyone want to make a hypothesis as to what will happen?"

No one said anything. Eric whispered, "fire."

"Ok, stand back a few steps. Ready? 1–2–3!" Ms. B. lit the charcoal lighter and flames shot up six feet into the air.

"Whoa!" Students had their cell phones taking pictures and video recording the rising flames. "Ms. B. Can you do it again? I wasn't ready. I want to take a picture."

"Ok. I can do it again, but before I do, look at the tub. What observations can you make from what you have just seen? The soap suds seem to have disappeared. Can anyone figure out why? What are you going to do with the video footage? How could the footage you record help you understand gas laws? Let me refill the tub again as you think about my questions and what you saw."

Moments later, Reese raised his hand. "The gas mixed with the soap suds. So when you lit the charcoal lighter into the tub, you added the ignition and more oxygen causing a combustion reaction or rather, the flames to shoot through the air."

"Reese, that was well said. Does anyone have more to add to Reese's explanation? Any other observations? Class, we will have a moment to write down our reflections after this last demonstration. What about video? How can we use video to help us understand gas laws?"

One student said, "I have my cell phone. I can record what happens or take pictures."

Another student said, "We can put the video on YouTube."

Elena said "I guess we can look at it and figure what happened and why."

"Great responses!" Ms. B. congratulated. "In your groups, if you can, I want you to try and film your experiment. What do I mean? All of you who have cell phones raise your hand. I want you, through pictures and video, to capture portions of your experiment. We will then look at Gay-Lussac's Law. What gas properties do you think we will investigate this time?" After a few moments several voices offered, "Pressure and temperature. Volume is constant."

The students went back to their lab tables and wrote their observations and reflections for the Methane Flame demonstration they witnessed. They took their notebooks and broke out into the same groups from the previous sessions to work on the Upside Down Can activity. Students had the following materials at their table: a soda can, a ring stand, a ring clamp, a 25-ml graduated cylinder, a wire mesh, a Bunsen burner, tongs, hand warmers, and a 1,000-ml beaker.

Students had the following question to investigate: What is the relationship between the pressure exerted by a gas and its temperature? The scientific practice of planning and carrying out the investigation will require students to clarify what counts as data and experimental variables while they are filming the investigation. They began to set up the lab equipment at each station. Reese took the ring stand and attached the ring clamp onto it. He then took the wire gauze to place the gauze onto of the ring clamp. He hooked the hose from the gas to the Bunsen burner and placed the Bunsen burner on the ring stand. He added 10 ml of water into the soda can and put the soda can on top of the wire mesh and heated the soda can for 20 minutes. Reese began to write the lab report as he and his lab partners waited for the water to evaporate from the soda can. One student took a picture of the setup.

After 20 minutes passed, Reese and his classmates saw steam coming from the soda can. Reese sensed something might happen. He told his lab partners to get ready. Students who had a video camera only had a few minutes they could record at a time. Reese filled the 1,000-ml beaker with 750 ml of cold water. Reese took the tongs and inverted the soda can into the beaker filled with water. POP! Someone gasped.

"Did you get it?" one student excitedly asked.

"Yes, I think so," another student replied.

The soda can crushed inside the water. Reese and his lab partners started writing. Ms. B. walked over to the table.

"Can anyone, other than Reese's group, tell me why the can exploded? I would like you to write out your observations and explanations and share them during oral presentations."

- *Tip for implementation: In this lesson, the opportunity to capture the science visually through video (though unscripted) could be used later as footage. Students could be guided to write a short script explaining the concept. Students could also find examples of the same concepts used in the real world, or in special effects Hollywood movies. Students may even compare the effects done in the movies with what was done in the lab. Students could also compare explanations with the movie plot elements to see if such fictional events are plausible scientifically.*

Incorporating Language Development, Chemistry and Print Media to Gases: Project Based Learning-Magazine Creation

The next class period, Ms. B. rolled into the room with a look of excitement on her face as Reese began to remove the items off the cart. He saw a bunch of colorful drawings. Students stopped chatting as Ms. B. arrived.

"Ms. B. What's this?" Reese inquired.

"Reese, I will let you and the class will know in a moment." Ms. B. wanted to explain whole class.

"Good afternoon class, I want to introduce a new project that I would like your feedback on. I want us to create a chemistry magazine." Ms. B. asked several students to hand out materials. First, she handed out magazines: *JET, Ebony, Home, Black Enterprise*, and *Entrepreneur*. Three magazines were placed at each lab table for students to view.

"Now, how many of you read magazines?" A few hands went up.

"How many of you read comic books?" More hands went up.

"How many of you read graphic novels?" Even more hands went up. "Why do you read them?"

"It's fun," one student responded.

"I learn new things," another student commented.

"Because I'm bored at school," another student confessed.

"Thank you for being honest." Ms. B said.

Ms. B. then asked the students to pass out a set of drawings that were bundled together and white copy paper.

"We are going to create a class magazine. I would like everyone to contribute. This will count as a homework grade. You can contribute extra pages as content for extra credit. Take a look at the magazines at your table. I want us to brainstorm all of the components of a magazine. Take six minutes to look the magazine and then talk to your neighbor. We will come back together as a class."

After a few minutes had passed, Ms. B. called the class together.

"Now that you have looked at the different items, let's brainstorm. What do you see are some of the components of a magazine?"

Ms. B. wrote on the whiteboard as students began to call out items: table of contents, editor page, advertisements, illustrations, articles were a few examples. Afterwards, silence filled the classroom.

"Don't different magazines have different things?" one student asked.

"They do in terms of content, but the format of a magazine is pretty basic," Ms. B. assured. "To create a class magazine or to create a kid's magazine, let's look at the following format." Ms. B. handed out the magazine format she tailored for the class (Appendix A).

"Now, since we have been talking about gases, how many people would like to create a magazine on gases?"

Several students raised their hands.

"If you read your format, you will see you can add articles, jokes, riddles, songs, or haikus like the ones we have been doing. You can also add illustrations, quizzes, concept maps, and short stories. I have some models for you to see. These models are old. I know that we can create a class magazine that is better than this."

Ms. B. issued the class a challenge. She wanted the class to do more independent writing. She wanted more creativity and wanted to see what her students could do if the students chose the science content. Ms. B. felt the students were up to the task.

Days later, students came to class with their sketch drawings. Some students drew from scratch while others used computer software to create their work of art. Some students decided they wanted to create their own magazine instead of just one or two pages. Some students decided they wanted to create more pages for articles, jokes, comics, or cover page, if the class magazine had missing parts. Students enjoyed creating the magazine so much for gases that they decided to create a magazine for the remaining chemistry topics: molarity, acids, and bases. Both sections of chemistry classes lamented they would have liked to have made magazines at the beginning of the semester. They felt it helped them understand the chemistry much more and it made the class fun.

When these same chemistry students took the final exams, 48% of them achieved a final grade of 80% or higher. Incorporating media into the class increased their level of understanding significantly.

IDEAS FOR ADAPTATION BY OTHERS

Before the lesson on gases, educators could invite newspaper editors, magazine editors or other media professionals to talk to the students about media literacy using scientific content. Explosions excite students so examples of newspaper content could feature gas explosions with the details of the explosion and the investigation. Educators could also invite police officers or other forensic experts to discuss forensics, arson and other causes of fires and explosions that they investigate. Experts could show students how they write reports and what they look for. Experts can compare the realities of their job to what's done in pop culture or on television. Educators could also invite screen writers to teach elements of screen writing and show students examples of scripts with scientific elements. The screenwriter could then model a sample public service announcement (PSA) using the footage from class. Educators could also go online and find scripts with scientific themes, scientific elements, and video footage and discuss with students how the science was captured on the screen. Students could then judge from their investigations if the same scientific concept was depicted accurately onscreen. Science educators may want to choose concepts that have been easily adaptable in movies to study.

Contests and competitions are another avenue and a great way to get students involved in video production. Collaborating with other subjects, such as science-history, science-art, or science-music is yet another way educators could assist students in bridging STEM literacy with media literacy. Science educators can contact local universities, local TV stations, and the media relations department to find professionals that can work with students. The science educator can also collaborate with media specialist, the journalism teacher, or the digital media teacher to plan lessons and projects that incorporate science content. The science educator teaches the science and can pose this question to students: How can you message the science you learn in a clear, fun way using art and technology? Flip phones, iPhones, Androids and other smartphones all have the capability to capture pictures and video that can be uploaded or placed in presentation formats such as Prezi or PowerPoint. Students can use social media such as Twitter, Facebook, and Instagram to showcase their science learning.

Career and Technical Education (CTE) programs that have film or mass media programs can be used to bridge vocational instruction, develop communication skills, and media literacy skills all at the same time. When science educators collaborate with the CTE instructors they can foster multiple skillsets in underserved students. Environmental Science students could be tasked to create a 30-second PSA about protecting their neighborhood from trash dumping. Students can draw, sculpt or paint themes containing scientific concepts and enter the designs in art shows, art exhibitions or contests, or media exhibitions.

The CTE or science educator could embed short media lessons that include techniques on messaging and the psychology of consumers; expose stereotypes and misconceptions and also to discover effective ways of conveying ideas and information. In doing so, students begin to learn how to absorb, critique and question media-conveyed news.

Students could create and produce newspapers, newsletters, or school-based TV programs that explore scientific issues in their neighborhood or in the hallways of their schools. Students could film lab experiments, dissections, and other laboratory procedures to help other students learn proper lab safety procedures, techniques, and engineering concepts.

Two former science students were invited to participate in a video documentary contest that spoke of their experiences in science. Students spoke candidly about their science classroom experience and what they learned. Students then demonstrated a physics experiment and explained the concept. Problems with editing and time commitments prevented the students from editing the clip and sending the entry into the contest. Although the project did not make the contest as an entry, the students learned a lot about using media to convey their message.

Producing content for media outlets helps students learn concepts in depth so that they can communicate those messages to general audiences to understand. Students become more vested into the media and begin to see the meaning behind the science or the "why" behind the science.

CONCLUSION

The National Science Board reported the following findings:

> The abilities of large numbers of potential future STEM innovators currently go unrecognized and are underdeveloped. Though cognitive ability is only one of many attributes of a future innovator, it is important. Identifying this ability as early as possible is critical for developing an appropriate educational intervention. Abilities may develop at different rates for different individuals, so educators must diligently seek out potential throughout the entire educational continuum. (2010, p. 62)

STEM educators have a prime opportunity to combine concepts of STEM literacy and media literacy to engage students in science. Combining STEM and media literacy strategies have the potential to elevate student achievement, student motivation and student interest in science. These skillsets can produce advocacy skills in students. These skills can also help diverse learners improve in science and language acquisition. They can also help build stronger communication skills with students with special needs. Ultimately, these new skills can prepare many students for careers in STEM fields that will foster creativity, innovation, and solutions that will directly address global challenges.

APPENDIX A

NAME_____ DATE_____ PERIOD____

MAGAZINE BASICS

PROCEDURES FOR CREATING A MAGAZINE
How do I Create a Kids' Magazine?

The Magazine must have:

- Must have **front and back cover**. **Cover page** must have title of magazine, author, and names of featured articles.
- Cover and back pages must be **illustrated**.
- Must have table of contents page with article and page number.
- Must include at least four (4) articles. Each article must include topic and author's byline and should be a minimum of two (2) paragraphs long.
- Articles can be informative (tell about something), narrative (tell a story) persuasive (present an idea or argument about something you feel strongly about and may include evidence to support your argument), or descriptive (describing something).
- Magazine can include poetry, interviews, short story, music or movie reviews, advice columns, puzzles, riddles, jokes (no inappropriate jokes such as "your momma" jokes or dirty jokes), recipes, fashion, etc.
- Magazine Authors must include the following items in their magazine: advertisements, short story about their topic; an interview of a scientist within their topic; a career choice using that topic, jokes, quiz (on that topic) You can also add puzzles, word searches, poetry, songs, advice columns.
- Magazines must also include advertisement. These advertisements must be original. (In other words, you must create the ads yourself). You may write only two full-page ads. Your ads may be included on a feature page.
- Magazine must include a page dedicated to the school year. This page must include highlights of the things you liked the most in _____. For instance, you may have excelled making concept maps. If so, include a concept map for your reader to solve.
- Magazine must be at least 10 pages and should be no more than 20 pages in length, including advertisements, articles, school year page, and front and back pages.
- Each page must be numbered.

APPENDIX B

NAME_____DATE_____PERIOD____

Scoring Rubric for MAGAZINE

Must have minimum 10 pages of material
_____/30 points

Cover has title of magazine, author, and names of featured articles
_____/30 points

Cover and back pages must be illustrated
_____/40 points

Table of contents page with article and page numbers
_____/20 points

Each article with topic and authors byline
_____/20 points

Each article at least two (2) paragraphs long
_____/20 points

Has at least two original advertisements
_____/20 points

Has at least 2 games, quizzes, and / or concept maps readers can work on
_____/20 points

TOTAL = _____/200 points
Grade = _____

NOTE

Since the first publication of this book in 2014, several organizations have sought to address the issue of communicating science using media to public audiences at the middle and high school levels. Two such promising examples, both come from the Public Broadcasting System (PBS). The PBS News Hour Student Reporting Labs Program (https://studentreportinglabs.org/reporting-labs-curriculum/), a project-based learning initiative funded by the National Science Foundation that seeks to engage middle and high school students in STEM fields by having them produce science video news stories. The curriculum guides students through journalism basics while choosing a STEM story topic within a larger context or theme. Students conduct their own research and can work with a scientist mentor who acts as an expert advisor. The curriculum is guided by the Next Generation Science Standards (NGSS) and is meant to engage students through interest - driven and locally relevant projects.

PBS also has a Learning Media curriculum platform which incorporates the NGSS. The new Science curriculum and Engineering and Technology hierarchy offers over 120,000 free digital resources. The digital resources are subject-specific and provide a skeleton that each resource hangs on within the catalog. The Learning Media resources are connected to national and state standards, various documents and other artifacts. This curriculum hierarchy helps teachers find digital resources faster using PBS Learning Media's platform in tandem with the advanced search, sort and filter capabilities. Teachers can explore topics where the results might include background essays, discussion questions and videos within just a few clicks. Teachers get a wonderful opportunity to deeply engage their students in STEM content while enhancing their teaching with digital media.

WORKS CITED

Andersen, N. (1992). Making a case for media literacy in the classroom: Impact of images: Education. *Media & Values, 57*. Retrieved October 13, 2013, from http://www.medialit.org/reading-room/making-case-media-literacy-classroom

Assessment & Teaching of 21st Century Skills (ATCS21s). (2013). *What are 21st century skills?* Retrieved from http://atc21s.org/index.php/about/what-are-21st-century-skills/

Aufderheide, P. (1993). *Media literacy. A report of the National Leadership Conference on Media Literacy*. Retrieved June 22, 2013, from http://www.eric.ed.gov/ERICWebPortal/search/detailmini.jsp?_nfpb=true&_&ERICExtSearch_SearchValue_0=ED365294&ERICExtSearch_SearchType_0=no&accno=ED365294

Bencze, L., Sperling, E., & Carter, L. (2011). Students' research-informed socio-scientific activism: Revisions for a sustainable future. *Research in Science Education, 42*(1), 129–148.

Berger, I. (2012). *STEM literacy is for everyone*. Retrieved October 13, 2013, from blog.irvingwb.com/blog/2012/09/stem-thinking-skills.html

Bybee, R., McCrae, B., & Laurie, R. (2009). PISA 2006: An assessment of scientific literacy. *Journal of Research in Science Teaching, 46*(8), 865–883.

Castano, C. (2008). Socio-scientific discussions as a way to improve the comprehension of science and the understanding of the interrelation between species and the environment. *Research in Science Education, 38*, 565–587.

Considine, D. M., & Haley, G. E. (1999). *Visual messages: Integrating imagery into instruction* (2nd ed.). Englewood, CO: Teacher Ideas Press. Dawson, V. M., & Venville, G. (2010). Teaching strategies for developing students' argumentation skills about socioscientific issues in high school genetics. *Research in Science Education, 40*, 133–148.

Evagorou, M. (2011). Discussing a socio-scientific issue in a primary school classroom: The case of using a technology supported environment in formal and non-formal settings. In T. Sadler

(Ed.), *Socio-scientific issues in the classroom: Teaching, learning and research* (pp. 131–160). New York: Springer.

Goteborg, UNESCO, & Nordicom. (2001). Recommendations addressed to the United Nations Educational Scientific and Cultural Organization. In *Education for the media and the digital age*. pp. 273–274.

Grace, M. (2009). Developing high quality decision making discussions about biological conservation in a normal classroom setting. *International Journal of Science Education, 31,* 4, 551–570.

Gresch, H., Hasselhorn, M., & Bögeholz, S. (2011). Training in decision-making strategies: An approach to enhance students' competence to deal with socio-scientific issues. *International Journal of Science Education, 35*(15), 2587–2607.

Hynd, C., Holschuh, J., & Nist, S. (2000). Learning science: What motivates students to learn complex scientific information. *Reading/Writing Quarterly, 16,* 23–57.

Klosterman, M., Sadler, T., & Brown, J. (2012). Science teachers' use of mass media to address socio-scientific and sustainability issues. *Research in Science Education, 42*(1), 51–74.

Michaels, S., Shouse, A. W., & Schweingruber, H. A. (2007). A new vision of science in education. In *Ready, set, science! Putting research to work in K–8 science classrooms (pp. 83–92)*. Washington, DC: National Academies Press.

Mueller, M. & Zeidler, D. (2010). Moral–ethical character and science education: Ecojustice ethics through socioscientific issues (SSI) In D. Tippins, M. Mueller, M. Van Eijck, & J. Adams (Eds). *Cultural studies & environmentalism: The confluence of ecojustice place based science education on indigenous knowledge systems.* Pp. 105–128. New York. Springer.

National Science Board. Committee of Education and Human Resources. (2010). *Preparing the next generation of STEM innovators: Identifying and developing our nation's human capital.* Charleston, SC: BiblioBazaar.

Next Generation Science Standards Release (2013). Retrieved November 09, 2013, from http://www.nextgenscience.org/development-overview.

Norris, S. P., & Phillips, L. M. (2003). How literacy in its fundamental sense is central to scientific literacy. *Science Education, 87,* 224–240.

Sadler, T. D. (2009). Situated learning in science education: Socio-scientific issues as contexts for practice. *Studies in Science Education, 45,* 1–42.

Sadler, T. D., Barab, S. A., & Scott, B. (2007). What do students gain by engaging in socioscientific inquiry? *Research in Science Education, 37,* 371–391.

Sadler, T. D., & Donnelly, L. (2006). Socioscientific argumentation: The effects of content knowledge and morality. *International Journal of Science Education, 28*(12), 1463–1488.

Sadler, T. D., & Zeidler, D. L. (2005). The significance of content knowledge for informal reasoning regarding socioscientific issues: Applying genetics knowledge to genetic engineering issues. *Science Education, 89,* 71–93.

Scientific Literacy. (1996). Retrieved October 13, 2013 from http://www.literacynet.org/science/scientificliteracy.html.

Von Aufschnaiter, C., Erduran, S., Osborne, J., & Simon, S. (2008). Arguing to learn and learning to argue: Case studies of how students' argumentation relates to their scientific knowledge. *Journal of Research in Science Teaching, 45*(1), 101–131.

What is media literacy? A definition ... and more. (2013). Retrieved June 22, 2013, from http://www.medialit.org/reading-room/what-media-literacy-definitionand-more.

Wu, Y.-T., & Tsai, C.-C. (2007). High school students' informal reasoning on a socioscientific issue: Qualitative and quantitative analyses. *International Journal of Science Education, 29*(9), 1163–1187.

Wu, Y.-T., & Tsai, C.-C. (2010). High school students' informal reasoning regarding a socio-scientific issue, with relation to scientific epistemological beliefs and cognitive structures. *International Journal of Science Education, 33*(3), 371–400.

Zeidler, D. L., Sadler, T. D., Applebaum, S., & Callahan, B. E. (2009). Advancing reflective judgment through socioscientific issues. *Journal of Research in Science Teaching, 46*, 74–101.

Zeidler, D. L., Sadler, T. D., Simmons, M., & Howes, E. (2005). Beyond STS: A research based framework for socioscientific issues education. *Science Education, 89*(3), 357–377.

Zohar, A., & Nemet, F. (2002). Fostering students' knowledge and argumentation skills through dilemmas in human genetics. *Journal of Research in Science Teaching, 39*, 35–62.

CHAPTER SIXTEEN

Discourse Strategies FOR English Learners IN THE STEM Classroom

JENAY SHARP LEACH

INTRODUCTION

I had spent nearly ten minutes explaining to thirty 4th graders how energy in a roller coaster changes from potential to kinetic. I thought I did all of the right things. I asked them if they had ever been to an amusement park. I pantomimed the ups and downs of the roller coaster. And for another ten minutes, I excitedly built to the climax of my presentation: unveiling the cool experiment they would conduct to investigate how changing the height of a hill affects the speed of a car. At that moment a little hand tentatively rose into the air.

"Yes?" I asked.

"What's a roller coaster?"

The reality of my teaching context washed over me like a tidal wave. I had made no attempt to determine the cultural or linguistic background knowledge these students brought into my classroom. Here I was droning on about potential and kinetic energy, using the roller coaster as a conceptual model, when some students had no idea what I meant.

English Learners (ELs) are defined by the U.S. Department of Education as students with limited English proficiency (Lhamon & Gupta, 2015). ELs are the fastest growing student population in the country (National Center for Educational Statistics, 2012) and all science, technology, engineering, and mathematics (STEM) teachers face the challenge of educating students with various language needs in the classroom. Inquiry-based activities can be challenging for ELs, requiring students to produce both academic and social language (Reiser,

Berland, & Kenyon, 2012). Research shows that students in science classrooms have difficulty navigating both academic and social discourses, where academic discourse is usually privileged over social discourse (Anderson, 2007; Carlsen, 2007; Moje, Collazo, Carrillo, & Marx, 2001).

While it may seem counterintuitive, inquiry-based activities can still be particularly powerful for ELs, as they provide authentic opportunities for language production, which happens when they participate in scientific and engineering processes that are intrinsically collaborative, hands-on, and language-driven (Lee & Luykx, 2006; Nutta, Bautista, & Butler, 2011; Santau, Maerten-Rivera, & Huggins, 2011). Inquiry-based investigations that provide language supports for ELs and other students with language needs offer all students access to rigorous STEM content and processes.

To bridge gaps between academic and social language, I implement discourse strategies in my classroom. I offer students intentional opportunities to talk about STEM content in a way that complements the inquiry process. Discourse strategies generate entry points for all students to participate in collaborative, hands-on investigations and they also address specific language needs of ELs. This chapter will present the challenges of the linguistically diverse STEM classroom, introduce inquiry as an authentic language-learning context for ELs, and discuss how to effectively implement discourse strategies in the STEM classroom.

BACKGROUND

English Learners and Science Instruction

Because ELs are now enrolled in three out of four public schools in the U.S., it is imperative that all teachers consider ELs' needs when planning and implementing instruction (NGSS Lead States, 2013). Science, in particular, presents specific difficulties for ELs as scientific knowledge is generated through language. Science is a set of practices that uses evidence to construct arguments to account for patterns in the natural world (National Research Council [NRC], 2007b). Thus, we create scientific knowledge by engaging in scientific practices, which require the production of language.

Following this consideration of the role of language in STEM, we must also consider the cultural norms and values inherent in the STEM classroom. Aside from the obvious English language considerations for ELs, science and engineering have their own norms, or rules, for social and cognitive participation, which typically differ from children's everyday experiences (NRC, 2007b). The discipline of science is vocabulary intensive, and like other STEM disciplines and special interests, it is traditionally communicated through academic registers of speech—linguistic styles shared by an organized group of people (Anderson, 2007). Classroom discourse tends to mimic the discourse registers of scientists and engineers in the field where the rules of argument, explanation, and evaluation of evidence differ from those in social settings (Anderson, 2007; NRC, 2007b). Students' comfort and familiarity with the norms of scientific and engineering practices vary, as their experiences vary "with their cultural, linguistic, and economic background" (NRC, 2007b, p. 186).

Recent education reports specifically call attention to discourse in the inquiry-based, integrated science classroom, especially in light of varying student experiences with scientific and engineering norms. The *Framework for K-12 Science Education* (NRC, 2011) urges science instruction that engages students as scientists and engineers, communicating the results of inquiry, collaboratively designing solutions, and using discourse to make sense of the ideas communicated. The NRC Report *Taking Science to School* (2007b) undergirds the *Framework*, providing four strands for student proficiency in science, including "participate productively in scientific practices and discourse" (p. 2). In order for students to participate in scientific practices and discourse, students need support to learn the "norms of scientific argument, explanation, and the evaluation of evidence," recognizing that these norms "differ from those in everyday life" (NRC, 2007b, p. 186). The *Next Generation Science Standards* (NGSS Lead States, 2013) recognize that engagement in scientific and engineering practices offers rich opportunities for language learning:

> When supported appropriately, these students are capable of learning science through their emerging language and by comprehending and carrying out sophisticated language functions (e.g., arguing from evidence, constructing explanations, developing models) using less-than-perfect English. By engaging in such practices, moreover, they simultaneously build on their understanding of science and their language proficiency (i.e., capacity to do more with language). (NGSS Lead States, 2013, p. 392)

PERSISTING OPPORTUNITY GAPS

Despite the urgings of recent standards and educational reform movements that have promoted inquiry-based approaches in the classroom since the early 1990s (American Association for the Advancement of Science [AAAS], 1993; NRC, 1996, 2000), there exists an achievement gap between English fluent students and students who speak English as a second language (Calderón, Slavin, & Sanchez, 2011; Lewis, Maerten-Rivera, Adamson, & Lee, 2011). The inquiry approaches advocated by the reform documents require students to take the linguistic initiative to ask questions and provide insights, which in turn deepen student understanding of science concepts and develop a student's sense of the nature of science and discipline of engineering (Lewis, Lee, Santau, & Cone, 2010). Many teachers lack the training to implement these reform practices (Santau et al., 2011), and "teachers of [ELs] face the added challenge of enabling these students to learn academic content and processes while also developing oral and written proficiency in English" (Lee & Buxton, 2013, p. 37).

Developing English proficiency constrains STEM classroom participation and creates opportunity gaps for ELs. While scientific and engineering practices provide opportunities for situational learning, the discourse demands are high, requiring students to create meaning and navigate sociolinguistic norms in order to successfully communicate with peers. This situation can be further complicated if teachers assume all students have high levels of English fluency. Furthermore, students sometimes execute sophisticated avoidance techniques in order to appear as English proficient, such as pretending to lack an understanding of the content, when the actual problem is an inability to express their understanding in English (Monzó & Rueda, 2009).

To further complicate matters, many teachers mistakenly assume that EL students must become English proficient before learning science (Lee, Maerten-Rivera, Buxton, Penfield, & Secada, 2009). While traditional content instruction assumes proficiency in English as a prerequisite for content learning, research suggests that ELs can learn to improve their English skills by engaging in inquiry practices (Lee & Luykx, 2006; Santau et al., 2011; Stoddart, Pinal, Latzke, & Canaday, 2002; Zwiep & Straits, 2013).

STEM CLASSROOM DISCOURSE

Advocating for inquiry instruction as a way for students to construct shared meanings, Lee and Buxton (2013) assert that content areas should provide authentic context for literacy development, which serves to eliminate opportunity gaps. English provides the medium for student engagement with the content and small groups provide ELs a safe environment to speak and learn from their classmates. Moreover, ELs benefit from hands-on inquiry instruction in particular (Lee, 2005; Lewis et al., 2010, 2011). Hands-on activities are less dependent on academic language than traditional lecture-based instruction and reduce the linguistic burden on ELs (Lee, Maerten-Rivera, Penfield, LeRoy, & Secada, 2008). Hands-on activities based on observable phenomena are "more accessible to students with limited science experience than is decontextualized textbook knowledge" (Lewis et al., 2011, p. 157). Hands-on inquiry activities can also facilitate students' communication of their understanding in a variety of modalities, including verbal, kinesthetic, pictorial, graphic, and written text (Lewis et al., 2011).

Integrating language and content must occur in such a way that allows students to discuss and debate ideas with evidence from their investigations, working collaboratively, and engaging in student-to-student dialogue (Zwiep & Straits, 2013). Simply providing social interactions is not enough for students "to fully comprehend how to use the language and linguistically negotiate within it" (Shea, Shanahan, Gomez-Zwiep, & Straits, 2012, p. 3). In order for ELs to engage in meaningful and purposeful language production, teachers should intentionally and skillfully provide discourse opportunities. For example, students should discuss ideas and hypotheses, use evidence to make logical claims, communicate their experimental and engineering designs, and connect their results to the larger body of scientific knowledge. Inquiry without discourse can create misconceptions about the true nature of science (Llewellyn & Rajesh, 2011), which is social, creative, subject to change, and non-authoritarian (AAAS, 1990). Lamentably, some teachers view inquiry "as synonymous with hands-on instruction and the importance of discourse and reflection is overlooked" (Stoddart et al., 2002, p. 684). Not only must the instructional model allow for collaboration and the social construction of ideas, but also, the teacher must explicitly implement discourse strategies.

Without the proper scaffolding, teachers' use of academic language can create barriers to learning for ELs and limit their participation in the STEM discourse community. Students require instruction to link their social discourses to academic discourses in order to enhance their understandings of scientific phenomena (Suarez & Otero, 2013). This linking is demonstrated by a case study of 13 third-grade EL students as they explored how the physical characteristics of strings affect the sounds they produce (Suarez & Otero, 2013). In this study, ELs engaged in discourse with their peers as they experimented with changing the tension, length, and frequency of vibration on guitar strings, describing the resulting changes in pitch and volume. As students described the sounds using familiar language, their own words became incorporated into the class discourse. Interestingly, the students also used academic language, such as the term "vibration," to explain the sounds they heard. This inquiry-based activity encouraged students to engage in meaningful sense-making through scientific discourse without the added burden of requiring students to use academic jargon or terminology. For example, students described "loose" strings as making a "tong-tong" sound and "tight" strings as making a "ting-ting" sound. The use of these words afforded students an entry point to academic concepts and an eventual use of the academic vocabulary of "tension" and "pitch." Suarez and Otero (2013) conclude that classrooms where students use familiar language to think about and explain observations promote the development of language skills and conceptual understanding.

Scaffolding Discourse

Without proper scaffolding, high-level reasoning through discourse is rare in STEM classrooms (Hardy, Kloetzer, Moeller, & Sodian, 2010). For instance,

Hardy et al.'s (2010) analysis of discourse from six elementary science classrooms found that students did not support their claims with evidence. Transcriptions of discussions about sinking and floating from six 3rd and 4th grade classrooms were analyzed for the quality of scientific reasoning. Results showed that whole-class discussions were dominated by unsupported claims and there was little evidence of scientific reasoning. The authors argue that teacher scaffolding is necessary to help students attend to the relevant aspects of the task, model more advanced thinking and understanding, and develop skill in scientific argumentation by linking evidence with claims. For example, I often prompt my students with the question, "How do you know?" when they make a claim, or provide a prompt such as, "I know that glass bends light *because* ..."

While Hardy et al.'s (2010) results showed that teacher prompts are associated with higher levels of student reasoning, other research demonstrates that such prompts limit students' creativity (Carlsen, 2007; Shemwell & Furtak, 2010). Teachers often limit students' discourse by providing scaffolds that lead all students to the "correct" answer (Shea et al., 2012; Zwiep & Straits, 2013). Proper linguistic scaffolding is scaffolding that is differentiated to respond to individual learners: teachers provide support for language use and vocabulary by adapting the linguistic complexity to help learners reach the next level in their development and then removing supports once the learner has achieved independence with a task (Baker et al., 2009; Gibbons, 2003).

A tension can exist between scaffolding students and allowing students to produce their own discourse. Hogan, Nastasi, and Pressley (1999) contend that teachers' control of discourse leads students to higher levels of thinking and higher-quality explanations, but student-driven discussion is more "generative and exploratory" (p. 379). In their discourse analysis of two 8th grade science classrooms, Hogan et al. (1999) examined discourse patterns and scientific reasoning in peer- and teacher-guided discussions as students constructed mental models of matter. Results indicated differences in students' collaborative discourse as they worked in small groups with and without teacher guidance. The presence of the teacher led students to generate higher quality explanations faster than the peer-only groups. However, the discourse in the peer-only groups was more diverse and some of these groups achieved high levels of reasoning without the assistance of the teacher.

The tension between providing too much and too little discourse support creates a "Goldilocks" situation for teachers in the STEM classroom. How can we know the "right" amount of scaffolding? While some scaffolding can lead students to higher levels of thinking and help them to make connections to the larger body of scientific knowledge, too much scaffolding can limit student creativity. In my view, it is our responsibility as STEM teachers to encourage students to explore and test the boundaries of what is possible. We must counter the notion that there is one correct answer; in the interdisciplinary STEM classroom, there are often many viable solutions to real-world problems. Further, when talking about scientific content, though students may make inaccurate statements, we must use caution not to impose a "correct" way of talking about content. For example, Moje's (1995) research described a classroom scenario in which the teacher preferred this kind of convergent thinking. In this situation, the teacher asked students to explain the function of a graph. Several students offered explanations about how graphs show how one variable affects another. The teacher dismissed all of these explanations, waiting for a response that included the term "relationships." While the students clearly understood the concept, she wanted them to explicitly say, "Graphs show relationships." We can assess a student's understanding of a concept to determine whether or not a student holds a misconception while still allowing the student to describe the concept without the proper vocabulary or register. Students can produce conceptually rich talk when discourse structures are not onerous, but rather adapt the linguistic complexity of the task to respond to students' individual language needs.

Classroom science discourse for ELs is influenced by many factors, as described above. Students' language proficiency, background knowledge, and skill in navigating between academic and social discourses all contribute to their ability to participate in STEM classroom discourse. The nature of the instructional activities and the sociocultural norms of the classroom are also important for providing a learning environment conducive to student talk. Further attention to the complex interplay of these factors, as well as how teachers consider these factors when planning and implementing STEM instruction is needed. Discourse strategies can be used to encourage open exploration for all students while still providing ELs with entry-points for participation.

BEST PRACTICES: DISCOURSE STRATEGIES

Discourse strategies are intentional opportunities for students to talk about STEM content. Discourse strategies give each student the opportunity to respond to a prompt, rather than the traditional sequence of a teacher posing a question to the whole class and having one student respond (Shea et al., 2012). "Discourse strategies" is an umbrella term that encompasses discourse *structures* and discourse *scaffolds*. Discourse structures are student tasks implemented by the teacher to encourage student-to-student conversations. Discourse scaffolds support student talk by providing tools and on-ramps to participation. Both discourse structures and discourse scaffolds enable students to participate in the spheres of scientific and engineering practices as prescribed by the NRC (2011): investigating, evaluating, and developing explanations and solutions. Discourse strategies mediate these spheres and allow students to build linguistic bridges between their home discourses and academic discourses (Figure 16.1).

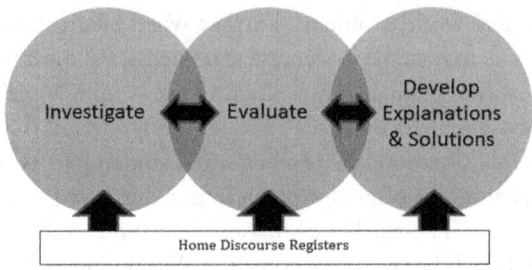

Figure 16.1. In STEM Classrooms, Students Investigate, Evaluate, and Develop Explanations and Solutions. Discourse allows students to use their home discourse registers to access these practices and work within and between the spheres.
(Source: Author.)

Discourse Scaffolds

I had taught in a non-linguistically diverse, upper-middle class community of Northern Virginia for a few years before moving to the school of the roller coaster incident. My new school was in the same district but ELs comprised 40% of the school's population when I started there. This proportion quickly grew to 60% within four years. I soon realized that what worked at the first school would not work at the second, but I was at a loss for what to do. While I knew that hands-on experiences would serve my learners well, I also knew that I wanted them communicating in English and increasing their academic literacy. As the EL population was quickly growing, my principal had the foresight to ask the English Speakers of Other Languages (ESOL) teachers to present at a staff meeting. The ESOL teachers introduced the concept of sentence frames and I absorbed the strategy like a sponge. The very next day, I tried using tiered sentence frames with my students and was met with higher levels of participation not only from my ELs, but all of my students.

What I realized was that sometimes students need support to be able to participate in discourse. Discourse for ELs needs to be scaffolded. These scaffolds can be more or less supportive, based on student needs, and can be removed once students are able to participate in STEM discourse without their use. Examples of discourse scaffolds are sentence starters (also called prompts) and tiered sentence frames.

Sentence starters

My ELs responded well to sentence starters because it gave them the words to discuss scientific concepts. For example, when studying circuits, my students added lightbulbs to a simple series circuit. To share their findings, I asked them to complete the sentence, "When I added bulbs to a series circuit, _____." Here, the students are given the word "bulb," so they can use it to explain that the bulbs dim or grow dark. I then prompted students to explain what evidence they used to help them complete the sentence: "When I added bulbs to a series circuit, they get dim because _____." The students' hands-on experience with circuits taught them that the brightness of the bulbs is an indication of the amount of current flowing through them. Therefore, my students talked about how the bulbs somehow slow or stop the current. I was able to link their observation to the word "resistance." Sentence starters enable students to sharpen and refine their thinking by using prompts to encourage the sharing of ideas and extending thinking. Students can learn the words (such as "resistance") while the sentence structures help them to think and communicate.

A word of caution is in order here with regards to using sentence starters properly. The starters should prompt student thinking and encourage exploration. Sentence starters should not promote convergent thinking or lead students to the "right" answer. For example, can you differentiate between the likely student responses to the following sentence starters?

A. The amount of space something takes up is _____.
B. Volume is _____.

In Sentence A, there is one correct answer: volume. In Sentence B, students can talk about capacity, three-dimensional space, how volume relates to other variables such as density or pressure, the units used to measure volume, or even make connections to the loudness of a sound. Sentence B promotes student-to-student discourse.

Tiered sentence starters

Though asking for a single correct response as in Sentence A above is not as rigorous as Sentence B, sometimes students need more scaffolding to respond to an open-ended sentence starter. In this case, the teacher can tier the sentence starters to prime a student's thinking and gradually increase the depth of response required of the student. Tiered sentence starters are leveled in complexity. For example, when learning about forces, I demonstrated pushing a door closed and pulling it open. Then, I assigned the term "force" to the concept by prompting students with, "A force is a _____ or _____." Students responded, "A force is a push or pull." After having the students work with wooden blocks, masses, and spring scales to make observations, I prompted the students to respond to "Forces _____." Here, I listened to how my students described forces. Some students continued to describe forces as pushes and pulls, while others discussed how forces can cause change or accelerations. Thus, I scaffolded for the first group of students by adding an additional tier, "Forces cause _____."

A student with nascent levels of English proficiency may only be able to respond to the lowest tier, but the teacher should gradually increase the linguistic complexity of the sentence starters as the student's English language skills improve. The lowest linguistic tiers are also the lowest cognitive tiers. I once observed a teacher whose ELs were able to repeat back to her the vocabulary that they learned about animal adaptations. She would prompt, "When an animal looks like another animal, that's called _____." The students would shout, "Mimicry!" but when they were later asked to identify a picture demonstrating mimicry, or explain why an animal would mimic another, they were unable to do so. Rather, the teacher could have used the lower tiered sentence frame to give students the words to be able to respond to the higher tiers, and thus help the students to achieve an understanding of the concept of mimicry (Table 16.1).

Table 16.1. Example Tiered Sentence Starters.

	Tier	Example	Additional Language Supports
Increasing Linguistic Complexity / Increasing Cognitive Load ↓	1	Monarch Viceroy image retrieved from: https://commons.wikimedia.org/w/index.php?curid=3611141 This is an example of _____. Or: The non-poisonous viceroy butterfly looking like the poisonous monarch butterfly is an example of _____.	Pictures can be used as additional scaffolding for students.
	2	Mimicry is an example of _____.	
	3	An adaptation is _____.	At the higher tiers, the student can respond using diagrams and gestures in addition to spoken discourse.
	4	Adaptations help animals to survive by _____.	

(Source: Author.)

Discourse Structures

Before I learned about discourse scaffolds I could not understand why students struggled so much to participate in academic conversation. While my elementary students enthusiastically responded to my request to turn and talk to a partner, my high school sophomores and juniors did not. I would ask students to turn and talk to a partner and they would stare blankly at me or mutter a word of response to their partner and stop without engaging in conversation. I realized that producing academic conversation between and among students must be taught. I have found that discourse scaffolds embedded into discourse structures give students the tools to participate in this conversation. Examples of discourse structures presented here include "talk partners," "scientist and engineers' meetings," "consensus placemats" (Virginia Department of Education, 2012), "three-way interviews" (Hopkins, 2005), and "one stays" (Murie, 2016). In each of these structures, discourse scaffolds should be used until students are able to discourse without their use.

Talk partners

A simple discourse structure is "talk partners." Students are taught how to communicate by asking elaboration and clarification questions that deepen content understanding. Teachers assign student pairs based on linguistic need. In my classroom, I paired students at similar linguistic levels so that the students felt comfortable working with each other and so that I could more easily differentiate instruction. In my opinion, talk partners should remain the same for a marking period in order for students to establish familiarity and comfort with their partners.

Talk partners should be used frequently throughout instruction to give students time for processing new information, making connections, and reflecting on their learning. In order to use this structure effectively, the teacher must carefully craft thought-provoking questions that drive students' understanding of the essential concepts or invite students to share their solutions to complex, multi-faceted challenges. I used talk partners to increase student engagement in the lesson. I liked to have students first discuss their answer with their partner, and then I asked some students to share with the whole class. Talk-partners give all students the opportunity to think and respond, rather than just calling on one student to answer a question. Talk partners also allow students to try-out their response before sharing with the whole class, which builds confidence, especially for ELs. I would often pause during direct instruction and ask students to tell their partner the answer to a quick knowledge-level question, repeat or paraphrase important information, or make a personal connection to the material. The knowledge-level questions I have described have their place, as I believe they increase student participation and engagement, but teacher questions should also get at the deeper, essential understandings of the larger unit and give students an opportunity to process their thinking, such as, "How has technology changed the way we live in the 21st century?" These questions can be revisited throughout the unit as students' understanding evolves.

When introducing this structure, I first demonstrated for students how to use the elaboration and clarification questions by having a model conversation with a student, my co-teacher, or the class. I posed the question and then asked my partner to respond. For less mature students, I find that it is helpful to assign Partner A and Partner B and alternate which partner responds first. I then asked a follow-up question using an elaboration/clarification prompt (Table 16.2). After my partner responded, I continued asking elaboration and clarifying questions until we were both satisfied, and then the roles reversed. I found it helpful to post the questions or provide laminated or cardstock copies to the students to access throughout class. I often noticed students referencing the questions during their partner-talk time. When used frequently, students began to use the questions without prompting from the teacher!

Table 16.2. Elaboration and Clarification Questions.

Elaboration questions	Clarification questions
Can you tell me more about _____?	A question I have is _____.
What is your evidence?	Why do you think that?
How do you know?	I think you said _____, is that correct?
Have you thought about _____?	Can you say it again in a different way?

(Source: Author, Collected from Zwiers & Crawford, 2011.)

If I had utilized student talk partners back when I was teaching potential and kinetic energy using the roller coaster example, I could have started with a picture of a roller coaster or a physical model, or presented a more familiar situation such as a car on a hill and asked the students, "Where can we find the energy in this situation?" Later in the unit, after students develop an understanding of the types of energy and the relationships among them, I could have revisited the same scenario and asked, "How does a roller coaster work?" This question is open enough to encourage discourse between students as they work through the relationships among the objects in the system with regards to conservation of energy.

Scientists' or engineers' meetings

Once using the elaboration/clarification discourse prompts become routine, they can be incorporated into Scientists' or Engineers' Meetings. At a meeting, students sit in a circle facing each other, either as a class or in smaller project teams. Students meet to discuss student-generated questions at each phase of the hands-on investigation: planning, investigating, and evaluating (Tables 16.3 and 16.4). Students can plan experiments, engineering designs, and solutions to problems through initial brainstorming sessions. I found the initial planning stages to be a prime time for a scientist meeting, as my students sometimes struggled with designing experiments. Using the wisdom of the class, students were able to identify and isolate variables through the whole-class discussion.

Table 16.3. Scientists' Meeting Discourse Prompts.

Investigation phase	Scientist prompts
Planning	• I think ____ will happen because ____. • I hypothesize ____ because ____. • One question I have is ____. • I think we should start by ____, then ____, then ____. • The independent variable will be ____. • The dependent variable will be ____. • What is the effect of ____ on ____?
Investigating	• The effect of ____ on ____ is ____. • A pattern we notice is ____. • Our results were similar to (another group) because ____. • Our results were different from ____ because ____. • I wonder if ____.
Evaluating	• I wonder if ____. • The effect of ____ on ____ is ____. • The relationship between the variables is ____. • I can infer ____ because ____. • We can conclude ____ because ____. • If we did this again, we would ____. • We agree with (another group) because ____. • We disagree with (another group) because ____. • I can connect this to ____. • In the real world, ____.
General Responses	
• I agree with ____ because ____. • I disagree with ____ because ____. • Can you tell me more about ____? • I like what ____ said about ____ because ____. • I was thinking something similar to ____ because ____. • To rephrase what you are saying, ____.	

(Source: Author, Collected from Fairfax County Public Schools, 2016; Michaels & O'Connor, 2012; Michaels, O'Connor, Hall, & Resnick, 2002; NRC, 2007a.)

Table 16.4. Engineers' Meeting Discourse Prompts.

Investigation phase	Engineer prompts
Planning	<ul><li>Something I know about this issue is _____.</li><li>I think _____ will happen because _____.</li><li>I hypothesize _____ because _____.</li><li>One question I have is _____.</li><li>I think we should start by _____, then _____, then _____.</li><li>The data we should collect are _____.</li></ul>
Investigating	<ul><li>_____ is working because _____.</li><li>_____ is not working because _____.</li><li>Our results were similar to (another group) because _____.</li><li>Our results were different from _____ because _____.</li><li>We should repeat _____.</li><li>I wonder if _____.</li></ul>
Evaluating	<ul><li>I wonder if _____.</li><li>A pattern I notice is _____.</li><li>The effect of _____ on _____ is _____.</li><li>I can infer _____ because _____.</li><li>We can conclude _____ because _____.</li><li>If we did this again, we would _____.</li><li>_____ worked because _____.</li><li>_____ did not work because _____.</li><li>We agree with (another group) because _____.</li><li>We disagree with (another group) because _____.</li><li>The best design was _____ because _____.</li><li>One solution would be _____.</li><li>I can connect this to _____.</li><li>In the real world, _____.</li></ul>
General Responses	
<ul><li>I agree with _____ because _____.</li><li>I disagree with _____ because _____.</li><li>Can you tell me more about _____?</li><li>I like what _____ said about _____ because _____.</li><li>I was thinking something similar to _____ because _____.</li><li>To rephrase what you are saying, _____.</li></ul>	

(Source: Author, Collected from Fairfax County Public Schools, 2016; Michaels & O'Connor, 2012; Michaels, O'Connor, Hall, & Resnick, 2002; NRC, 2007a.)

Once the students have collected data, or tested their engineering designs, the meetings can be a powerful way for students to learn from each other, particularly when there are discrepancies between groups' results. In this structure, the teacher acts as a facilitator, encouraging students to share their results, identify commonalities, and make inferences. I also found this discussion to be invaluable for me as the teacher, because it helped me to pinpoint areas where my students held misconceptions.

The prompts should be made available for students to reference during meetings. Encourage your students to write their own prompts and add them to the list, and re-word the prompts based on your students' linguistic needs. For example, I first started with the prompt, "I think _____ will happen because _____," to help my students make predictions. This prompt became "I hypothesize _____ because _____" once my students associated the concept of making predictions with the word "hypothesize."

ADOPTING DISCOURSE STRATEGIES IN YOUR CLASSROOM

In my experience, the most difficult aspect of using discourse strategies is giving students' control of their learning. While this seems like a desirable outcome, it can be very difficult to be patient while students wrestle with concepts or create an inefficient or flawed design. I found it difficult to remain quiet during scientists' meetings, as the students often looked to me for affirmation after venturing a statement. It can be tempting to re-word or expand upon their answers, but it should be your aim to prompt further thinking through questioning, or even better, teach the students to ask the follow-up questions. When my students discourse with each other, they seem to create more meaning because they are forced to internalize the material. There are other strategies that can help to encourage student-to-student dialogue, including "consensus placemats," "three-way interview," and "one strays".

Consensus Placemats

This strategy is used to help groups of students reach a consensus and can be used at any stage of the inquiry or design process, or be used to help students reflect on deeper concepts. For example, students may need to reach a consensus about the definition of a term such as, "technology." Each student would work alone to write or draw their own definition, and record their ideas on their designated area of the placemat. Then, students are asked to discuss their own definitions and come to a group consensus or generalization to record in the middle (Figure 16.2). Terms that work well for this exercise should be central to your course, be open to interpretation, and require student reflection. These exercises can be a great way to build classroom culture as students must come to an agreement.

Another use for consensus placemats is to help students decide the best way to go about doing something, such as developing their own procedure for an experiment or challenge problem. For example, if students are challenged to build a bridge that can support the most weight, each student can individually brainstorm their design and then come to a consensus about what to build as a team. Students

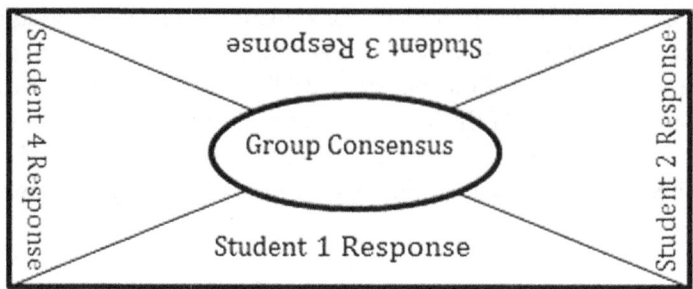

Figure 16.2. Consensus Placemat.
(Source: Author.)

can then revisit their design and suggest modifications after testing using the same procedure.

Three-way Interview

The "three-way interview" is an alternative to consensus placemats and can be used for the same instructional reasons. Students in groups of four pair off and take turns responding to a question, challenge, or prompt. Student 1 and Student 2 interview each other while Student 3 and Student 4 interview each other. Then Students 1 and 3 interview one another while Students 2 and 4 interview one another. Finally, the group members discuss their ideas and reach a consensus. I recommend having students record their own ideas first, then take notes during each interview, and finally, record the group consensus statement (Figure 16.3).

My Ideas	
Interview 1:	
Interview 2:	
Group Consensus:	

Figure 16.3. Graphic Organizer for Three Way Interview.
(Source: Author.)

One Strays

This discourse structure is meant to encourage between-group discourse. Once student groups have formulated their own ideas (possibly using the consensus placemat or three-way interview), three group members remain with their original group and one group member "strays" to another group. The stray then listens to the new group's ideas and then shares his or her group's ideas. The new group should compare and contrast the two ideas, using the discourse prompts provided above. The teacher should use discretion as to how many rotations the stray should do before returning to his or her home group. If the students' ideas are highly varied, more rotations should be completed. Once returning to his or her home group, the stray communicates what he or she learned by visiting the other groups and the home group modifies their ideas accordingly.

CLASSROOM MANAGEMENT CONSIDERATIONS

Efficient and effective STEM classroom discourse can only be realized by establishing discourse routines and procedures. I established discourse norms with students by first discussing how classroom discourse should look and sound. My students needed to be taught active listening skills such as turning to face speakers and giving speakers their full attention. Be aware of students' home cultural norms that may influence this discussion. For example, in some cultures, it is acceptable to talk over other people as doing so demonstrates interest and engagement. The teacher and students should agree to the discourse norms and they should be posted in the classroom.

The teacher may also want to establish procedures for communication, as discourse is highly student-centered and it can be difficult for the teacher to communicate with each group or the class as a whole as students engage in discourse. In my classroom, I established group "spokespersons," a role that rotated with each new class project. The spokesperson is responsible for raising her hand to signal when her group has a question and for communicating her group's question to me. If noise levels become an issue, the teacher may want to have a "noise level meter" displayed at the front of the room where the teacher signals to the students their current noise level compared to acceptable levels. This can be as simple as a stop light poster with Red, Yellow, and Green to symbolize "Too Loud," "Caution," and "Acceptable."

Discourse configurations should change depending on the structure being used. It is helpful to train students to move into various configurations (i.e., partners, team, whole class circle) to reduce class time spent on moving chairs and desks. I was fortunate to have a large enough classroom where I could have an open space for scientists' meetings that was separate from the group lab tables. If you do not have such a space, provide students with diagrams or tape indicators on the floor and give students time to practice moving into different configurations. Challenge your students to move within one minute. Student partners and groups should also change depending on the discourse structure. Consider using flexible grouping strategies to change group members depending on the task at hand or changing student groups as students' linguistic abilities improve and change over the course of the school year.

In addition to classroom norms and routines, I find that it is extremely important to normalize error (Lemov, 2010). Discourse requires students to take risks in the STEM classroom. Normalizing error means that students understand that mistakes are expected and are a normal part of learning in an inquiry-based, collaborative classroom. When a student suggests something incorrect or implausible, I model for students how to respond respectfully to each contribution. I encourage student ideas and help students to feel comfortable. This can be especially challenging in the linguistically diverse classroom, as students may be hesitant to use their developing English. Equal discourse participation must be established as an expectation in the collaborative, inquiry-based STEM classroom. It can help to partner students of similar linguistic levels. I begin the school year expecting only partner and small group discourse from ELs and then gradually increase expectations for ELs to participate in whole-class discourse in scientist and engineers' meetings.

CONCLUSION

Perhaps the 4th-graders I referenced at the beginning of this chapter did not know the English word for "roller coaster," or perhaps they had never seen one, but either way, a hands-on experience with a model or toy roller-coaster coupled with discourse strategies would have provided a much more meaningful and effective learning experience. Discourse strategies structure hands-on investigations so that students must use higher-level thinking skills to explain their reasoning, while at the same time, they address the language needs of ELs by develop their academic English in a scaffolded way. Discourse strategies can help to lessen the language load by providing terms and allowing the use of more familiar language, and then scaffolding to build up to more academic language. Hands-on experiences give students opportunities to explore concepts, and discourse strategies give students the words and formats to name and discuss these concepts.

STEM classrooms that utilize discourse strategies are engaging for students because all students participate. Not only are students cognitively engaged through the active construction of their knowledge, but students also learn that they have a voice and that their voice is valued. Discourse strategies form a linguistic bridge between students' home discourses and the academic discourse of the STEM classroom and give students opportunities to increase their academic literacy in a comfortable, exploratory environment. STEM instruction designed to engage ELs provides students with opportunities to take initiatives during the inquiry process, and ensures equitable learning opportunities.

Initially, putting students in charge of their own learning can be a challenge for both the teacher and the students. It can be difficult for teachers to hand over the reins to the students. We teachers mean well and desperately want our students to learn, so we often try to "ensure" our students' learning by controlling the classroom discourse. I have been amazed by how much more my students can learn when I step out of their way!

Just as teachers can be hesitant to give the reins, students can be hesitant to take them. Some of my own students expressed initial doubts and complained that I was not doing my job as teacher. However boring it may be, students take comfort in their familiar role as passive recipient of knowledge and teacher as "sage on the stage." I explained to my students that I work hard to construct opportunities for them to find their own answers while providing coaching and guidance. I smiled thinking how much easier it would be to stand up and deliver a lecture. It takes courage and persistence from both the teacher and students to continue working through the teacher's and students' new discourse roles, but after the students learn what is expected, they come to see the worth of their own discoveries: students take ownership of their learning, become more curious, and learn the content on a much deeper level. Discourse helps all students to become better thinkers, and to me, developing better thinkers should be our ultimate goal.

WORKS CITED

American Association for the Advancement of Science. (1990). *The nature of science*. Retrieved from http://www.project2061.org/publications/sfaa/online/chap1.htm

American Association for the Advancement of Science. (1993). *Benchmarks for science literacy*. New York: Oxford University Press.

Anderson, C. (2007). Perspectives on science teaching. In S. Abell & N. Lederman (Eds.), *Research on science education* (pp. 3–30). New York: Routledge.

Baker, D., Lewis, E, Purzer, S., et al. (2009). The communication in science inquiry project (CISIP): A project to enhance scientific literacy through the creation of science classroom discourse communities. *International Journal of Environmental and Science Education, 4,* 259–274.

Calderón, M., Slavin, R., & Sanchez, M. (2011). Effective instruction for English learners. *The Future of Children, 21*(1), 103–127. doi: 10.1353/foc.2011.0007

Carlsen, W. (2007). Language and science teaching. In S. Abell & N. Lederman (Eds.), *Research on science education* (pp. 57–74). New York: Routledge.

Gibbons, P. (2003). Mediating language learning: Teacher interactions with ESL students in a content-based classroom. *TESOL Quarterly, 37,* 247–273. doi: 10.2307/3588504

Hardy, I., Kloetzer, B., Moeller, K., & Sodian, B. (2010). The analysis of classroom discourse: Elementary school science curricula advancing reasoning with evidence. *Educational Assessment, 15*(3–4), 197–221. doi: 10.1080/10627197.2010.530556

Hogan, K., Nastasi, B. K., & Pressley, M. (1999). Discourse patterns and collaborative scientific reasoning in peer and teacher-guided discussions. *Cognition and Instruction, 17,* 379–432.

Hopkins, B. (2005). 3 way interview. Retrieved from https://billsteachingnotes.wikispaces.com/file/view/COMMUNICATION+3+WAY+INTERVIEW.pdf

Lee, O. (2005). Science education and English language learners: Synthesis and research agenda. *Review of Educational Research, 75,* 491–530. doi: 10.3102/00346543075004491

Lee, O., & Buxton, C. A. (2013). Integrating science and English proficiency for English language learners. *Theory into Practice, 52*(1), 36–42. doi: 10.1080/07351690.2013.743772

Lee, O., & Luykx, A. (2006). Science education and student diversity: Race/ethnicity, language, culture, and socioeconomic status. In K. Sawyer (Ed.), *The Cambridge handbook of learning sciences* (pp. 171–197). New York: Cambridge University Press.

Lee, O., Maerten-Rivera, J., Buxton, C., Penfield, R., & Secada, W. (2009). Urban elementary teachers' perspectives on teaching science to English language learners. *Journal of Science Teacher Education, 20*, 263–286. doi: 10.1007/s10972=009-9133-z

Lee, O., Maerten-Rivera, J., Penfield, R., LeRoy, K., & Secada, W. (2008). Science achievement of English language learners in urban elementary schools: Results of a first-year professional development intervention. *Journal of Research in Science Teaching, 45*, 31–52.

Lemov, D. (2010). *Teach like a champion: 49 techniques that put students on the path to college (K-12)*. San Francisco, CA: John Wiley & Sons.

Lewis, S., Lee, O., Santau, A., & Cone, N. (2010). Student initiatives in urban elementary science classrooms. *School Science & Mathematics, 110*(3), 160–172. doi: 10.1111/j.1949-8594.2010.00018.x

Lewis, S., Maerten-Rivera, J., Adamson, K., & Lee, O. (2011). Urban third grade teachers' practices and perceptions in science instruction with English language learners. *School Science and Mathematics, 111*(4), 156–163. doi: 10.1111/j.1949-8594.2011.00073.x

Lhamon, C., & Gupta, V. (2015). Dear colleague letter: English learner students and limited English proficient parents. Retrieved from http://www2.ed.gov/about/offices/list/ocr/letters/colleague-el-201501.pdf

Llewellyn, D., & Rajesh, H. (2011). Fostering argumentation skills: Doing what scientist really do. *Science Scope, 35*(1), 22–28. Retrieved from ERIC database.

Michaels, S., & O'Connor, C. (2012). *Science talk primer*. Boston, MA: TERC.

Michaels, S., O'Connor, M. C., Hall, M. W., & Resnick, L. (2002). *Accountable talk: Classroom conversation that works*. Pittsburgh, PA: University of Pittsburgh.

Moje, E. (1995). Talking about science: An interpretation of the effects of teacher talk in a high school science classroom. *Journal of Research in Science Teaching, 32*, 349–371. doi: 10.1002/tea.3660320405

Moje, E., Collazo, T., Carrillo, R., & Marx, R. W. (2001). "Maestro, what is quality?": Language, literacy, and discourse in project based science. *Journal of Research in Science Teaching, 38*, 469–495.

Monzó, L., & Rueda, R. (2009). Passing for English fluent: Latino immigrant children masking language proficiency. *Anthropology & Education Quarterly, 40*(1), 20–40. doi: 10.1111/j.1548-1492.2009.01026.x

Murie, C. (2016). The effects of communication on student learning. Retrieved from http://www.kaganonline.com/free_articles/research_and_rationale/communication_effects.php

National Center for Educational Statistics. (2012). *The condition of education: English language learners*. Retrieved from https://nces.ed.gov/programs/coe/indicator_cgf.asp

National Research Council. (1996). *National science education standards*. Washington, DC: National Academy Press.

National Research Council. (2000). *Inquiry and the national science education standards: A guide for teaching and learning*. Washington, DC: National Academy Press.

National Research Council. (2007a). *Ready, set, SCIENCE!: Putting research to work in K-8 science classrooms*. Washington, DC: The National Academies Press. doi: 10.17226/11882

National Research Council. (2007b). *Taking science to school: Learning and teaching science in grades K-8*. Washington, DC: The National Academies Press.

National Research Council. (2011). *A framework for K–12 science education: Practices, crosscutting concepts, and core ideas*. Washington, DC: National Academies Press.

NGSS Lead States. (2013). *Next generation science standards: For states, by states*. Washington, DC: The National Academies Press.

Nutta, J., Bautista, N., & Butler, M. (2011). *Teaching science to English language learners*. New York: Routledge.

Reiser, B., Berland, L., & Kenyon, L. (2012). Engaging students in the scientific practices of explanation and argumentation. *Science & Children, 49*(8), 8–13.

Santau, A., Maerten-Rivera, J., & Huggins, A. (2011). Science achievement of English language learners in urban elementary schools: Fourth-grade student achievement results from a professional development intervention. *Science Education*. Retrieved from www.wileyonlinelibrary.com. doi: 10.1002/sce.20443

Shea, L., Shanahan, T., Gomez-Zwiep, S., & Straits, W. (2012). Using science as a context for language learning: Impact and implications from two professional development programs. *Electronic Journal of Science Education*, (16), 2.

Shemwell, J. T., & Furtak, E. (2010). Science classroom discussion as scientific argumentation: A study of conceptually rich (and poor) student talk. *Educational Assessment, 15*, 222–250. doi: 10.1080/10627197.2010.530563

Stoddart, T., Pinal, A., Latzke, M., & Canaday, D. (2002). Integrating inquiry science and language development for English language learners. *Journal of Research in Science Teaching, 39*, 664–687. doi: 10.1002/tea.10040

Suarez, E., & Otero, V. (2013). 3rd grade English language learners making sense of sound. *AIP Conference Proceedings, 1513*(1), 406–409. doi: 10.1063/1.4789738

Virginia Department of Education. (2012). Consensus placemat activity. Retrieved from http://www.doe.virginia.gov/instruction/science/professional_development/2012/institute_6-8/08_Consensus_Placemat_Directions.pdf

Zwiep, S., & Straits, W. (2013). Inquiry science: The gateway to English language proficiency. *Journal of Science Teacher Education, 24*, 1315–1331. doi: 10.1007/s10972-013-9357-9

Zwiers, J., & Crawford, M. (2011). *Academic conversations: Classroom talk that fosters critical thinking and content understandings*. Portland, MA: Stenhouse Publishers.

CHAPTER SEVENTEEN

Increasing Literacy Skills IN THE STEM Classroom

APRIL LANOTTE

INTRODUCTION

Literacy in science, technology, engineering, and math (STEM) has a variety of meanings to educators. While many agree it is important for students to be literate in STEM, educators often do not have a solid grasp of what literacy means to them, what it means to others, and what the implications are for student learning. In order to better solidify definitions of STEM literacy in a variety of contexts, and to use these understandings to develop and increase STEM literacy skills in students, two of the most often-cited perspectives are presented.

Perhaps the most common viewpoint for looking at STEM literacy is to focus on the ability to understand STEM content itself. *Science Literacy: Concepts, Contexts, and Consequences* (National Academies, 2016) defines science literacy as the knowledge needed to communicate science: "Access to science—whether using knowledge or creating it—necessitates some level of familiarity with the enterprise and practice of science" (p. 1). Other similar definitions of STEM literacy state that students and the general public need a basic understanding of STEM content such as genetics, Newton's Three Laws, how electricity works, and how to reduce fractions, for example. This type of STEM literacy allows people to carry on discussions about STEM content or about issues that include STEM content (Meeder, 2014; US Department of Education, 2014; Meeder, 2014). The ability to read or properly write about STEM is not a part of this description.

The second viewpoint for looking at STEM literacy, and the focus of this chapter, is to study a person's ability to *communicate* STEM. This thrust makes sure students know how to critically and accurately read STEM content and be able to meet language expectations, how to write in formats that are acceptable in STEM content areas, and how to communicate appropriately both for STEM audiences and about STEM to the general public (using each technique appropriately). While both viewpoints about STEM literacy are equally valid, this second definition looks at some of the deeper reasons many of our students struggle with STEM content: students simply don't know how to read, write, or communicate through the specific language demands expected in science, technology, engineering, and math. Once students can meet these demands, they can better understand the STEM concepts presented to them. In addition, better language skills allow students to enjoy the content more.

When it comes to teaching literacy, STEM teachers often say, "It's not my job. Literacy is the English teacher's job," looking at literacy as an *us vs. them* fight. Science teachers teach science content and English teachers teach communications skills, some say. While it might not be the math teacher's job to teach Renaissance poetry, or the technology teacher's job to help students analyze *The Great Gatsby*, it is every teacher's job to ensure students are literate in different contexts. Since literacy demands in one content area look very different than literacy demands in another, all educators need to be the guides for their specific areas of expertise.

This chapter highlights applicable pedagogical strategies and techniques that can increase student understanding of written and oral materials in STEM content areas. Authentic student literacy skill development can be a part of all content instruction without the fear of the classroom becoming an English class. Since expected communication styles can seem like a foreign language to those not already familiar with the ways STEM experts communicate, STEM teachers who are already fluent in their content area's accepted forms of communications can help their students become better readers, writers, and communicators of this content more so than any other teacher. Specialized skills such as interpreting algebraic equations or developing sound arguments around climate change debates are skills an English teacher cannot be expected to excel in but are areas of expertise for STEM content teachers.

BACKGROUND

Discussions, books, and conference workshops for elementary school and English teachers frequently focus on increasing student literacy skills, yet nearly disappear at the secondary level, especially outside of the language arts world. Quick surveys of STEM conferences such as National Science Teachers Association,

National Council of Teachers of Mathematics, and International Technology and Engineering Educators Association routinely support workshops and sessions on STEM literacy focused on the elementary level, but rarely include offerings for secondary STEM teachers. As a result, students enter secondary education faced with new ways of learning more complex content, yet have very little literacy support beyond elementary school.

It is often assumed that secondary students enter the classroom prepared to read, write, and communicate appropriately for their grade level, and that students readily transfer knowledge from one subject area to another (such as the proper way to write a research report or summarize an article). However, that is not the case. In addition, STEM students have specific literacy skill needs (Zollman, 2012). Without literacy support in each classroom, students may not acquire content-specific communications skills or learn how to transfer skills between subject areas. Toulmin and Meghan (2007) found that STEM literacy also requires, more than other subjects, overlapping interdisciplinary literacy skills, concepts, and processes.

The argument has been made that there is not enough time to add literacy into an already-packed STEM curriculum. However, utilizing literacy skill-building techniques in a classroom should be seen as a way of teaching and not an addition to existing content. In many cases, changing the classroom environment to be more inclusive of both reading about and participating in STEM content through activities or labs increases both literacy skills and content knowledge. Swan (2003) showed that students who observed and interacted with scientific phenomenon *and* had access to related texts about the same phenomenon were more engaged than those students who did not have the same literacy connection. This two-tiered approach to teaching content adds to student content understanding.

LITERACY EXPECTATIONS

When students enter a secondary classroom, they need to be able to read text that matches the needs of the content they will be studying, and will continue this need in higher education and beyond. At the secondary level, this involves reading predominantly expository text versus narrative text. However, studies referenced in the Common Core State Standards (CCSS) have found that students read very little expository text, especially in elementary and middle school, and that this type of text is more difficult for students to read. The shift in complexity and style generally occurs around 4th grade when content delivery changes from a more narrative style to an informational one. Technical terms become more of a focal point than previous readings. Texts become longer, sentence and structural complexity increases, graphical representation becomes more important, and the variety of text styles begins to vary widely depending on specific content area (DeBoer,

Carman, & Lazzaro, 2012). In high school, students are mainly expected to read expository texts, but only on a superficial level. With science and math textbooks containing more expository text than most other textbooks, students need practice and instruction on the best ways to read this kind of text (CCSS, 2010).

Many content teachers leave the teaching of literacy skills to the English teacher, yet these skills are most appropriately taught in *non*-language arts classes. Connections need to be made in STEM classrooms between literacy skills and content to increase student interest and the potential for them to select a STEM field for a career. Ainley and Ainley (2011) found a direct relationship between adolescents' enjoyment of science and their continuing interest in learning science. A common roadblock for this interest and content understanding lies in the difficult vocabulary of STEM-focused texts, since many words utilized in these classes are not commonly used elsewhere.

In the United States, the average student learns at least 50,000 vocabulary words by the time they finish 12th grade, or 3,000–4,000 words per year (Graves, August, & Mancilla-Martinez, 2012). Most people need to interact with a word or phrase 10–15 times before it becomes a regular part of their vocabulary (Skirble, 2010). For English Language Learners (ELLs) or students with learning disabilities, that number is sometimes as high as 60–70 times per word. With this level of required repetition, only the most relevant or most often-used words are retained. With statistics such as these, teaching strategies that help students increase vocabulary understanding is crucial.

Added to the challenge of vocabulary is the complexity of texts. Over the past twenty years, readability levels of science texts have increased in complexity, while students' reading ability has decreased. The National Assessment for Educational Progress (NAEP) reading results for 2015 show that only 37% of the approximately 18,700 twelfth graders who took the test scored at or above the proficient level (NCES, 2015). Stewart (2006) found that text comprehension was more difficult in biology, chemistry, and geology than in physics. Biology and geology texts tend to have long, unfamiliar vocabulary words, and chemistry texts contain more abstract concepts in addition to unfamiliar vocabulary words that all add to complexity. Physics textbooks contained more formulas and abstract concepts, in general, than other sciences, but the words themselves were found to be more familiar, therefore more readable. With many schools using Project Lead the Way (PLTW) curriculum materials for engineering courses, the expectation for courses such as these is that students are able to read and write at the appropriate grade level and are aligned to Common Core Standards as such (PLTW, n.d.).

Math textbooks can be complex for different reasons, including unfamiliar structure, non-narrative format, and sometimes, poorly written materials. These will be covered in more depth later in the chapter. Written materials in a technology class, since there may not be a textbook, often face similar difficulties as math textbooks when teachers use technical manuals and other written materials such as guides and instructions students may not be familiar with (DeWitt, 2013).

Another reason students may struggle with reading include what they are, and are not, reading. Since 2008, Renaissance Learning (Renaissance Learning, n.d.), responsible for the widely accepted and used Accelerated Reader (AR) reading comprehension program, has gathered data on what books U.S. students are reading. Many schools, particularly elementary and middle schools, require students to read books from AR book lists and take comprehension tests in order to earn points. Teachers and/or librarians set point goals for students to earn in a set amount of time. Students can earn more points at one time for reading more complex books, or can read many smaller-point books to reach the same point total. The flexibility of this point system also makes it possible, if students are not properly tracked, for students to remain literarily unchallenged, not needing to read increasingly complex text in order to earn their necessary points.

Accelerated Reader's compiled lists of most-read books during the 2013–2014 school year included data collected from 9.8 million students from 31,000 schools across the country. Their data reflected 318 million book titles (Renaissance Learning, 2014). Results showed that while there were a large *number* of books being read, the reading level of books was not increasing in kind as students progressed through grade levels (see above about the ability for students to avoid increasingly complex texts). For example, in grades 7–12, *The Hunger Games* was the most read book for each grade during 2014, with the exception of 11th grade, where the same book was ranked #2. *Diary of a Wimpy Kid: The Third Wheel* started to appear on the most commonly read book list in grade 3, remained at the top between grades 3 through 6, then continued to appear on the twenty most-read books through grade 9.

The grade equivalency of a text equals the grade and month in school; a 5.6 means an average student should be able to read at this level during the sixth month of fifth grade. Lexile levels are a more complex measure of reading ability, but research that compares Lexile levels to approximate grade level equivalencies shows that Lexile Levels between 420 and 650L approximate 2nd grade levels, and 925–1,070L are around 6th grade (Johnson, 2016. Both grade level equivalency and Lexile levels are not indications of text quality, content complexity, or maturity; instead they reflect sentence length, average numbers of syllables per word, and general ability to read text. *The Hunger Games* is rated at a grade equivalency of 5.3, with a Lexile level of 810L, while *Diary of a Wimpy Kid* is at the 5.6 grade level equivalency, with a Lexile level of 1,060L (Scholastic Books, nd).

Many middle and high school students, it appears, regularly read books at fifth grade reading levels, which will not be enough for them to succeed in postsecondary education and/or the workplace. For a comparison to reading materials beyond the classroom, GED test materials are written at 1,060L, SAT/ACT test materials and military training manuals are 1,180L, and university textbooks average 1,395L (Massengill, 2013). Many technical manuals that may be used in technology classes are also written at higher levels. Apple's MacBook Pro manual, for example, scores above a twelfth grade reading level (Weins & Bluff, 2015). The author's calculation of LEGO's EV3 User Guide also scores well above a twelfth grade reading level due to sentence length, as well as larger vocabulary words. If students are spending much of their time reading vast amounts of books with lower reading complexity (such as a 10th grader reading the text complexity of what is expected in 5th grade), it is unreasonable to assume that they will enter STEM classrooms, which usually contain complex, technically difficult text, and be able to read text without some intervention or support.

In addition, many students do not regularly discuss, read, or write about STEM content outside of the classroom setting (Kenney, Hancewicz, Heuer, Metsisto, & Tuttle, 2005). Familiarity with standard communications styles, the ability to analyze STEM content, and comfort levels working with this language only comes from repeated use and practice. If that is not happening regularly outside of the classroom, it will be difficult when in the classroom. STEM teachers must realize that extra time and attention needs to be devoted within the classroom in order to develop content-fluent students.

INCREASING TEXT UNDERSTANDING

When adding or increasing literacy support for students in a STEM classroom, it is more important for teachers to have the proper mindset than it is for teachers to be experts in language arts. As nearly any reading expert will tell other educators, support should be provided to students before they read, while they read, and after they read (Colorado Department of Education, 2008). Support does not need to be given in all categories each time a student reads. Much like the scaffolding when introducing new STEM content, when students are first introduced to accepted means of communications in a content area, they should receive more support, with a gradual release of responsibility as reading, writing, and communications skills progress.

One way to shift the STEM education mindset is to look at reading, writing, and communicating as *active* processes. Math teachers would rarely expect students to know how to solve linear equations without guiding students through the process first. This is an acquired skill, and teachers take an active role in the learning process. In that same manner, STEM teachers need to look at how they, themselves, read, write, and communicate and model these behaviors with students. By checking to see if students understand what they read, literacy components become much more active and meaningful.

Meeting the requirements of reading and communicating STEM content can be like a foreign language for many. Schwartz (2008) proposed that people tend to think about math content in two major categories: mathematical nouns and mathematical verbs. Mathematical nouns are objects such as numbers, measurements, shapes, etc. Mathematical verbs are actions applied to problem-solving and reasoning. Verbs can be broken down into four categories: modeling and formulating; transforming and manipulating; inferring; and communicating. Ensuring a good mix of both objects and actions is a way to teach students more than a list of vocabulary words, and makes it easier to look at how material should be presented to students.

Just as it is with learning any language, learning nouns and verbs in isolation does not make someone fluent in that language. For that, Schwartz (2008) refers to two other necessary categories: *content* and *process*. Content, often referred to as *topics*, contains both objects and actions. Process is a more content-general set of procedures followed in mathematics practice and is not limited to the specific topic or skill being learned, but is applicable on a wider scale. Fluency comes from utilizing all categories.

Successful Versus Struggling Readers

Perhaps the most overarching topic within literacy research is the list of qualities that successful readers possess (RAND Reading Study Group, 2002). Different research studies highlight different traits (anywhere from four to twenty); this chapter will focus on six qualities. These skills speak to more than reading and are skills that can be developed in any content area. RAND's research shows that successful readers can summarize, question, predict, visualize, self-monitor, and evaluate. Many STEM educators already work on these skills with students in relation to learning STEM content, but may not actively apply them to the literary components of their classes.

Successful readers question text (Rafael, 2010). Critical thinking and questioning are skills that STEM teachers foster, so asking students to do the same with the text they read complements existing teaching strategies. Struggling readers tend to approach all written text in the same way (Tankersley, 2005). These students may read a poem the same way they read a scientific article. Guidance that includes asking pre-reading questions about text format, purpose, and other considerations can help them place text in the correct context. Asking students to consider the source and viewpoint or to look at the article's purpose prior to reading can make students critical readers. While most experts already do this before they read an article, many students will not analyze where the article is coming from or why it is being written.

There is little a teacher can do to change how prepared students are when they first enter their classroom. However, helping these students read successfully in class through modeling can break a cycle of non-reading. Tankersley (2005) states, "we have to help them see reading as a bridge to learning more about the things that matter to them" (p. 7). Many adults who label themselves readers first became readers through the interest generated during a content class and not in an English class. McRae and Guthrie (2009 documented that readers who are intrinsically motivated spend 300% more time reading than those who have low intrinsic motivation for reading. Compared to other motivations, intrinsic motivation for reading was most highly associated with whether or not readers read widely and frequently on their own accord.

McRae and Guthrie (2009 found that most English teachers support their texts, meaning they either read text aloud and/or use text with students in the classroom before students are expected to work on their own. In comparison, they found that text is used but is mostly unsupported in the science classroom-it may be used as a basis for homework, yet is rarely incorporated into the classroom setting. Actual text from a math book is hardly used; instead, textbooks are primarily resources for math problems and do not often require students to use text to guide them through the process of understanding mathematical content. As a result, students do not have the literacy skills to use the textbook for guidance when needed.

BEST PRACTICES

As an educator with over twenty years of teaching experience in both secondary English and secondary science, I have a passion for English and science content areas, and perhaps a more global perspective on education in general. I currently teach the *Reading in the Content Area* course for my university's secondary science, math, and engineering teacher preparation program, which allows me to work with future STEM teachers and also pull from my own classroom experiences incorporating literacy techniques into a science classroom, as well as my experiences utilizing STEM content in my English classes.

When I first started to teach *Reading in the Content Area*, I was surprised by the resistance pre-service teachers had about integrating literacy strategies into STEM lessons and activities (and have since discovered the same to be true of many in-service teachers I work with, as well). What I initially thought would be a course spent energizing students and empowering them to add cross-curricular teaching strategies to their lesson plans and activities turned into a battle to convince them that literacy was a crucial skill students needed and did not have. I also had to convince them that STEM teachers were the only teachers truly qualified to develop these skills in students.

During that first semester, I found that my students and I were not speaking the same language; the literacy I was thinking about was not the literacy they were thinking about. I also made a mistake in assuming interest, which many of us do in our content areas, no matter the subject. Now I spend the first part of the course introducing students to what is meant by STEM literacy. Once we speak the same language, once they understand that they are the *most* qualified person to teach content literacy to their students, they realize that they should teach literacy. And many of them want to, since the course also gives them the tools to do so.

What students found from course content, some of which is included here, is that solutions are not as time-consuming or as difficult as they may seem, and increasing literacy in the classroom is not an "all or nothing" venture. Many literacy strategies can be implemented one at a time, when appropriate. Not everyone who teaches a STEM class comes from an English background, as I did. Yet the following literacy strategies do not require such a background.

WHAT I LEARNED ALONG THE WAY

When I taught secondary students, I used to dislike using the textbook. I avoided it whenever possible and found it far more interesting to use current events articles and research, videos, labs, active lessons, and notes when necessary. In physics classes we predominantly worked on projects. We were always in the lab in chemistry class, and for biology classes, it was rare to find us in the classroom. For all of my classes, we were literally out in the field. In my English classes, I pulled heavily from articles, stories, and other non-textbook readings, preferring to engage students through real-life examples and primary source materials. Many of my students enjoyed the class structure, and for a large part, students succeeded in my classes. Interest levels were high, along with standardized test scores. Administrators rarely had to deal with student behavioral issues since there really wasn't much time for students to do anything except work.

I don't regret much about how I structured my classes, except when I reflect on the preparation—or lack of preparation—that took place in getting my students ready for certain aspects of college. Some of these students went straight from my active classroom into a lecture-based introduction to chemistry course in college and were unprepared. They suddenly had to read several chapters of text at a time, listen to droning lectures, then take tests on that material. Some had no idea what to do. If thinking about a student's ability to utilize a textbook, I also think about a student's lack of ability to use their math textbook, where they are often referred to only for problem samples. Instead, occasionally asking students to use the instructional portions of the math text while in class, rather than walking students through specific instructions on the board, for example, can help them learn to use the book. Then, if a student is absent or is having trouble with homework, they have tools to support their own learning. Since more technical textbooks can be particularly difficult to read, added support in class will help them outside of class.

Now, I approach the secondary classroom with a mixture of strategies. I still prefer an active classroom, full of inquiry and roll-up-your-shirtsleeves kinds of adventures. However, I also realize that I am not their only source of information and content and that they need to learn other skills, too. More than anything, they need literacy skills. They need to be able to locate quality sources of information, critically read text, and appropriately communicate STEM content. Students need to be more self-sufficient, and integrating literacy skills in the STEM classroom can do that.

Keep in mind that the level of content knowledge a student possesses does not necessarily relate to how well a student can read, write, or communicate about STEM content. An honors trigonometry student is not necessarily a strong reader and writer. Likewise, an all-star English student may not be able to read and write at the same level in a science class as they do in their English class since the types of communication for each subject area is very different. The English teacher will be guiding students through literature; STEM teachers should be doing the same with their content. Earlier in my science teaching career, I made the mistake of assuming that some of my more advanced students would automatically transfer their knowledge of writing research papers in English class, for example, into my biology class, where they had to prepare a research paper on sustainable energy. I remember one honors student, in particular, who literally copied and pasted information from several internet sources directly into a research paper. He did not even bother to change the font of each source. When questioned, he seemed confused. "But this is a paper for biology class, not English class," he explained. That was the first time I really began to think consciously about the need to model what was expected of my science research assignments. By simply pointing out my expectations and how those expectations aligned with, and diverged from their English class assignments, many frustrating hours of grading poor quality research papers were saved. As a result, quality increased, and I suspect, student frustration levels decreased since they knew what was expected.

A Few of My Favorite Things

Over the years, I've found and adapted several of my favorite, go-to techniques and strategies that I find myself using in many situations, with each of these showing the best results from my students. These are great places to start when first incorporating literacy into a STEM classroom. Further on in this chapter, in the More Literacy Techniques and Strategies section, additional ideas are included to allow educators to add more literacy based on their comfort level, incorporating strategies ranging from the introductory literacy level through more advanced techniques and strategies.

Word walls and word maps

Cronsberry (2004) has shown that word walls have the ability to increase student independence in reading and writing skills in the secondary classroom as well as in elementary classrooms, where they are more common. Word walls are filled with vocabulary words students need to know and provide visual reference for words they have already been taught. A variation on word walls is the word map, where words—either on paper or on the wall—are arranged in a specific way in order to convey word associations.

There are different ways to display word walls in the classroom. Oftentimes, words are written or printed on cards and reflect current content vocabulary. Cards may be color-coded to help organize words, and may also be arranged as word maps (Photo 17.1) to help students establish relationships between words. Maps can also become frameworks for larger writing projects. While word walls may be teacher- or student-generated, for maximum effectiveness, student involvement or student-led word walls are recommended. In my own classroom, I often found that students would look back at sections of the wall where words used to be, or even when these words had to be covered over during standardized testing.

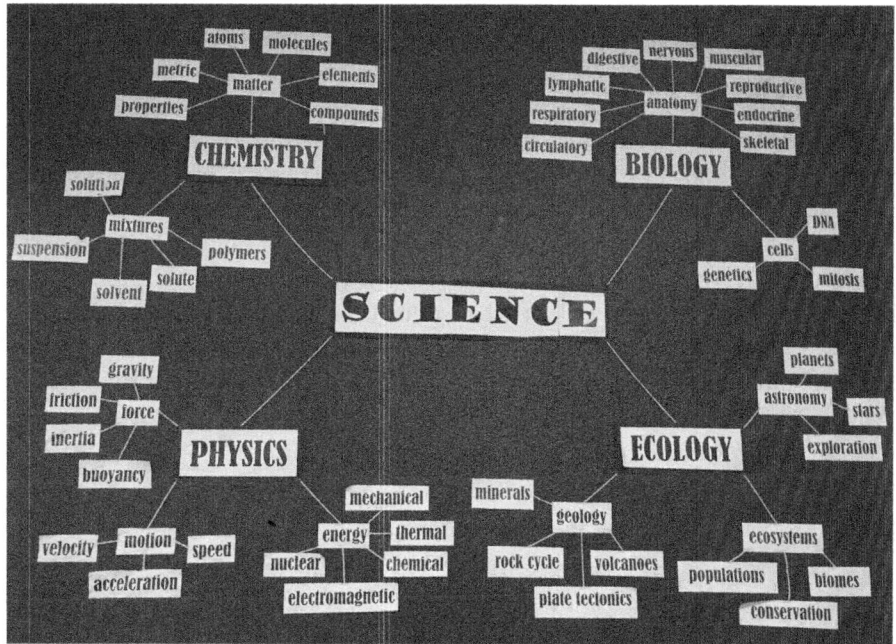

Photo 17.1. Word Wall in an 8th Grade Science Teacher's Classroom.
(Photo by April Lanotte.)

Word maps require students to think about and describe words in several ways rather than just writing one definition. The Frayer Model (Figure 17.1) is a commonly used map.

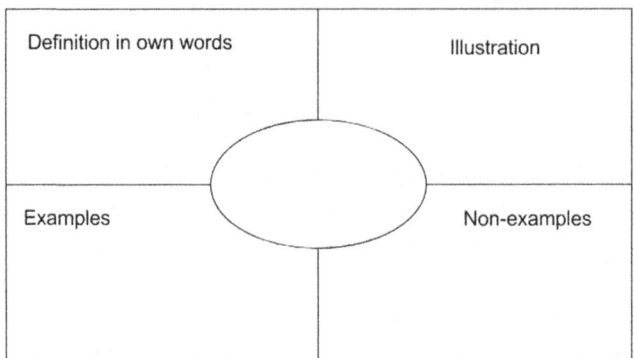

Figure 17.1. Frayer Model.
(Source: Author.)

Boxes allow multiple ways to express a definition (such as drawings, examples, or synonyms). The *non-examples* category can be particularly powerful since knowing what a word does *not* mean is sometimes more difficult than a definition alone.

Readability tests

Sometimes one of the struggles students have in a STEM classroom is that the material they are expected to read has too high a reading level. Conducting a simple readability test, such as the Raygor Readability Test, allows educators to calculate the approximate grade level equivalency of a text. A word of caution, however: tests measure general sentence and word lengths, but do not measure context such as the background knowledge (or lack of) that students bring to the classroom, student interest level in a topic, colloquial language or nuances in word usage that may affect comprehension, or appropriateness of the topic for a given grade level. Readability tests are helpful for written text, but there is no standardized readability test that can calculate grade equivalency for mathematical formulas (Johnson, 2016).

Modeling

As content experts, STEM teachers are already familiar with ways to properly interact with technical text. By modeling what we do as active readers in our content areas, we can significantly help students understand how they should approach reading STEM content-either in class or outside of class. Being aware of and pointing out some of the many potential pitfalls in technical text can help students understand what they are reading by following a teacher's example. In particular, modeling how we approach and deal with common textual problems will show students what to do when they, too, encounter these challenges. Some of these challenges are:

- overly complex sentences and unfamiliar vocabulary
- words that have very different meanings in the scientific world than in everyday life
- unfamiliar page layout (separate graphs and pull-out areas on a page that may confuse a student rather than enrich their understanding)
- faulty arguments in an article

Math texts, in particular, often suffer from:

- confusing writing styles or poor writing
- more dense sentence structures than other types of writing
- lack of student interest in content

Learning charts

Charts that help students keep track of what they know, what they want to know, and what they have learned (K-W-L chart) or similar variations can allow students to become more active readers and can set the stage for students, helping them focus on aspects of what they are about to read (Ogle, 1986; Sejnost, 2007). In the KWL example (Table 17.1), five columns are included that can be used in part, or in their entirety, depending on the classroom situation.

Table 17.1. Modified KWL Chart Sample.

What do I KNOW about the topic or problem?	What do I WANT to know about the problem or topic? OR What does the problem want me to find?	What included information do I NOT need?	What STRATEGY will I use to solve the problem?	What have I LEARNED? OR what is the solution?

(Source: Author.)

ADDITIONAL LITERACY TECHNIQUES AND STRATEGIES

Each of the teaching techniques and communication strategies provided in this chapter helps improve student literacy skills. How much literacy to add depends on each situation and on the teacher's comfort level. Additional strategies and techniques in this section, along with the strategies already mentioned, can be added over time or when appropriate. With so many different literacy strategies and activities to choose from, it can be helpful to look at activities as pre-, during-, and post-activities, making reading, writing, and communicating more active and a more valuable part of learning content at each stage.

Pre-reading Writing Strategies and Activities

Pre-reading and writing techniques can help students before they begin to read. If text is too difficult for students, they will need extra support as they read, and teachers should not expect students to read too much on their own. If a text is too easy for students, additional or different materials can be given to challenge students. Three strategies are creating a Content Area Reading Inventory (CARI), making predictions, and Write-to-Learn.

A CARI can assess individual students' abilities to interact with text. While a readability test provides general guidelines about what average students in a certain grade level *should* be able to read, the CARI assesses where specific students *are* in terms of understanding a text. A CARI is not used to assess a student's comprehension of content, but is used to assess whether they can *read* a specific text. To create a CARI, students read a short passage from the text, then answer 10–15 questions based on the text and text structure. Items found in the index, table of contents, graphs, and other sections can show a student's ability to locate information. Questions to assess students' abilities to make sense of text can also be used (Sejnost, 2007).

Successful readers make predictions about what they will be reading, which allows them to start thinking about the text even before a page is read (Benchmark Education, nd, Brown and Briggs, 1989). Asking students what the text section might cover, where they would look for specific information, and asking them how they might approach reading a specific text provides context. Teachers can also ask about key areas and guide students through their own reading of lab procedures or instructions rather than reading to students. For articles, teachers can ask students to predict what the story will cover. Since many struggling readers approach all text in the same way, asking students to think about a reading prior to reading may help them approach the text differently.

A Write-to-Learn (WTL) activity is a form of writing that can help students begin to think through information and develop their thought processes through writing. In a WTL, students use the process of writing to unlock and/or synthesize information through less formal, typically ungraded assignments. This changes writing to a learning tool versus an assessment tool, and as a result, can remove some of the stress certain students feel when needing to write as a performance assessment. While writing, students work their way through content, further developing their thoughts as they write. Examples of WTL strategies include: asking students to be a drop of water and to write about going through the water cycle; writing unsent letters to famous mathematicians; creating journal entries or warm-ups to get students to actively think about what they do or do not know about a topic, taking notes, writing blogs, making drawings, or creating fictitious tweets about a topic area. Asking students to write about their past struggles with factoring or with balancing chemical equations can help them begin to think specifically about their skills prior to working with them in class, without stressing about the writing itself.

During-reading, Writing, and Listening Strategies and Activities

Providing support while students read, write, or listen helps them become more conscious and active about what they are reading, writing, or listening to. Any teacher who has asked students to read a passage then discuss its content, only to face a sea of blank faces during the discussion can tell you that many students appear to be engaged, yet cannot recall what they just read. Oftentimes, by adding writing components or providing guidance while they read or listen can go a long way in improving skills. For many educators, teaching vocabulary, expecting students to make observations or take notes, and asking students to read material are already a part of the classroom routine. Adding more substance to these tasks can increase student literacy skills. In addition, introducing current events articles and using a trade book or primary source material to teach content outside of the textbook format can increase student interest, particularly students who may otherwise feel disengaged from STEM content.

Since vocabulary instruction is an enormous component of guiding students through STEM content, framing the types of vocabulary words students are expected to know can help teachers figure out what, and how, to teach or clarify a variety of words.

STEM content teachers will not need to focus on teaching students basic words; focus will instead be on words that have content-specific applications. Some of these words can be somewhat universal in meaning, such as *analyze, describe, and graph*, yet their appropriate usages intersect and diverge with other content areas in certain contexts, so it is important to point this out to students. *Theory* and *hypothesis*, when used in a science context, for example, take on different meanings than they do in everyday language. Other, more specific content-specific words require guidance and instruction that content teachers are uniquely qualified to teach.

Literacy research by Catherine Snow shows nine basic principles that can help with vocabulary instruction (Snow, 2010). Within these principles, strategies include: picking words that are part of word families, contain prefixes and suffixes, and can be taught in groups vs. one word at a time; providing multiple opportunities to use words; providing learner-friendly definitions; cultivating word awareness; and showing how words relate to other words. Utilizing a mixture of any of these strategies creates a classroom environment where students have the opportunity interact with vocabulary in more complex ways than simple memorization.

Schwartz and Kenney (2008) noted in their research on literacy and mathematics that learning vocabulary out of context does not make one fluent in a language. In order to add more context in the classroom, there are many different strategies that can help with this. For example, rather than using a standard fill-in-the-blank vocabulary activity, ask students to more actively use vocabulary words in context. To deepen the usefulness of vocabulary notecards, students can also use the cards to create a graphic organizer, arranging words to illustrate where these words belong in relation to one another. Students might create a hierarchical structure of words, a flow chart, or a compare/contrast graphic.

In a graphic word organizer (Figure 17.2), middle school students may be learning about atoms. If graphically organizing vocabulary, they may choose to organize the list of words, and then explain their organization, allowing for assessment of their understanding.

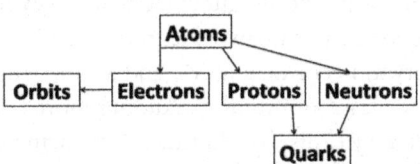

Figure 17.2. Graphic Word Organizer Example.
(Source: Author.)

Making written observations and taking notes improves both writing and observation skills. Observations do not always need to be narrative; labeled illustrations also require many of the same communications skills as writing in a more traditional format. Taking notes during class is also an effective way to enhance student writing skills, particularly if students create their own notes and do not copy teacher-created notes. Cohen, Kim, Tan, and Winkelmes (2013) found that when students synthesize their own notes, they make information their own since they are processing, restructuring, and presenting information in their own ways, which are all crucial components of learning material. Asking students to draw about what they just read can also help them visualize complex concepts.

As students read material, STEM educators need to expect the most of their students' reading. Close reading, the opposite of skimming material, requires students to develop deeper understandings and to analyze complex texts. Students critically think about what they are reading, first reading, then re-reading a text passage to find multiple layers of information within one section. Using probing questions that require students to look back at a text they've already read forces students to reconnect with text and analyze content. Teacher-led questions should require that students re-read that section, focusing on a specific task such as looking for implications of a research study, implied meanings, or even ambiguities in the passage (Moss, Lapp, Grant, & Johnson, 2015).

Current events articles offer students the opportunity to read about timely STEM content and typically have higher student interest than textbook information alone. Many student-based current events resources also have supplementary lesson materials available. The Appendix contains suggestions for reputable article sources. Two strategies, guided reading and leveled readers, can provide support for students as they read current events materials.

Most students are not used to reading current events articles, textbooks, and many other sources of informational text; they need guidance. It is easy for students to become distracted by pictures, accompanying videos, or any number of items within an article. Providing guidance through worksheets that help students work through an article can help them focus on specific aspect(s) of the article. Guided reading worksheets are different from comprehension questions at the end of a chapter or article; guided worksheets are completed *as* students read text to help support desired focal points. Because of its somewhat scripted nature, guided reading is most helpful for struggling readers.

Since students read and comprehend at different levels within one classroom, it is important to differentiate but still cover the same basic content. Several websites, included in the Appendix, provide leveled readers, which offer different reading levels for the same article. Each level covers the same story, but has more or less detail and differing writing complexity depending on the level(s) chosen. The class can then have a collective discussion about the content without the worry of leaving students behind or not challenging other students.

Trade books are novels or books, fiction or nonfiction, that tie to content in a less formal way than a textbook. Books such as *The Immortal Life of Henrietta Lacks* by Rebecca Skloot (2010) can be used to teach students about cells in culture, as well as historical context for scientific research. Fictional stories like *The Phantom Tollbooth* (Juster, 1961) can give middle school students the opportunity to play with numbers. Biographies, as well as books such as *Spare Parts: Four Undocumented Teenagers, One Ugly Robot, and the Battle for the American Dream* (Davis, 2014) and *The Martian* (Weir, 2014) can be used to supplement engineering and technology classes. For many of these books, lessons, activities, and even entire units have already been created by other teachers and can be found online. Like current events articles, the appeal of trade books is the increase in interest level for students.

Two cautions about the use of trade books involve readability levels and appropriateness of content. A quick readability test can let the teacher know how difficult the text will be to read from a grade level standpoint. Books may also need to go through an approval process for appropriateness of content.

Utilizing primary sources in a STEM classroom is also a natural fit. There are many primary sources that are invaluable for STEM content research, and with the rise in technology, many of these sources are available for students to access and use via the Internet. The Library of Congress has been digitizing much of their material and is continuously compiling sets of primary sources for teachers that can be accessed by topic or theme, along with online professional development that shows teachers how to access and use primary sources in the classroom. For example, a themed set of Wright brothers materials includes lesson plans, photographs, letters between the Wright brothers and other aviation pioneers of the time, newspaper articles, drawings, and telegrams. Links to this and other primary source materials available through organizations like Newseum and Smithsonian Institution are provided in the Appendix.

Post-reading Strategies and Activities

After students read content, it is important for them to question what they just learned, summarize and analyze that content, synthesize their thoughts, and present their material. Formal writing assignments are ways to do this. In addition, alternative assignments and summarization strategies can also help teachers assess how well students learned specific material. Multiple forms of assessment allow different types of students to communicate their learning in ways that work for them.

Students should question text before, during, and after they read, just as they should question any information they learn. Asking *How* and *Why* types of questions to students after they read requires them to think more critically about what they've just read. *What if?* questions can also force them to think deeper about the text. Many students spent early elementary school years questioning text, yet become more resistant to this as they get older. Prompting students to do this once again can help them better understand what they've read.

Many teachers assign a formal research paper in their STEM classes. Formal writing assignments evaluate student understanding of content knowledge as well as students' abilities to communicate in an academic style. So how does a STEM teacher assign a formal writing assignment without becoming the grammar teacher? One solution is to focus primarily on the paper's purpose, deciding what the major goals are for the paper. While grammar, writing style, proper citation, etc. are also valuable components of a research paper and should be evaluated, content teachers should be careful not to spend the bulk of their time and energy searching for grammatical errors. Focus should be on the accuracy of content, validity of claims, use of supporting evidence, and other factors that are crucial to STEM experts. Mirroring what would happen if a paper was submitted for publication in a professional journal and contained poor grammar, the paper can be returned with instructions to correct these errors. Problematic sections can be highlighted or circled, providing general guidance such as "your citations need to be improved." Allow students to take responsibility to correct their own writing. Grading rubrics should include a category or two for grammar, writing, proper citations, etc., which will reinforce the need to include all of these elements; however, the percentage of the overall grade for these categories should not overshadow content.

Teachers need to give students guidance and provide expectations for research and final products when writing STEM-focused research papers, including ways to find credible sources and showing students professional STEM publications or samples of former student work. While the teacher may have read over 100 papers, most students have only seen their own, and perhaps one or two others.

Expectations vary in different situations, and students cannot be expected to automatically know what is required. Honors English students will not automatically apply the same research and writing skills to a STEM class. Simply pointing out that the same expectations apply in STEM classes, or highlighting differences, can eliminate frustrating hours of grading papers that do not meet teacher expectations.

There are other ways to assess student understanding beyond the formal writing assignment. Alternative writing formats can take focus away from the stress of writing, oftentimes improving student writing and enjoyment. Writing activities include writing unsent letters to mathematicians, scientists, engineers, or others; creating comics to summarize or explain a concept; and summarizing a concept on one side of a sticky note. Students can make trading cards of mathematicians, engineers, or even concepts or items such as the Pythagorean Theorem or the Hubble telescope. Other assignments can include conducting interviews with famous people, or even interviewing concepts such as sines and cosines. All provide different ways to show understanding and potentially interest different students.

ADAPTING YOUR STEM CLASSROOM TO INCLUDE LITERACY

Introducing literacy skills in a STEM content class is not an *all or nothing* venture. STEM teachers should spend time thinking about their own specific literacy skills and the ways they interact with the STEM content they read, write, and communicate. This can help develop ways to model literacy skills to students. Establishing a proper mindset for including literacy opens the classroom for more student success. Focusing more on literary approaches to learning content does not necessarily mean taking much more time away from content, but is more about reframing how material is presented. Thinking about pre-, during-, and post-reading strategies will improve student literacy skills and there will be less time spent reviewing or re-teaching material.

When first introducing new literacy-focused approaches to teaching STEM content, some students may not respond positively at first, while others will be more engaged than ever. Students, like adults, get used to the way they have been taught, and may not like change. Don't give up. The material presented in this chapter asks more of students, and places more responsibility in their hands. The teacher is the guide, setting the stage and modeling how reading, writing, and communicating is done by those who know STEM content. After that, students are responsible for their learning and their communication skills. In time, they will see that they are in charge of what they learn, and that they have tools to do so.

What NOT to Do

Not all literacy techniques are helpful. There are a few ineffective, popular, and sometimes detrimental literacy strategies that should be avoided. These include popcorn reading, reading large text passages, and placing a disproportionate focus on grammar during the writing process.

Popcorn reading is a technique where one student reads a section of text, stops reading at any point, and randomly calls on another student to pick up where they left off, continuing until the text selection is finished. While this might force students to pay attention, Finley (2014) has shown this to be completely an ineffective and detrimental reading strategy because it stigmatizes poor readers, weakens comprehension, and sabotages fluency and pronunciation. Cope also notes that "unrehearsed oral reading is the single most negative experience reported by adolescents about their entire school experience" (as cited in Tankersley, 2005, p. 3). For English Language Learners, it can be especially detrimental, because they need to hear how text is *supposed* to be read, and this technique breaks up sentences and leads to mispronounced words and halting sentence flow. As an alternative, a strong reader can read the passage aloud, an audio recording can be used, or students can read to a partner then switch. The most effective option, research shows, is for students to silently read to themselves, then discuss the reading as a group (Finley, 2014).

Assigning large chunks of text for homework, particularly with little to no direction, is not productive. With these vague reading assignments, students often return to class the next day and have little to contribute about what they read. Instead, keep reading assignments smaller and more focused. Before reading, provide the context for reading, with a few questions to guide students or something to think about while they read.

When it comes to writing, nothing takes the interest away from writing or literature faster than an obsession with grammar. Leave deeper discussions about prepositional phrases to the English class. Instead focus on critical content analyses that pertain to STEM content expertise.

CONCLUSION

The *Reading in the Content Area* course in our university's secondary teacher preparation program was a long-dreaded course. Former students now return to campus and share the ways they are using the course material, some of which is contained here, in their own classes. One former student was really excited to integrate literacy into her own class and could not wait to try it out. Like me, she was excited to inspire students with the addition of literary activities, only to be met with sour opposition from skeptical students. "This isn't English class," they grumbled, complaining about daily warm-up questions that required writing and personal reflection. After a while, students stopped complaining as much, and one day, when a large project was getting too far behind and the teacher put daily warm-up questions on hold for a few days, she found herself answering to the resistance students had about *not* writing. Daily writing exercises had become a part of the class, and they wanted to keep it up. They had a reason to write, and they liked it.

Not all students love to read and write. For that matter, not all teachers like to read and write. Some students have not been expected to write very much, particularly in math and science classes, and when they have done so in these classes, it has usually been in an evaluative context. What STEM teachers often forget, in contrast, is that other students dread math and science classes because they *do not* get to write in these classes. Not all of these students are excited to learn math and science content (or they do not *think* they want to learn that content, perhaps because they haven't been able to show their knowledge or learn content in a way that appeals to them). No matter if students, or teachers, want to read, write, and communicate STEM, they are essential skills. Integrating more STEM literacy into classes will not only improve student skills in these areas, it will also have the secondary effect of increasing student critical thinking skills overall.

It is clear that there is a need for *all* teachers to teach literacy, and that for certain skills, the content teachers are best prepared for that job. STEM teachers have more literacy skills than they realize; they use them in their own lives, oftentimes without thinking about them. Even STEM teachers who do not consider themselves good readers or writers find, upon reflection, that they have their own strategies for comprehending new vocabulary, strategize before they read text, are critical of information sources and quality, and make appropriate adjustments to their audience when communicating STEM content. Being models for students to follow is one of the easiest ways to begin incorporating these skills into the classroom. There is little fear of students reading, writing, and communicating *too much*, especially if that practice is deliberate and builds on both literacy and its application in content areas. With STEM teachers being the best prepared to teach students STEM content, who better to teach students STEM literacy skills than the STEM teacher?

APPENDIX

The following literacy resources have either been discussed in the chapter or are good resources to use when including literacy techniques in the classroom.

Leveled Reading

NASA's "Cosmic Times" series of astrophysics lessons and activities utilize three reading levels for content: http://cosmictimes.gsfc.nasa.gov

Newsela is a current events resource where stories are available at a variety of reading levels as well as a Spanish version of the article. Comprehension questions and writing prompts are also included: https://newsela.com

Primary Source Materials

The Newseum is a museum that focuses on the history of news, including newspapers and other media. They also have teacher and student resources for primary sources: https://newseumed.org

The Library of Congress offers many primary source materials, including the sources themselves as well as teacher supplementary materials: http://www.loc.gov/teachers/classroommaterials/primarysourcesets/

The Smithsonian Institution also offers primary source materials, such as documents and photos, lesson plans, and videos. Items can be found individually and are also compiled by topics such as *Invention, Transportation,* and *Colonial America*: http://www.smithsoniansource.org/display/primarysource/search.aspx

Pedagogy Textbooks That Support Reading and Writing in Content Areas (Books I Use in the *Reading in the Content Area* Course I Teach)

Kenney, J., Hankewicz, E., Heuer, L., Metsisto, D., & Tuttle, C. (2005). *Literacy strategies for improving mathematics instruction.* Alexandria, VA: Association for Supervision and Curriculum Development.

Sejnost, R., & Thiese, S. (2007). *Reading and writing across content areas* (2nd ed.). Thousand Oaks, CA: Corwin Press.

WORKS CITED

Ainley, M., & Ainley, J. (2011). Student engagement with science in early adolescence: The contribution of enjoyment to students' continuing interest in learning about science. *Contemporary Educational Psychology, 36*(1), 4–12. doi: 10.1016/j.cedpsych.2010.08.001

Benchmark Education. (n.d.). *Comprehension skills, strategies, and best practices.* Retrieved 2016, September 28, from http://www.benchmarkeducation.com/best-practices-library/comprehension-strategies.html

Brown, D., & Briggs, L. D. (1989). Success in reading: Four characteristics of strategic readers. *Reading Horizons, 30*, article 4. Retrieved from http://scholarworks.wmich.edu/cgi/viewcontent.cgi?article=1650&context=reading_horizons

Cohen, D., Kim, E., Tan, J., & Winkelmes, M. (2013). A note-restructuring intervention increases students' exam scores. *College Teaching, 61*(Summer), 95–99.

Colorado Department of Education (2008). *Understanding the relationship between reading comprehension skills and strategies* [PDF document]. Denver, CO: Colorado Department of Education.

Cronsberry, J. (2004). Word walls: A support for literacy in secondary school classrooms [PDF document]. Retrieved from http://www.readingrockets.org/content/pdfs/World_Walls_-_A_Support_for_Literacy_in_Secondary_School_Classrooms.pdf

Davis, J. (2014). *Spare parts: Four undocumented teenagers, one ugly robot, and the battle for the American dream.* New York: Farrar, Straus, and Giroux.

DeBoer, G., Carman, E., & Lazzaro, C. (2012). The role of language arts in a successful STEM education program. [Powerpoint slides]. Retrieved from https://research.collegeboard.org/sites/default/files/publications/2012/7/presentation-2010-12-language-arts-stem-k12-implmentation.pdf

DeWitt, P. (2013, March 14). Do textbooks still have a place in schools? [Web log post]. Retrieved from http://blogs.edweek.org/edweek/finding_common_ground/2013/03/do_textbooks_still_have_a_place_in_schools.html

Finley, T. (2014). 11 Alternatives to "round robin" (and "popcorn") reading. [Web log post]. Retrieved from http://www.edutopia.org/blog/alternatives-to-round-robin-reading-todd-finley

Graves, M., August, D., & Mancilla-Martinez, J. (2012). *Learning words in a second language.* In Teaching vocabulary to English language learners. TESOL Press/Teachers College Press. Retrieved from http://www.tesol.org/docs/books/bk_teachingvocab_750.pdf?sfvrsn=2 (accessed 2016, September 23).

Johnson, K. (2016). *Measuring the reading age of books and other reading matter* [PDF document]. Retrieved from http://timetabler.com/reading.html

Juster, N. (1961). *The phantom tollbooth.* New York: Scholastic.

Kenney, J., Hancewicz, E., Heuer, L., Metsisto, D., & Tuttle, C. (2005). *Literacy strategies for improving mathematics instruction.* Alexandria, VA: Association for Supervision and Curriculum Development.

Library of Congress. (n.d.). *Primary source sets.* Retrieved from http://www.loc.gov/teachers/classroommaterials/primarysourcesets/

Massengill, G. (2013). College and career readiness: through the lens of Lexiles. *The TTAC Telegram, 17(3),* 5. Retrieved December 30, 2016, from https://ttac.gmu.edu/telegram/issue-3

McRae, A., & Guthrie, J. T. (2009). Promoting reasons for reading: Teacher practices that impact motivation. In E. H. Hiebert (Ed.), *Reading more, reading better* (pp. 55–76). New York: Guilford Press.

Meeder, H. (2014). What is "STEM literacy"? [Web log post]. Retrieved from http://nc3t.com/stem-literacy/

Moss, B., Lapp, D., Grant, M., & Johnson, K. (2015). *A close look at close reading: Teaching students to analyze complex texts, grades 6–12.* Alexandria, VA: Association for Supervision and Curriculum Development.

National Academies of Sciences, Engineering, and Medicine. (2016). *Science literacy: Concepts, contexts, and consequences.* Washington, DC: The National Academies Press. doi: 10.17226/23595

National Center for Education Statistics. (2015). *The nation's report card: 12-Grade reading and mathematics, 2013.* Washington, DC: National Center for Education Statistics. Retrieved from http://www.nationsreportcard.gov/reading_math_g12_2015/#reading

National Governors Association Center for Best Practices, Council of Chief State School Officers. (2010). *Common core state standards for English language arts and literacy in history/social studies, science, and technical subjects.* Washington, DC: National Governors Association Center for Best Practices, Council of Chief State School Officers.

Ogle, D. (1986, February). The K-W-L: A teaching model that develops active reading of expository text. *The Reading Teacher, 39,* 564–570.

Project Lead the Way. (n.d.). PLTW's Alignment to Standards. Retrieved from https://www.pltw.org/our-programs/standards-alignment

Rafael, T. (2010, April). *Reading comprehension* [PDF document]. New York: McGraw-Hill: Wright Group LEAD21.

RAND Reading Study Group. (2002). *Reading for understanding: Toward an R & D program in reading comprehension.* Santa Monica, CA: RAND.

Renaissance Learning. (n.d.). Accelerated reader. Retrieved from https://hosted23.renlearn.com/69533/

Renaissance Learning. (2014). *What kids are reading: The book-reading habit of students in American schools.* Wisconsin Rapids, WI: Renaissance Learning.

Scholastic Books. (n.d.). Classroom books: Browse by level | Scholastic. Retrieved from http://teacher.scholastic.com/products/classroombooks/browse_level.asp

Schwartz, J., & Kenney, J. (2008). *Tasks and rubrics for balanced mathematics assessment in primary and elementary grades.* Thousand Oaks, CA: Corwin Mathematics Press.

Sejnost, R., & S. Thiese. (2007). *Reading and writing across content areas* (2nd ed.). Thousand Oaks, CA: Corwin Press.

Skirble, R. (2010, April 27) *In schools, a way to keep language from getting in the way of science.* Voice of America News. Retrieved from http://learningenglish.voanews.com/a/in-schools-a-way-to-keep-language-from-getting-in-the-way-of-science-92244839/117660.html

Skloot, R. (2010). *The immortal life of Henrietta Lacks.* New York: Crown Publishers.

Snow, C. (2010). SERP | Word generation - proven v. myths. Retrieved September 13, 2016, from http://wordgen.serpmedia.org/proven_v_myths.html

Swan, E. A. (2003). *Concept-oriented reading instruction: Engaging classrooms, lifelong learners.* New York: The Guilford Press.

Tankersley, K. (2005). *Literacy strategies for grades 4–12: Reinforcing the threads of reading.* Alexandria, VA: Association for Supervision and Curriculum Development.

Toulmin, C. N., & Meghan, G. (2007). *Building a science, technology, engineering and math agenda.* Washington, DC: National Governor's Association.

US Department of Education. (2014). *STEM literacy.* Retrieved December 31, 2016, from https://y4y.ed.gov/learn/stem/introduction/stem-literacy/

Weins, K., & Bluff, J. (2015). *Tech writing handbook* [PDF document]. Santa Barbara, CA: Dozuki.

Weir, A. (2014). *The Martian.* New York: Crown Publishers.

Zollman, A. (2012), Learning for STEM literacy: STEM literacy for learning. *School Science and Mathematics, 112,* 12–19. doi: 10.1111/j.1949-8594.2012.00101.x

CHAPTER EIGHTEEN

Promoting Science Literate Identities Through the Use of Trade Books

PAULO A. OEMIG

INTRODUCTION

When I think about teaching science, I think of inquiry-based instruction—*doing* investigations and *writing* lab reports. Teaching science should be conducive to developing deep appreciation for and understanding of the subject matter, both of which can lead to the formation of science literate identities. This requires providing students with opportunities to engage in the *disciplinary literacy* of science—the unique ways of expressing and representing information in the field—and creating an active classroom environment (Shanahan & Shanahan, 2008).

Learning in a discipline entails authoring particular identities (Gee, 2003). A science literate identity reflects the cultural and communicative practices shared among members of the scientific community. In school, this means fomenting practices, such as asking questions, using models, engaging in argument from evidence, and obtaining, evaluating, and communicating information (National Research Council (NRC), 2012). These practices can be made accessible to students through the use of inquiry and trade books. Careful incorporation of trade books not only benefits all students, but it makes content accessible to English learners (ELs) as well. Literacy becomes intertwined with inquiry to make information applicable and real.

According to Bangart-Drowns and Bankert (as cited in Hattie, 2009), inquiry-based instruction promotes critical thinking. Some of the greatest positive effects of this type of instruction have been seen on non-mainstream students (see Appendix A for discussion of non-mainstream usage) whose backgrounds may not have previously been valued or have been overlooked (p. 209). Inquiry not only contextualizes practices, but it can make content core ideas more meaningful. The goal is to "prepare students with sufficient core knowledge so that they can later acquire additional information on their own" (NRC, 2012, p. 31). The complexity of science rests in that it ultimately entails a way of knowing. In science we come to know through evidence, reasoning and testing. Appropriating science as such does not happen overnight; it demands incorporating a way of processing information that becomes part of a student's identity. Science literate identity, then, involves the literacy practices associated with science *and* its multiple forms of linguistic expression and representation. Science literate identity is the authoring or enactment of oneself to reflect such literacy practices, ways of knowing, doing, thinking and acting (Gee, 2012).

The need to promote science literacy in all adolescents is ever increasing to meet the demands for knowledge production and innovation (Carnegie Council on Advancing Adolescent Literacy, 2010). In the midst of implementing the new Common Core State Standards (CCSS) and the Next Generation Science Standards (NGSS), many states are stressing a vision of literacy that closely connects the different processes of communication (e.g., reading, writing, speaking, and listening) and an interdisciplinary approach based on informational texts (CCSS Initiative, 2010; NGSS Lead States, 2013; NRC, 2012). Furthermore, some highlight the importance of science literacy as a way to strengthen economic prosperity and power; science literacy is seen as a critical component of human capital competencies (Congressional Research Service, 2012; Klein, Rice, & Levy, 2012). Results from the National Assessment of Educational Progress (NAEP), however, show that most adolescents lack the science literacy skills needed for knowledge production and innovation (National Center for Education Statistics, NCES, 2013). More troubling is the continuing reduced performance in science among Latino/a students compared to their White European American peers (NCES, 2013), as well as the underrepresentation of Hispanics in science, technology, engineering, and mathematics (STEM) professions (Landivar, 2013).

Science literacy has been understood as a means for students to acquire the knowledge to lead productive lives and make informed decisions in a society heavily influenced by science and technology. *Benchmarks for Science Literacy* (Project 2061, 1993) defined science literacy not so much as the ability to do science, but to "enhance the ability of a person to observe events perceptively, reflect on them thoughtfully, and comprehend explanations offered for them" (p. 322). Today, science literacy is associated with the *practices* of science, in particular, with "obtaining, evaluating, and communicating information" (NRC, 2012). *A Framework for K–12 Science Education* takes an active approach to literacy where practices involve engaging students directly in carrying out investigations, and communicating and evaluating information. "Reading, interpreting, and producing text are fundamental practices of science" (NRC, 2012, p. 74); as suggested by the *Framework* accomplishing this is heightened when inquiry-based instruction is integrated with literacy.

Since the school in general and the science classroom in particular exhibit a particular community of practice (Lave & Wenger, 1991), distinctive behaviors are expected in order to succeed, behaviors that may inhibit minority students' science learning. As I have worked with Latino/a students and English Learners (ELs), examining their identities in relation to science and its literacy, I have observed that learning science becomes more difficult for these students. They face not only overcoming content-knowledge barriers, but also language barriers. In a science classroom all students are (or are expected to be) in the process of acculturating the practices of science.

Promoting science literate identities, however, cannot wait until students reach high school or they master the English language. Students' self-concepts—how they view themselves in relation to science—have to be fostered throughout schooling. Teaching science needs to involve meaningful investigations, but to teach science literacy, students must also engage in reading experiences that make science comprehensible. Reading in the discipline of science requires skills that differ from reading for pleasure. Reading in science means problem-solving to figure the meaning of text. To do this, students must not only be able to decode and recognize words, but understand a wide range of symbols and images and how these assist problem-solving and interpretation. Science textbooks, unfortunately, often present content in a dulled manner and vocabulary is defined superficially. Trade books offer an excellent context for learning language and literacy. The use of trade books is not intended to replace textbooks or other print, but rather to supplement these. Through inquiry, meaning can be elicited from text, and concepts and investigations contextualized.

In this chapter, I share my practice of incorporating trade books in the teaching of science through inquiry-based instruction. The format of trade books offers unique advantages over textbooks for engaging students in a science lesson. Trade books can be excellent springboards for students' understanding of basic science concepts. Unlike science textbooks, which can be challenging especially for students with reading difficulties and ELs, trade books tend to focus on fewer concepts and better contextualize information. Incorporating the use of selected trade books to complement the science textbook can greatly enhance learning in the classroom. It has been my experience that trade books stimulate students intellectually and emotionally. They are engaging because students readily connect with the illustrations and storylines.

BACKGROUND

The notion of literacy has changed over time. Once it was thought to be a simple set of skills needed for reading and writing; today, however, "literacy is not a single or monolithic entity. Rather it is a set of multi-faceted social practices that are shaped by contexts, participants, and technologies" (National Council of Teachers of English, 2011, p. 1). Those social practices refer to particular ways of acting in a classroom to succeed academically. In the science classroom learning cannot take place passively; we need to show students how to engage in the literacy practices and discourse that constitutes science. Similarly, Gee (2002) and the New London Group (1996) suggest that for students to become competent in the discourse of the science classroom, they must appropriate certain practices, such as those for evaluating claims and representing information. We know that science as a discipline favors inquiry, but how does this relate to science education and language learning with ELs?

The Chèche Konnen project (Warren, Ballenger, Ogonowski, Rosebery, & Hudicourt-Barnes, 2001) has conducted case studies for over three decades on low-income students from Latino and African American backgrounds in both bilingual and monolingual settings. They approach science through open inquiry, requiring identification of variables and experimentation as an exploratory process for meaning making. Instead of teachers delivering information to their students, they provide opportunities for students to engage in the discourse and literacy of science. The project focuses on the resources non-mainstream students draw from to evaluate observations. The findings indicate that non-mainstream students are capable of conducting inquiry and using scientific reasoning through open-ended inquiry and promoting scientific literacy. Over the years, the Chèche Konnen project has found that non-mainstream students' cultural practices can be utilized to access science literacy and argumentation. Science investigation is guided by students' own observations and questions—one question leading to other questions and a new round of inquiry. Students' first languages are seen as a resource for making sense of scientific practices, which contributes to the overall goal of building a science learning community.

In other research, Lee and Luykx (2006) underscore the importance of developing instructional congruence in the science classroom. Their work centers on building a natural link between students' cultural expectations and rules of classroom interactions, and also between students' linguistic experiences and the discourse of the science classroom. As an example of the effects of building instructional congruence, Brown and Ryoo (2008) gave a class of fifth graders an assessment of photosynthesis using scientific language. Students who were taught by explaining scientific concepts in everyday language before being introduced to scientific terminology performed better than students who were taught by using scientific terminology from the start of the unit. Support for this type of instruction also comes from Lemke (1990) and Gibbons (2015). Both recommend that teachers use informal or colloquial speech initially for students to understand information more easily and provide students with opportunities to speak to one another while making sense of science.

MISCONCEPTIONS IN TEACHING CONTENT AREA LITERACY

In teaching content area literacy, several misconceptions have been identified, which make the teaching of science to non-mainstream students more complex (Bean, Readence, & Baldwin, 2011; Manzo, Manzo, & Thomas, 2005). Some of these include the following:

- Students have learned to read in elementary school, so they have sufficient knowledge to engage effectively with the information encountered in textbooks.
- The mental information processes involved in reading textbooks are the same as those taught and used in elementary school.
- Content reading means teaching phonics and other basic skills not directly related to a teacher's subject areas.
- Teachers are seen as information dispensers.

Most secondary teachers see themselves as subject specialists and content experts whose main obligation then is teaching that content knowledge (Stoddart, Pinal, Latzke, & Canaday, 2002). As such, secondary teachers believe that students have already been taught literacy skills prior to arriving to middle or high school. However, as students progress from elementary into secondary school and from general language arts and mathematics to disciplinary contents, there is a clear transition in discourse and literacy skills necessary to appropriate new specialized subjects (Queen, 2013). When the discourse and literacy of science have the potential to act as gatekeepers for many students, especially non-mainstream, a general academic identity might not be enough to shore up a science literate identity. Keeping in mind these misconceptions, it is not surprising that in their attempts to learn science, many non-mainstream students might experience identity conflicts.

Some science education researchers have explored ways to bridge the apparent gap between learning to use the language of science while maintaining cultural identity. For instance, Reveles and Brown (2008) studied how EL students learned to accommodate identities that supported the situated learning of science. The authors use the metaphor of *contextual shifting* (e.g., changing ways of speaking and acting) to stress the role of teachers in facilitating student understanding of scientific literate practices in relation to school science. An underlying assumption in such studies is the notion that a student's home language, in relation to science discourse, can serve as a gatekeeper for student learning. When students are provided with opportunities to explain and talk about science concepts in their everyday language and then guided from conceptual instruction to language instruction, they are more prone to appropriating science literacy.

COMPREHENSIBLE INPUT, SOCIOCULTURAL PERSPECTIVES AND TRADE BOOKS

Speaking too rapidly or having students read texts that are far above their reading abilities tend to result in many students, including ELs, not understanding the content. Comprehensible input refers to communicating in a manner that can be understood by students despite them not comprehending every word used. According to Krashen's (1985) theory of language acquisition, providing learners this input helps them acquire language naturally. Trade books do exactly that; they provide relevant and easily accessible context.

In addition to making input comprehensible, effective teaching and learning of science should also take into account sociocultural perspectives on literacy. Sociocultural perspectives focus on "the description of the 'ways of knowing' unique to particular social, cultural, and educational groups" (Alexander & Fox, 2013, p. 17). It is the student in relation to others who constructs knowledge by sharing a group history (e.g. gender, age, ethnicity), and/or place (e.g., a science classroom community), unfolding over time that influences identity formation. In short, learning is socially mediated. It is this sociocultural perspective that provides the most appropriate springboard for examining the needs of culturally and linguistically diverse students. This perspective shifts the conception of non-mainstream students as lacking skills or having deficits into viewing their literacies as assets. Literacy comes to be understood not as something singular, such as science literacy, but as content made of multiple knowledges/experiences. As such the intent is on reconciling schooled and unschooled knowledge; taking into account students' prior knowledge to guide their understanding and use of language (e.g., spontaneous v. scientific concepts, Vygotsky, 1986). Learning becomes situated as language assumes a particular *Discourse* (see Appendix B for Gee's understanding of discourse) in a particular situation (Gee, 2012).

To illustrate the value of adopting a sociocultural perspective of learning consider the children's book, *Fish is Fish*. In the book Fish is Fish the author Leo Lionni describes the friendship between a tadpole and a minnow as they grow into frog and fish. One day the frog leaves the pond to explore the world beyond water, upon returning he recounts to his friend the things he has seen. As the frog recounts his sighting of birds, the fish imagine them as "large feathered fish." The fish reinterprets every account through the experience of a fish (e.g., a cow as a black and white spotted fish, humans as cladded fish walking upright).

Lionni's story captures a fundamental insight about learning: new understandings are constructed upon a foundation of existing understandings and experiences. In the case of the fish, his only frame of reference is locomotion through water. It is not enough for the frog to provide factual and accurate information; the fish is unable to organize knowledge conceptually. For instance, walking or breathing air remain abstract concepts to the fish. Similarly, the understandings children carry with them into the classroom significantly shape how they make sense of what they are taught. Just as the fish constructed an image of a human as a modified fish, children use what they know to shape their new understandings. Students' prior knowledge and experiences should not be ignored. The frog in the story provides information to the fish about humans, birds, and cows that is accurate and relevant, yet obviously insufficient. Feathers, legs, udders and sport coats are surface features that distinguish each species. But if the fish is to understand how land species are different from fish and different from each other, this factual information will not be of much help. The concepts take on meaning in the knowledge-rich contexts in which they are applied. "The goal of learning," Alexander and Fox (2013) affirm, is "no longer seen as the development of an individually held body of knowledge, but rather the creation of a mutual understanding arising in the social interaction of particular individuals in a particular context at a particular time" (p. 16).

The fish in Lionni's story accepts the account from the frog quite passively. Had the fish been actively comparing the information with what he already knew, he might have noticed that putting on clothes would limit the swimming ability of fish. This is an important example of regulating one's learning through metacognition (i.e., thinking about your thinking). The notion of metacognition relates to being able to self-monitor one's thinking; how one goes about processing information. Guided inquiry, where the teacher provides the steps and reasoning for an investigation, serves as a way to self-monitor in science. Similarly, in reading trade books, or informational text, the teacher can model how to make sense of meaning by thinking aloud. Proficiency in reading requires one or more metacognitive strategies to comprehend text. The use of such strategies develops over time as students appropriate the ones better suited in comprehension (Pressley, Wharton-McDonald, Mistretta-Hampston, & Echevarria, 1998).

When science literacy is conceived as a kind of inquiry, literacy practices such as asking questions, setting purposes, clarifying ambiguities, making inferences from given information, and constructing arguments can be used to support scientific inquiry rather than substitute it (Krajcik & Sutherland, 2010; Pearson, Moje, & Greenleaf, 2010). Pearson et al. (2010) argue that text driven inquiry is not about passively receiving information, but rather an active process of meaning making. These types of inquiries can be seen as investigations in their own right. Without a careful incorporation of reading and writing in subjects like science, students can graduate from high school with a precarious understanding of the role literacy has in science. Similarly, if students are not engaged in the scientific practices of the laboratory and inquiry, such as obtaining, evaluating, and communicating information (NRC, 2012), they will have a limited understanding of the nature of science.

SO WHY TRADE BOOKS?

In children's literature, trade books are defined more by format than by content. These include either chapter books or picture books. Chapter books tend to be text heavy whereas picture books use illustrations and images as much as words to express meaning. Even though picture books are associated with children, they can be valuable for a wide range of students. They can be used to share information in a more accessible manner, introduce a concept, or model a reading strategy. Considering that science textbooks contain complex vocabulary and cover a lot of topics, students may struggle understanding them, particularly ELs and those with reading difficulties. When students are reading aloud, they contribute to listening comprehension and vocabulary acquisition (Garan, 2007). Trade books also make excellent complementary texts, are written by independent authors, as opposed to publishers who seek to sell the same textbook to as many states as possible and thus providing a limited viewpoint, and are readily available and accessible.

Rice's (2002) comprehensive evaluation on the use of trade books to teach science cites several benefits: (a) trade books focus on fewer concepts and provide more in-depth treatments, (b) are more engaging and less confusing than texts, (c) the storylines help students better understand and remember concepts than textbooks that tend to portray science as facts to be memorized, (d) and their illustrations often make abstract ideas more understandable. Trade books have been used successfully to introduce the scientific method and the excitement of discovery, as well as enhance creativity and thinking skills. Although there is no single universal scientific method, I use it here to encompass a variety of inquiry approaches to generate scientific knowledge based on evidence. Rice also points out that many of the advantages of using trade books in teaching science are not only good for the elementary grades but also for secondary level students. Given these benefits, it is not surprising that trade books have the potential to improve students' attitudes toward science.

BEST PRACTICE

Using trade books in my eighth grade science classes, and with my university pre-service teachers, originated from the need to reach English learners. There are many trade books which could be introduced to students at this level but I would like to share several specific trade books I introduced to students which yielded successful results. *Carlos y el Zorrillo (Carlos and the Skunk)* by Romero Stevens (1997) allows me to revisit the scientific method. *Full Speed Ahead! How Fast Things Go* (Crushiform, 2013) is used as an opportunity for students to write their own stories on motion-related themes. *Things That Float and Things That Don't*

(Adler, 2013) is a trade book that can engage students in the topics of density, buoyancy, and engineering design.

Carlos y el Zorrillo

Every year I perform different demonstrations and ask students to describe what they have seen. I want my students to interpret data, look for patterns, and explain observations by using evidence. I also want my students to collaborate, so small groupings provide these opportunities. These demonstrations are often used prior to my students performing full-fledged laboratory investigations. Their purpose is twofold: introduce a related scientific concept to the investigation and pique students' interest.

In one demonstration, the diving or dancing raisins, as anticipation to the scientific method, my students test the effect of a solution on the floatability of raisins. I do not tell my students what we are testing, nor do I mention concepts such as floatability or the scientific method. Unlike what is common practice in language arts, in inquiry-based teaching there is no frontloading, that is, introducing vocabulary prior to the instructional activity. The idea is for students to use their own language and for me to assess and guide them in their observations. With my guidance the students discover concepts, which makes these moments more memorable. In the diving raisins demonstration, I ask students to pour 120 mL of water and 20 mL of vinegar into a 250 mL beaker. After mixing the solution well, I ask them to add one teaspoon of baking soda and then a few raisins. I ask students to organize their observations in their notebooks and create a data table to test different questions (e.g., what would happen if more vinegar is used?). After students have tested their hypotheses, variables, and recorded their observations, but prior to going over their findings, I read aloud *Carlos y el Zorrillo*. Since reading aloud is a performance, I make sure my voice and facial expressions match the emotions of the text and often pause to stress a point or make comparisons to the raisins demonstration.

The raisins demonstration is a segue that introduces my students to the scientific method or, as Albert Einstein referred to it, the scientific cycle (Kemeny, 1959). *Carlos y el Zorrillo* is a "factional" story that is a blend of fiction and facts. It takes place in New Mexico and it is about a kid who is sprayed by a skunk and how he goes about to get rid of the smell on his body. After I read the book aloud I underscore how Carlos tested different liquids (e.g., water and tomato juice) to get rid of the skunk smell and how my students tested different amounts of vinegar and water to, for instance, keep the raisins afloat the longest. This book illustrates the concept of the scientific method well and validates students' lives as the story takes place in the state of New Mexico where I teach. Moreover, it can be read in Spanish, as I have, recognizing many students' linguistic backgrounds as a valuable educational resource.

Full Speed Ahead! How Fast Things Go

Promoting science literate identities is facilitated when the experiences students bring into the classroom are taken into account. In activating prior knowledge, it is useful to provide opportunities for students to share their ideas related, in this case, to motion. A graphic organizer, such as a concept web, can help with organizing ideas. Students can think-pair-share or work in small groups to discuss motion. For instance, after students have time to share in pairs their ideas about motion, I start a whole class discussion. I address their thoughts on a concept web, where the main word, motion, is written in the middle of the board and student contributions are written around it, making the individual student contributions visually accessible to the entire class. This is a time for students to discover connections with the general topic and for me to assess their knowledge and understanding from the unit on motion and Newton's laws.

After reviewing the students' ideas on motion, I introduce the book, *Full Speed Ahead! How Fast Things Go*—a National Science Teachers Association (NSTA) science outstanding trade book. I ask students to pay attention to the images and factual information regarding the speed of all items included within its pages. Asking students to write down surprising facts in their science journals as they read (or are read to in a read aloud scenario) elicits curiosity. The text in this trade book is limited to identifying the items shown on one page and their common speed on the opposite page. For instance, on one page a depiction of Earth is shown next to a bullet (not to scale) on the opposite page the objects are identified and in big letters their speed is displayed in the center of the page: "1,700 Km/h—1,056 mph." Invariably, such comparisons always raise expressions of surprise from students. The reason for level of engagement lies in connecting something students may have a general understanding of (e.g., the tremendous speed of a bullet) with something they may not yet grasp (e.g., the tremendous speed of the Earth).

I also use this trade book as an entry point for students to create their own trade books on a topic related to the unit we completed. The goal is for students to write, illustrate and create stories either by themselves or in small groups. The completed book must have a narrative that is engaging; within it there must be at least one science concept addressed as per our unit just studied and include at least three illustrations that support the text. Completion of their books may take up to three weeks, while I provide time in class for them work. I do not dedicate a whole class period to the task—most of it is done outside the classroom. I also recruit the help of the Language Arts teacher. Student trade books combine a story with informational elements. One eighth grade student who had moved with his family from Mexico a year and a half earlier wrote and illustrated a story in Spanish. He recounted how his dad raised horses for local races in their hometown of Parral and how they calculated the speed of the horses. Once student books are finished, I take my students on "tour" for them to read their creations to sixth- and seventh-graders. With the assistance of parents, one year I took a group of eighth graders to the elementary school from which most of our students come to read their trade books to first and second graders. Figure 18.1 shows part of one of my students created book "Family Matters: The Story of Mine," describing the alkali family of the periodic table of the elements.

Things That Float and Things That Don't

Things That Float and Things That Don't, another NSTA outstanding science trade book, introduces density as a key concept to understand the buoyancy of different objects. Before reading the book, I solicit students' understanding with a question, such as, "Why do some objects float and others don't?" or "Why do ice cubes float in water?" It is important to record students' responses and make them visible for the class by either writing on the whiteboard or using a document camera. The collective thinking of the class is displayed for building on each other's ideas. Depending on the grade level, the word "density" may or may not come up, but very likely the word "heavy" will be mentioned.

I introduce the book by showing its cover and rephrasing the title of the book as a question: "What are things that float and things that don't?" The book becomes the hook to grab students' attention and to frame conceptual understanding through the inquiry activities adapted from it. As I read aloud, I pause at different places and ask students to make predictions and answer questions. It is important for students to interact with the book's content, to be active listeners by making connections or anticipating what is next through my questions or prompts.

At the same time, I want to make sure that I keep the flow of the reading. In order to facilitate this, my students must respond to three prompts as I read: What did you learn? What questions do you still have? What idea and/or image stuck with you? I have found these prompts enable students to listen with intention, anticipate, and infer. Reading aloud also generates an opportunity to model reading strategies. For instance, I pause after this passage: "A large boat that might weigh thousands of pounds will float. A whole crowd of people could be on that boat, and it would still float … [but a] pebble would sink quickly." I ask students: "How is this possible if a pebble is a lot lighter than the big boat?"

Things That Float introduces many technical terms that middle school students may not be familiar with. For example, they may know *displacement* in describing motion, but not in finding out volume. Re-reading helps to emphasize important sections and check for students' understanding. The use of realia (household or everyday accessible objects) can be used to clarify the meaning of displacement in this context. Dropping a rock or other sinkable object into a transparent cup half-filled with water and marked on its side to show the water/air interface, illustrates

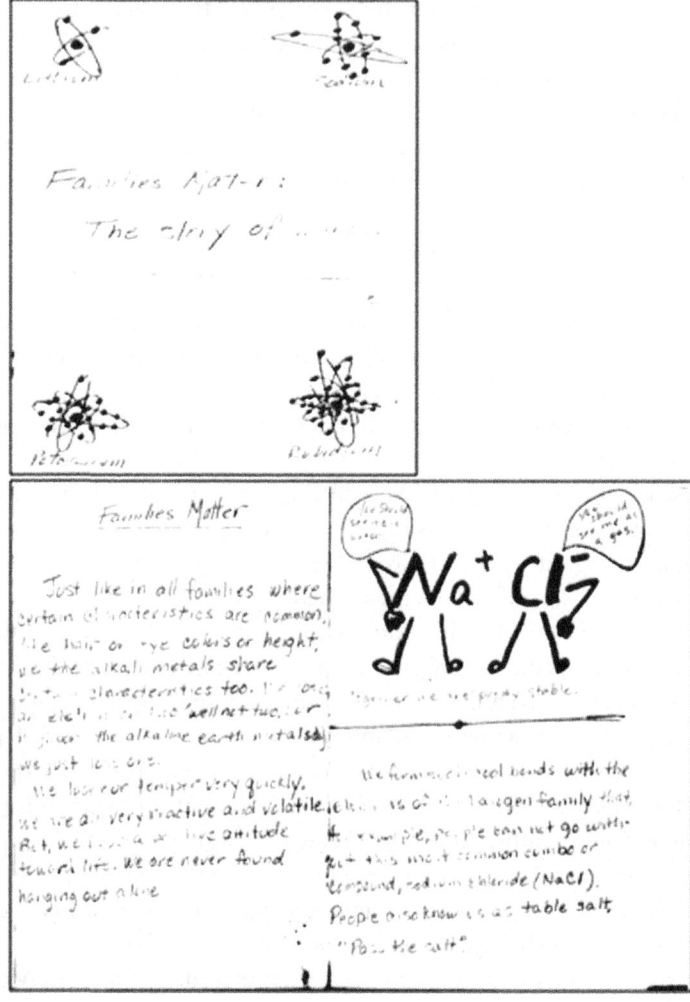

Figure 18.1. Student-Created Trade Book.
(Source: Author.)

the phenomenon of displacement. Not only do English learners benefit from such a practice, but all students come to appreciate the story and concepts presented in such context.

One part in this book that I revisit is the description of how to make modeling clay float. A ball of clay sinks in water, but if shaped in a particular way, it will float. I adapt this activity as an engineering challenge but instead of using clay I have students use aluminum foil. I tell students that they will be building boats, but that, in small groups, they will make them out of aluminum foil. The aim is for their boats to hold as much weight as possible in the form of pennies (or washers) when placed in water. Students must sketch a blueprint and write their procedures in their science notebooks, as well as build a prototype out of paper that I provide. Once students have their sketches, procedures, and prototype ready, I provide each group with a square piece of aluminum (25 by 25 cm) to build their boats. I invite one group to test their designs by placing the empty boat in the water and add one penny at a time until it sinks. On a data table students record their results. Once all groups have tested their designs, the design that held the most pennies is compared with the rest. I prepare a class data table on the whiteboard for groups to enter their boat's name and weight held (or amount of pennies if scale is not available). After all groups have had an opportunity to test their designs, I read the book again, this time telling students to pay attention specifically for clues on factors to increase their boats' capacities. As part of the engineering design process, students have another opportunity to improve their original designs. I add another column to the table for new weights. We discuss how the latest results compare to the original. Giving students the opportunity to revisit the class data table reinforces the notion of data literacy that is, interpreting and organizing data in a meaningful context. Finally, in another data table, features such as perimeter, surface area, volume, and height of walls of the different boats are compared (Table 18.1). By doing this, students are able to refine their understanding by showing a logical relationship between experience, claims, and evidence. The book becomes a medium that facilitates understanding.

Table 18.1. Possible Features to Include in the Things That Float and Things That Don't Data Table.

Team's name	Number of pennies or weight	Surface area (cm^2)	Perimeter (cm)	Wall's height (cm)	Time floating

(Source: Author.)

These are just three examples of trade books that I have incorporated during inquiry-based instruction. Understanding literacy as something static and limited to text makes reading science passive. Reading, writing, and doing science with the support of trade books makes it an active act. Students question, make connections, infer, evaluate the importance of information, collaborate, and test their ideas. By letting students explore and revisit concepts, they learn from each other and become more involved in their own learning.

Literacy through science makes discourse relevant via tangible connections. In order to become proficient in the discourse features of a discipline, acting on content in fields is needed, including science (Moje, 2008). That is to say, to become fully literate in science necessitates acting like a scientist. The text cannot be disconnected from scientific practices and vice versa. Science content and literacy learning is developed by students when their overall objective is to learn science (Guthrie & Wigfield, 2000). In promoting science literate identities, text is a mediational tool to learn and discover the natural world.

ADAPTING THE PRACTICE

In teaching science, trade books can be used prior to, during, or after having completed a unit on a particular concept. Trade books complement other forms of print such as textbooks or digital media. Trade books can make concepts more comprehensible than textbooks alone. The key rests in knowing how to select the right narrative and deciding on how best to use it. Locating resources for finding appropriate trade books and guidelines for choosing these books (with inquiry-based instruction in mind) can be challenging so the following suggestions are offered.

How to Select Trade Books

An excellent resource to look at is the National Science Teachers Association (NSTA). A list of outstanding science trade books is available annually in the March issues of the NSTA journals, *Science and Children, Science Scope,* and *The Science Teacher*. A list of NSTA science trade books published in 2015 and previous years is also available online. Other sources for finding outstanding children's trade books include the Young Adult's Choices Reading List and Teachers' Choices Reading List. Another list, created by a librarian, is the database of award-winning children's literature: Database of Award-Winning Children's Literature (DAWCL). This site makes possible searches at different levels, such as age of reader, publication date, science topic, and awards won. These lists provide a starting point where to look for trade books. Links to all these sources are provided in Appendix C. Also, included in this appendix is a link to a short article on *STEM Resources and Materials for Engaging Learning Experiences,* which describes how children's literature can support STEM and how accessible resources can facilitate exploration and inquiry.

In order to decide on how to select the book that best aligns with the curriculum, Altieri (2016) offers an evaluation guide. Her guide, covering the areas of science content, images, writing, informational text features, and overall design is fairly comprehensive. Some questions regarding the science content include the following: "Does the author have expertise related to the information discussed in the text?" "Has the author consulted experts in the field?" It is important to note that although a topic may not be discussed in detail, consideration must be given to evaluate whether simplification of the topic results in misleading information.

Since stereotypes found in texts influence science performance, both text and images should be examined to ensure that gender and ethnic stereotypes are not present. Although some progress has been made, depictions of male White European American students in science textbooks continue to be prevalent. The development of science literate identities is thwarted for non-mainstream students when they do not see themselves represented in science textbooks (Altieri, 2016). Regarding images, these need to be realistic. Questions such as, "Do the photographs support the text and are they accompanied by credits or captions?" and "Are the illustrations accurate in scale?" should be kept in mind while deciding on a science trade book.

Although the writing style must contain precise scientific terminology, it also needs to be engaging. Here are writing style questions to consider: "Is there a logical presentation of concepts?" "Is the narrative engaging for students?" Trade books may also contain text features, like diagrams, sidebars, captions, insets, etc. It is important to note the features in the books selected to make students aware of them. You can use these questions to evaluate the use of text features: "Does the book contain useful captions, sidebars and/or charts and keys?" "Is there a glossary?" "Does the book use different typefaces?" And regarding the overall design of science trade books, the following questions should be kept in mind: "Are the layout and format appropriate for the content?" "Is the layout appealing to students?"

How to Use Trade Books

Regardless of whether trade books are incorporated before or during a lesson, prior to reading, assess students' understanding of the topic. Finding out students' overall knowledge and comprehension on a topic can provide information on how much students know and/or how well our teaching has helped them advance understanding. For this purpose, a concept map like the one I used with *Full Speed Ahead*, or another graphic organizer, such as a K-W-L (i.e., what I *know*, what I *want* to learn, what I *learned*) could work. With students' prior-knowledge in mind, it is easier to know what parts of the reading aloud to emphasize (or omit). Before reading, remind students that as they listen, they will have to respond to the prompts on the organizer.

During reading, since it is a performance, it is important to read with expression befitting of the story. Every so often make eye contact with students; you are determining their attention levels and showing that you care about their reactions. Share the illustrations with the students as appropriate to support a particular passage. If you have a document camera, use it to amplify the images. Without sacrificing the flow of the reading, model reading strategies; every now and then pause, and as applicable, model how you make sense of the information. Phrases that start with, "I wonder if" and "this reminds me of" can make your thinking visible to students. After reading, give students time to reflect on their "take-aways," and go over their responses to the prompts.

One way to accomplish student sharing can be through "think-pair-share". Individually students think about and review their responses. Students are then paired, or can also be arranged in small groups, in order to share their ideas. A classroom-wide discussion could follow. As you facilitate whole-class sharing you gain a sense of students' knowledge and the type of evidence they use to describe their understanding of the reading. This can help you decide whether or not to re-read the whole trade book at a later time or address particular sections right away. The trade book is a flexible medium to either introduce or reinforce a concept, and a model for students to write their own trade book to teach peers as in the case I described with *Full Speed Ahead*.

Inquiry and Trade Books

As meaning-making and problem-solving are central to literacy, the incorporation of inquiry provides a natural framework to integrate both. In using inquiry aligned to attain NGSS learning goals, four principles of instructional design must be kept in mind: (a) instructional materials must be considered in light of clear performance expectations (i.e., learning goals), (b), learning experiences need to be logically sequenced to maximize instruction, (c) learning experiences need to be planned to maximize conceptual understanding (e.g., both disciplinary core ideas and crosscutting concepts) with opportunities to perform the practices of science and engineering, and (d) students need to be given opportunities for reflection, discussion, and argumentation from evidence (Bybee, 2015, p. 68).

Trade books and inquiry complement each other; both support the doing of science. Introduce the lesson with an open-ended question or demonstration that relates to the trade book, as opposed to starting with a lecture. Collect students' responses and record their questions without providing detailed responses or directions. Read aloud the trade book following the guidelines described. Have students work in small groups to perform an investigation and gather data. If appropriate, revisit the trade book and allow students to re-do their experiments. Allow students to communicate their findings by either presenting to the class or writing their own trade books.

According to the knowledge students possess, inquiry can assume different modes (Banchi & Bell, 2008). In verification inquiry, students are given the question, procedures, and the expected outcome of the investigation in advance. In structured inquiry, students are given the question and the procedures, but students discuss their results in light of the evidence they uncover. In guided inquiry, students are given the question but they brainstorm the procedures and explain the results. In "open" inquiry, the questions to investigate originate from the students themselves, as well as the procedures and the communication of results.

The best practices discussed in this chapter are intended to be flexible and adaptable to various settings. The use of trade books can be incorporated at any grade level and the guidelines presented here make the selection of trade books possible. The level of inquiry to implement with trade books will depend on students' depth of knowledge and familiarity with carrying out investigations. The earlier inquiry-based literate practices are implemented in the lives of students, the greater the likelihood of developing science literate identities.

The trade books discussed here are not solely intended to emphasize science. The use of trade books through an inquiry-based approach facilitates the understanding of the STEM fields as integrated. The interconnectivity among the different subject areas can be better served through real-world scenarios where student have to engineer solutions to problems. For instance, in using the book *Things that Float and Things that Don't* students are not only to design the best boat, but as part of that design elaborate on the mathematics of assessing different designs. Trade books such as *Papa's Mechanical Fish* and *Roller Coaster* can also be used to illustrate the engineering design process, mathematics and science. Students use tools and other technologies to build, for instance, a roller coaster out of foam insulation and express their findings through multiple modalities.

Novel Engineering (NE), a Tufts University Center for Engineering Education and Outreach (CEEO) project, have identified several trade and chapter books which provide the context for students to address engineering processes (see Appendix C for this resource link). Students design solutions by brainstorming technological affordances. Technology is understood here not as digital technologies but as applied science (e.g., the design of a wheelbarrow to facilitate transportation of wood). This understanding of technology to serve practical purposes aligns well with the concept of computational thinking; that is reflected as a thought process toward problems and their solutions. NE's approach is humanistic in nature as it argues for including social studies and humanities in engineering and thus making STEM more relevant to students (Hynes & Swenson, 2013). In the real world, more often than not, STEM fields are integrated, and promote collaborative work in teams by finding best solutions to known problems.

CONCLUSION

Selecting and incorporating trade books in the science classroom might seem unnecessary when textbooks dominate the curriculum. Textbooks, however, are mass-produced for a vast number of students who may not share many things in common (Altieri, 2016). Additionally, the writing style in textbooks is often dull and even though the terminology may be defined, it is often decontextualized in a glossary or lacking a narrative that makes it accessible (Lent, 2012). In my own research, I have found that high school students have become quite adept at skimming for answers instead of reading for conceptual understanding (Oemig, 2015). Supplemental reading options can complement textbooks well with up-to-date information and appealing layouts. Supplemental reading does not need to be directly related to the topic in question. The more exposure students have to reading similar texts or texts about similar subjects, the more their likelihood of conceptual understanding increases.

While we engage our students in doing investigations, we have to ensure that they are also engaged in all aspects of the science classroom. Rather than treating literacy as ancillary to science learning, encouraging reading, writing, communication, and collaboration must be fully embedded in the teaching of science and take on a role to the par of inquiry-based activities. Furthermore, students should be encouraged to read a wide range of materials; trade books are just one option. Trade books, and particularly picture books, are excellent sources to pique students' interest at all grade levels. Introducing appropriate informational texts to young students is beneficial in developing a strong science foundation and affinity toward science. If students are not given opportunities to experience informational texts in elementary school, they will struggle in school and in successfully participating in the various content areas of secondary school. Fisher and Frey (2011) maintain that "the solution isn't as easy as assigning more expository texts … Students have to be taught how to read, write, speak, listen, and think in each of the disciplines which they are expected to perform" (p. 344).

Integrating science and literacy is challenging, but challenges often reinforce learning by the opportunities they provide. To reach all of our students, we need to familiarize ourselves with diverse educational practices. We must modify approaches and strategies so that all students can reach their potential, including English Learners and those who struggle with reading or have other learning challenges. We should encourage our students to both develop and represent their knowledge in a variety ways. By recognizing this need, we can successfully build our students' interest in and passion for science.

APPENDIX A

I chose to make a distinction between mainstream and non-mainstream students to do away with the term minority as it has the connotation of referring to a numerical minority, which is not always accurate. Instead, non-mainstream refer to those students whose families do not possess the institutionalized discourses of the school or are familiar with those institutions' expectations (e.g., students from low income families, English learners, etc.).

APPENDIX B

James Gee makes a distinction between everyday language within society in general, that is, discourse with a lower case "d" and Discourses indicative of specific spheres in life and values and ways of thinking expected in a particular Discourse, such language in use at school.

APPENDIX C

List of National Science Teachers Association science trade books:
http://www.nsta.org/publications/ostb/ostb2016.aspx.
Young Adults' Choices Reading List:
https://www.literacyworldwide.org/get-resources/reading-lists/young-adults-choices-reading-list
Teachers' Choices Reading List:
https://www.literacyworldwide.org/get-resources/reading-lists/teachers-choices-reading-list.
Database of Award-Winning Children's Literature:
http://www.dawcl.com/
Novel Engineering:
http://www.novelengineering.org/books
STEM Resources and Materials for Engaging Learning Experiences:
http://www.naeyc.org/yc/stem-engaged-learning

WORKS CITED

Alexander, P. A., & Fox, E. (2013). A historical perspective on reading research and practice, redux. In D. E. Alvermann, N. J. Unrau, & R. B. Ruddell (Eds.), *Theoretical models and processes of reading* (6th ed.) (pp. 3–46). Newark, DE: International Reading Association.

Altieri, J. L. (2016). *Reading science: Practical strategies for integrating instruction.* Portsmouth, NH: Heinemann.

Banchi, H., & Bell, R. (2008). The many levels of inquiry. *Science and Children, 46*(2), 26–29.

Bean, T. W., Readence, J. E., & Baldwin, R. S. (2011). *Content area literacy: An integrated approach* (10th ed.). Dubuque, IA: Kendall Hunt.

Brown, B. A., & Ryoo, K. (2008). Teaching science as a language: A "content-first" approach to science teaching. *Journal of Research in Science Teaching, 45*(5), 529–553.

Bybee, R. W. (2015). *The BSCS 5E instructional model: Creating teachable moments.* Arlington, VA: NSTA.

Carnegie Council on Advancing Adolescent Literacy. (2010). *Time to act: An agenda for advancing adolescent literacy for college and career success.* New York: Carnegie Corporation of New York.

Common Core State Standards Initiative. (2010). *Common core state standards for English language arts & literacy in history/social studies, science, and technical subjects.* Retrieved from http://www.corestandards.org/assets/CCSSI_ELA%20Standards.pdf

Congressional Research Service. (2012). *Science, technology, engineering, and mathematics (STEM) education: A primer.* Retrieved from http://www.stemedcoalition.org/wp-content/uploads/2010/05/STEM-Education-Primer.pdf

Fisher, D., & Frey, N. (2011). Best practices in content-area literacy. In L. M. Morrow & L. B. Grambell (Eds.), *Best practices in literacy instruction* (pp. 343–360). New York: Guilford Press.

Garan, E. (2007). *Smart answers to tough questions.* New York: Scholastic.

Gee, J. P. (2002). Identity as an analytic lens for research in education. In W. G. Secada (Ed.), *Review of research in education: Vol. 26.* (pp. 99–125). Washington, DC: American Educational Research Association.

Gee, J. P. (2003). *What video games have to teach us about learning and literacy* (2nd ed.). New York: St. Martin's Press.

Gee, J. P. (2012). *Social linguistics and literacies: Ideology in discourses* (4th ed.). New York: Routledge.

Gibbons, P. (2015). *Scaffolding language, scaffolding learning: Teaching second language learners in the mainstream classroom* (2nd ed.). Portsmouth, NH: Heinemann.

Guthrie, J. T., & Wigfield, A. (2000). Engagement and motivation in reading. In M. J. Kamil, P. B. Mosental, P. D. Pearson, & R. Barr (Eds.), *Handbook of reading research* (pp. 406–424). Mahwah, NJ: Erlbaum.

Hattie, J. C. (2009). *Visible learning: A synthesis of over 800 meta-analyses relating to achievement.* New York: Routledge.

Hynes, M., & Swenson, J. (2013). The humanistic side of engineering: Considering social science and humanities dimensions of engineering in education and research. *Journal of Pre-College Engineering Education Research, 3*(2), 31–42.

Kemeny, J. G. (1959). *A philosopher looks at science.* New York: Van Nostrand Co.

Klein, J. I., Rice, C., & Levy, J. (2012). *U.S. education reform and national security.* New York: Council on Foreign Relations.

Krajcik, J. S., & Sutherland, L. M. (2010, April 23). Supporting students in developing literacy in science. *Science, 328,* 456–459.

Krashen, S. (1985). *The input hypothesis: Issues and implications.* London: Longman.

Landivar, L. C. (2013). *Disparities in STEM employment by sex, race, and Hispanic origin.* American Community Survey Reports, ACS-24. Washington, DC: U.S. Census Bureau.

Lave, J., & Wenger, E. (1991). *Situated learning: Legitimate peripheral participation.* Cambridge: Cambridge University Press.

Lee, O., & Luykx, A. (2006). *Science education and student diversity: Synthesis and research agenda.* Cambridge: Cambridge University Press.

Lemke, J. L. (1990). *Talking science: Language, learning, and values.* Norwood, NJ: Ablex.

Lent, R. C. (2012). *Overcoming text fatigue: 21st century tools to revitalize teaching and learning.* Alexandria, VA: ASCD.

Manzo, A. V., Manzo, U. C., & Thomas, M. M. (2005). *Content area literacy: Strategic teaching for strategic learning* (4th ed.). Hoboken, NJ: Wiley.

Moje, E. B. (2008). Foregrounding the disciplines in secondary literacy teaching and learning: A call for change. *Journal of Adolescent and Adult Literacy, 52*(2), 96–107.

National Center for Education Statistics. (2013). *The condition of education 2013.* Retrieved from http://nces.ed.gov/pubs2013/2013037.pdf

National Council of Teachers of English. (2011). *Literacies of disciplines* [Policy brief]. Retrieved from http://www.literacyinlearningexchange.org/sites/default/files/cc0211policy-2.pdf

National Research Council. (2012). *A framework for K-12 science education: Practices, crosscutting concepts, and core ideas.* Washington, DC: National Academies Press.

New London Group. (1996). A pedagogy of multiliteracies: Designing social futures. *Harvard Educational Review, 66*(1), 60–93.

NGSS Lead States. (2013). *Next generation science standards: For states, by state*s. Washington, DC: National Academies Press.

Oemig, P. A. (2015). Bringing science into the elementary school through literacy. *The New Mexico Journal of Reading, 35*(3), 14–20.

Pearson, D. P., Moje, E., & Greenleaf, C. (2010). Literacy and science: Each in the service of the other. *Science, 328,* 459–463.

Pressley, M., Wharton-McDonald, R., Mistretta-Hampton, J., & Echevarria, M. (1998). Literacy instruction in 10 fourth- and fifth-grade classrooms in upstate New York. *Scientific Studies of Reading, 2*(2), 159–194.

Project 2061 (American Association for the Advancement of Science). (1993). *Benchmarks for science literacy.* New York: Oxford University Press.

Queen, J. A. (2013). *Student transitions from middle to high school: Improving achievement and creating a safer environment.* New York: Routledge.

Reveles, J. M., & Brown, B. A. (2008). Contextual shifting: Teachers emphasizing students' academic identity to promote scientific literacy. *Science Education, 92,* 1015–1041.

Rice, D. C. (2002, March). Using trade books in teaching elementary science: Facts and fallacies. *The Reading Teacher, 55*(6), 552–565.

Shanahan, T., & Shanahan, C. (2008). Teaching disciplinary literacy to adolescents: Rethinking content area literacy. *Harvard Educational Review, 78*(1), 40–59.

Stoddart, T., Pinal, A., Latzke, M., & Canaday, D. (2002). Integrating inquiry science and language development for English language learners. *Journal of Research in Science Teaching, 39*(8), 664–687.

Vygotsky, L. S. (1986). *Thought and language* (A. Kozulin, Trans.). Cambridge, MA: MIT Press. (Original work published 1934)

Warren, B., Ballenger, C., Ogonowski, M., Rosebery, A., & Hudicourt-Barnes, J. (2001). Rethinking diversity in learning science: The logic of everyday language. *Journal of Research in Science Teaching, 38*(5), 529–552.

Children's Literature Cited

Adler, D. A. (2013). *Things that float and things that don't* (A. Raff, Illustrator). New York: Holiday House.

Crushiform. (2014). *Full speed ahead! How fast things go.* New York: Abrams Books.

Fleming, C. (2013). *Papa's mechanical fish* (B. Kulikov, Illustrator). New York: Ferrar Strauss Giroux.

Frazee, M. (2006). *Roller coaster.* New York: Harcourt.

Lionni, L. (1970/1998). *Fish is fish.* New York: Dragonfly.

Romero Stevens, J. (1997). *Carlos and the skunk/Carlos y el zorrillo* (J. Arnold, Illustrator). Flagstaff, AZ: Rising Moon.

CHAPTER NINETEEN

Building Community Partnerships AND Integrating Arts AND Social Studies TO Strengthen STEM Learning

JOHN F. SMITH AND JUNE TEISAN

INTRODUCTION

Tamika and her friends head to science class at the start of the school day. Tamika is pretty sure her teacher mentioned they'd begin a new unit today and a friend reminds her it has something to do with plants. Tamika thinks, "Who really cares about plants, anyway?" As she approaches the classroom she hears a crescendo of voices. She quickens her pace and enters a space that looks more like a greenhouse or plant store than her classroom. Flowers in jars and vases top each desk, table and counter. The "Do Now" directions on the front board direct Tamika to find a flower she likes, sketch it in her notebook, and guess what part of the world the plant came from. Step two of the directions asks her to think of questions she might ask a plant expert. Tamika wonders if the lady standing next to her teacher might be some kind of plant scientist. Tamika smiles and starts sketching a bright orange flower with a twisty shape.

Students blossom in learning spaces that are connected to real world contexts and concerns. Whether teachers are devising unit introductions that draw in students or designing extension activities that fuel imagination and innovation, teachers strengthen student comprehension, creativity, skills, and senses of agency through thoughtful planning and preparation. In the case of Tamika's middle school science class, thoughtful planning and preparation included combining science, art, social studies, and community partnerships to make learning irresistible to Tamika and her peers. Her teacher arranged for leftover flowers to be delivered from a local funeral home in advance of a botany unit to engage students (Photo 19.1). Further,

Tamika's teacher arranged a guest speaker to present in an atypical way, asking students to come up with questions of interest to them.

Photo 19.1. Floral Displays in June's Classroom, Donated by Local Funeral Homes, Ready to Enrich the Planned Botany Studies.
(Photo credit: June Teisan.)

Schools are increasingly seeking ways to strengthen and integrate learning across STEM (science, technology, engineering, and math) subjects (National Academy of Engineering [NAE] & National Research Council [NRC], 2010). Beyond integrating STEM subjects, though, are opportunities to connect art and social studies with STEM courses and curricula. Including art and social studies alongside STEM is a response to instruction in STEM subjects that may lack context and connections to the community (Melaville, Berg, & Blank, 2006), as well as moral and ethical considerations (Sadler, 2004). Linkages among the arts, social studies, and STEM subjects are readily apparent in the world outside of school: engineering challenges requiring creative and equitable solutions, socially responsible scientific research aiming to improve the human condition, and technological innovations springing from an understanding of societal and global concerns with an eye toward aesthetics. While educators have created a number of different acronyms to describe the integration of additional disciplines with STEM, chapter authors will refer to the integration of art and social studies respectively as STEAM and STEMS^2 (See Table 19.1 for additional STEM-based acronyms for inspiration).

To fully realize the possibilities presented by STEAM and STEMS^2 learning, teachers can partner with individuals and institutions in the community that surround the school. It is one thing for a textbook or a slideshow with photos to show examples of how math matters when making a mural or how science concepts inform public policy. It is another thing for a local veterinarian to discuss the history and ethics of animal research as a guest speaker or for students to explore how art

Table 19.1. STEM-based Acronyms.

STEM	Science, technology, engineering, and mathematics
STEAM	Science, technology, engineering, art, and mathematics Science, technology, engineering, and applied mathematics
STREAM / STR^{2}EAM	Science, technology, research, engineering, art, and mathematics Science, technology, writing, engineering, art, and mathematics Science, technology, reading, writing, engineering, art, and mathematics Science, technology, research, writing, engineering, art, and mathematics
STEMS^2	Science, technology, engineering, mathematics, and social studies
STREAMS^2	Science, technology, reading, engineering, mathematics, and social studies
STEM+C	Science, technology, engineering, mathematics, and computer science
STEMx	Science, technology, engineering, mathematics, and anything else!

and engineering converge at a paper packing manufacturing plant in their neighborhood. Community partnerships that include people, places, and processes outside of school push STEAM and STEMS^2 learning to be more than just another curricular fad or a reshuffling of existing learning goals. Community contexts and connections provide spaces for students to learn content, practices, and traditions of STEM disciplines while supporting students as they explore STEM as a vibrant and vital part of a world without distinct disciplinary boundaries. Further, these partnerships open up possibilities for students to develop cognitive competencies like creativity and innovation as well as interpersonal and intrapersonal competencies, such as those outlined in the National Research Council's report *Education for Life and Work* (2012a). Knowledge, practices, and competencies do not exist in a vacuum but instead have purposes and associations that are revealed by enacting STEM learning that includes the arts, social studies, and community partners.

By reaching out to the community and infusing the arts and social studies into STEM, educators assume a vital role as connector. Who better than a teacher to forge partnerships that link caring adults beyond the classroom to young people in the classroom and, in doing so, situate learning in meaningful, real-life contexts? As the experts in teaching and learning and with a deep connection to individual students, teachers can tap into networks and resources that enliven learning, empower caring adults to get involved, and ensure students' needs are met.

For years in their teaching practices, June and John (who goes by "Trey") implemented STEAM and STEMS^2 learning through partnerships across a wide array of community networks. They have seen the benefits of STEM learning increase when they build community partnerships that incorporate the arts and social studies. In this chapter they will share successes with infusing the arts, social studies, and community collaborations into STEM learning experiences. The chapter will conclude with a compendium of tips and insights for teachers to carry out similar work. Avenues and tools for integrating arts, social studies, and community partnerships are truly limitless, powered by teacher creativity and a willingness to weave together multiple facets that may not, at first glance, appear connected to the topics under study. The outcome is a beautiful bouquet of learning experiences.

BACKGROUND

Learning connected to community and real world concerns offers opportunities to mend the disconnectedness that marks modern schooling. First, students often do not feel school connects to their lived experiences (Bridgeland, DiIulio, & Morison, 2006) and "perceive school learning as separate from life learning" (Bouillion & Gomez, 2001). Curricula include abstract ideas and terms that are often disconnected from communities, home, work and everyday life (Bouillion & Gomez, 2001; Root-Bernstein & Root-Bernstein, 2013; Swango & Steward, 2003; Tan, Calabrese Barton, Turner, & Gutiérrez, 2012). Second, subjects within the school are often taught separately from each other (Bean, 1995), and teachers themselves typically operate in what Wise (2012) calls "egg-crate" schools, walled off from one another (p. 24). Textbooks, school-day schedules, and transcript requirements reinforce the separation between subjects. Third, within a given subject like science, teachers may address concepts like chemical bonding and biogeochemical cycles in isolation from one another; they also teach terms in isolation from the practices like building and testing models that constitute the discipline as a way of thinking and coming to know (Dewey, 1910; NRC, 2012a). Similarly, math class becomes a set of formulas to be memorized. Faced with so much disconnectedness, students rightly ask teachers how things they learn in school are relevant or why they matter. Further, as they move through each grade, students may become less interested in STEM subjects (Sadler, Sonnert, Hazari, & Tai, 2012). Tying learning to community contexts and breaking down disciplinary barriers offers opportunities for students to reconnect and find relevance in and purposes for what they are learning.

From Separate Subjects to STEM to STEAM and STEMS[2]

The disconnectedness of schooling is complicated by diverse opinions of how and why students should learn STEM subjects. For instance, arguments for improving science education draw from two competing, yet at times complementary, visions: preparing students to be professionals in STEM fields and ensuring students are scientifically literate to participate as citizens in the broader society, even if they do not pursue STEM careers (Roberts, 2007). The distinction can be thought of as learning to do science versus learning to use science (Hazen & Trefil, 2009). In mathematics education, stakeholders might argue for a traditional view of math education in which students learn math because it is elegant and prepares them for careers in STEM fields while other stakeholders promote math's applicability in daily life and usefulness in building a more just society (Felton, 2014). In technology education, rapid changes in and dependence on technological knowledge, products, and processes have not been accompanied by increases in technological literacy (NAE & NRC, 2002).

Amidst disagreement about why and how students should learn about STEM subjects individually, efforts to integrate STEM as a connected set of disciplines have taken hold. In *STEM Integration in K–12 Education* (NAE & NRC, 2010), a range of possibilities exist for STEM integration in classrooms and schools. In some cases, multiple teachers might teach each STEM subject separately. In other cases, teachers and schools might ignore disciplinary boundaries and create curricula that integrate all of the disciplines. The NRC's *Framework for K–12 Science Education* (2012b), for instance, makes an effort to integrate engineering principles into traditional science learning, and emerging models successfully couch science within engineering design challenges (Berland, 2013). In other cases still, a teacher might integrate engineering with science while math is left for another class (and technology is not explicitly addressed). As the *STEM Integration in K–12 Education* (NAE & NRC, 2010) report and others point out, there is little agreement about the best path forward for designing integrated STEM learning (Breiner, Harkness, Johnson, & Koehler, 2012).

The addition of arts and social studies to STEM learning further complicates, but also responds to, the discussion about why and how students should learn science, technology, engineering, and math. Just as arguments for improved STEM education often center on creating a productive and innovative workforce (e.g., Carnevale, Smith, & Melton, 2011), so too do arguments for integrating art with STEM include mention of molding a creative and innovative workforce (Eger, 2013; Henriksen, 2014; Piro, 2010). In comparison, arguments for including social studies in STEM focuses on responding to vexing challenges that cannot be solved through technology alone (Hardin, 1968) and latching on to the human need for stories and narratives (Berkowicz & Myers, 2015). Beyond these arguments for integrating arts and social studies with STEM is the key notion that disciplines have only recently become so disconnected (Edelstein, 1999; Root-Bernstein & Root-Bernstein, 2013). STEAM and STEMS^2 become necessities, not just because they might provide a more creative workforce, but also because these approaches may well allow us to understand—and change—the world in new and important ways. Further, as Trey recently explored as science teacher-in-residence at the Library of Congress during the 2015–2016 school year, historical events and STEM learning are deeply intertwined as parts of a complex and collective human experience (Bell, 2016; Smith, 2015).

Building Community Partnerships for Context and Connections

Both June and Trey have seen first-hand the positive effects of coupling learning with community partners: families, professionals, institutions, and industry. Efforts to build community partnerships in schools make use of diverse set of educational literatures, and those literatures reveal that partnerships can take many forms. A report from the Coalition for Community Schools (Melaville et al., 2006) compares and contrasts (1) civic education, (2) environment-based education, (3) place-based learning, (4) service learning, and (5) work-based learning (further described in Table 19.2). Some of these terms may be more familiar than others. Each pedagogical approach is a model for thinking about the ways in which communities and schools should interact and influence curricula. What ties all of the aforementioned approaches together is that they seek to "fully engage young people, by harnessing their natural interest in where and how they live and by using their own community as a source of learning and action" (Melaville et al., 2006, p. 2). When we use the term "community partnerships" in this chapter, we are referring to this diverse range of pedagogical possibilities.

Related educational approaches like problem-based learning (PBL) have emerged to complement community-based learning efforts. For instance, it is not uncommon to infuse an environmental education unit with problem-based learning structures. Problem-based learning involves posing a real-world challenges that require solutions and from which all subsequent learning follows out of necessity (Allen, Duch, & Groh, 1996; Savery, 2005). As Grow and Plucker (2003) point out, it is possible for teachers to implement PBL without redesigning the entire curriculum. This is not to say that transitioning from teaching a single subject to STEM, STEAM, or STEMS^2 is easy (O'Neill, Yamagata, Yamagata, & Togioka, 2012). Teachers can identify real-world problems that have been used in lessons in the past and that can serve as catalysts for further learning. While not all PBL units involve community partnerships or require social studies connections, community-based and civic-inspired learning experiences complement PBL structures.

Teachers may also be familiar with project-based learning, another kind of PBL. In project-based learning models, students may respond to real-world problems by creating a product, making the process virtually indistinguishable from problem-based learning (Larmer, & Mergendoller, 2015). In other cases,

Table 19.2. Possible Pedagogical Approaches to Community-based Learning and Partnerships.

Civic education	Emphasizes preparing students to participate in a democratic society; links academic content to civic and political engagement
Environment-based education	Uses the spaces surrounding the school and in the broader community to elicit student interest and prior knowledge; connects further learning to these spaces
Place-based learning	Builds curriculum based on local histories, cultures, environments, and economies; frames student learning tasks in terms of community needs
Service learning	Connects academic learning goals with community service possibilities; offers opportunities for students to define and celebrate service goals
Work-based learning	Locates student learning in career contexts; attempts to connect K-12 with post-secondary training and jobs; may involve mentoring and role models

Note: Melaville, Berg, and Blank (2006) also describe a sixth example that pertains mainly to students and professors at colleges and universities that is not quite as relevant to this chapter.

(Source: Adapted from Melaville et al., 2006.)

project-based learning may focus on the creation of a product within a given set of parameters that are more confined than the pathways involved in problem-based learning (Savery, 2005). Small distinctions aside, the core feature of both examples of PBL is that they provide opportunities for students to find meaning and motivation in what otherwise might be abstract learning that lacks relevance and a relationship to students' lived experiences. Both examples of PBL find their ways into community partnerships and complement a third kind of PBL: place-based learning. This third version of PBL, which we mentioned at the beginning of this section as one model for community-based learning, might involve a problem or a project or may simply be an investigation couched in a local context with which students are familiar, creating an extra layer of purpose and engagement for students (Sarkar & Frazier, 2008; Zimmerman & Land, 2014).

While practices and research related to building community partnerships may have specific features that keep them distinct from each other in educational literature, teachers often take action without adhering strictly to one pedagogical framework or another. Based on the specific and diverse needs of students in classrooms in a given year, teachers may need to adjust plans and programs over time. Both theory and practice related to building community partnerships through approaches like environment-based education, which may attempt to break down perceived barriers between humans and natural systems (Tidball & Krasny, 2011), or service learning (e.g., Stanton, Giles, & Cruz, 1999) provide a menu of options from which teachers can make decisions to meet the needs of learners. Further, one might think of these approaches within particular community contexts as "STEM learning ecosystems" (Traphagen & Traill, 2014). The examples from June's and

Trey's classrooms in this chapter stretch across labels like service learning and PBL and remain grounded in a key idea: no matter which theories or practices teachers tie their work to, context, connections, and community are central components.

Challenging Standardized Learning and the Narrative of Failure

Restoring balance to the curriculum is another important benefit of instruction that incorporates STEAM, STEMS^2, and community collaborations. Teach-and-test curricula have dominated schooling in the past two decades, affecting not only how much time students spent learning science but also the nature of the learning (Jorgenson & Vanosdall, 2002). Haberman (2010) warns against engaging in the "pedagogy of poverty," which is constituted by teacher-centered instruction focused on basic skills, compliance, and standardized learning and testing. With the reauthorization of the Elementary and Secondary Education Act of 1965 in the United States known as the No Child Left Behind Act of 2001 (NCLB), states largely defined student and teacher success by a single metric: student performance on annual standardized tests. The resulting emphasis on annually tested subjects like English Language Arts and math set in motion a cascade of consequences. A review of data from the U.S. Department of Education's Schools and Staffing Survey revealed that over the past two decades, elementary school students spent less time on science (Change the Equation, 2014). Increased pressure to prepare students to pass standardized state tests required under federal accountability frameworks affected the arts and social studies even more than science and pushed them out of some elementary schools (Fitchett & Heafner, 2010; Spohn, 2008). At the high school level, while the percentages of students taking advanced math and science courses increased during the first decade of NCLB, the percentages of students graduating with credits in computer science, engineering, and technology barely budged (Nord et. al., 2011). The latest federal reauthorization, the Every Student Succeeds Act of 2015 (ESSA) seeks in part to restore a more well-rounded education with language in the legislation to directly address the dearth of arts and social studies.

When educators build community partnerships and implement STEAM and STEMS^2 learning, they can restore the balance across the curriculum and offer students multiple entry points into what needs to be learned. Indeed, standardized tests and related instruction typically focus on "what" needs to be learned. Some students find motivation in exploring how what needs to be learned can be used, why it matters that we learn about it, and how we might change, refine, or adapt what we learn to a new context (McCarthy & McCarthy, 2006). Tapping into art, social studies, and community connections situates STEM learning in holistic and humane contexts that meet the diverse needs of learners and provide multiple entryways into learning. These multiple entryways challenge one-size-fits-all curricula and learning outcomes.

Community connections with STEAM and STEMS[2] address another issue related to standardized instruction and testing and the related narrative of failure: school districts sometimes suffer from public relations problems. Media stories often focus on a narrow band of school news from sporting events and human interest stories to more alarming reports of low test scores and school violence. Families with children in school receive firsthand accounts of the day-to-day work in classrooms, but citizens and communities learn about education largely from these headlines, news articles, and sound bites (Swango & Steward, 2003). When teachers purposefully link curricula to community resources, institutions, and individuals beyond the school they showcase student strengths, talents, and passions to community members. This form of civic engagement channels hours of productive effort toward a greater good and adults outside of schools can meet with students face to face, forge positive relationships, and engender faith in the next generation.

BEST PRACTICES

Inside and outside of their classrooms, June and Trey have worked to improve student learning by building community partnerships that incorporated the arts and social studies. In doing so they have added variety to the basic "meat and potatoes" curriculum, cultivated relationships beyond the classroom that enriched learning, linked classroom learning to critical issues in the community, and empowered students for positive social impact.

Community Connections with Art and Social Studies Can Enliven Learning

Enhancing STEM instruction by infusing art, social studies, and community partnerships has been a highly successful pedagogical strategy. In June's middle school science classes she often thought of it as offering a buffet of experiences so that a wider array of student tastes could be accommodated. For example, during the required botany unit, June launched into the subject with a zany "Chia pet" activity using old hosiery, washed and donated by families and staff. The bottom 8 or 9 inches of the nylons were cut away and the toe area filled with about a tablespoon of grass seed. Soil was then added to make a potato-shaped head. The open end would be knotted, the whole head moistened then turned upright in a plastic cup (seed side up).

Once the "hair" began to poke through the toe of the nylon, students were offered googly eyes and felt to cut into facial features like eyebrows and lips. They used hot glue guns to affix the decorations. Students raced into the room daily to visit their Chia heads, asking to trim the hair or grow it long enough to braid.

Soon conversations sprung up comparing growth rates between the array of specimens, rife with observations like "Jamar's is in the sun all day, but mine was in the darker corner of the window; I think that's why his is taller." Questions about how to make the hair/plants grow faster fostered discussion about fertilizers. This low-budget high-impact activity can be extended when students ask to try other types of seeds or simply want to make sillier designs by adding sideburns or beards to a new creation. Imagine students asking to come in early to engineer the best way to provide light to the "facial hair" of their botany faces while still fostering optimal growth of the plants on the top of each head! This infusion of art into a traditional general science lesson ramped up student engagement and interest in the content.

The energy and excitement during botany studies continued to build when June's students would arrive to find the classroom filled with floral displays—blossoms of every shape and color covering desks and counters. Each student was directed to find a flower or two that they liked, to make a pencil sketch of those blooms, then rub the petals and leaves across their sketches to paint color on their drawings with the plant pigments. Conversation flowed as the artwork took shape and students shared advice—which petals gave the deepest hue or how to rub onto the paper but not through the paper. Soon, while the artists worked, stories would be shared—of grass stains on pants during epic sporting events or berry juice on fingers from harvesting fruit in grandma's backyard—dialog that uncovered prior knowledge in a relaxed, informal flow of conversation that enriched the learning experience for all. June was able to guide the discussion, capitalizing on teachable moments, like when the word chlorophyll popped up or a pupil wondered aloud about using fall leaves instead of the green ones available in the classroom. June suggested activities that led students to delve more deeply into studying the chemistry of plant pigments; to perhaps grind up some plant parts in a blender with isopropyl alcohol, pour the goo into beakers, then drape strips of coffee filters over the rim of the beakers and into the goop and watch what happens. The floral art was always proudly displayed throughout the school and the resulting investigations that came from student questions proved more memorable than any textbook-directed activities. The stunning array of blooms was donated every year by a local funeral home, floral displays that might otherwise have landed in the dumpster. The funeral directors were delighted to see the flowers put to good use and the only cost was the time June spent packing up her van for multiple trips transferring the flowers to her classroom.

As the culminating project for the botany unit, June coordinated an annual outdoor event with her seventh grade students serving as nature guides at a local park for all of the district second graders. Preparing for the day-long "Teach Me About Trees" venture, middle school students harnessed their new-found knowledge of flowers and leaves, photosynthesis and plant pigments to design and refine botanical lessons targeted to their elementary audience. The middle school scientists were in charge of creating engaging lessons, designing hands-on activities, arranging for materials and supplies to support the learning, and ultimately delivering robust instruction to groups of squirmy 7-year-olds in the great outdoors. This long-running community-based program yielded multiple benefits:

- Parents and extended family members serving as chaperones on the day trip were gratified to see their children connecting with older students from the school community.
- Elementary educators were freed up from instructional responsibilities for one glorious day and could enjoy observing their pupils soak in the subject matter.
- The younger kids heard solid content from new faces and voices, adding a special spark to the plant studies when the "big kids" did the teaching.
- Many of the seventh grade "instructors" remembered being the awed second graders five years earlier and felt a special connection to our school community.
- Year after year park service staff and building principals witnessed a positive multi-generational experience that was tightly connected to content and local resources.

Linking Learning to Critical Issues in the Community

STEM instruction can also link to critical community issues and the everyday lived experiences of citizens beyond the classroom. June found that transforming her standard microbiology unit into a study of food safety was an important means of fostering personal well-being for students and their families (Byrd-Bredbenner, Abbot, & Quick, 2010). Teens, particularly in an urban setting, are often in charge of caring for elderly family members or younger siblings, two age groups highly susceptible to foodborne illness.

In addition to studying the basic state-mandated standards pertaining to microbiology, June had students interview family members to ask them about previous food poisoning experiences, discuss safe food handling procedures with area restaurant managers, or create handwashing safety posters to post in the school bathrooms. Illustrating "Wanted" posters featuring the nefarious bacteria that could be contaminating food is a much more lively endeavor than labeling a bacteria diagram on a worksheet. Designing engaging posters and brochures opens the door to creative, artful expression and can have a significant positive power when shared with fellow students, school staff, and audiences beyond the school.

June created budget-friendly in-school field trips for the microbiology investigations. Each of her classes visited the school kitchen to learn safe food handling practices in a working commercial kitchen. Trained food-service staff were delighted to host small groups of students over a series of days to explain the precautions enacted every day that keep microbes at bay. June also transformed the entire school facility into a research site. Armed with petri dishes and cotton swabs, teams of student scientists selected areas around the school to swab for bacteria. Would the phone in the principal's office harbor more bacteria than the handle on a gym locker? Would the bacteria count from a faucet be drastically different between the men's and women's restroom? The roving research teams received plenty of interest and support from the greater school community during their studies. As part of a monthly science speaker series, June also invited scientists from the local health department and university laboratories to delve more deeply into the science of food safety with her classes. Engineering advances in food safety, like high pressure processing and complicated manufacturing procedures, were revealed using online videos and web research.

Investigations like these focused on real-world issues in commonplace locations—be it a family kitchen or local eatery. These investigations also created a lifelong awareness of STEM subjects, in this example, specifically microbiology. Students were motivated by these authentic contexts to learn about the differences between bacteria and viruses and often the conversation naturally expanded further to explore antibiotic resistance, vaccines, and global epidemics. June found that by framing microbiology studies within the context of food safety, the impact of teaching and learning transformed what had been a somewhat dry academic exercise into a dynamic ongoing investigation with potential for saving lives from food-borne illness. The impact of increased student knowledge and skills reached beyond the classroom and into the community by offering safer meals prepared by the students, whether at home or at their jobs in the foodservice industry.

Community Connections Coupled with Math, Art, and Technology

Trey, who taught biology, chemistry, and engineering classes and chaired the science department in his high school, was working with students as part of an after-school engineering club to prepare for local and national math-related events. At the time, very few local math competitions existed for students in Philadelphia. He reached out to a local educational advocacy organization, the Philadelphia Math + Science Coalition, and a national online professional development site for math teachers headquartered in Philadelphia, The Math Forum, to create a series of out-of-school time, competitive and collaborative mathematics and engineering events collectively called the Philadelphia Engineering and Math Challenge (EMC). The EMC allowed his students to learn alongside and compete with students from other schools across the city. One component of the EMC was inspired by "Three Act Math Tasks" described by Dan Meyer (2011) and after a couple of annual iterations evolved into a citywide, online competition. As part of the competition students create videos about mathematics in the everyday world. As part of the video challenge, teachers and youth in Philadelphia public schools not only received pedagogical support from educators Max Ray-Riek and Suzanne Alejandre at The Math Forum but also connected their mathematics learning with topics in the community.

The competition blended mathematics, technology and art, as students focused not only on highlighting and communicating mathematical ideas but also on creating digital content. In the first year of the EMC math video event, Trey's students created math videos about stacking books to fill up a room and about proportions involved in making macaroni and cheese. The videos connected to immediate concerns both at school and at home, both somewhat outrageous and quite practical. Students also planned out storyboards and debated the ways in which they would visually represent the scenarios they presented. Would speeding up the footage make the video more engaging for the viewer? Would various angles provide greater visual interest? Aesthetic questions in the video productive process were present alongside the mathematical ideas. In subsequent years, students created and posted videos on the web about mathematical ideas in their lives that had even greater social relevance. One student group made a video about working and saving to buy a car. Another student group made a video about statistics related to policing and the Black Lives Matter movement. What became clear from the videos that the young people created was that the issues and topics they cared about had a place in the math curriculum. Further, student videos showcased differences in style, sound, and movement, which are important aesthetic considerations in video production.

Taking Learning Beyond the School

Community partners and STEM, STEAM, and STEMS^2 learning can also involve taking field trips, engaging in community-based projects, and meeting STEM professionals outside of the school. These efforts can connect to units and learning goals but also push beyond key terms and course content to include and embrace regional and community issues and concerns. For instance, in Philadelphia, Trey connected with local professionals and organizations to organize monthly field trips and monthly guest speakers to engage students in his biology and chemistry classes. One such trip took students to Bartram's Garden, one of the oldest botanical gardens in the United States, to collect data as part of a Schuylkill River water study. The site of the study was surrounded, on both sides of the river, by former industrial and commercial activities that neighbored Bartram's Garden, reflecting the changing history of the city. The site was also part of more recent efforts to reimagine community connections to the river and establish trails for recreation.

Students measured turbidity as well as different levels of dissolved chemicals as part of a larger effort to understand the impact the surrounding sites had on the health of the river.

Five minutes down the road from the school in the other direction, Trey's students from various grades and science classes who reported that they were interested in engineering visited a major paperboard packaging manufacturer in the region. Most of the students did not know the business existed in the neighborhood. Discussions with the professionals in the plant and with the plant's CEO highlighted the connections among technology, engineering, mathematics, and art in printing and assembling paperboard products for packaging. Students marveled at the rate of production of everyday paper items and were exposed to the various processes involved in creating items that we might take for granted.

Although June teaches in Michigan, the Great Lakes State, she was disheartened by the lack of curricular focus on water science and environmental stewardship. To address this void, she created the Student Water Initiative in Michigan, also known as the "SWIiM Team," an elective, place-based STEM research and fieldwork art program. Funded by grants she authored, and enacted with a rotating set of students in an elective class called "FutureThink", June's students constructed and deployed water-quality buoys, connected with water stewardship organizations to investigate threats to water quality, and actively explored wetlands through science and art, sketching wildlife in field journals, collecting leaves to press, identify and display, etc. These middle school student scientists won awards and earned thousands of dollars in prize money for their research, campaigned for water protection through art-infused promotions, and presented reports to government agencies at the local and national level. Key to the success of the program were partnerships June forged with community organizations: Michigan Sea Grant, Wild Birds Unlimited, Huron-Clinton Metro Parks, The Education Foundation of the Edsel and Eleanor Ford Estate, and many more. Each partner organization offered unique support for student learning, a distinctive puzzle piece that contributed to a comprehensive learning endeavor.

June's students who had grown up less than a mile from the local lake sailed on the lakes, many for the very first time, aboard the Michigan Sea Grant school ship. Amateur birders recruited by the owner of the retail store, Wild Birds Unlimited, led bird-watching walks on the grounds of the Edsel Ford estate so students could witness firsthand the beauty and diversity of birds in their natural shoreline ecosystem (Photo 19.2). Naturalists at the Huron-Clinton MetroParks assisted in site selection for deploying water quality buoys within the park waterways, led discussion of rich Native American historical lore, and hosted nature walks that included artistic tips for student sketches in field journals. Researchers from the Great Lakes Environmental Laboratories mentored students, helping with data analysis and submitting research to online science competitions. The powerhouse

Photo 19.2. Great Lakes Studies Included Bird Watching Hikes Led by Staff at Wild Birds Unlimited and the Michigan Audubon Society on the Grounds of Edsel and Eleanor Ford Estate. (Photo credit: June Teisan.)

combo of place-based instruction, intensive, targeted support from community experts, and authentic action to address meaningful local issues proved irresistible to June's middle school students. After participating in this intensive "Future-Think" elective class, many of these urban learners went on to pursue advanced studies in STEM fields.

Empowering Students for Positive Social Impact

A vital goal of education is to nurture and strengthen students' sense of purpose and individual agency related to community betterment. Linking such worthy civic-minded purposes to STEM studies is a natural fit since the fields of science, technology, engineering and math have brought myriad advances that positively impact humanity. Urban centers in particular can be areas of greatest need for purposeful support and meaningful community engagement. June teaches in a district that borders Detroit, a vast urban center with a proud heritage as a titan in engineering and technology. However Detroit has been crippled by unemployment, poverty, racial tension, crime, and neighborhood blight. June felt it was important to enlist her students as active agents of change in Detroit and began in an unlikely way: exposing students to art. Steeped in innovation, Detroit of the 1930s was a world-class STEM capital, and became known around the world as "The Motor City."

Most students in June's care know too well the struggles of present-day Detroit but lack context of the city's rich history of innovation, how STEM played a role in making Detroit a global capital "back in the day," and work that is being done to reverse Detroit's decades-long slide. June's theory was that students could learn of Detroit's rich heritage, grasp more clearly the intricate challenges of the city today, and cast a vision for how they could engineer brighter future by using their time and talents. *Detroit 1933/2033* (dubbed "D-33" by the students) involved extensive programming over a year-long unit: students delved into the engineering motifs and social justice themes behind masterful work of Diego Rivera's "Detroit Industry" murals housed at the Detroit Institute of Arts (Photo 19.3); trekked to modern day auto factories as well as recreated assembly lines from the past to explore the marvels of engineering that built the Motor City; created STEM-connected art tied to the technology revitalizing the city today; interviewed city residents steeped in the history, challenges, triumphs, and vitality of Detroit; met with graffiti artists and spoken word performers whose voices were reshaping Detroit's narrative; studied the chemistry of fresco art and foundry work and created art in studio sessions at the DIA; formulated student action plans casting a vision to reinvigorate the city through targeted, small-scale community action. Some of these activities were enacted before or after school with students who chose to participate, but the vast majority of the studies took place during class or on trips during the school day. Curriculum connections were clearly spelled out; the science of simple machines (what it takes to build a complex machine like an automobile) and geology (earth materials needed for car production), as well as cross-curricular tie-ins with June's social studies colleagues helping to shape and teach the historical and sociological aspects of the Detroit 1933/2033 unit. The end result was rich STEM learning with a STEAM/STEMS[2] twist that empowered students as concerned citizens and informed community advocates.

HOW OTHERS CAN IMPLEMENT

Drawing from our experiences building community partnerships and facilitating STEM, STEAM, and STEMS[2] learning in two different urban regions of the country, we have discovered a number of practices that support these kinds of efforts.

Photo 19.3. Students Explore the STEM and Social Justice Themes in Diego Rivera's "Detroit Industry" Murals at the Detroit Institute of Arts as Part of the "Detroit 1933/2033" Project. (Photo credit: June Teisan.)

Map Out Community Characteristics with Arts and Social Studies Connections in Mind

Begin by reflecting on what your community and region have to offer. Pay special attention to opportunities to highlight the arts and social studies.

1. Brainstorm with colleagues and students and talk to other community members to create a list of the institutions and industries that currently and

historically have had a role in shaping your region. In the Detroit area, June connected lessons to the vital automobile industry and its history, enlisted support from Great Lakes non-profit environmental groups, and utilized resources from a local funeral home to bolster STEAM learning experiences. Trey tapped into a growing computer coding scene in Philadelphia to bring professionals who build mobile apps to speak to students as part of a monthly speaker series. No organization is too large or too small.
2. Consider the diverse ecologies and environments that make up your region. In some cases, these ecologies and environments may be hyperlocal, like a specific city block, while others may pertain to major geological features like lakes or rock layers. Both Trey and June tapped into local bodies of water to explore the interactions among history and STEM.
3. Learn more about local histories, students' lived experiences, and community points of pride. The key is to determine and document what characteristics make your region unique and to build a set of contacts and learning experiences based on these characteristics.

Build a Network of Community Contacts

Based on the list you create about community characteristics, begin to network with members of the community who can strategically help you connect with myriad local contacts and infuse arts and social studies to your STEM classroom.

1. Check websites for email addresses and phone numbers and seek out social media pages and handles, provides possibilities for making initial formal connections. Trey found more than one local guest speaker using connections initiated via Twitter.
2. Be prepared to discuss the work that you do while commuting on public transportation and at the grocery store. Attend community events and open houses. Create business cards and carry them with you. In each of these instances, use an elevator pitch to share your vision for teaching and learning and how community partners can support this vision. Use the business card and ask for their business cards so you can follow up. Trey once used his pitch in an actual elevator and connected with an entomologist who was visiting a local science museum.
3. Tap into the families at your school and the connections to STEM, art, and social studies they might bring to the table. At back-to-school night, find out which family members work in STEM-related fields or in spaces where a future field trip might be scheduled. Some family members might be artists, activists, app developers, historians, public officials, or engineers. Others may be interested in sharing how STEM, STEAM, and STEMS^2

ideas connect to hobbies or work in the home. Friends of students' families may also catch wind of your efforts to connect learning with the community: a student's mother at Trey's school introduced him to an acquaintance from her church who travels the world overseeing ethical research that involves primates.
4. Share example stories from other community partnerships to paint a picture for potential partners. In some cases, you may contact a community partner with an idea in mind; in other instances, you may simply want to introduce yourself and suggest that you are looking to connect your students with rich educational experiences. Trey had success creating a series of fliers and a website describing various existing partnerships, which he shared with colleagues, families, and friends, thus generating further interest from other community partners.

Maintain and Strengthen Partnerships

Once you have established valuable connections outside of your classroom, do not let them fade. In our work, we have learned ways to keep the hard-won relationships vibrant.

1. As you make connections, maintain a database of community partners. Keep track of who you contact, the ways in which they have contributed or partnered with your school, and information related to when you followed up. This follow-up might take the form of thank you notes with photos of students in action, of printed certificates of appreciation, or of a simple note via email. Enlisting students to compose thank you notes is a natural extension of their learning.
2. When inviting external partners to work with your classes, be prepared to offer tips on connecting with students. Experts in fields outside of education can benefit if you help them tailor their jargon-rich/highly technical language for a younger audience. It can be intimidating for even the most seasoned professional to volunteer to speak with groups of students, so our advice has been: ditch the powerpoint, offer activities, incorporate visuals, and strive for "talking with" instead of "talking to." In our experience, guest presenters appreciate the input and the entire experience is uplifted when teachers strategically share their pedagogical wisdom in advance!
3. Branding can be a powerful strategy for efficiently and effectively sharing your vision so others learn about your cause and choose to join in. In today's fast-paced world, creating a tagline, a logo, an interview-ready "elevator pitch" are all ways to increase visibility of your program. In our work we've found that administrators will support these branding efforts and

see their value as strategic community-facing mechanism to share news of great programs and build momentum across the community.

Overcoming Obstacles

Building community partnerships and integrating arts and social studies into STEM classrooms may come with a few challenges. Some of the tips below may help you address these obstacles.

1. In some cases, booking buses and finding funding for field trips and additional supplies may serve as barriers to visiting learning spaces outside of school and taking advantage of community connections. With the help of your administrators, families, and colleagues both at school and in teacher networks, seek out funding from foundations, professional organizations, businesses, crowd-funding sites (like DonorsChoose and GoFundme), and community stakeholders.
2. Sometimes you might hear "no" from a principal, community partner, students, your own social network, or funders. Do not be afraid to push back and seek out the "why" behind the resistance. You are the expert in your subject, in pedagogy, in knowing your students. This is a way to advocate and lead. It may take some persuasion, further explanation of your vision for improved outcomes, and as you walk through the details you may chart an even better course. Circle back, ask and ask again.
3. If your ideas are met with reluctance or resistance, start small. Test some ideas in afterschool spaces or as part of an elective class. June piloted several of her STEMS^2 and community collaborations with a subset of students a year or two before she integrated the practices into her science courses. In this way she created a proof of concept that assured school leaders of the efficacy of the programs.
4. Time is always a teacher's nemesis; finding time to address all of the standards is a chief concern. Know that the work of integrating STEAM and STEMS^2 and community partnerships can be accomplished within the curriculum. June has found ways to make room by committing to a philosophy of depth over breadth. Her perspective has been that we are choosing to paint with more colors and during the creative process one may need to make space for a vibrant splash of inquiry so we hold off on a less-than-dynamic drab section. June's students achieved a deep understanding of the role of pigments and the incredible complexity of plant parts during the botany activities described earlier. She found the time to enact those vibrant experiences by eliminating some of the uninspired explorations prescribed in the textbook.

5. Throughout all of this, do not forget to enlist colleagues. Opening the curriculum and breaking down barriers between traditional school learning and these more meaningful practices can complicate the task of teaching. This complication for us, though, was more than worth the effort and many of our colleagues were excited to join in when they too witnessed the positive impact on students and the broader community. Building better STEAM and STEMS^2 learning with the help of community partnerships refueled and nourished our educator selves and souls, and that's too good to keep to ourselves!

CONCLUSION

Integrating the arts, social studies, and community partnerships into STEM learning offers opportunities to connect learning within disciplines, across disciplines, and with real world contexts. These linkages optimize the benefits of STEM education and push back on tendencies to standardize learning or compartmentalize it into subject silos that strip away richness and context. In our own classrooms we have seen the positive impact of integrating the arts, social studies, and community partnerships. Students who needed an extra nudge to participate more fully in STEM subjects benefited from expanded entry points, pathways, and opportunities for expression. Students who already embraced STEM subjects were pushed to consider complex ways that science, technology, engineering, and math interacted with other subjects and in community contexts. In all, we believe incorporating STEAM, STEMS^2, and community partnerships into teaching is a healthy, holistic, powerful practice that honors the diverse interests and passions of students and connects them to a wider family of supportive adults beyond the classroom.

WORKS CITED

Allen, D. E., Duch, B. J., & Groh, S. E. (1996). The power of problem-based learning in teaching introductory science courses. *New Directions for Teaching and Learning, 68*, 43–52.

Beane, J. A. (1995). Curriculum integration and the disciplines of knowledge. *Phi Delta Kappan, 76*(8), 616–622.

Bell, D. (2016, September 1). Looking back: Outstanding blog posts from former teacher in residence Trey Smith. Teaching with the Library of Congress [blog]. Retrieved September 20, 2016, from http://blogs.loc.gov/teachers/2016/09/looking-back-outstanding-blog-posts-from-former-teacher-in-residence-trey-smith

Berkowicz, J., & Myers, A. (2015, January 11). Social studies and STEM. *Education Week*. Leadership 360 Blog. Retrieved August 19, 2016, from http://blogs.edweek.org/edweek/leadership_360/2015/01/social_studies_and_stem.html

Berland, L. K. (2013). Designing for STEM integration. *Journal of Pre-College Engineering Education Research, 3*(1), 22–31.

Bouillion, L. M., & Gomez, L. M. (2001). Connecting school and community with science learning: Real world problems and school–community partnerships as contextual scaffolds. *Journal of Research in Science Teaching, 38*(8), 878–898.

Breiner, J. M., Harkness, S. S., Johnson, C. C., & Koehler, C. M. (2012). What is STEM? A discussion about conceptions of STEM in education and partnerships. *School Science and Mathematics, 112*(1), 3–11.

Bridgeland, J. M., DiIulio Jr, J. J., & Morison, K. B. (2006). *The silent epidemic: Perspectives of high school dropouts.* Washington, DC: Civic Enterprises.

Byrd-Bredbenner, C., Abbot, J. M., & Quick, V. (2010). Food safety knowledge and beliefs in middle school children: Implications for food safety educators. *Journal of Food Science Education, 9*(1), 19–30.

Carnevale, A. P., Smith, N., & Melton, M. (2011). *STEM: Science technology engineering mathematics.* Washington, DC: Georgetown University Center on Education and the Workforce.

Dewey, J. (1910). Science as subject-matter and as method. *Science, 31*(787), 121–127.

Edelstein, W. (1999). The cognitive context of historical change: Assimilation, accommodation, and the segmentation of competence. *New Directions for Child and Adolescent Development, 1999*(83), 5–17.

Eger, J. (2013). STEAM ... now!. *The STEAM Journal, 1*(1), 8.

Every Student Succeeds Act of 2015, Pub. L. No. 114–95 § 114 Stat. 1177 (2015–2016).

Felton, M. (2014, June 23). Why teach mathematics? *Mathematics teaching in the middle school* [blog]. National Council for Teachers of Mathematics. Retrieved August 12, 2016, from http://www.nctm.org/Publications/Mathematics-Teaching-in-Middle School/Blog/Why-Teach-Mathematics_

Fitchett, P. G., & Heafner, T. L. (2010). A national perspective on the effects of high-stakes testing and standardization on elementary social studies marginalization. *Theory & Research in Social Education, 38*(1), 114–130.

Grow, P. L., & Plucker, J. A. (2003, December). Good problems to have: Implementing problem-based learning without redesigning a curriculum. *The Science Teacher, 70*(9), 31–35.

Haberman, M. (2010). The pedagogy of poverty versus good teaching. *Phi Delta Kappan, 92*(2), 81–87.

Hardin, G. (1968). The tragedy of the commons. *Science, 13*(162), 1243–1248.

Hazen, R. M., & Trefil, J. (2009). *Science matters: Achieving scientific literacy.* New York, NY: Anchor Books.

Henriksen, D. (2014). Full STEAM ahead: Creativity in excellent STEM teaching practices. *The STEAM Journal, 1*(2), 15.

Jorgenson, O., & Vanosdall, R. (2002, April). The death of science?: What we risk in our rush toward standardized testing and the three R's. *Phi Delta Kappan, 83*(8), 601–605.

Larmer, J., & Mergendoller, J. (2015). Why we changed our model of the "8 essential elements of PBL." The Buck Institute for Education. Retrieved August 16, 2016, from http://www.bie.org/blog/why_we_changed_our_model_of_the_8_essential_elements_of_pbl

McCarthy, B., & McCarthy, D. (2006). *Teaching around the 4MAT cycle.* Thousand Oaks, CA: Corwin.

Melaville, A., Berg, A. C., & Blank, M. J. (2006). *Community-based learning: Engaging students for success and citizenship.* Washington, DC: Coalition for Community Schools.

Meyer, D. (2011, May 11). The three acts of a mathematical story. *dy/dan* [blog]. Retrieved October 20, 2016, from http://blog.mrmeyer.com/2011/the-three-acts-of-a-mathematical-stry

National Academy of Engineering and National Research Council. (2002). *Technically speaking: why all Americans need to know more about technology*. Washington, DC: National Academies Press.

National Academy of Engineering and National Research Council. (2010). *STEM integration in K-12 education: Status, prospects, and an agenda for research*. Washington, DC: National Academies Press.

National Research Council. (2012a). *Education for life and work: Developing transferable knowledge and skills in the 21st century*. Washington, DC: The National Academies Press.

National Research Council. (2012b). *A framework for K-12 science education: Practices, crosscutting concepts, and core ideas*. Committee on Conceptual Framework for the New K-12 Science Education Standards. Washington, DC: National Academies Press.

No Child Left Behind (NCLB) Act of 2001, Pub. L. No. 107–110, § 115, Stat. 1425 (2002).

Nord, C., Roey, S., Perkins, R., Lyons, M., Lemanski, N., Brown, J., & Schuknecht, J. (2011). *The nation's report card: America's high school graduates* (NCES 2011–462). U.S. Department of Education. National Center for Education Statistics. Washington, DC: U.S. Government Printing Office.

O'Neill, T., Yamagata, L., Yamagata, J., & Togioka, S. (2012, September). Teaching STEM means teacher learning: Three veteran teachers learn that teaching STEM courses requires more than just changing course content. *Phi Delta Kappan, 94*(1), 36–40.

Piro, J. (2010). Going from STEM to STEAM. *Education Week, 29*(24), 28–29.

Roberts, D. A. (2007). Scientific literacy/science literacy. In S. K. Abell & N. G. Lederman, (Eds.), *Handbook of research on science education* (pp. 729–780). Mahwah, NJ: Lawrence Erlbaum.

Root-Bernstein, R. S., & Root-Bernstein, M. M. (2013). *Sparks of genius: The thirteen thinking tools of the world's most creative people*. Boston, MA: Houghton Mifflin Harcourt.

Sadler, P. M., Sonnert, G., Hazari, Z., & Tai, R. (2012). Stability and volatility of STEM career interest in high school: A gender study. *Science Education, 96*(3), 411–427.

Sadler, T. D. (2004). Moral and ethical dimensions of socioscientific decision-making as integral components of scientific literacy. *Science Educator, 13*(1), 39–48.

Sarkar, S., & Frazier, R. (2008). Place-based investigations and authentic inquiry. *The Science Teacher, 75*(2), 29–33.

Savery, J. R. (2005, May). Overview of problem-based learning: Definition and distinctions. *Interdisciplinary Journal of Problem-Based Learning, 1*(1), 9–20.

Smith, J. F. (2015, October 15). Primary sources in science classrooms: Paint, poisoning, proportions, and public health and policy. Teaching with the Library of Congress [blog]. Retrieved September 20, 2016, from http://blogs.loc.gov/teachers/2015/10/primary-sources-in-science-classrooms-paint-poisoning-proportions-and-public-health-and-policy

Spohn, C. (2008). Teacher perspectives on No Child Left Behind and arts education: A case study. *Arts Education Policy Review, 109*(4), 3–12.

Stanton, T. K., Giles Jr, D. E., & Cruz, N. I. (1999). *Service-learning: A movement's pioneers reflect on its origins, practice, and future*. San Francisco, CA: Jossey-Bass.

Swango, C. J., & Steward, S. B. (2003). *Help! I'm teaching middle school science*. Arlington, VA: NSTA Press.

Tan, E., Calabrese Barton, A., Turner, E., & Gutiérrez, M. V. (2012). *Empowering science and mathematics education in urban schools*. Chicago, IL: University of Chicago Press.

Tidball, K. G., & Krasny, M. E. (2011). Urban environmental education from a social-ecological perspective: Conceptual framework for civic ecology education. *Cities and the Environment, 3*(1), 1–20.

Traphagen, K., & Traill, S. (2014). *How cross-sector collaborations are advancing STEM learning*. Los Altos, CA: Noyce Foundation.
Wise, A. E. (2012). End the tyranny of the self-contained classroom. *Education Week, 31*(18), 24–25.
Zimmerman, H. T., & Land, S. M. (2014). Facilitating place-based learning in outdoor informal environments with mobile computers. *TechTrends, 58*(1), 77–83.

CHAPTER TWENTY

Zoology Brüt

Using Backward Design to Explore the Sixth Extinction Through Art, Architecture and Appetite

MELISSA GEORGE

INTRODUCTION

Recent press, both scientific (Ceballos et al., 2015) and popular (Kolbert, 2014; Stevens & Ahnemann, 2015), warns that humans are swiftly altering the biodiversity of the planet. The Earth's species are declining at 1,000 to 10,000 times the normal background rate of one to five per year. This increase is primarily due to human activities driving habitat loss, exotic species introduction, and climate change (Chivian & Bernstein, 2008). Emphasizing that the K-12 educational system must prepare students for college, career, and citizenship to address such problems, the *Next Generation Science Standards* (*NGSS*) outline science, technology, engineering, and math (STEM) performance expectations (PEs) that address human impacts on Earth's systems (Achieve, 2013a). Although these standards integrate the three dimensions of quality STEM education (scientific and engineering practices, crosscutting concepts, and disciplinary core ideas), the impetus to embrace them often focuses narrowly on producing more scientists and engineers to compete in a globally competitive marketplace, rather than inspiring a diverse population of citizen scientists to become knowledgeable and active in their own communities.

Looking towards the future, *A Framework for K–12 Science Education* (National Research Council [NRC], 2012) suggests that further research should focus on (a) understanding students' interests and backgrounds, especially those of underserved and underrepresented groups; (b) using this information to structure learning goals that harness the *NGSS* (Achieve, 2013a, 2013b); and (c) tailoring assessment and learning activities to meet these goals. This chapter recounts my story as a teacher-researcher responding to these challenges through the use of an instructional design technique known as *backward design* (Wiggins & McTighe, 2005).

BACKGROUND

Defining STEM literacy goals and delineating ways to achieve them for the diverse American K–12 student population are major thrusts of the standards movement (Achieve, 2013a, 2013b; NRC, 2012; President's Council of Advisors on Science and Technology [PCAST], 2010, 2012). In particular, *A Framework for K–12 Science Education (Framework)* anticipates that students who "actively engage in scientific and engineering practices and apply crosscutting concepts to deepen their understanding of the core ideas" (NRC, 2012, pp. 8–9) will see how science and engineering are instrumental in addressing major challenges posed by global environmental change. While the documents emphasize student outcomes, they simultaneously underscore the crucial role of the teacher-researcher, "... most important, research is needed on classroom-level contexts ... that engage and support a wider range of students in high-quality teaching and learning experiences with the concepts, ideas, and practices" (p. 325). Acting on this impetus, I addressed the *Framework*'s research challenges, namely to understand students' interests and backgrounds, especially those of underserved and underrepresented groups, and used this information to design a course based on *backward design* (Wiggins & McTighe, 2005) that would harness the three dimensions of *NGSS* (Achieve, 2013a, 2013b; NRC, 2012). *Backward design* requires teachers to think first about structuring specific learning goals and gathering assessment evidence for those goals. Thus, prior to planning instruction, teachers consider their students' needs, the learning sought and the evidence for that learning.

Understanding Students

A defining feature of the *NGSS* (Achieve, 2013a) is the emphasis on equitable learning environments to draw upon the talents of all Americans, including students of diverse backgrounds, particularly those belonging to accountability groups defined by the *No Child Left Behind (NCLB) Act of 2001* and the reauthorized *Elementary and Secondary Education Act* (United States Department of Education, 2010). These groups include students who are economically disadvantaged, members of underperforming racial and ethnic groups, disabled, or have limited English proficiency. According to a 2015 study by the National Center for Education Statistics, low income students (eligible for free or reduced-price lunches) now make up the majority, 51%, of the students across the nation's public schools (Suitts, Barba, & Dunn, 2015). Similarly, the student population in the United States is increasingly more racially and ethnically diverse. The *Condition of Education* (Kena et al., 2016), indicates that from 2003 to 2013 the percentage of white students enrolled in public schools dropped from 59 to 50%, while the number of Hispanic students rose from 19 to 25%. During the 2013–2014 school year, the percentage of English Learners (EL) was 9.3%. Students receiving special education services rose to 13% during the 2013–2014 school year. My classroom composition mirrored the statistics on ethnic minorities (50%) and EL (11%), but a higher percentage of my students received either free and reduced lunch (70%) or special education services (20%).

Research has uncovered some of the strategies that are important for broadening participation in these groups. Such strategies include: (a) connecting science to students' sense of place, (b) supporting learning by offering resources and multiple modes of knowledge representation, and (c) encouraging belief in one's own abilities (Achieve, 2013b; Chemers, Hu, & Garcia, 2001; Lee & Luykx, 2007). Inclusion of these elements helps to sustain student interest (Basu & Barton, 2007), develop ties to the community (Turner, Gutierrez, Simic-Muller, & Diez-Palomar, 2009), address social justice issues (Gutstein, 2003), and validate cultural traditions (Bricker & Bell, 2014; Vomvoridi-Ivanović, 2012).

Establishing a sense of place

The *NGSS* identifies one of the key strategies to support students from diverse backgrounds as "connecting science education to students' sense of place" (Achieve, 2013b, p. 31). In his book *Place-Based Education* (2005), David Sobel provides evidence that "place" provides a rich avenue for goal setting, learning and assessment. Place, as an educational construct, signifies "a unique and bounded biophysical and cultural environment" (Greenwood, 2013, p. 93) that can range from a small classroom to an expansive watershed. Sobel states that the solutions to many of our ecological problems lie in an approach that celebrates, empowers, and nurtures the cultural, artistic, historical, and spiritual resources of place. Key features of the place-based model are: (a) respecting the cultural experiences and prior knowledge that students bring to the classroom (Achieve, 2013b); (b) understanding that knowledge is constructed in cultural contexts, socially negotiated, and distributed among people and tools (Brown, Collins, & Duguid, 1989; Lave & Wegner, 1991; Putman & Borko, 2000; Salomon, 1993); and (c) utilizing place-based learning, a dynamic approach that centers on the active exploration of real-world challenges and problems (Krajcik & Blumenfeld, 2006).

Using multiple modes and representations

In addition to providing a sense of place, when students can interact with multiple learning *modes*, sense with various *modalities*, and communicate their understanding through various types of *representations*, their learning is enhanced (Achieve, 2013b; Ainsworth, 2006; Dunlosky, Rawson, Marsh, Nathan, & Willingham, 2013; Mayer & Moreno, 2002, 2007; Nixon, Smith, & Wimmer, 2015). Presentation *mode* refers to how the material to be learned is presented. Modes can include spoken words, written words, images, gestures, diagrams, or 3D objects (Kress, 2010; Lemke, 1998). Modalities refer to the sense systems used by the learner to receive the material (visual, auditory, tactile, and so on). Accordingly, the most effective learning environments are those that use mixed-modes to appeal to various modalities. Engaging students in this way is different than matching instruction with learning style; the first has extensive research support, the latter does not (Cuevas, 2015; Pashler, McDaniel, Rohrer, & Bjork, 2008). Representations, however, refer to how students display their knowledge construction (Sherin, 2000). While learning, students are exposed to a plethora of instructional modes and process these through various modalities. To communicate their understanding, students should be encouraged to create representations of their ideas.

Sherin (2000) relates how middle and high school students invented scientific representations during a "Motion Picture" activity. After watching computer simulations, students depicted elements of various objects' motion through drawing images, temporal sequences, line segments and graphs. Implementing technological tools expanded the quantity and variety of modes and modalities that students used to interact, interpret, and design. Sherin (2000) states that by "giving students the opportunity to express their creativity, we can provide an entry into the scientific endeavor for students who might otherwise not be attracted to science" (p. 438). Similarly, Nixon et al. (2015) found that middle school science students who were explicitly taught multiple modes to represent their ideas were better able to express scientific knowledge. In particular, Hispanic and female students expressed their ideas more clearly using modes other than, or in addition to, words.

Encouraging self-efficacy

Self-efficacy is defined as a "judgment about one's ability to organize and execute the courses of action necessary to attain a specific goal" (Rittmayer & Beier, 2008, p. 1). Students' self-efficacy beliefs are based on four sources of information: mastery experiences, vicarious experiences, social persuasion, and physiological reaction (Bandura, 1997). Mastery experiences are typically hands-on activities such as lab work, design-projects, or experiments that are structured to be challenging, but not impossible. Engaging in these experiences boosts students' knowledge and confidence. When students have limited mastery experience, which is often the case with diverse learners, they infer the capability level based on the performance of others perceived as similar; thus, teacher provision of experiences such as group mentoring, field trips, or guest speakers allows students to interact with role-models who may be closer in age and background to themselves (Zeldin & Pajares, 2000).

Understanding by Design

Understanding students' needs is forefront to designing curricula and assessment that focuses on developing and deepening students' knowledge of important scientific ideas. To understand is to "make connections and bind together our knowledge into something that makes sense of things" (Wiggins & McTighe, 2005, p. 7). In *Understanding by Design*, Wiggins and McTighe (2005) provide a framework, the *backward design* (BD), for constructing curricular units that focus on in-depth development of big ideas, exploration of essential questions, and authentic applications. The BD framework proceeds in three stages from what is typically identified as the opposite end of the planning process:

- Stage 1: Identify the desired results by considering the learning goals for the course.
- State 2: Determine acceptable evidence that students are making progress towards goals.
- Stage 3: Plan learning experiences that facilitate students' construction of meaning.

The three stages are linked; the desired results of Stage 1 dictate the type of assessment evidence needed in Stage 2 and suggest the types of instruction and activities planned in Stage 3.

Stage 1: Identifying desired results

The process of BD is a deliberate approach used to avoid two common mistakes in instructional design: the aimless coverage of content and the use of isolated activities (Wiggins & McTighe, 2005). Stage 1 calls for thinking deeply about the specific learning sought. There are several considerations. The first is to *identify established goals* that are formal and long-term, such as state content standards, district program goals, departmental objectives, and exit-level outcomes. The second consideration is to *specify the essential questions* that highlight the "big ideas (that) are chosen especially for their power to explain phenomena" (Wynn & Wiggins, 1997, p. v). Another consideration is *targeting student understandings* that link the content to students' lives and "spark meaningful connections, provide genuine inquiry and deep thought, and encourage transfer" (Wiggins & McTighe, 2005,

p. 28). The final consideration is to *outline the key knowledge and skills* students will acquire as learners. This includes stating both the declarative knowledge (basic concepts, vocabulary, factual information) and procedural knowledge (basic know-how or discrete skills) that will be learned by the unit's end. The key is to focus on the knowledge and skills that are essential for meaning-making and transfer and are explicitly taught and assessed (Wiggins & McTighe, 2005).

Stage 2: Determining acceptable evidence

After clarifying learning goals in Stage 1, the next step of BD is to determine evidence of student understanding. According to Wiggins and McTighe (2005), understanding is the "ability to *transfer* what we have learned to new and sometimes confusing settings" (p. 40). Evidence that students are able to transfer understanding includes the following six facets: explaining concepts, interpreting information, applying in new contexts, demonstrating perspective, displaying empathy, and having self-knowledge.

Determining appropriate evidence of understanding requires a match between the format of the assessment and the desired results. If learning basic facts is the goal, then quizzes provide adequate and efficient measures. However, if the goal is deeper understanding, performance tasks and projects that are complex, open-ended, and authentic are more appropriate.

Stage 3: Planning learning experiences and instruction

After identifying desired learning goals and determining acceptable evidence, the next stage of BD is to consider the instructional strategies and learning experiences needed to achieve the desired results. McTighe and Wiggins (2004) use the acronym "WHERETO" (p. 14) to guide effective and engaging learning plans; accordingly, teachers should:

> W—know *where* they are going content-wise, *why* they are learning the content, and *what* is required as evidence of transfer;
> H—hook the students with engaging activities;
> E—provide opportunities to explore and experience;
> R—schedule time for students to rethink and revise their work based on feedback;
> E—encourage self-evaluation;
> T—tailor the plan to address the needs and interests of the students; and,
> O—organize and sequence to maximize engagement and effectiveness.

ZOOLOGY BRÜT: A BEST PRACTICE DESIGN

Over my 20 years of teaching in Indiana, classroom diversity has increased; in response I continually alter course curriculum to provide an equitable learning environment. From 2013 to 2016, I taught a yearlong zoology course in a large urban high school. Reflecting on my students' needs, I integrated the *NGSS* (Achieve, 2013a, 2013b) into my school district's curriculum, and designed *Zoology Brüt*, a course that embraced the term *art brüt* coined by French artist Jean Dubuffet (Piery, 2006) referring to knowledge created outside the boundaries of the official culture. I taught my course with this same philosophy *broadening participation* for my diverse students, the majority of whom were "outside" of the boundaries of the college science track, to construct ideas about animals' roles in the ecosystem and to express these developing ideas in non-traditional ways.

In this section, I highlight the three year development of this course using BD as a means of teaching for understanding and embracing place-based projects. The projects focused on: (a) organism study of structure and function through multiple representations (art); (b) ecosystem modeling and monitoring facilitated by student-designed devices (architecture); and (c) evaluation of the culturally embedded culinary use of animals (appetite). The course capitalized on the school community's strengths (strong fine arts, performing arts, and culinary arts departments as well as engineering and building trade programs) and expanded into the civic community through watershed education resources, university outreach, and field trips to local zoos. These avenues provided students opportunities to "dig deeper" and explore personally relevant connections between materials, design, community, and the environment.

Phase 1: Covering Content or Creating Context? (2013–2014)

Designing a course is an iterative process. When I began teaching zoology, I had access to a curriculum, digital files from a former teacher, and a college-level textbook. The curriculum was a traditional comparative anatomy scope and sequence focused on learning the structures and functions of a representative animal in each phylum. The digital files included PowerPoint lectures, a lab manual with dissection activities, homework, quizzes, and tests. The textbook was significantly beyond the reading level of most, if not all, of the students. I relied on the materials for guidance, but quickly realized they would need to be revamped to fit with my style as a teacher and the students' learning needs. In response, I organized the first quarter into several mini-units, each focusing on a different invertebrate phylum and including a variety of activities. Like any teacher new to a curriculum, I was training myself and my students through routine, adapting to my students' needs by: (a) offering a smorgasbord of activities (teacher-directed, small group, individual); (b) providing the materials, tools, and time to complete the tasks; and (c) using specific instructional strategies to address various learning modalities and needs.

The students were engaged; however, I believed I was committing what Wiggins and McTighe (2005) refer to as the "twin sins of traditional design," (p. 16), activity without explicit sense-making and curriculum coverage without overarching goals. I saw the fallacy in my methods in student responses to the course narrative review. Although they expressed enjoying the class and learning a lot of facts, they were not able to answer deeply a relatively simple question: *What is the role of animals in maintaining ecosystem biodiversity?* The narrative review also revealed that few students had first-hand experiences with ocean animals or animals in a natural setting. Students familiar with animals were pet-owners or participants in 4-H fairs. Thus, the students' role in protecting the biodiversity of the planet was an abstract concept.

In the spring semester, since I had a better understanding of students' learning modalities, outside-of-class interests, and background experiences, I attempted to connect zoology with other aspects of their lives including the unique offerings of our high school curriculum: culinary arts, visual and performing arts, and building trades. Attending an *Arts in Education* workshop injected confidence in my idea that the integration of art-based activities would enhance my students' learning. I tested optional enrichment opportunities: a life-sized student designed and painted *Tree of Life* on the front wall of the classroom, an in-class fabricated and fried chicken cooking demonstration, an entry in the local watershed rain barrel contest, and the building of traps to catch (and release) animals from the adjacent Durkee Run Storm Park for study. I received a small grant sponsored by a local business to fund these projects. As the year ended, I knew that the projects increased student excitement about learning zoology; student evaluations were positive, and enrollment was high. Additionally, the school principal was pleased with the active learning in my classroom. However, I was concerned that I was not explicitly communicating a guiding intellectual purpose to the course. I asked myself: *Was I teaching for understanding?*

Phase 2: Identifying Results, Determining Evidence, Planning Experiences (2014–2015)

Preparing for the next year, I thought deeply about my goals for student understanding. I taught in Indiana, a state that was not on board with *NGSS* (Achieve, 2013a), in a district that used student data compiled from semester-end multiple choice tests to assess teachers' effectiveness; however, I decided to think beyond these constraints. I wanted the first semester's goals, assessment, and instruction to scaffold second semester, allowing for increased student ownership and independence as the year progressed; I utilized a key feature of BD, the *Understanding by Design* (UbD) template (Wiggins & McTighe, 2011), to guide planning of Stages 1, 2, and 3 of my *Zoology Brüt* course.

Stage 1: Identifying desired results

I took several steps in Stage 1; the first was to *identify established goals* by aligning the district curriculum with the three dimensions outlined in the *Framework* (NRC, 2012) and the *NGSS* (Achieve, 2013a). The methods that scientists use as they investigate and build models of the world and engineers employ as they design and test systems are acquired through Dimension 1, *scientific and engineering practices* (SEP). These practices frame students' efforts to understanding core content by focusing on investigating the world, developing explanations and solutions, and evaluating their ideas. Similarly, Dimension 2, the *crosscutting concepts* (CC) are unifying themes that help students to construct and connect knowledge across disciplines. Dimension 3, *disciplinary core ideas* (DCIs), are the organizational structures of the scientific content knowledge that have broad importance across multiple disciplines, provide a key for investigating more complex problems, and relate to students' interests and societal concerns. I combed the DCIs and realized that the core ideas in the *Zoology Brüt* course bridged Life Sciences, Earth and Space Sciences, and Engineering Design.

Second, I *specified the essential questions* that focused on the course's big ideas and then translated them to reflect the *NGSS* (Achieve, 2013a). The essential questions were: How did life on Earth originate and develop? What pressures exist on life on Earth? Why is Earth currently in an era that is *Racing Extinction*? I translated them to reflect the *NGSS* (Achieve, 2013a) DCIs. For example, questions such as: How do the traits of populations of organisms change over time? (HS-LS4), What factors account for both species' unity and diversity? (HS-LS4), and How do elevated carbon dioxide levels influence the co-evolution of Earth's surface and the life that exists on it? (HS-ESS3) emphasize the importance of: appreciating the Earth's biodiversity, realizing the human impacts, and designing solutions to address these impacts.

Another consideration was *targeting student understandings* that link content to students' lives; I wanted students to understand biodiversity as well as critically examine its current state and their role in its future. In *Geology of Mankind*, Dutch chemist Paul Crutzen (2002) termed our present geological era the *Anthropocene* (Kolbert, 2014) because of the overwhelming evidence of human effects on the planet. The expanding human population and its consumption of resources (land, water, energy, and living organisms) threaten biodiversity. Direct and indirect threats resulting from human activity are habitat destruction, habitat fragmentation, habitat degradation (by pollution), global climate change, overexploitation of species for human use, introduction of invasive species, the spread of disease, and synergistic effects between these threats (Primack, 2008; Primack & Bacon, 2011). The biosphere has been stressed beyond its ability to regenerate itself leading us to the current era of the Sixth Extinction (Ceballos et al., 2015; Kolbert, 2014; Stevens & Ahnemann, 2015). The Anthropocene (Crutzen, 2002) is a provocative topic that generates inquiry and deep understanding about the *Zoology Brüt* courses' established goals and essential questions. According to Wiggins and McTighe (2011), *transfer* is the "long-term aim of all education" (p. 14), when students transfer information they take what they have learned in one context and use it in another. Transfer depends upon *meaning-making*, that is, reflecting on and analyzing one's learning to make "an important generalization, a new insight, (or) a useful generalization that makes sense out of a prior experience or learning that was either fragmented or puzzling" (Wiggins & McTighe, 2011, p. 14).

Finally, I *outlined the key knowledge and skills* students use for meaning making and transfer. For example, during the first semester (invertebrate zoology) students: dissect to uncover the anatomy of a squid (skill) and learn its structures (knowledge); recognize the variation of squids' observed traits depends on both genetic and environmental factors (meaning); and map these variation to locations on Earth (transfer). During the second semester (vertebrate zoology), students design solutions (skill) to problems of human impact that limit organism populations (knowledge) and their interactions with each other (meaning) to consider the challenges facing biodiversity (transfer).

Stages 2 & 3: Determining evidence and planning experiences

After clarifying desired results, I followed Wiggins and McTighe's (2005) recommendation to think like an assessor by focusing on these questions: What kinds of evidence will determine student understanding? How will this evidence guide instruction? Effective assessment is a collection of evidence from simple checks for understanding to traditional tests, quizzes, and open-ended prompts to complex performance based projects. Assessments should vary in "scope (simple to complex), time frame (short to long term), setting (decontextualized to authentic content), and structure (highly directive to unstructured)" (Wiggins & McTighe, 2005, p. 152). The various assessments compile a "scrapbook" that tells a story of progressive understanding through the unit. I designed learning activities that would promote "understanding, skill, student interest and excellence" (p. 193) using the WHERETO acronym discussed earlier to plan engaging and effective lessons.

Again, the content of the fall semester, invertebrate zoology, posed a steep learning curve for my students in terms of content, skills, and understandings. My students lived in land-locked Indiana and traveled infrequently; thus, prior knowledge of marine ecology and experiences with ocean animals was limited. Additionally, in this advanced course, students were challenged for the first time to apply concepts learned in biology (hierarchical organization, ecosystem interactions, characteristics determined by DNA, evidence for evolutionary relationships) and to develop lab and research techniques (dissection of animals, preparation of microscopy samples, preparing artifacts). I structured a variety of performance tasks to address these considerations by scaffolding learning and assessment to meet the identified *NGSS* PEs (Achieve, 2013a). Additionally, I wanted the first semester's goals, assessment, and instruction to prepare students for more ownership and independence as the year progressed.

I revised the mini-units, nine focusing on an invertebrate phylum and one outlining the culminating project. Each unit included a variety of assessments: (a) informal checks for understanding during group and lab activities, (b) quizzes and tests of structure and function, and (c) academic prompts focused particularly on the identified PEs that required students to evaluate claims regarding ecosystem interactions and clarify relationships between DNA and expressed traits. The semester culminated in presentation of a performance task, the *Ocean Biodiversity Research Project*, which included model building.

Weekly, students addressed prompts, open-ended questions or problems, that required them to work in small groups to analyze, synthesize, and evaluate. For example, I began the Fall semester with a model building activity which served two purposes: to elicit student ideas about invertebrate structure and to review characteristics of cooperative group learning. Students were given a Zometool kit (Zometool, 2016) to build a model of an invertebrate animal; Zometools are various colored struts and connectors that can be fashioned into geometric shapes. Model building required both creativity and critical thinking; student presentations provided data on their initial ideas about invertebrate animals.

During the mollusk mini-unit, I used a problem-based lesson from the Howard Hughes Medical Institute's (HHMI) *BioInteractive* series entitled *Exploring Biodiversity: The Search for New Medicines* (HHMI, 2016) that explores the venomous carnivorous cone snails found in the Philippines. The activity used video clips and activities to introduce students to: Dr. Olivera, a native Filipino who studies drug development related to cone snail venom; the reliance of the Filipino cultural on mollusks for food, hobby, and medicine; and the principles of taxonomy. I focused the problem on the sorting of seashells by morphological characteristics, constructing a cladogram and later a phylogenetic tree based on these characteristics, and then using DNA sequences to evaluate and revise the tree. Like the Zometool activity, the cone snail sorting activity required students to work in small groups to analyze, synthesize, and evaluate.

The performance task for Fall 2014 was the *Ocean Biodiversity Research Project*. I structured this project differently than 2013–2014 project to address student challenges (deciding on a project, understanding the tasks, using time wisely) and my own concern (mastering *NGSS* content). First, I hooked the students by showing the video *Plastic Paradise: The Great Pacific Garbage Patch* (Soto & Sun, 2013). The video chronicles a young journalist, Angela Sun, as she travels to Midway Atoll to uncover the mystery of the garbage patch ultimately shedding light on how our society's vast consumption of disposable plastic is affecting our oceans.

After discussion of the video's message, students chose to research an ocean location keeping in mind geographical and climatic features that influence animal habitat. Second, as we progressed through the nine mini-units focused on invertebrate phyla, students used the *Marine Species Identification Portal* (ETI BioInformatics, 2016) as a starting point for research. This portal streamlined the process; students were able to search by phylum and ocean location. They compared their animals and appreciated both the unity and diversity of those belonging to the same phylum. I revised the PowerPoint slide bank template from the previous year to simplify and focus student research on these features: (a) classification, (b) description, (c) habitat, (d) distribution, (e) interesting facts, and (f) captioned photos. Progressing through the study of invertebrate phyla, students added representative organisms to their slide bank. At the end of the semester, they summarized the information and presented it to their classmates.

This research was integrated into the mini-units as we progressed through them. For instance, during the porifera mini-unit, our lab experiences were limited to viewing whole samples and prepared slides; consequently, I noticed students were having trouble understanding the poriferan cell specialization. Thus, we created models of the sponges researched. Piggybacking on the message of *Plastic Paradise* (Soto & Sun, 2013), students chose recyclable containers that matched the shape of their sponge. We used flour, water, glue and newspapers to make paper maché and other recyclables to represent various structures, including specialized cells. The classroom became a community space where students collaborated to represent their sponge's features and construct understanding of structure and function.

Another integration of our ocean biodiversity research occurred during the mollusk unit. We had dissected both a clam and a squid; subsequently students researched other mollusks juxtaposing their anatomy to the dissected creatures. During this research, students found that many ocean mollusks were food sources. Reminding them of the maritime Filipino dependency on mollusks, I offered the students a challenge: to determine if the mollusk in their ocean location was used for food. This was a high interest activity, and since most of the students had not eaten mollusks, we prepared clam chowder for students to sample during their marine invertebrate presentations. This became a team effort: finding a recipe, determining how much to make, shucking the clams, and preparing the soup.

At the semester's end, student evaluations showed that besides dissections, the most engaging activities were those that incorporated creating, building, and cooking. Thus, I decided to write an Indiana Public Schools Foundation Grant through Tippecanoe County. My grant, entitled *Zoology Brut: Exploring Animals Using Art, Architecture, and Appetite,* was funded and spanned two school years (2014–2015, 2015–2016). The goal of the grant was to increase high-school students' awareness of invertebrate and vertebrate animals in local ecosystems using non-traditional avenues while addressing learning goals outlined in district curriculum as well as the *NGSS* (Achieve, 2013a, 2013b). These goals were met during the next three semesters.

In Spring 2015, in accordance with the grant activities, I encouraged the students to think of different ways to represent their developing understandings. I knew that students who truly understand would exhibit the following six facets at the term's end:

- Facet 1: Justifies powerful *explanations*
- Facet 2: Offers meaningful *interpretations*
- Facet 3: Provides contextual *applications*
- Facet 4: Sees a variety of *perspectives*
- Facet 5: Demonstrates *empathy*
- Facet 6: Reveals *self-knowledge* (Wiggins & McTighe, 2005).

First, I differentiated the product choices so students could meet the mini-unit requirements. For example, one pre-dissection requirement was completing a coloring sheet which identifies the structure and function of each representative animal's anatomy. During the semester's first mini-unit, *Class Agnatha: Jawless Fishes*, students color-coded a labeled diagram of the lamprey in its juvenile and adult stages. I allowed students to use their smartphones to photograph various aspects of animal dissection. Students compared their standardized drawings with live dissections and posted annotated photographs on the class Facebook Page, *Dr. G's Zoology Page* (2016). Additionally, using their coloring sheets, journals, dissection notes, and photographs, they compared systems of fish groups by diagramming different systems on whiteboards and presenting them to their peers. These activities enhanced understanding (Facets 1 through 4, above) by requiring evidence based explanations, interpretations, and applications forcing students to appreciate many perspectives.

Second, I added an *environmental issues* component to each mini-unit that explicitly addressed the phyla under study. For example, during the *Class Agnatha: Jawless Fishes* mini-unit, I showed the *Silent Invaders: Sea Lamprey* (Breitenstein, 2010) episode which portrays: the invasion of the exotic species, *Petromyzon marinus*, from the Atlantic Ocean to the Great Lakes in the 1920s; the damage done by the parasitic fish to the freshwater fishery; and the variety of mechanisms made to control the invaders as well as the controversy associated with these mechanisms. We also discussed how various cultures depend upon the lamprey as a food source. These activities deepened the facets of student understanding by: providing contextual applications (Facet 3), presenting a variety of perspectives (Facet 4), providing a vehicle to experience empathy (Facet 5), and developing self-knowledge about the issues (Facet 6). Tweaking the mini-units by adding photography and linking environmental issues set the stage for the first phase of the grant.

The first set of grant related projects, *Zoology Brüt Phase 1: Fish Edition*, began early in the spring semester. They included: (a) expanding studies of marine invertebrates and their ocean locations; (b) focusing on the three classes of fish that inhabit these ocean ecosystems; and (c) exploring how humans depend upon fish and marine ecosystems as resources, yet are adversely affecting them through unsustainable practices. The projects were designed to address *NGSS* disciplinary core ideas, crosscutting concepts, or scientific/engineering practices (Achieve, 2013a) by focusing on one or more of the guiding questions below:

- How do the structures of organisms enable life's functions?
- How do humans depend on Earth's resources?
- How can humans design and evaluate solutions to real world problems that take into account considerations such as cost, safety, reliability, aesthetics, as well as social, cultural, and environmental impacts?

Each class brainstormed possible projects that related to art, architecture, and appetite and then decided on research topics about fish that were related to one or more of the essential questions. A list on a large classroom white board displayed all classes' ideas. Students worked alone, in pairs, or in small groups for six weeks to research their ideas and prepare an explanatory product. Class time each week was dedicated to project work and all materials were provided by the grant. I laid out the plan for the projects: (a) Week 1: brainstorm projects, research three topics, decide on one; (b) Week 2: begin specific research, draft materials list; (c) Weeks 3, 4, and 5: continue research, begin and complete project design; (d) Week 6: plan presentation; and (e) Later weeks: present findings.

Many zoology students were concurrently enrolled in art or journalism classes, thus integrated ideas across disciplines. Art projects included children's books (fish ecosystems, fish anatomy, fish species), drawing/paintings (tropical fish, deep sea fish), fictitious newspaper articles (fish of Hawaii), mobiles (fish of Indiana), models (tropical ocean diorama), mosaics (tropical fish), PowerPoints (sharks, sport fishing), and sculptures (sharks). Students were required to prepare an artist's statement for their project which relayed their inspiration, research, and message.

Architecture projects included engineering breeding tanks for zebrafish and a larger-than-life plastic bottle fish sculpture. Students from each class established tanks at the back of the classroom and tested various designs for fish breeding chambers. When results were not as planned, I arranged a field trip to the Lueng Lab at Purdue University which uses zebrafish as retinal degeneration models (Leung, 2015). My students learned about the disease causing gene network for retinal degeneration as well as how the model fish were bred and raised. After the visit, the students compared their breeding strategies to those used by scientists and realized how the conditions of our school tanks differed; students successfully revised their designs, grew zebrafish embryos, and shared these with their classmates for microscope viewing.

A fish sculpture was inspired by a community art project designed by Kasey Jones (National Aquarium, 2014), a large outdoor turtle sculpture created with plastic trash bags. Our version used PVC pipe and wire to create an eight foot long frame. Students laid out a paper design, colored the bottles by pouring paint into them, and threaded them through thick wire. The goal of the design was to bring awareness to the harmful effects of plastic on ocean animals.

Appetite projects focused on easy to find fresh fishes in Indiana, namely tilapia and salmon. Students identified the anatomy, cultural significance, and sustainable fishing practices. This information, along with recipes, was included in a book. Students from each class section participated in a cook-off and voted on a winner.

The grant project established a place-based community. All the supplies were provided for the students, and most of the work was completed in the classroom, either during in class "project blocks" or optional after school work sessions. Rubrics standardized expectations. Students presented a variety of projects to their respective classes and self-evaluated their work. According to the students, the strengths of the *Zoology Brüt Fish Projects* were: (a) choosing the topic and presentation mode, (b) learning from the variety of work done by classmates both in and outside their class block, and (c) creating the "workshop" atmosphere of the classroom learning community. Challenges of the projects were deciding on a topic and using class time wisely. I agreed with their evaluations and had additional concerns regarding the depth of understanding of the human relationship to using and protecting Earth's resources.

Phase 3: Reflecting and Revising (2015–2016)

Stage 1: Desired results

In the Fall of 2015, my zoology course load increased to six sections.. Many students were encouraged to sign up for the class by either their friends, due to the community building grant-related projects, or counselors who saw the relevance of the course's integrated themes. I capitalized on the strengths of the previous year's course, both in the day to day activities structured by the mini-units and in the semester long performance task. I explicitly focused on the concern that arose at the end of the previous year's *Fish Edition Projects*, namely, the depth of student understanding about the relationship of how humans use and protect Earth's resources. I introduced this theme and wove it through the mini-units and projects. I presented the idea of the Anthropocene (Crutzen, 2002; Kolbert, 2014) by having students read and discuss the prologue and first chapter of the *Sixth Extinction* (Kolbert, 2014).

These pages began unravelling the evolutionary history of extinction; stories recounted the disappearance of creatures in various parts of the world. The sad tale of the demise of the charismatic Panamanian golden frog population introduced several ideas: the unique lifecycles of various amphibian species; the global extent of the diminishing amphibian population due in part to the chytrid fungus; and the collaboration of various scientists and local citizens to establish conservations centers. This story of the increasingly fragmented rainforest motivated students to dig deeper into other issues such as the effects of ocean acidification on the Great Barrier Reef. Additionally, I established an online book club with guiding chapter themes for students who wanted to continue exploration of the *Sixth Extinction* (Kolbert, 2014). As the semester progressed, I emphasized threats to marine ecosystems and the variety of ways the diverse scientific community studies them. *The Shape of Life* video series (Sea Studios Foundation, 2015) provided connections to contemporary scientists and their studies.

Students applied their understandings of how and why scientists study marine invertebrates in *Zoology Brüt Phase 2: Marine Invertebrate Edition*. Because of the preface provided by the *Sixth Extinction* (Kolbert, 2014) discussions, the goals of the *Ocean Biodiversity Research Project* seemed clearer to these students than those of previous years. I kept the PowerPoint task the same as the Fall of 2014, but expanded the model building project to include all invertebrates. Rubrics guided both PowerPoints and models; this scaffolding helped with decision making, wise use of time, and quality of science displayed in the projects.

The students were tasked with choosing one invertebrate from the early part of their research and designing a model of that animal from recycled materials. They were to identify, represent, and explain the specific structures and how they helped the animal perform its life functions. All of this work was done in class (model building time was structured into each week's schedule), although students were encouraged to stay for after-school work sessions and allowed to take models home to work on them. This activity generated a workshop atmosphere where approximately 126 invertebrate models were constructed. Each student had a shared lab station. Students were inspired to create because of the buzz of activity during each class period and the variety displayed across classrooms.

Once the designs were completed by the six classes, we set up a classroom museum with specific areas for the different marine invertebrate model categories. Each student also created a legend for their model that included ten structures and functions and how they were identified by the recycled materials. Students decided criteria and voted on the *best* (as operationalized by the students) models in each category. Voting occurred during an informal museum gallery walk as students ate clam chowder prepared as in the previous year. The *best* models were displayed in a large glass case in the entryway to the high school building.

Stages 2 & 3: Determining evidence and planning experiences

In the Spring of 2016, vertebrate zoology included revised mini-units that emphasized student choice, explicit connections to animal biodiversity, and creative expression. The grant's final phase, *Zoology Brüt Phase 3: Racing Extinction Edition,* was facilitated during this semester.

Animal biodiversity and conservation issues were explored through excerpts from the non-fiction work, *The Sixth Extinction* (Kolbert, 2014), and the video *Racing Extinction* (Stevens & Ahnemann, 2015). Both media emphasize that human activities are changing the Earth. The video challenged many students to think about conservation issues that affect vertebrate animals such as overfishing, poaching exotic animals for clothing and pets, and consuming animals as meat. The take home message in *Racing Extinction* is to "#StartWith1Thing" (Stevens & Ahnemann, 2015) to make a lasting impact on the planet.

Because these messages were so powerful to the students, I focused the *Zoology Brüt* projects on the idea that we, as a planet, are *Racing Extinction* (Stevens & Ahnemann, 2015). Students were challenged to: choose an *issue* of concern, *investigate* the issue, and *inform* others through an art, architecture, or appetite project. Using inspiration from the *Racing Extinction* teacher resources, I made a list of possible issues (overuse of plastics, exotic animal trade, conservation of native ecosystems), investigative questions, and informational products; students were tasked with choosing their *three I's* (issues, investigative questions, and informational products) from my list or designing their own.

The semester project was divided into several segments: (a) a week spent brainstorming issues; (b) three weeks devising questions and researching them; (c) four weeks creating a product to inform others, and; (d) a week presenting the product. Besides preparing a class presentation, many students also set up an exhibit to teach others in a community outreach fair that I organized at the citywide *Paint the Rain Barrel* contest. Several aspects of this project were significant. First, there was a sense of urgency in the student projects because they were adhering to the *Racing Extinction* motto "#StartWith1Thing" (Stevens & Ahnemann, 2015). Students were clearly motivated to: find ways to reuse plastic, investigate the declining populations of amphibians in Indiana, and explore vegetarianism. For example, two students worked on creating a loom and weaving plastic newspaper bags into rugs that could be used as mats for naptime in all-day kindergarten classrooms. Second, there was increased awareness of animal use as a commodity in other cultures for food sources, sacred objects, medicinal remedies, and materials for clothing and jewelry. One group wrote a "What Not to Cook Cookbook" describing recipes of how various endangered animals were used as food sources, and then suggesting possible alternatives. Another student tried a vegetarian diet for two weeks and recorded each meal as an Instagram post (Bisner, 2016). Each of her vegetarian choices was related to a sustainable practice such as composting food scraps, conserving water and land through avoiding livestock, and so on. Another student created a large multimedia sculpture of bones wrapped with fake fur and horns caught in a trap. Her artistic statement said:

> The title of my piece is "Ew, What is That?" I researched what animals were being hunted for fur/horn in six continents (North America, South America, Europe, Africa, Asia, and Australia). A multitude of species are threatened by hunting, most of which people have never heard of so they need to become more aware. I wanted my piece to look like a pile of skinned and dehorned animals in the back of a poacher's truck … that would move people to realize that without any change in our behavior, many more species will become extinct. (Isabella, Artist's Statement, 5/9/16)

A significant aspect of the project was the sense of place within the classroom across the six sections of zoology, outside the classroom in the local Durkee Run Storm Park, and extending into the larger community through field trips and the Paint the Rain rain barrel contest. Like variations of the project during earlier semesters, both physical and temporal space supported students work and facilitated cross classroom collaboration. Several student groups worked on animal projects: raising native amphibians and reptiles that we gathered from the storm park, ordered online, and rescued in the wild. Students designed appropriate structures to house and breed the animals which were watched intensively by all classes; they began to distinguish between the type of care required for native organisms, lab models, and pets. Field trips to the neighborhood Columbian Park Zoo allowed students to physically connect with four exotic creatures (an eight banded armadillo, a woma python, a red-tailed African grey parrot, and a hellbender salamander) and the threats they face due to human decisions. An additional field trip to the Indianapolis Zoo provided further exploration into local and global conservation issues. Lastly, several groups of students investigated water conservation issues. Eight groups designed and painted rain barrels and entered them in a local contest to bring awareness to water conservation issues (Photo 20.1). Other determined the quality of water from community parks and waterways. Since the Durkee Run Storm Park (adjacent to our high school) was the contest location, we used this avenue to share the *Racing Extinction* projects and teach the community.

I conducted a survey for the final project phase, *Zoology Brüt Phase 3: Racing Extinction Edition*, through SurveyMonkey (2016). The major highlights at the end of the project were: 87% of the students understood that human choices influence the Earth's systems and may accelerate our extinction as a species, and 73% of the students claimed that the topic of *Racing Extinction* should be explored in future Zoology classes.

Photo 20.1. One of Many Rain Barrels Entered in Community Water Conservation Contests. (Photo by Melissa George.)

USING BACKWARD DESIGN FOR COURSE PLANNING

The story told in the previous section depicts my journey to design and enact a placed-based curriculum that adhered to the vision: (a) delineated in recent reform documents; (b) described in the BD literature; and, (c) embodied (by me) as a teacher researcher addressing classroom constraints and accommodating student needs. Teacher design of lessons, units, or courses of instruction, is paramount to the craft of the teaching profession. Teachers realize that design is an iterative process that may take several years to "get right" because ultimately it involves reflecting, extending, and re-evaluating. While the three stages of BD are: (a) identifying desired results; (b) determining acceptable evidence; and (c) planning experiences; building a course is "more like painting from a black canvas that painting by numbers" (McTighe & Wiggins, 2004, p. 4). I streamline the strategies for BD that were discussed in previous sections using the UbD template; keep in mind, however, the path is circuitous.

Stage 1: Identifying desired results

This stage focuses on four considerations addressed in earlier sections: (1) identifying established goals; (2) specifying essential questions; (3) targeting student understandings; and (4) outlining key knowledge and skills. The thrust of this stage is clarification of priorities to arrive with a product that embodies the six facets of understanding: explanation, interpretation, application, perspective, empathy, and self-knowledge (Wiggins & McTighe, 2005). Sifting through established content standards (national, state, district), reviewing district curriculum expectations, and determining overlap is a good place to start *identifying established goals*. How does this process help? First, broadly highlighting the key verbs and recurring nouns in district curriculum delineates procedural and conceptual knowledge, respectively. The purpose is to identify core concepts, principles, theories, and processes that will serve as the focal points for your design. Second, matching this information to the both the *NGSS* (Achieve, 2013a) procedural standards, *scientific and engineering practices* (SEPs), as well as the conceptual standards, *crosscutting concepts* (CCs) and *disciplinary core ideas* (DCIs), enables an alignment check. This process is a bit messy, but paint with broad strokes; sticking to the main headings in the curriculum and standards may help with this. For example, a district curriculum learning objective might state: "Students will be able to compare the anatomy of animal phyla." This maps to *NGSS* LS1.A, "Systems of specialized cells within organisms help them perform the essential functions of life" (Achieve, 2013a, p. 105).

The next considerations, *specifying essential questions* and *targeting student understandings*, allow goals to be unpacked into transferable understandings that give meaning and importance to facts. One way to do this is to look for patterns in your map that represent core ideas as an "outgrowth of related and suggestive pairs" (Wiggins & McTighe, 2005, p. 74). In the preceding example, the relationship between *structure and function* is implied in both the district curriculum and standard LS1.A (Achieve, 2013a), suggesting both a targeted understanding: *Animals in each phyla have structures that enable them to carry out life's functions* and an essential question: *How does organism structure enable life's functions?* Essential questions typically frame and guide inquiry into the topic of understanding.

Once the established goal is unpacked, delineate what students will know (knowledge) and be able to do (skills) in order to demonstrate this understanding.

While knowledge includes terminology, definitions, and critical relationships, skills can be related to thinking (comparing, analyzing, and interpreting), communicating (listening, speaking, demonstrating), and researching (inquiring, investigating, dissecting). Relating to the above example, students will know the functions of various animal structures (knowledge) and use dissection to compare them (skill).

Table 20.1 shows how I utilized the UbD template (Wiggins & McTighe, 2011) to structure Stage 1 of *Zoology Brüt*. On the left hand side of the table are the established goals I wrote in the form of essential questions based on the alignment of district curriculum objectives and *NGSS* DCIs (Achieve, 2013a). On the right hand side, I list how students will transfer understanding and acquire

Table 20.1. Understanding by Design Template, Stage 1.

Invertebrate/Vertebrate Zoology Brüt Course Stage 1—Desired Results	
Established Goals	Transfer of Understanding
Students will investigate these *NGSS* (Achieve, 2013a; NRC, 2012) essential questions based on Dimension 3 (below) and expanded through Dimensions 1 and 2 (right). HS-LS1: How does organism structure enable life's functions? HS-LS2: How do organisms interact with each other and their physical environment? HS-LS3: How are the traits of one generation of organisms related to the previous generation? HS-LS4: What factors account for both species' unity and diversity? HS-ESS2: How do elevated carbon dioxide levels influence the co-evolution of	*Students will use their understanding to:* • Demonstrate that systems of specialized cells within organisms help them perform essential functions. • Compare invertebrate/vertebrate organisms based on the structures they use to carry out life's processes in various locations on Earth. • Represent how organisms interact with each other and their terrestrial or oceanic environment. • Evaluate claims supporting that changes in environmental conditions may result in: (a) increases in the number of individuals of some species, (b) the emergence of new species over time, and (c) the extinction of other species. • Propose that the variation of organisms' observed traits depends on both genetic and environmental factors. • Argue that adaptive changes due to natural selection contribute to the planet's biodiversity. • Consider the challenges facing animal biodiversity.
	Acquisition of Knowledge and Skill
	Students will know: • The structures of representative organisms in each animal phyla that allow them to carry out life's functions.

(*Continued*)

Table 20.1. (*Continued*)

Invertebrate/Vertebrate Zoology Brüt Course Stage 1—Desired Results	
Established Goals	Transfer of Understanding
Earth's surface and the life that exists on it? HS-ESS3: How can resource management protect biodiversity? HS-ETS1: How can engineering design be used to enhance the sustainability of Earth's resources?	• Ecosystems limit animal populations through the availability of resources as well as the challenges of competition, predation, disease, and human influences. *Students will be skilled at:* • Using dissection and research skills to uncover and compare representative organisms' anatomy. • Designing solutions to human impacts on biodiversity. • Expressing their findings orally, in writing, in diagrams, using technology, and by modeling.

(Source: Author, Collected from Wiggins & McTighe, 2011.)

knowledge and skills using "performance verbs" (Wiggins & McTighe, 2011, p. 100) that indicate the six facets of understanding.

Stage 2: Determining acceptable evidence

Having just identified desired results in Stage 1, it is now time to think about the evidence that would indicate successful learning of these goals. Wiggins and McTighe (2005) state that in Stage 2, one must "think like an assessor, not a teacher" (p. 148). The assessment evidence focuses on the outcomes sought (Stage 1), rather than a mechanism for generating grades or planning activities. Three questions can help frame the assessment design:

- What types of evidence indicate student understanding?
- What characteristics of this evidence should be examined?
- Does the evidence enable us to infer a student's knowledge, skill, or understanding?

We will consider each of these questions.

Effective assessment requires gathering a range of evidence: informal checks for understanding, observations, tests and quizzes, academic prompts, and performance tasks. Performance tasks are the most authentic because they provide evidence that students are able to transfer information in context. For example, Table 20.2 lists performance tasks I designed for Stage 2 of *Zoology Brüt*. One

Table 20.2. Understanding by Design Template, Stage 2.

Invertebrate/Vertebrate Brüt Zoology Unit
Stage 2—Determining Evidence

Evaluative Criteria	Performance Tasks
Students will meet these NGSS (Achieve, 2013a; NRC, 2012) performance expectations: HS-LS1-2: Develop and use models to illustrate the organization of systems (Achieve, 2013a, p. 263). HS-LS2-6: Evaluate evidence that the stability of ecosystem interactions depends upon ecosystem conditions (Achieve, 2013a, p. 267). HS-LS3-1: Ask questions to clarify relationships between DNA and inherited traits (Achieve, 2013a, p. 270). HS-LS4-1: Communicate that multiple lines of evidence support biological evolution (Achieve, 2013a, p. 272). HS-ESS2: Develop a model to demonstrate carbon cycling (Achieve, 2013a, p. 122). HS-ESS3: Evaluate solutions that reduce human impacts (Achieve, 2013a, p. 125). HS-ETS1: Analyze global challenges to propose solutions (Achieve 2013a, p. 129).	*Students will show their learning by evidence of:* • Displaying appropriate appreciation and technique during dissection of various organisms. • Describing representative organisms' structures and corresponding functions in an illustrated (or photographed) journal summary (or Facebook post). • Comparing the systems of animal phyla through small-group annotated diagrams and *Gallery Walks*. • Documenting representative invertebrates from each phylum that live in a specific ocean location in a PowerPoint presentation. • Creating models to illustrate the structural organization of a representative marine invertebrate. • Discussing the challenges of scientists to understand the history and complexities of ecosystems. • Conducting research on the range of conditions that influence the stability of animal ecosystems. • Deciphering cladograms that show the evolutionary relationships between animals. • Drawing, presenting, and evaluating cladograms that show evolutionary relationships between animals. • Researching global challenges to sustainability and proposing, designing, evaluating, and presenting solutions through art, architecture, or appetite.
	Other Evidence
	Students will show their learning by evidence of: • Responding orally and/or in writing to prompts requiring evaluation of animal structure and function, scientific research, and evolutionary relationships. • Demonstrating proficiency on tests and quizzes about animal structure and function and evolutionary relationships with reinforcement from on-line quiz applications.

(Source: Author, Collected from Wiggins & McTighe, 2011.)

of the tasks is *researching global challenges to sustainability and proposing, designing, evaluating, and presenting solutions through art, architecture, or appetite.*

Wiggins and McTighe (2005) have created a design tool using the acronym GRASPS to guide the construction of performance tasks. The letters correspond to design elements: "Goal, Role, Audience, Situation, Performance, Standards" (p. 157). We will walk through the *Zoology Brüt* performance task above using GRASPS.

- Goal: *The challenge is* to identify an issue threatening the planet's sustainability, investigate the issue, and decide on a way to inform others.
- Role: *You are a* sustainability advocate who is translating his/her message through creative means.
- Audience: *Your audience is* the general public. *You need to convince them* that humans are negatively impacting the biodiversity of the planet.
- Situation: *The context you find yourself in* is an environmental advocate.
- Performance: *You will create* a product focusing on art, architecture, or appetite *to* communicate awareness of the issue.
- Standards: *Your product must* meet the standards outlined in the rubric.

Ideally, the characteristics of student responses, products, or performances would indicate one or more of the six facets of understanding (Wiggins & McTighe, 2005) discussed previously. For example, the *Zoology Brüt* performance task outlined above requires students to explain, interpret, apply, show perspective, empathize, and have self-knowledge (Facets 1–6) about the issue they choose. I designed both holistic and analytical rubrics to describe both the characteristics of the products and the degrees of understanding along a continuum. Midway through the project work, I utilized the holistic rubric to evaluate the overall impression of student work and to provide feedback to steer the project in a focused direction. At the end of the project, I used an analytic rubric to evaluate different dimensions of quality and proficiency.

Devising a rubric for evaluation is again an iterative process. The first time through, a teacher may provide general guidelines, but as the products are turned in, a range of understanding and proficiency is exhibited. Arter and McTighe (2001) propose a six-step process for analyzing student work and creating rubrics from this analysis:

- Step 1: Gather examples of student work.
- Step 2: Sort into different stacks and evaluate reasons for sorting.
- Step 3: Cluster the reasons into characteristics of performance.
- Step 4: Write a definition of each characteristic.
- Step 5: Select student samples that exemplify each characteristic.
- Step 6: Continually refine.

After deciding on evidence and defining its characteristics, we must check to see that the evidence gathered is valid and reliable. Validity means whether an assessment measures what is intended. When examining rubrics, a teacher must ask: Could the criteria be met without understanding? Could the criteria not be met, but the student shows understanding? These are important checks; answering in the affirmative to either of them indicates an invalid assessment. Besides measuring what is intended, we also need to know that our assessments will provide consistent results, or show reliability. Reliable evidence will indicate a credible pattern or trend; using a mix of assessments that tap into the various facets of understanding is most appropriate. For instance, in the *Zoology Brüt* performance task, the hands-on application (Facet 3), also requires a statement of purpose (Facet 4) and a self assessment (Facet 6).

Stage 3: Planning experiences

After clarifying desired results and delineating appropriate assessments comes planning engaging and effective learning activities. In Stage 3 we focus on who the students are and what they will need to achieve the desired results of Stage 1 and perform well at the tasks outlined in Stage 2. McTighe and Wiggins (2004) use the WHERETO acronym outlined earlier to design a student-centered learning plan. As a reminder, each letter of the acronym represents a key element of the planning design (in italics below); however, the elements do not need to appear in order.

Students should know *where* they are going to content-wise. This involves including activities that: diagnose where students are coming from (their prior knowledge, misconceptions, learning modalities, and special needs or talents); outline relevant and beneficial learning goals that take into account student questions, personal quests, and community connections; and present culminating task requirements. Often this explanation of *where* may begin with a *hook* to pique and *hold* student interest. For example, starting with a mystery, a challenge, a personal experience, or an emotional connection often engages students to construct meaning. From there, a teacher should think about how to facilitate content *exploration* so that students are *equipped* for expected performances. Content exploration will involve: direct instruction to help students find information, monitor their understanding and self evaluate; coaching to provide feedback and practice; and facilitative instruction to assist students in constructing, examining, and extending meaning. Examples of facilitative instruction include: cooperative learning, guided inquiry, problem-based learning, and open-ended questioning. Content exploration might also include place-based experiences such as the school garden, other departments within the school (culinary arts, engineering, art), field trips, and community liaisons.

Understanding develops as a result of *rethinking, revising, reflecting* and *evaluating*. Various strategies can be employed to help students rethink such as constructing arguments for debates, especially those that require students to shift perspective. Likewise, editing sessions, peer critiques, and rehearsals provide opportunities for students to revise; and, journals, think-alouds, and metacognitive prompts encourage reflection. The second *E* in WHERETO proposes that teachers build in opportunities during lessons for students to *self-evaluate*. This might involve: setting aside time at the beginning or end of a lesson to consider conclusions as well as unresolved questions; requiring written self-assessments be attached to every product or performance; and including quick-writes to summarize main points and ask questions.

Finally, teachers should *tailor* their design to address student differences in background knowledge, skill levels, talents, special needs, and learning modalities and *organize* the learning activities accordingly. Some topics are most effectively covered in a linear and didactic fashion; others are more appropriately "uncovered" in an inquiry-oriented, experiential manner. McTighe and Wiggins (2004) note that an appropriate "sequence of teaching and learning experiences will equip students to engage with, develop and demonstrate desired understandings" (p. 226). Table 20.3 outlines my planned learning events in *Zoology Brüt* in terms of the WHERETO acronym.

Table 20.3. Understanding by Design Template, Stage 3.

Invertebrate/Vertebrate Brüt Zoology Unit Stage 3—Learning Plan	
	Learning Events *Student success at transfer, meaning, and acquisition depends on...*
W *Where* are we going? *Why*? *What* is expected? *What* prior knowledge exists?	• Presentation of essential questions and expectations (goals). • Provision of course rationale and goal application (relevance). • Explanation of key assignments/assessments and involve students in determining evaluation criteria (expectations). • Assessment of prior knowledge and misconceptions (diagnosis).
H How do we hook and *hold* student interest?	• Hooking: Beginning with essential questions that pique student interest in environmental issues related to animal biodiversity. • Holding: Allowing student choice of learning activities within each mini-unit and culminating projects.

	Invertebrate/Vertebrate Brüt Zoology Unit Stage 3—Learning Plan
	Learning Events *Student success at transfer, meaning, and acquisition depends on...*
E How do we *equip* students for expected performances?	• Direct instruction: interactive lectures, graphic organizers, demonstration/modeling, evaluating video clips. • Facilitative teaching: cladogram building, anatomical diagramming, animal dissecting, readings on environmental issues, lab experimentation, vocabulary practice. • Coaching: authentic assessments (projects), feedback.
R How will students *rethink* and *revise*?	• Reconsideration of structure/function through animal phyla from simple to complex. • Scaffolding of biodiversity threats from natural to manmade.
E How will students self-*evaluate* and reflect on learning?	• Employment of metacognitive strategies: think-pair-share, jigsaw, quick-write, survey and exit tickets, performance profile, Socratic seminar, project-based learning. • Requirement of self-assessment after formal products
T How will we *tailor* learning to interests, needs, and styles?	• Content: Open-ended essential *NGSS*-based questions; targeted instruction matched to prior knowledge and skills. • Process: Address learning modalities with variety of resources. • Product: Choice for mini-units and project-based learning.
O How will we organize and sequence the learning?	• Organization: Immersion in mini-units with an overarching project to connect and add depth. • Sequence: Recursive learning/doing/reflecting; focus on intriguing issues first, then make sense of experiences

(Source: Author, Collected from Wiggins & McTighe, 2011.)

CONCLUSION

One of the guiding principles of the recent reform documents is to engage students with sustained opportunities to develop a thorough understanding of scientific ideas (Achieve, 2013a; NRC, 2012). *Backward design* addresses this principle by structuring curricula, assessment, and instruction to develop and deepen understanding of important ideas by keeping the "end in mind" (McTighe & Wiggins, 2004, p. 25). Using three stages, the model encourages designers of curriculum to: focus on the appropriate goals and targeted understandings (Stage 1), provide appropriate and varied assessments for evaluation (Stage 2), and plan effective and engaging learning experiences (Stage 3).

Although the UbD template (McTighe & Wiggins, 2004) helps to organize the product (see Tables 20.1–20.3), the process typically unfolds in an iterative and unpredictable way as was told in my story of the *Zoology Brüt* design. My quest to create an *NGSS*-based course design (Achieve, 2013a) hinged on: understanding the vast needs of the students I taught, capitalizing on the expectations and resources of the school and local communities, and revisioning the role of education in a global society. Although my quest was lofty, I was able to achieve it because I planned with specific goals in mind. Too often STEM teachers begin and remain focused on textbooks, favorite lessons, or engaging activities rather than the desired result: understanding.

This chapter tells my story of curriculum design and provides many insights into the use of the UbD guidelines such as clarifying content priorities, fashioning assessments that reflect the six facets of understanding, and using the WHERETO (McTighe & Wiggins, 2004) acronym to guide lesson planning. The Association for Supervision and Curriculum Development (ASCD) website, http://www.ascd.org/research-a-topic/understanding-by-design-resources.aspx# online, offers further assistance. The website gives an overview of the UbD framework; provides articles, sample chapters, and learning videos; and offers support in the way of professional development courses, webinars, and an on-line learning community.

The best practices of STEM education help students to become thoughtful about and productive with both the procedural and conceptual goals of the *NGSS* (Achieve, 2013a). Such goals steer away from content coverage and aimless activities, and towards facilitating deep thinking to prepare all students for the world beyond school to apply what they have learned to issues and problems they will face in the future. I encourage STEM teachers to embrace *backward design*, a method that provides a way to craft curriculum to reflect *learning for understanding*, which helps students: (a) acquire important information and skills, (b) make sense of content, and (c) transfer their learning to new situations within school and beyond it.

WORKS CITED

Achieve. (2013a). *The next generation science standards, Volume 1: The standards–arranged by disciplinary core ideas and topics*. Washington, DC: The National Academy Press.

Achieve. (2013b). *The next generation science standards, Volume 2: Appendices.* Washington, DC: The National Academy Press.

Ainsworth, S. (2006). Deft: A conceptual framework for considering learning with multiple representations. *Learning and Instruction, 16,* 183–198.

Arter, J. A. & McTighe, J. (2001). *Scoring rubrics in the classroom: Using performance criteria for assessing and improving student performance.* Thousand Oaks, CA: Corwin Press.

Bandura, A. (1997). *Self-efficacy: The exercise of control.* New York: W. H. Freeman and Company.

Basu, S. J., & Barton, A. C. (2007). Developing a sustained interest in science among urban minority youth. *Journal of Research in Science Teaching, 44*(3), 466–489.

Bisner, B. (2016). Whatbayleighate. Retrieved from https://instagram.com/whatbayleighate/

Breitenstein, J. (Producer). (2010). *Silent invaders: Sea lamprey.* Retrieved from https://www.youtube.com/watch?v=EqskoRWMXbA

Bricker, L. A., & Bell, P. (2014). "What comes to mind when you think of science? The perfumery!": Documenting science-related cultural learning pathways across context and timescales. *Journal of Research for Science Teaching, 51*(3), 260–285.

Brown, J. S., Collins, A., & Duguid, P. (1989). Situated cognition and the culture of learning. *Educational Researcher, 18*(1), 32–42.

Ceballos, G., Ehrlich, P. R., Barnosky, A. D., Garcia, A., Pringle, R. M., & Palmer, T. M. (2015). Accelerated modern human–induced species losses: Entering the sixth mass extinction." *Science Advances, 1*(5), e1400253–e1400253. doi: 10.1126/sciadv

Chemers, M. M., Hu, I., & Garcia, B. F. (2001). Academic self-efficacy and first-year college student performance and adjustment. *Journal of Educational Psychology, 93*(1), 55–64.

Chivian, E., & Bernstein, A. (Eds.). (2008). *Sustaining life: How human health depends on biodiversity.* Center for Health and the Global Environment. New York: Oxford University Press.

Crutzen, P. (2002). Geology of mankind. *Nature, 415,* 23.

Cuevas, J. (2015). Is learning styles-based instruction effective? A comprehensive analysis of recent research on learning styles. *Theory & Research in Education, 3*(13), 308–323.

Dunlosky, J., Rawson, K. A., Marsh, E. J., Nathan, M. J., & Willingham, D. T. (2013). Improving students' learning with effective learning techniques: Promising directions from cognitive and educational psychology. *Psychological Science in the PublicInterest, 14*(1), 4–58.

ETI Bioinformatics. (2016). Marine species identification portal. [Data file]. Retrieved from http://species-identification.org

George, M. (2016). Dr. G's zoology page. Retrieved from https://www.facebook.com/Dr-Gs-Zoology-Page-371860326284441/

Greenwood, D. A. (2013). A critical theory of place-conscious education. In R. B. Stevenson et al. (Eds.), *International handbook of research on environmental education.* New York: Routledge.

Gutstein, E. (2003). Teaching and learning mathematics for social justice in an urban, Latino school. *Journal for Research in Mathematics Education, 34*(1), 37–73.

Howard Hughes Medical Institute. (2016). *Exploring biodiversity: The search for new medicines.* Retrieved from http://www.hhmi.org/biointeractive/exploring-biodiversity-search-new-medicines

Kena, G., Hussar, W., McFarland J., de Brey, C., Musu-Gillette, L., Wang, X., Zhang, J., Rathbun, A., Wilkinson-Flicker, S., Diliberti M., Barmer, A., Bullock Mann, F., & Dunlop Velez, E. (2016). *The Condition of Education 2016* (NCES 2016-144). Washington, DC: U.S. Department of Education, National Center for Education Statistics. Retrieved from http://nces.ed.gov/pubsearch

Kolbert, E. (2014). *The sixth extinction.* New York: Picador.

Krajcik, J. S., & Blumenfeld, P. (2006). Project-based learning. In R. K. Sawyer (Ed.), *The Cambridge handbook of the learning sciences.* New York: Cambridge.

Kress, G. (2010). *Multimodality: A social semiotic approach to contemporary communication.* New York: Routledge.

Lave, J., & Wegner, E. (1991). *Situated learning: Legitimate peripheral participation.* Cambridge: Cambridge University Press.

Lee, O., & Luykx, A. (2007). Science education and student diversity: Race/ethnicity, language, culture, and socioeconomic status. In S. K. Abell & N. G. Lederman (Eds.), *Handbook of research on science teaching.* New York: Routledge.

Lemke, J. (1998). *Teaching all the languages of science: Words, symbols, images, and actions.* Retrieved from http://academic.brooklyn.cuny.edu/education/jlemke/papers/barcelon.htm

Leung, F. Y. (2015). Leung Lab Blog [Web log post]. Retrieved from https://www.bio.purdue.edu/lab/leung/blog/?p=1828

Mayer, R., & Moreno, R. (2002). Aids to computer-based multimedia learning. *Learning & Instruction, 12,* 107–119.

Mayer, R., & Moreno, R. (2007). Interactive multimodal learning environments. *Educational Psychology Review, 19,* 309–326. doi: 10.1007/s10648-007-9047-2

McTighe, G., & Wiggins, J. (2004). *Understanding by design: Professional development workbook.* Alexandria, VA: Association for Supervision and Curriculum Development.

National Aquarium. (2014). Community art project: Calypso structure. Retrieved from https://www.youtube.com/watch?v=wbinQpfAEnM

National Research Council. (2012). *A framework for K-12 science education: Practices, crosscutting concepts, and core ideas.* Washington, DC: National Academies Press.

Nixon, R. S., Smith, L. K., & Wimmer, J. J. (2015). Teaching multiple modes of representation in middle school science classrooms: Impact on student learning and multimodal use. *School Science & Mathematics, 115,* 186–199. doi: 10.1111/ ssm.12119

No Child Left Behind (NCLB) Act of 2001, Pub. L. No. 107–110, § 115, Stat. 1425 (2002).

Pashler, H., McDaniel, M., Rohrer, D., & Bjork, R. (2008). Learning styles: Concepts and evidence. *Psychological Science in the Public Interest, 9*(3), 105–119.

Piery, L. (2006). *Art brut: The origins of outsider art.* Paris: Flammarion.

President's Council of Advisors on Science and Technology (PCAST). (2010). *Prepare and inspire: K-12 education in science, technology, engineering, and math (STEM) for America's future.* Retrieved July 1, 2016, from https://www.whitehouse.gov/sites/default/files/microsites/ostp/pcast-stem-ed-final.pdf

President's Council of Advisors on Science and Technology (PCAST). (2012). *Engage to excel: Producing one million additional college graduates with degrees in science, technology, engineering, and mathematics.* Retrieved July 1, 2016, from https://www.whitehouse.gov/sites/default/files/microsites/ostp/pcast-engage-to-excel-final_2-25-12.pdf

Primack, R. B. (2008). *Essentials of conservation biology.* Sunderland, MA: Sinauer Associates.

Primack, R. B., & Bacon, E. E. (2011). Evolution and biodiversity conservation. In M. Hutchins (Ed.), *Grzimek's animal life encyclopedia* (pp. 335–345). Detroit, MI: Gale. Retrieved August 1, 2016, from http://libraries.state.ma.us/login?gwurl=http://ic.galegroup.com/ic/scic/ReferenceDetailsPage/DocumentToolsPortletWindow?displayGroupName=Reference&action=2&catId=&documentId=GALE%7CCX1919700046&zid=41b9728b5415d7a54697d-164572866c3#grzv_0000_0001_0_img0249&source=Bookmark&u=mlin_w_forbes&jsid=237821702217b5c8ff01cb2ee167995f

Putman, R. T., & Borko, H. (2000). What do new views of knowledge and thinking have to say about research on teacher learning? *Educational Researcher, 29*(1), 4–15.

Rittmayer, A. D., & Beier, M. E. (2008). Overview: Self-efficacy in STEM. Retrieved from http://www.AWEonline.org

Salomon, G. (1993). No distribution without individuals' cognition: A dynamic interactional view. In G. Salomon (Ed.), *Distributed cognitions: Psychological and educational considerations* (pp. 111–138). New York: Cambridge University Press.

Sea Studios Foundation. (2015). *The shape of life: Story of the animal kingdom*. Retrieved from http://www.shapeoflife.org

Sherin, B. L. (2000). How students invent representations of motion: A genetic account. *Journal of Mathematical Behavior, 19*, 399–444.

Sobel, D. (2005). *Place based education: Connecting classrooms and communities*. Barrington, MA: The Orion Society.

Soto, T. L. (Producer), & Sun, A. (Director). (2013). *Plastic paradise: The great American garbage patch* [Motion picture]. New York: Virgil Films & Entertainment.

Stevens, F., & Ahnemann, O. (Producers), & Psihoyos, L. (Director). (2015). *Racing extinction* [Motion picture]. Broomfield, CO: Ripple Effect Productions, LLC.

Suitts, S., Barba, P., & Dunn, K. (2015). A new majority: Low income students now a majority in the nation's public schools. Retrieved from Southern Education Foundation website: http://www.southerneducation.org/getattachment/4ac62e27-5260-47a5-9d02-14896ec3a531/A-New-Majority-2015-Update-Low-Income-Students-Now.aspx

SurveyMonkey. (2016). Retrieved from https://www.surveymonkey.com

Turner, E. E., Gutierrez, M. V., Simic-Muller, K., & Diez-Palomar, J. (2009). "Everything is math in the whole world": Integrating critical and community knowledge in authentic mathematical investigations with elementary Latina/o students. *Mathematical Thinking and Learning, 11*, 136–157.

United States Department of Education, Office of Planning, Evaluation and Policy Development. (2010). *A blueprint for reform: The reauthorization of the Elementary and Secondary Education Act*. Washington, DC: Author.

Vomvoridi-Ivanović, E. (2012). Using culture as a resource in mathematics: The case of four Mexican–American prospective teachers in a bilingual after-school program. *Journal of Mathematics Teacher Education, 15*(1), 53–66.

Wiggins, G., & McTighe, J. (2005). *Understanding by design* (Expanded 2nd ed.). Alexandria, VA: Association for Supervision and Curriculum Development.

Wiggins, G., & McTighe, J. (2011). *The understanding by design guide to creating high-quality units*. Alexandria, VA: Association for Supervision and Curriculum Development.

Wynn, C. M., & Wiggins, A. W. (1997). *The five biggest ideas in science*. New York: John Wiley & Sons.

Zeldin, A. F. (2000). Against the odds: Self-efficacy beliefs of women in mathematical, scientific, and technological careers. *American Educational Research Journal, 37*, 215–246.

Zometool. (2016). ZT education. Retrieved from http://www.zometool.com/pages/zt-education.html

CHAPTER TWENTY-ONE

Using Self-Regulated Learning Processes to Support Scientific Thinking

ERIN PETERS-BURTON

INTRODUCTION

Students in grades K–12 are growing up in a world where they have almost limitless access to information—an ideal environment to foster life-long learning. Although some of this information is reliable, much of it is not. It is increasingly important to teach students how to think logically so that they can sort through information to make knowledgeable choices for themselves and their community. With this in mind, an important practice in the science, technology, engineering, and mathematics (STEM) classroom is to teach students how to be "self-regulated" learners, learners who can monitor their own progress in the context of scientific thinking.

As educators, many of us have worked with students who have skills that result in high performance, are excited by challenging coursework, and can modify their own behaviors to keep them on the path to success. Students who exhibit these strategic behaviors, high levels of self-awareness, and well-organized reflection skills are typically referred to as sophisticated *self-regulated learners* (Zimmerman, 2000). Unfortunately, many other students often display poor motivation in school and struggle to effectively manage and control their learning progress, even when they receive positive messages and encouragement from teachers. To complicate matters, teachers can also become discouraged and frustrated when students struggle because they may lack insight into what students are thinking about during a particular class activity. Fortunately for teachers, self-regulated learning (SRL) strategies are not innate in a learner. Teachers can support students with SRL strategies while learning so that students become more proficient in STEM knowledge and skills. Students who engage in advantageous SRL strategies can become independent learners. When struggling students receive appropriate guidance, they develop greater levels of agency to improve their competencies on future STEM projects.

BACKGROUND

The struggles of U.S. students in STEM courses have been well documented. Recent reports indicate that the U.S. is not only falling behind in science performance in its K–12 education system, but also international rankings on science tests tend to get worse as students mature (i.e., 4th to 8th grade testing; National Center for Educational Statistics (NCES), 2013; President's Council of Advisors on Science and Technology (PCAST), 2010). Data also show that U.S. students possess relatively low levels of interest in STEM subjects and exhibit deficiencies in various types of STEM practices, such as designing investigations to reliably collect data and then using data to support logical conclusions (NCES, 2013). Although there are clearly a variety of variables (e.g., quality of instruction, nature of feedback motivation, self-regulation) that may contribute to this persistent pattern of science and mathematics underachievement, researchers have increasingly turned their attention to the importance of students' regulatory processes. In fact, across virtually most academic skill areas and content domains, including reading, mathematics, writing, and science, the literature has clearly established SRL processes as key determinants of students' achievement (Butler, Beckingham, & Lauscher, 2005; Cleary & Platten, 2013; Graham & Harris, 2005; Guthrie & Wigfield, 2000; Sinatra & Taasoobshirazi, 2011). Specifically in STEM education, effective instruction should not only increase learning, it should help nurture the motivation, strategic behaviors, and metacognitive skills of students, particularly as they progress through middle and high school contexts (Millar & Osborne, 1998; Peters & Kitsantas, 2010; Schraw, Crippen, & Hartley, 2006; Sinatra & Taasoobshirazi, 2011). In fact, several studies have investigated the effects of self-regulatory interventions incorporating key SRL strategies on the science achievement of school-aged students (Cleary & Labuhn, 2013; Peters, 2009; Peters & Kitsantas, 2010; Peters, 2010; Peters-Burton & Botov, 2016). Collectively, these studies show that when SRL strategies are incorporated into classroom or tutoring activities, they are effective in developing STEM content knowledge and STEM thinking practices.

Self-regulated learners progress through three phases of a learning cycle when they approach a learning task: forethought, performance, and self-reflection

(Zimmerman, 2000). The "forethought" phase (i.e., *before* acting on the learning task) refers to motivational and analytical processes that precede efforts to act, and set the stage for action such as considering prior experience with similar learning tasks and setting process-oriented goals. The "performance" phase includes processes that occur *during* the learning task, such as focusing attention towards implementation of the task, as well as self-monitoring. The "self-reflection" phase (i.e., *after* the learning task) refers to the processes that influence a person's individualized response to the performance task, such as the use of standards to make self-judgments about the performance. As learners cycle through these SRL feedback loops, they develop more sophisticated learning processes, progressing through each phase as long as they attend to strategies connected to positive outcomes. Conversely, learners who cycle through SRL feedback loops with maladaptive strategies may decrease achievement in their performance around the learning task.

There is accumulating evidence that incorporating SRL-related supports in science teaching can help students develop deeper thinking about science. Studies have shown that students engaged with scientific thinking using an SRL coaching technique have improved scientific thinking and content knowledge over comparison groups (Peters, 2012; Peters & Kitsantas, 2010). For example, seventh- and eighth-grade students who participated in a cycle of SRL-based instruction demonstrated increased connections among ideas about empirical evidence, practices, and tentativeness (Peters-Burton, 2015b). SRL theory has also been helpful in teaching in-service teachers about the nature of science and inquiry-based teaching approaches (Peters-Burton, 2013, 2015a). SRL is useful in many contexts because it is a tool to illuminate and examine learning processes very closely—akin to using a microscope to examine things not apparent to the naked eye.

Forethought

The forethought phase of self-regulated learning occurs after a learner understands what is expected in a learning task but before the learner attempts to accomplish the learning task. This phase encompasses processes leading to the learning task, including self-efficacy, task value, goal setting and strategic planning. Each of these processes is explained in detail below.

Self-efficacy

Self-efficacy is defined as a learner's confidence in their ability to complete a particular task (Bandura, 1986). Self-efficacy has four known sources: prior experiences, vicarious experience, verbal persuasion and physiological reactions such as excitement (Bandura, 1986). Because self-efficacy is a person's confidence level for a particular task, every time a learner encounters this defined task, their self-efficacy can change. Take, for instance, a learner attempting to learn how to solve one-variable algebraic equations. If the first time the learner encounters the need to solve a one-variable algebraic equation they experience or perceive success, they build higher self-efficacy in that domain. Alternatively, if they experience failure in this domain, it tends to lower self-efficacy. Having a high self-efficacy is helpful to learners because students with high self-efficacy tend to take more risks and push themselves to learn with more competence and confidence. Learners with low self-efficacy tend to avoid those types of learning tasks or behave so that outward appearances look successful, but the learner has not mastered the skill or knowledge.

Task value

Task value, as the name suggests, is the learner's perception of how useful the targeted knowledge or skill is to their learning repertoire (Deci, 1975), or future goals. Students who have high task value for the targeted learning will persist and push themselves to master the knowledge or skills. Students with low task value will at best have superficial learning—perhaps enough to pass a test but not much more.

Goal setting and strategic planning

Goal setting and strategic planning are key strategies in mastering a learning task. Goal setting is most advantageous when learners set short-term goals that are measurable and attainable, and build to long-term goals that have meaning for the learner (Zimmerman & Martinez-Pons, 1986). Strategic planning is the process of planning discrete steps to attain the short-term and, ultimately, the long-term goals, much like the computational thinking STEM habit of mind: decomposition and algorithm design. Students who have advantageous goal setting and strategically plan to achieve those goals tend to accomplish more than those who do not have appropriate goal setting and strategic planning techniques. Students who have poor goal setting techniques tend to stumble through learning without an end goal or a way to chart their way to the completed task.

Performance

Self-regulated students will first take into account the processes of learning before they attempt a learning task (forethought) and then proceed to the performance phase. During this phase, students can regulate their learning progress through attention focusing and metacognitive monitoring techniques. The explanation of each process below can help teachers understand what students do while they are engaged in a task.

Attention focusing

During the performance phase, self-regulated students will find out what the key learning objectives are for accomplishing a task, and if that information is not immediately available, self-regulated learners know how to seek help. On the other hand, students who lack SRL skills are unaware there are key skills and knowledge on which to focus attention and may attend to irrelevant facts or skills that create misconceptions. Often students who lack self-regulated learning skills are afraid or do not understand how to seek help when struggling with identifying what is important to learn. It is imperative that teachers point out the most important information to students. Because students are naïve learners and are not familiar with the material being learned, they may focus their attention on irrelevant facts and skills, thus spending valuable learning time on unproductive tasks.

Metacognitive monitoring

Metacognitive monitoring refers to one's ability to understand what they know and what they do not know, as well as acting on that knowledge. A prerequisite for metacognitive monitoring is the ability to recognize the key information in the learning task and to have a standard for competency by which to compare their performance. It follows that if a student is unaware of the information on which to focus attention, then they will have difficulty attempting metacognitive monitoring. Students who are adept at metacognitive monitoring perform a learning task with an exemplar on hand that they can refer to in order to check on their progress toward accomplishing the learning task. By checking against the exemplar, the student is aware of the information on which to focus attention and has the ability to examine their progress in terms of the exemplar.

Self-Reflection

Self-reflection occurs when students have attempted the learning task, the learning task has been assessed, and feedback has been provided to the learner. This feedback can be given by the teacher, as in a traditional assessment, but this feedback could also be from the physical world as in the example of trying to complete a robotics course. Students receiving feedback from a robotics course would immediately understand if their programming was successful or not based on the motion of the robot. The self-reflection phase includes student perception of success in the learning task (self-evaluation and self-reaction), as well as learner perception of the reasons for success or failure at a learning task, and the learner's ability to change strategies for techniques that were unsuccessful at reaching the learning task. The resulting perceptions a learner takes away from a self-reflection phase will feed into the subsequent forethought phase when the learner attempts a similar learning task in the future.

Self-evaluation and self-reaction

When a learner experiences self-evaluation and self-reaction, they interpret the feedback they receive and reconcile the feedback with their perception of their performance (Zimmerman, 2008). Skillful self-regulated learners tend to have a close alignment of their own perception with their actual performance, meaning that if they feel they were successful at the learning task it is likely that an outside observer would agree that they are successful. Students who are poor self-regulated learners tend to have a different perception of their skills than the feedback suggests, which often frustrates these types of learners because they cannot understand or do not trust the feedback given by outside observers.

Attribution

Attribution refers to learner perceptions of the causes of success or failure in a learning task. There are certain attributions that are more helpful to learners than others (Zimmerman & Kitsantas, 1997). For example, if a learner attributes their success or failure to external factors, such as the teacher not liking them, there is little that can be done to adjust strategies to be a more adept learner. However, if success or failure is attributed to internal factors, such as not spreading out study sessions and cramming for a test, much can be done to adjust learning strategies to be more successful. Other factors such as stability or ability to control their successes or failures in a learning task contribute to ways students can be self-regulated learners.

Adaptivity

Adaptivity refers to a learner's ability to change their technique for learning based on perception of success and attribution. Students who have ways to adapt their strategies have more flexibility to be productive learners than those who do not have ways to change their strategies. Even if students are at a loss for ways to change their strategies, they can be adaptive if they know how to find reliable resources that can help them build their repertoire for learning techniques.

SELF-REGULATED LEARNING IN PRACTICE

My experiences with designing and implementing self-regulated learning supports, such as using metacognitive prompts with 8th grade physical science and 12th grade physics students learning electricity and magnetism, has helped students think more like scientists. The metacognitive prompts are a strategy that generally addresses SRL by first guiding students to think like scientists and then asking students to talk about ways they think about science. The SRL-infused curriculum uses SRL in a different way from the metacognitive prompts by intentionally planning for each SRL process (i.e., self-efficacy, self-monitoring, adaptivity) in lessons. Although the metacognitive prompts and SRL-infused curriculum take different approaches, they are both ways that the theory of learning supported by SRL can be used to diagnose student difficulty and help students progress in learning STEM practices more efficiently.

Metacognitive Prompting

Metacognitive prompts are supports that are embedded into lessons so that students have some extra instruction guiding them to do STEM practices. The prompts are a general way to solicit self-regulated learning and guide the growth of thinking as students learn how to perform a STEM practice while learning about content. The prompts are operationalized in these ways: the modeling phase consists of teacher demonstration, the emulation phase consists of a group of bullet points that act as a checklist, the self-control phase consists of a shortened version of the emulation checklist and several questions about justifying choices, and the self-reflection phase consists of a few questions about rationalizing choices.

The intention of the metacognitive prompts can be explained by the analogy of an adult teaching a child to ride a bike. When the adult first introduces the child to bike riding, the adult rides the bike to demonstrate where hands and feet go and how to make the bike move forward (modeling). Then the child sits on the bike and attempts to mimic the adult, and the adult supports the child by holding tightly to the bike because the child will most likely fail to succeed without help (emulation). As the child becomes more proficient in riding the bike, the adult provides less support by only holding on to the bike intermittently and the child has some success in riding the bike on their own (self-control). Finally, when the child is proficient at riding a bike, the adult does not need to support the child and can focus on providing feedback to fine tune the child's bike riding skills (self-regulation). Once the child reaches the self-regulation phase of the coaching, the adult is confident that the child can ride the bike independently.

The intention of the metacognitive prompts is to support students in learning STEM practices in an academic setting using the same coaching strategy. Like the bike example, the metacognitive prompts can support students to use empirical evidence to support claims. First, the teacher can provide an example of a claim, such as the length of a standing wave for a tuning fork, showing the class how they used a tube inside a cylinder of water to gather evidence related to the claim. Then, the teacher can give a different frequency tuning fork to students from which to gather evidence to make a claim, supporting student attempts with a checklist pointing out the most important processes. When students perform another investigation, teachers can fade support by shortening the checklist and asking students to explain their choices. When students are fully self-regulated learners, they can make claims that are backed by evidence without the support of prompts.

The metacognitive prompting in my classroom was comprised of prompts (i.e., checklists and questions) that coach students to become aware of the strategies they use to learn STEM practices. The overall purpose of the prompts is to

a) model to students how to do the skill,
b) allow students to try the skill with a large amount of teacher support,
c) gradually fade support for the student to only key elements, ask students to try the skill again and ask questions about why they did certain things to learn the skill while the teacher gives feedback, and
d) allow students to try the skill on their own while explaining how they are acting and thinking like STEM professionals.

The STEM practices that I have supported with this coaching strategy include the following: using empirical evidence to make claims, making observations, recording data, analyzing data, adapting conceptual models of phenomena, and using heuristics to solve physics problems.

Modeling

The intention of this phase is for teachers to present behavior for students to later emulate. I selected one STEM practice to focus on (e.g., making observations), and I showed students how I would approach a problem and apply the practice as a STEM professional would. For example, I might set up a demonstration so that students can observe factors that change reaction rates, I "think out loud" about ways I measure reaction rate (e.g., how I isolate variables, what I was observing, and what I wrote down), and I explain to students how I am paying attention to the way I connected my background knowledge (i.e., knowing how a chemical reaction occurs) to the rate of chemical reaction, the reasons I used standard observations, and how I explained enough detail in my observations on the rates to be replicable for others.

In an example of how scientists use conceptual models and adapt them based on evidence in an 8th grade class, I focused on how permanent magnets work. I introduced the idea of domains of electrons lining up to make the property of magnetic field. I wanted students to use their imaginations to do a "think experiment" and anticipate what might happen to the magnetic strength of a permanent magnet if it were continually dropped. First, I asked them to draw models of domains for a strong magnet and a weak magnet. I drew my model and explained what I used in my drawing to represent each piece of what I thought about domains.

Emulation

During the emulation phase, students are expected to perform the targeted STEM practice on their own, but with plenty of support from the teacher. I asked students to take a strong permanent magnet and measure how many paperclips it could hold, using a way to measure this without introducing surface area. When setting up learning for this phase, I gave my students a task to accomplish that was similar to the one that I just modeled and passed out a checklist of important strategies to take into account. The checklist is a feature of the emulation phase that provides support to students who have little to no experience with the particularly practice in question. The checklist for the emulation phase on conceptual modeling included these points:

- When I was doing this lab, I thought about times when I saw something similar to my results.
- I looked for patterns in my results as I gathered data.
- I thought about many different conclusions that my results could explain and chose the one that made the most sense to me to modify the drawing of my conceptual model.
- The conclusion I chose made sense compared to the ways scientists explain related phenomena.

Then, I asked students to drop a permanent magnet from a given height and measure the number of paperclips it could hold after each drop and record until they completed 25 drops. Students drew a conceptual model reflecting any changes they saw in the data. An appropriate drawing would be a drawing of a permanent magnet before it was dropped that showed most sections of the domains lining up the electron spin in the same direction. Also, the appropriate conceptual model drawing would have a picture of a permanent magnet after it was dropped with few sections of domains with their electron spin lined up, demonstrating the cause for the weakened magnetism. Students used the prompts and went back to their data to look for changes, discussing how these changes might have happened. Students changed their conceptual models based on the discussion and the ways the model mechanisms connected to changes in the data.

Self-control

In the self-control phase, students tried the same STEM practice they observed in the modeling phase and tried for the first time in the emulation phase. The teacher monitored the quality of the student performance. To support students, the teacher provided another checklist for students, but because students had more experience with the practice, the checklist for the self-control phase was much shorter. In addition to the shortened checklist, the teacher asked students to respond to a few questions.

The next time my students completed the targeted STEM practice, "observing", I gave them a short list of essential strategies with questions about why they made certain decisions about the scientific process. For example, I shortened the checklist to only the first two bullets, "When I was doing this lab, I thought about times when I saw something similar to my results" and "I looked for patterns in my results as I gathered data" and then asked students a question; "How do you know your observations could be understood by someone who was not here to participate?"

Self-regulation

In the modeling phase, the student observes the teacher talk out loud about how to do the science practice while performing the science practice. In the emulation phase, the student tries to do the same science practice as the teacher did in the modeling phase, but because the student is new at it, the teacher supports the student with a checklist of important strategies. In the self-control phase, the student is more proficient at the science practice and the teacher lets them try the science practice again with less support, which includes a shortened checklist with questions for the student to answer. In the self-regulation phase, the student has had enough practice and support to do the science practice correctly and independently. Teachers should allow the student perform the targeted science practice again and assess the quality of the student performance by asking questions so the student can explain their rationale behind accomplishing the science practice.

When I applied the metacognitive prompts for the self-regulation phase, I removed all of the checklist supports and asked students questions to justify why they performed a particular practice. For the observation example, I asked students "Would someone else be able to replicate your observations? Explain what you think could be replicated by someone with your level of expertise and what would be replicated by someone with a lower level of expertise?"

I have seven banks of checklists and questions for different STEM process skills that I used throughout the year. I found it better to focus on only one skill per unit because students can get confused about strategies if many different STEM practices are expected. I asked students to graph the number of times they used the statements in each checklist after each unit, and throughout the year, the students noticed that they used more and more statements.

Effectiveness of the Metacognitive Prompts

I conducted a quasi-experimental study over three years in an 8th grade class to confirm my anecdotal evidence that the metacognitive prompts worked well (Peters, 2012; Peters & Kitsantas, 2010a, 2010b). In the two group comparative study, one group had the metacognitive prompting intervention (i.e., treatment group) that supported their guided inquiry, and the other group (i.e., comparison group) did not have the metacognitive prompting intervention, but rather spent equal time reading about magnets. I found that the metacognitive prompts helped students learn significantly more about the nature of science than the classes without the prompts. The effect size for the content measure was $d = .50$ (medium) and for the nature of science measure was $d = .80$ (large). Also, a surprising thing that I found was that even though both groups spent the same amount of time in the class, the group with the prompts learned more about the content, even though they spent time with prompts which took away from their time with content. Students felt supported with the prompts and said that they were clearer about how they were acting like scientists. Students who were prompted also had more respect for evidence. Students who received the prompts said when there was a conflict in analyzing data, they would look at the data again, whereas the comparison group would ask the teacher or "do what the smartest kid in the group would do." Over time, the treatment group did not need the prompts to think like STEM professionals because they became independent, self-regulated learners.

SRL-based Curriculum

The metacognitive prompting interventions described above used SRL in a general way to coach students to become more proficient in learning STEM practices. However, as described earlier in this chapter, SRL has specific processes in the forethought, performance, and self-reflection phases. Although the metacognitive prompting intervention was effective in helping students use STEM practices, it implements SRL holistically, so I could not tell what SRL processes were being used by students and how well the students applied each process. In response to the need to examine SRL processes more closely, I worked with two other teachers to put together a whole-year curriculum, using all of the self-regulated learning processes described in the research literature. We were intentional in our planning to support as many SRL processes among the three phases (forethought, performance, and self-evaluation) for students as we could such as self-efficacy, goal setting, metacognitive monitoring, self-evaluation, and adaptivity (Peters-Burton, 2015a).

Over the year, we worked with three units that lasted approximately 12 weeks each. In the first unit, students designed and implemented independent research projects. The second unit consisted of an exploration of plate tectonics through a historical approach, and students examined scientific arguments as the theory was developed. The third unit was one where we had a USGS scientist train the students in population counts of horseshoe crabs so students could contribute to a citizen science project.

Learning goals

The learning goal for the unit on independent research was for students to work as scientists to answer a question of their choosing using empirical evidence (Figure 21.1). After I gave the students the assignment, but before learning started (i.e., forethought), I conferenced with them to help them set goals. Students were asked to choose their own topic, which created student value in the task and self-efficacy in their ability to complete the learning. The assignment also encouraged students to be mastery learners since it was their own question they were answering. Mastery learners focus on learning the material so they know it deeply, even if they make mistakes along the way. During all parts of the investigation (i.e., performance) we interrupted the students and asked them to explain to us how they were acting like scientists, which helped us determine if students were paying attention to the correct skills and knowledge in their independent work, as well as helped us support student self-monitoring of their progress. Students addressed self-reflection after they explained the validity of their work to the group and received comments from their peers and teachers. They did this by addressing how well they achieved their goals (self-reaction), what they thought made them successful or not successful (attribution), and asked what they might do in the future if they do another investigation (adaptivity).

The learning goal for the second unit on the historical approach to science was for students to understand more about how ideas and models about the formation of continents changed over time, becoming more sophisticated (Figure 21.2). Students were given journals in which they could write their thoughts, ideas, and feelings. We used these journals to document student SRL processes. Each class meeting during this part of the course involved telling a story about the ways people thought about the shape and movement of the continents on the Earth. The story began with mapmakers in the early 1700s noticing that the edges of the continents fit like puzzle pieces. Then, the story proceeded to the way the discovery of the continental shelf edges improved the fit of the puzzle pieces. Following that, the story unfolded the discoveries of surface minerals matching on opposite coasts of continents, and the discoveries of stratigraphic column matching on opposite coasts. Finally, the story ended with the discovery of mid-ocean ridges, which help to explain the mechanism

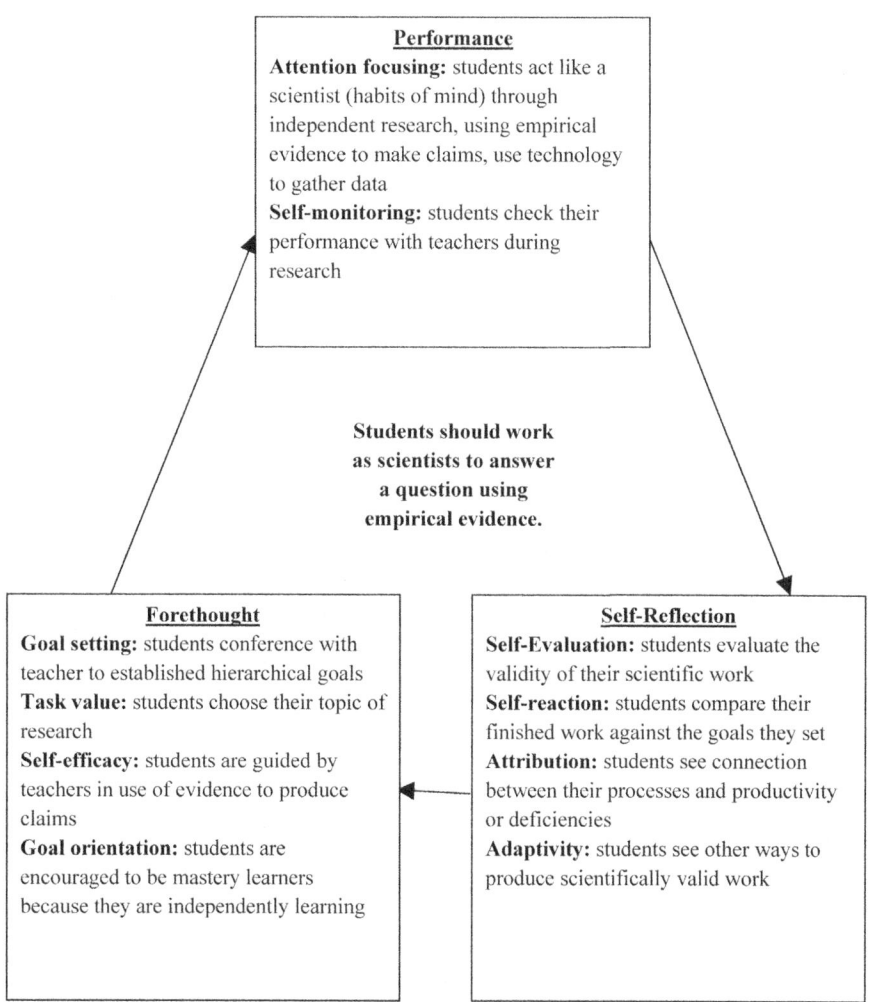

Figure 21.1. Phases of SRL in the Independent Research Project. This figure illustrates how each subprocess in SRL was enacted during the independent research project. (Source: Author.)

behind the movement of the plates. As each piece of the story was revealed, students also learned through primary resources how scientists thought about competing theories called lifters and drifters, and how scientists incorporated the new knowledge they learned through the process of argumentation. Students were asked to act like scientists during that time by generating a model for how the continents moved. Teachers asked students to explain their thinking and to answer questions about student's models and how they supported lifting or drifting theory.

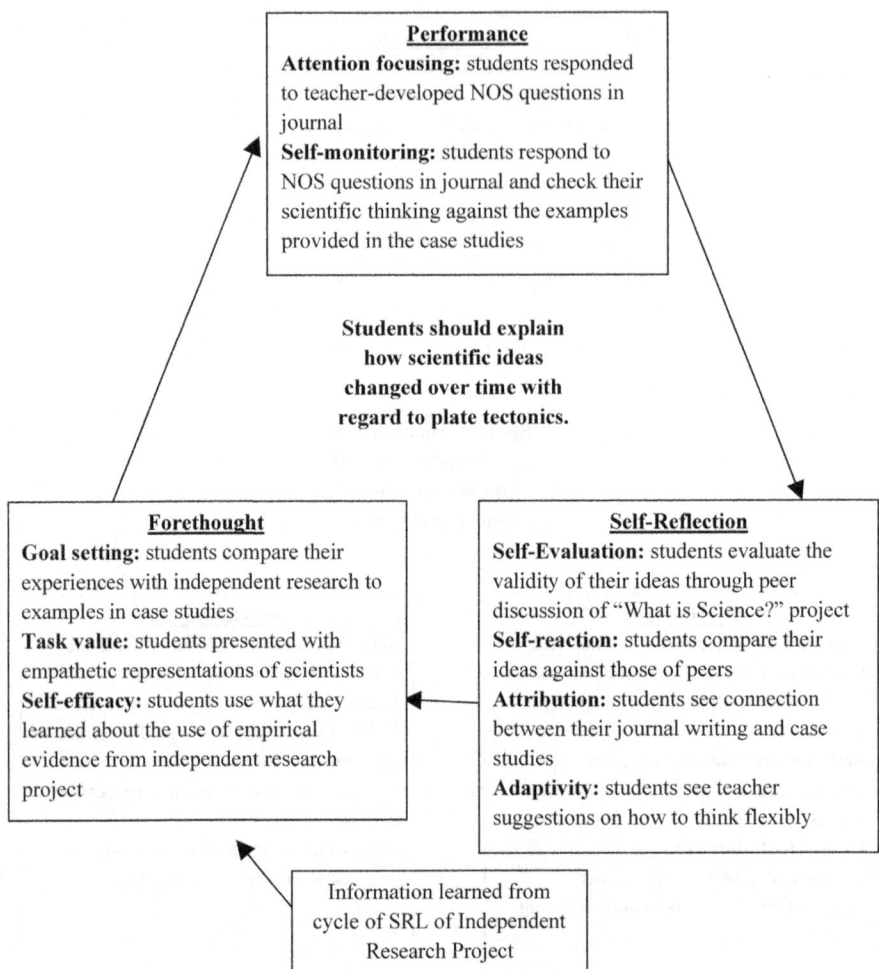

Figure 21.2. Phases of SRL in the Knowledge Building Section of the Class. This figure illustrates how each subprocess in SRL was enacted during the knowledge building section. (Source: Author.)

We supported student forethought processes in the knowledge building segment of the course by emphasizing how scientists can make mistakes and how they learn from them (task value). Often when students make mistakes, they feel as though they are not capable. Instead, we emphasized that by making mistakes, learning occurs, and students should value the opportunity to revise their thinking because it improves learning. We also supported student forethought processes by demonstrating to students that the skills they build in the independent research segment of the class were similar to the ones scientists used in building

Table 21.1. Rubric for Class Discussion.

Characteristic	Emerging (1)	Proficient (2)	Exemplary (3)
Follows guidelines of intellectual discussion and is civil	Criticizes other people personally instead of being critical of ideas; doesn't use appropriate language	Challenges the idea but without reason; uses appropriate language	Challenges the idea with solid reasoning; uses appropriate language; diverts any unproductive discussion
Makes claim	Claim unoriginal AND indirectly related to topic	Claim original AND indirectly related to topic	Claim original AND directly related to topic
Uses reliable sources for evidence	Uses unreliable resources (such as Wikipedia or blog)	Only uses textbook as resource	Uses outside reliable resources (such as a scientific journal or .gov or .edu website)
Appropriate level of evidence	Opinion-based evidence	One piece of researched evidence	More than one piece of researched evidence
Responds to the content of the discussion	No response or unrelated to claim	Response is indirectly associated with claim	Response is aligned with claim
Connects with what prior person says	Unrelated to current discussion	Stay on topic, but makes no connection with person before them	Acknowledges prior person's idea and elaborates on what previous person says
Able to defend their claim/ rebuttal	Has no response	Has a response but cannot back up response	Has a response and is able to back up response with further evidence
Uses appropriate reasoning	Reasoning is disconnected from claim	Reasoning is superficially connected to claim	Reasoning directly connects claim to evidence

(Source: Author.)

knowledge for plate tectonic theory (self-efficacy and goal setting). This builds self-efficacy by pointing out that students had successfully acted like scientists in the first part of the class. This also aided in student goal setting by providing a model for building ideas from evidence from the first part of the class for what they needed to accomplish in this part of the class. We supported student performance processes by asking students specific questions that focused their attention on understanding how scientists build knowledge over time and asked them to reflect on how their own knowledge was building (self-monitoring). This supports self-monitoring because it reminds students to see if their performance matches their goals. For example, we asked students how their ideas were built on evidence and if students were *not* able to explain how an idea developed from evidence, they were effectively identifying a gap to address. In this way, they were monitoring their performance to see if it matched their goals. To support student self-reflection, we asked students to participate in a whole class discussion guided by the question, "What is science?" and asked students to refer to their journals to back up their claims (self-evaluation). This was a culminating activity that was intended to guide students to go back to the ways they recorded their ideas to examine their progression of thinking. During the discussion, students were given rubrics (Table 21.1) to guide them to compare their ideas to their peers' ideas (self-reaction) and asked students to read their entire journal and reflect on what they felt went well and what needed improvement throughout the unit (attribution). The rubric supported self-reaction because students could measure their performance against a standard and rate themselves. The student reflection supported attribution by reflecting on the reasons for their successful model-building strategies and their no-so-successful model-building strategies. Finally, we asked students to explain what they might do differently the next time they encounter a unit like this one, supporting adaptivity so that students can recognize more than one way to learn. Overall, students responded that during the first unit, they were often discouraged if they didn't get the results they were expecting in their independent project. However, after this unit, students saw that they were making predictions to test them with empirical evidence, not to be "correct" with their prediction.

The learning goal for the third unit on horseshoe crabs was for students to identify interconnections of living things and resources in an environment and learn skills that scientists use to do population studies (Figure 21.3). This unit started with a study of the ecosystem in which horseshoe crabs live in Delaware, including the physiology and life cycle of horseshoe crabs. The unit also included why horseshoe crab population studies are needed, since the crabs are experiencing changes in population due to the unique coagulation properties of their blood. Following four weeks of researching background knowledge, the students were then trained by a U.S. Geological Service (USGS) ecologist on how to systematically to do a horseshoe crab count. We took the students to a Delaware beach on the last full moon of May, when the horseshoe crabs come to the water's edge to mate, and students counted horseshoe crabs for the USGS survey. We supported students' forethought processes by setting up experiences for authentic data collection (task value) and how to collect data that scientists would use (goal setting), which was similar to the skills the students learned in the prior two units (self-efficacy). Students who have high value for a learning task will persist in the task, even if they initially fail. When students have high self-efficacy, they are more likely to challenge themselves in their learning during the performance phase. Likewise, students who set goals that align with the teacher's goals and set goals that are achievable, yet challenging, are more prepared to use productive learning strategies during performance. We supported student performance processes by emphasizing the protocols set up by the USGS for data collection (attention focusing and self-monitoring). Once students completed their performance, we asked students to engage in self-reflection of their performance, a step in learning that is often overlooked. We supported student self-reflection processes by asking them to list the ways they were behaving like scientists with confirmation by the USGS ecologist (self-evaluation and self-reaction). Similar to the prior unit on knowledge building, we asked students to explain what worked well in the variety of ways they learned about ecology and what did not work well (attribution) and ways they might use different strategies in the future (adaptivity). Students responded that they were able to connect they ways they did their independent projects with the protocols they were given by the USGS, and as a result they felt that they were reinforcing their knowledge about scientific thinking.

Student outcomes

This 8th grade course was extremely effective in teaching students to embrace STEM practices (Peters-Burton, 2015a). Prior to the course, students generally thought that scientists used experimentation to answer questions and that science was something that was done external to themselves. Students felt that they were consumers of science, but that science was not part of their lives prior to the course. Several of the students communicated naïve understandings of science practices prior to this unit when they explained that they thought scientists went to work each day with a list similar to a cookbook lab and when they completed the procedure and wrote the conclusions, they went home for the day. After the course was over, students had a more sophisticated understanding of science practices because they adopted the view that scientists have a variety of methods they employed to answer questions and communicated that science was something that they could produce, not just consume. We felt that student growth in views of science practices was due in part to the way we supported their learning through SRL and how we started the unit with overlap in their beginning views of science practices (experimentation in the independent research segment). We then expanded their understandings as we took them through the knowledge building and the citizen science segment. This success was encouraging and led us to thinking about ways to help teachers learn how to support SRL.

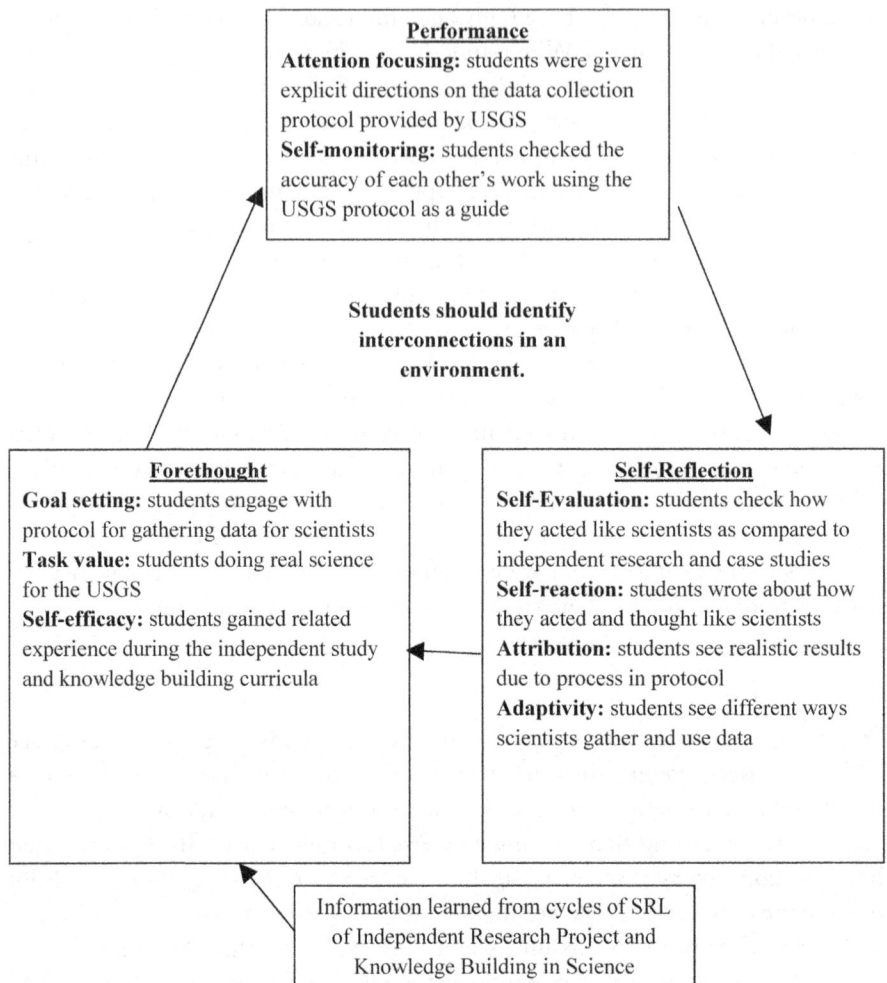

Figure 21.3. Phases of SRL in the Citizen Science Section of the Class. This figure illustrates how each subprocess in SRL was enacted during the citizen science section. (Source: Author.)

HOW OTHERS CAN SUPPORT SELF-REGULATED LEARNING IN PRACTICE

Metacognitive Prompts

Some students may seem unmotivated to learn science, when in reality students may appear uninterested because they do not want to appear as though they do not know what to do. They also might be unmotivated because they are often unsuccessful in attempts to learn science. Metacognitive prompts can help all students feel like part of the community of scientists because the prompts explicitly provide ways to think and act like scientists. This section describes the architecture of the metacognitive prompts so that teachers can construct them for their own classrooms.

As explained previously, metacognitive prompts are used as a coaching strategy and have four phases: modeling, emulation, self-control, and self-reflection (Zimmerman, 2000). The prompts are operationalized in these ways: the modeling phase consists of teacher demonstration, the emulation phase consists of a group of bullet points that act as a checklist, the self-control phase consists of a shortened version of the emulation checklist and several questions about justifying choices, and the self-reflection phase consists of a few questions about rationalizing choices. Ways to build prompts for each of these phases are explained below. However, before the prompts can be created, teachers must first define the STEM practice that is the target for student learning. Once the STEM practice learning target is articulated, teachers should plan for the students to engage with the targeted STEM practice multiple times so the students can be coached and progress through the stages with each experience. The minimum times students should engage using metacognitive prompts is four times (Peters & Kitsantas, 2010). Once the STEM practice is targeted and a plan for at least four student experiences is established, teachers can create the following coaching technique to guide students to master the STEM practice and the content in the lessons. To demonstrate how metacognitive prompts can be extended beyond science practices and into STEM practices, I will use an example of computational thinking in this section.

Modeling

In this phase, the teacher should plan to model the targeted STEM practice. There are different ways teachers can model skills for students, and some of them are more helpful for struggling students than others. Mastery modeling is the type of modeling where the teacher shows the students the correct way to perform the STEM practice. Although this may seem intuitively the best way to model, coping modeling is more helpful for learners who are learning new material or skills. Coping modeling occurs when the teacher demonstrates the STEM practice, but instead of demonstrating it perfectly, the teacher models ways learners typically make mistakes and talks out loud about how they realize they are making a mistake and how to backtrack to correct the mistake. Coping modeling helps students in two ways: first, it helps students to see both exemplars and non-exemplars to form full conceptions of the targeted STEM practice, and second, it helps students feel comfortable with making mistakes and demonstrates how to learn from those mistakes. An example of coping modeling when teaching how to use the computational thinking practice of decomposition to connect evidence to claims is demonstrated by making a claim that may be only partially backed up by evidence. The teacher then separates each piece of evidence to show how they may or may not support the claim, notices out loud that a portion of it is not backed up, and reconciles the counter claims with more evidence to back it up, adding it to the statement.

Emulation

Emulation occurs when the student performs the STEM practice for the first time, after viewing the modeling by the teacher. The teacher fully supports the student by providing checklists so that students can focus their attention on the most important aspects of learning. Teachers can create the checklists by documenting the key characteristics of the STEM practice. For example, in writing prompts for the STEM practice "using computational thinking when analyzing data", teachers can focus on the practice of decomposition, which is the breaking down data, processes, or problems into smaller, manageable parts; pattern recognition, which is observing patterns, trends, and regularities in data; abstraction, which is de-identifying the general principles that generate these patterns; and algorithm design, which is developing the step by step instructions for solving this and similar problems. Teachers should write at least three bullet points related to each STEM practice. For example, teachers can support students' skills in decomposition of data by writing the following three bullet points for a checklist for emulation:

- I have read the research question and broken down the topic by all of the different variables that are related to the research question.
- I have written a plan for the steps I need to take to calculate the derived variables needed to answer the research question (e.g., for speed I wrote down how to calculate speed from distance and time).
- I have reviewed my procedure and have noted when each variable in the data table has been measured so I can enter it into my data table.

Teachers can then write out at least three bullet points for each STEM practice and combine them all into an emulation checklist for students to use. Once students feel comfortable understanding what is expected of them for the STEM practice and the teacher has assessed the student and determined that they have acquired the skill of the targeted STEM practice, the student can progress to the next phase.

Self-control

In the self-control phase, students engage in the targeted STEM practice for the third time and teachers, having determined that students have acquired a beginner's level knowledge of the STEM practice, begin to fade support so the students can take more responsibility for learning. The teacher chooses one key bullet point per STEM practice to provide students with the support they may need as they gain more experience with their skills. For the example of computational thinking for data analysis, the teacher would choose one bullet point for the checklist from each of the STEM practices of decomposition, pattern recognition, abstraction, and algorithm design. In addition to the shorter checklist, teachers now shift student performance toward active explanation for their choices. When teaching computational thinking for data analysis, for example, teachers teaching in the self-control phase will ask students how they attempt to break up the problem into smaller parts, how they go about looking for patterns, how they move from specific information (data) to generalizations (conclusions), and how students create systematic processes to go about data analysis. Teachers can determine from student answers if students are internalizing and transferring the skills and knowledge for the STEM practice or if they are still following teacher direction. It is when students can answer the justification questions in the self-control phase reliably that they can move on to the self-regulation phase.

Self-regulation

In the self-regulation phase, students are expected to demonstrate that they can perform the STEM practice independent of teacher support. Teachers can determine if students can accomplish the STEM practice independently by providing students an additional chance at applying the STEM practice and by asking students to rationalize the choices they may during this attempt. For the example of computational thinking, teachers can ask students:

- Why do you break up the problem into smaller parts?
- What is the purpose of looking for patterns in data analysis?
- Why is it important in STEM to move from specific data sets to generalizations?
- What was your step-by-step process to solve this problem?

Students who can answer these questions accurately have most likely mastered the STEM practice and should be able to transfer their understanding of how to do this STEM practice to other investigations. Most importantly, students who can answer these questions understand why the STEM practice is important in the field and have become participants in the ways of knowing in STEM.

Organizing a SRL-based Curriculum

Teachers who wish to design such a SRL-based curriculum should break down their lessons into a series of well-defined learning tasks. Each of the learning tasks in the lessons should have supports from all three SRL phases: forethought, performance, and self-reflection. Luckily, for STEM teachers, using a 5e curriculum design (i.e., engage, explore, explain, elaborate, evaluate) aligns well with the SRL phases. Table 21.2 displays the alignment of the goals of a 5e curriculum design with the forethought, performance, and self-reflection processes of SRL.

5e Curriculum Design

The 5e Curriculum Design model (Bybee et al., 2006) helps students developmentally learn skills and concepts by guiding students through a series of activities that build student interest and knowledge. Students are first hooking the students to motivate them to learn more about the topic in the Engage component. The purpose of the Engage component aligns well to the processes in the forethought phase of SRL theory because it provides an anticipatory set to the performance of the learning task. Students in the Explore component are confronted with important concepts/issues and develop important aspects of the target concepts and skills for themselves. The Explain component consists of student generation of the explanation of the concepts and skills themselves based on their exploration, and important ideas originate with the students. The Elaborate component of 5e curriculum design consists of students extending their concept development. The Evaluate component of the 5e curriculum design includes multiple assessments that permit students to use multiple approaches to solutions and incorporates formative assessment for each approach. Table 21.2 demonstrates how SRL processes align with the activities in the 5e model.

Forethought

- Engage—build student self-efficacy and task value by connecting it to students' lives and prior experiences; give students short, non-evaluative experiences
- Explore—help students set goals for the learning task based on the experience the students gained in the engage portion of the lesson
- **Performance**
- Explain—show students exemplars for STEM practices to help focus their attention on the key characteristics of learning and help students monitor their own learning from this exemplar
- **Self-reflection**
- Elaborate and Evaluate—give frequent and timely feedback on their performance

Table 21.2. Intersections Between SRL Phases and 5e Model of Instruction.

	Engage Short activities that help students to:	Explore Laboratory-based activities that help students to:	Explain Opportunities for students to demonstrate	Elaborate Additional activities that help students:	Evaluate Assessments for students to:
Forethought	○ promote interest ○ expose prior connections ○ make connections between past and present learning experiences ○ Organize student thinking toward the learning outcomes	○ Provide students with common based of experiences ○ Elicit relevant content knowledge and process skills			
Performance		○ Generate new ideas related to content ○ Explore possibilities ○ Design and conduct a preliminary investigation	○ Explain their understanding of the concept or skill	○ Apply understanding to new contexts or more complex situations	
Self-reflection			○ How feedback from teacher or curriculum guides toward deeper understanding	○ Evaluate the utility of currently understood concepts and skills in new or more complex situations	○ Self-assess the links between prior and new knowledge and skills ○ Consider feedback from teacher or curriculum in their ability to acquire new knowledge and skills ○ Catalog what worked and what did not work in attempts to generate new knowledge and skills from labs

(Source: Author.)

CONCLUSION

Not only is the subject of science a venue for learning about the world around us, it is also a vehicle to teach students scientific ways of knowing. Although science has been in the forefront of important topics since public schooling was created in the United States, learners have faced challenges in its mastery; many have also been challenged learning mathematics. Self-regulated learning theory provides concrete ways educators can communicate to reticent students or students who feel marginalized by STEM that learning STEM subjects have value because of the important content and skills embedded within. Students can gain value for the learning tasks if STEM subjects are well designed, and well designed curriculum provides students more authentic experiences in STEM. SRL theory can be a template for well designed, meaningful, and effective learning experiences.

Traditionally, science subjects have been taught in ways that are competitive, leaving out students who do not feel part of the scientific community. However, SRL gives teachers ways to structure explicit explanations of how STEM knowledge is generated and how scientists, technologists, engineers and mathematicians think. Students who need more support in understanding the ways STEM professionals think (Peters & Kitsantas, 2010a) can participate more fully when they are supported with the structures provided by self-regulated learning.

WORKS CITED

Bandura, A. (1986). *Social foundations of thought and action: A social cognitive theory*. Englewood Cliffs, NJ: Prentice-Hall.

Butler, D. L., Beckingham, B., & Lauscher, H. J. N. (2005). Promoting strategic learning by eighth-grade students struggling in mathematics: A report of three case studies. *Learning Disabilities Research and Practice, 20*, 156–174.

Bybee, R., Taylor, J., Gardner, A., Van Scotter, P., Carlson, J., Westbrook, A., & Landes, N. (2006). *The BSCE 5E instructional model: Origins and effectiveness*. A report for Office of Science Education National Institutes of Health. Retrieved from http://science.education.nih.gov/houseofreps.nsf/b82d55fa138783c2852572c9004f5566/$FILE/Appendix?D.pdf

Cleary, T. J., & Labuhn, C. (2013). Application of cyclical self-regulation interventions in science-based contexts. In H. Bembenutty, T. J. Cleary, & A. Kitsantas (Eds.), *Applications of self-regulated learning across diverse disciplines: A tribute to Barry J. Zimmerman* (pp. 89–124). Charlotte, NC: Information Age Publishing.

Cleary, T. J., & Platten, P. (2013). Examining the correspondence between self-regulated learning and academic achievement: A case study analysis. *Education Research International*. doi: 10.1155/2013/272560

Deci, E. L. (1975). *Intrinsic motivation*. New York: Plenum Press.

Graham, S., & Harris, H. R. (2005). Improving the writing performance of young struggling writers: Theoretical and programmatic research from the center on accelerating student learning. *The Journal of Special Education, 39*, 19–33.

Guthrie, J. T., & Wigfield, A. (2000). Engagement and motivation in reading. In M. L. Kamil, P. B. Mosenthal, P. D. Pearson, & R. Barr (Eds.), *Handbook of reading research* (3rd ed., pp. 403–422). New York: Longman.

Millar, R., & Osborne, J. F. (Eds.). (1998). *Beyond 2000: Science education for the future*. London: King's College London.

National Center for Educational Statistics (NCES). (2013). *The condition of education 2013*. Washington, DC: Author. Retrieved from http://nces.ed.gov/pubs2013/2013037.pdf

Peters, E. E. (2009). *Thinking like scientists: Using metacognitive prompts to develop nature of science knowledge*. Saarbrücken: Verlag Dr. Müller Aktiengesellschaft & Co. KG Publishers.

Peters, E. E. (2010). Shifting to a student-centered science classroom: An exploration of teacher and student changes in perceptions and practices. *Journal of Science Teacher Education, 21*(3), 329–349.

Peters, E. E. (2012). Developing content knowledge in students through explicit teaching of the nature of science: Influences of goal setting and self-monitoring. *Science & Education, 21*(6), 881–898.

Peters, E. E., & Kitsantas, A. (2010a). The effect of nature of science metacognitive prompts on science students' content and nature of science knowledge, metacognition, and self-regulatory efficacy. *School Science and Mathematics, 110*, 382–396.

Peters, E. E., & Kitsantas, A. (2010b). Self-regulation of student epistemic thinking in science: The role of metacognitive prompts. *Educational Psychology, 30*(1), 27–52. doi:10.1080/01443410903353294

Peters-Burton, E. E. (2013, April). *Microanalysis of self-regulatory processes of elementary teachers learning to teach earth science through inquiry*. Paper presented at the annual conference of the American Education Research Association, San Francisco, CA.

Peters-Burton, E. E. (2015a). Outcomes of a self-regulatory curriculum model: Network analysis of middle school students' views of nature of science. *Science & Education, 24*, 855–885.

Peters-Burton, E. E. (2015b, January). *Incorporation of scientific argumentation into instruction: Results from a professional development for high school teachers*. Paper presented at the Annual Meeting of the Association for Science Teacher Education, Portland, OR.

Peters-Burton, E. E., & Botov, I. S. (2016). Self-regulated learning microanalysis as a tool to inform professional development delivery in real-time. *Metacognition and Learning*. doi: 10.1007/s11409-016-9160-z

President's Council of Advisors on Science and Technology (PCAST). (2010). *Prepare and inspire: K–12 education in science, technology, engineering, and math (STEM) for America's future*. Washington, DC: Author.

Schraw, G., Crippen, K., & Hartley, K. (2006). Promoting self-regulation in science education: Metacognition as part of a broader perspective on learning. *Research in Science Education, 36*, 111–139.

Sinatra, G. M., & Taasoobshirazi, G. (2011). Intentional conceptual change: The self-regulation of science. In B. J. Zimmerman & D. H. Schunk (Eds.), *Handbook of self-regulation of learning and performance* (pp. 203–216). New York: Routledge.

Zimmerman, B. J. (2000). Attaining self-regulation: A social-cognitive perspective. In M. Boekaerts, P. Pintrich, & M. Zeidner (Eds.), *Handbook of self-regulation* (pp. 13–39). San Diego, CA: Academic Press.

Zimmerman, B. J. (2008). Investigating self-regulation and motivation: Historical background, methodological developments, and future prospects. *American Education Research Journal, 45*, 166–183.

Zimmerman, B. J., & Kitsantas, A. (1997). Developmental phases in self-regulation: Shifting from process to outcome goals. *Journal of Educational Psychology, 89*, 1–10.

Zimmerman, B. J., & Martinez-Pons, M. (1986). Development of a structured interview for assessing student use of self-regulated learning-strategies. *American Educational Research Journal, 23*(4), 614–628.

CHAPTER TWENTY-TWO

Teaching Students Metacognition Through Discipline-Based Research AND Technology

REBECCA VIEYRA

INTRODUCTION

Metacognition is often described as "thinking about thinking," and is a vital skill that science, technology, engineering, and math (STEM) educators can teach their students. Applicable in many learning contexts beyond STEM, metacognition is an essential element of critical thinking that helps students differentiate between what they know and understand and what they do not know or do not understand. STEM is an ideal field in which to practice this skill, as STEM concepts can often be broken down into base units of knowledge, such as the laws of conservation of mass, energy, and momentum in science, and principles of logic in mathematics.

STEM teachers frequently use strategies with students that implicitly encourage metacognition. For example, students of science are routinely asked to make predictions and hypotheses, and to evaluate them against actual experimental outcomes. Students of mathematics are asked to show their work and to provide logical proofs. However, techniques that only implicitly use metacognitive strategies are often not enough to overcome pervasive naive conceptions about the world or students' perceptions of themselves as competent, capable learners of STEM. Students who understand the power of metacognition adopt a growth mindset and recognize that they can improve their learning, regardless of where they start. When STEM teachers help students to see themselves as malleable learners, as opposed to individuals with fixed levels of intelligence, they are more likely to engage willingly, with enthusiasm and success in the learning process (Blackwell, Trzesniewski, & Dweck, 2007).

Because teaching is not simply knowledge dissemination, but requires students to be active agents in developing their own understandings about the world, it is important to dissect how students visualize their understandings. Students' understanding is typically defined by cognitive scientists as a mental model of the world, a "collection of mental patterns people build to organize their experiences related to a topic" (Redish, 1994, p. 797). Physics education has made significant strides in discipline-specific research about change in mental models and what struggles they experience while doing so. Redish (1994) suggests that teachers need to be sensitive to four principles of concept change: (1) Construction: learners must build mental models for themselves; (2) Assimilation: learners must incorporate new ideas into their mental models; (3) Accommodation: learners must be willing to modify their mental models when new and/or conflicting observations are made; and (4) Individuality: learners have subjective personal experiences that shape their mental models. Encouraging students to communicate within themselves and to others about the struggles they encounter in any of these stages of conceptual growth is necessary. To advance their own learning, students must be willing to review, revise, and refine their own understanding about STEM concepts.

This chapter will provide an overview about how I used explicit metacognitive strategies and tools with my high school physics students, including the development and inclusion of mobile technology. My action research over the course of multiple years demonstrated that my students continually struggled with understandings about force and motion, both for those who had no prior knowledge, as well as for those who came in with naive conceptions. I opted to turn to educational technology as a support to help my students confront their thinking about difficult physics concepts through experimentation and the use of data collection tools that could provide an "objective view" of the world that students needed to consider as they developed their mental models. By developing an app for students' personal smartphones and tablets, students were able to better understand physics concepts with sensors that connected their movement and sensations of forces with objective data visualizations.

BACKGROUND

Research on Metacognition and Self-Regulation

Broadly, metacognition is a process of self-reflection on one's thinking that can entail checking one's assumptions, knowing one's weaknesses, and knowing when to adapt (Lovett, 2008). Metacognition has also been described as an aspect of the learning process that entails (1) believing that learning ability is not fixed, (2) planning and goal-setting for learning, and (3) monitoring and reflecting on one's learning (Science Education Resource Center, 2016). Metacognition has been the subject of educational psychology research in the past few decades, although the formalized practice of analyzing personal understanding dates back to the times of Socrates. The National Academies have convened a number of times to investigate cognitive research, much of which is summarized in their reports, *How People Learn* (Bransford, Brown, & Cocking, 1999), its science sequel *How Students Learn: Science in the Classroom* (Donovan & Bransford, 2005), and *Discipline-Based Education Research* (Singer et al., 2012). A project currently underway, *How People Learn II: The Science and Practice of Learning* (National Academy of Science, 2016), will synthesize additional recent findings. These strategies apply to all forms of learning, but are absolutely essential to STEM as students attempt to make objective models in math, evidence-based claims free from assumptions in science, and to appropriately use technology or make design changes in engineering.

Students can engage in metacognition in any context where they interact with their ideas by analyzing their reading, writing, speaking, and listening skills in the learning context (Wenning & Vieyra, 2015). Although teachers frequently employ metacognitive strategies with their students through various practices, such as Socratic dialogue and using exit slips, it is easy for teachers to forgo actually talking about the process of learning with students. Instead of merely demonstrating understanding of a concept, metacognitive questions might entail students responding to something like the following: "What do I already know about this topic?," "What is easiest/hardest for me to understand about this topic?," and "Under what circumstances can I apply what I know about this topic?" Self-regulation refers to the practice of responding and adjusting oneself based upon the answers to those questions. It refers to taking action to remediate by asking for help, trying new approaches to learning, and tracking learning progress.

Even with metacognitive tools, learning can require significant intellectual effort, especially when students arrive to the classroom with pre-existing ideas about the world that might be incomplete or not aligned with expert beliefs. As demonstrated by Redish (1994), learners are strongly subject to their own preconceptions, biases, and attitudes. This is particularly true for STEM fields, in which educational research has identified persistent, clear-cut naive conceptions around well-defined concepts (Hieggelke, Kanim, Maloney, & O'Kuma, 2013; Hieggelke, Maloney, & Kanim, 2011; Hieggelke, Maloney, O'Kuma, & Kanim, 2005; Knight, 2002; O'Kuma, Maloney, & Hieggelke, 2003). Naive conceptions are incredibly pervasive and persistent among learners—and, frequently, among teachers—even after making observations about phenomena that do not agree with their worldview (Harvard-Smithsonian Center for Astrophysics, 2016).

Metacognitive Strategies and Tools from General and Physics Education Research

Multiple efforts have been made by researchers to bring effective metacognitive strategies to K-12 classrooms. Marzano's (2000 & 2009) Nine Essential Strategies are widely used by teachers and professional development programs, and have been shown through a significant amount of educational research to enhance student achievement. Marzano suggests nine strategies to help students reflect on their learning process and to self-regulate by evaluating themselves against learning goals. Marzano's strategies include (1) identifying similarities and differences, (2) summarizing and note-taking, (3) reinforcing effort and providing recognition, (4) homework and practice, (5) nonlinguistic representations, (6) cooperative learning, (7) setting objectives and providing feedback, (8) generating and testing hypotheses, and (9) cues, questions, and advanced organizers. Likewise, Project CRISS (Creating Independence Through Student-owned Struggles) has a professional development curriculum filled with metacognitive practices supported by over a decade of educational research (Project CRISS, 2016). Project CRISS embraces principles and philosophies that build on (1) metacognitive practices: identifying background knowledge, purpose setting, active learning, organization, and intuition, (2) modeling, and (3) teaching for understanding of content knowledge. Although both Marzano and Project CRISS are discipline-agnostic, they share many similarities with discipline-specific approaches that have also been developed.

Although broad strategies described by Marzano and CRISS are effective both in and beyond STEM courses, discipline-based education research has shined some light on the struggles students have when learning specific topics, such as how to describe motion, or how to identify forces acting on an object. Although discipline-based educational research can take many forms, physics education has historically followed a pathway that focuses on discrete physics concepts and students' struggles in organizing their knowledge around central ideas (McDermott, 1984 & 1991). Notably, educational research has both developed and utilized *assessments* to define and measure student understanding before and after instruction, and specific *research-based teaching methods* to address known student learning challenges (Wenning, 2008). What follows are some examples of the outcomes from physics education research that are widely implemented in the high school physics teaching community to promote metacognitive practices.

Assessments

Concept inventories are one type of assessment developed from intensive interviews with students about their thinking, which uncovers both the driving ideas behind their thinking and naive conceptions they have. Concept inventories break down ideas into their fundamental pieces and are developed after identifying highly-attractive naive conceptions displayed by students, that are then formulated as potential distractors. With over twenty years of validation and reliability, the questions and answers generated in the Force Concept Inventory (Hestenes, Wells, & Swackhamer, 1992) demonstrate the pervasive naive conceptions students have about force and motion. Different from concept inventories, which are meant to give a teacher a holistic view of students' understanding, diagnostic tools are questions or sets of questions that have been developed through research to document concept change throughout learning cycle. Diagnoser Tools (Diagnoser Tools, 2016) is an interactive website that allows students to respond to very brief multiple-choice and short-answer questions. Responses allow teachers and students to directly connect their answers with pre-coded common facets of learning (correct concepts and naive conceptions) identified through research. Similarly, Tasks Inspired by Physics Education Research, also known as TIPERs (Heiggelke, et al., 2005, 2011, 2013), are performance-based tasks developed through research that require students to respond to questions in multiple formats, including ranking tasks, selecting the better of two explanations, and identifying what is wrong in an explanation.

Research-based teaching methods

Research has resulted in development of over 55 methods in physics education alone that are shown to improve students' content knowledge, skills, and attitudes (PhysPort, 2016). One such approach is the Modeling Method of Instruction (MMI) in Physics (Wells, Hestenes, & Swackhamer, 1995), which has now expanded to chemistry, biology, astronomy, and integrated physical science and engineering (American Modeling Teachers Association, 2017). MMI was developed from over 20 years of educational research born at Arizona State University, and ongoing research using the method demonstrates the power of classroom discourse as a metacognitive method to help students build evidence-based mental models through sense-making (Megowan-Romanowicz, 2016). Students constructively develop knowledge through Socratic dialogue, defend their conclusions based upon empirical evidence, and build consensus through whole class whiteboarding discussions. Additional in-depth information about whiteboarding can be found in Chapter 14 of this volume (Cushman-Patz, 2014).

Role of Technology in Metacognition

Fundamental to much of physics education research on metacognition has been technology. In the mid-1980s, microcomputer-based laboratories (MBLs) (Laws, 1988; Thornton, 1989; Thornton & Sokoloff, 1990) became a reality, allowing students to attach sensor probes to their graphing calculators and desktop computers to collect data on motion, force, sound, light intensity, magnetic field strength, and more. To meet the demand for teaching techniques that took advantage of technology, science education probeware companies such as Vernier (2016) and PASCO (2016) were born, founded by a high school physics teacher and a physicist, respectively. (Out of these companies also evolved newer and cheaper tools for collecting data relevant to all STEM fields, including pH, O_2 and CO_2 concentration, heart rate, EKG, humidity, etc.) This new technology allowed students to experiment more quickly and efficiently than ever before, collect data that might have otherwise been inaccessible in K–12 schools, and provided new ways to objectively observe and visualize data on a screen. The positive impact of methods employing MBLs are among the most frequently documented in physics education research with educational gains according to PhysPort's research validation summaries for MBL-based approaches (PhysPort, 2016). Meta-analyses (Vernier, 2007) have also shown significant positive differences on assessments of science content for those who used data-collection technology versus those who did not.

In the past ten years, a new evolution has taken place in technology: the advent of the mobile device as scientific probeware. Research is scarce in regard to using personal mobile devices as opposed to commercial probeware. However, many of the touted benefits of MBLs would seem to hold true for personal mobile devices, such as smartphones and tablets, as well, simply because they reproduce the same technological capabilities, yet in a smaller, more mobile format. Using sensors that are internal to most smartphones, students and teachers can download a variety of free applications (apps) that allow students to visualize and export data associated with acceleration, net force, angular velocity, magnetic field strength, sound intensity and frequency, among others. The popular adoption of mobile devices as sensor tools led the American Association of Physics Teachers to institute an *iPhysics* column dedicated to smartphone physics in their *The Physics Teacher* journal (AAPT, 2016). The column's editors, Khun and Vogt, suggest that "using such devices as experimental tools could foster conceptual learning" (Maciel, 2015). Because of the availability of smartphones as potential tools for metacognitive practices in STEM, they should not be overlooked by the education community for classroom experimentation and formal cognitive research.

BEST PRACTICE

Starting from my first year of teaching, I regularly administered the Force Concept Inventory (FCI) to my physics students at the beginning and end of the first semester. Although I had extensive teacher preparation in metacognitive strategies associated with mechanics in physics, I was unaware of how prone my students were to misunderstandings about Newton's Laws. Despite growing improvements each year, I found that my students continued to struggle with the basic tenets of Newton's 2nd law: a net force on an object results in acceleration of the object, and greater net forces results in greater accelerations, while objects with greater masses are more difficult to accelerate. Although students had few issues with the algebraic relationship between net force, mass, and acceleration, F_{net} = ma, students regularly defaulted to naive understandings when asked to apply the concept to real-world situations. This was especially true because students had so many apparently conflicting personal and sensory experiences with motion. Although true weightlessness for an object is not possible in the presence of a gravitational field, students have experienced "apparent weightlessness" on amusement park rides and have seen astronauts "float" around the International Space Station on video feeds. With these seemingly conflicting experiences, students failed to distinguish between actual weight, applied forces, and physical sensations such as feeling heavy at the bottom of a loop of a roller coaster and lighter near the top.

Most classical mechanics in physics, particle interactions in chemistry, and real-world applications of technology and engineering are dependent upon a good understanding of the relationship between forces and motion. While student struggles with Newton's 2nd law might seem to be only a small piece in a physics curriculum, Newton's laws make up the foundation for all of mechanics and understandings associated with motion, force, the major conservation laws of linear and angular momentum and energy, and its engineering applications. Understanding Newton's laws is one of the three main Disciplinary Core Ideas as set forth by the Framework for K-12 Science Education (National Research Council, 2012) and re-echoed in the Next Generation Science Standards (NGSS Lead States, 2013). Helping students to develop robust mental models associated with Newton's laws is vital to students in the physical sciences and to all students for college and career STEM-readiness.

However, this topic was difficult for many of my students because the simple algebraic relationship did not necessarily match up with their own sensory experiences. I soon realized that I needed an additional tool to help students self-check their thinking. While there were a variety of commercial science education probeware tools, such electronic scales and force plates, my students needed something more personalized to help change their very visceral misunderstandings about actual and apparent weight. To encourage my students be metacognitive about Newton's 2nd Law, I turned to my husband, a software developer, to create a smartphone app that could measure acceleration and the ratio of normal force to weight (frequently called "g-force"). What resulted was a robust set of resources that helped my students to use explicit metacognitive practices throughout the year to master Newton's laws in a way they had been unable to do before.

The conceptual learning sequence for Newton's 2nd law that I used with my students started with horizontal motion and forces and an experience pushing on a bowling ball. Students progressed to learn about force systems and would then attempt to transfer their understanding about horizontal forces to vertical ones. However, vertical force systems, especially those involving traditional "elevator-type" problems, are notoriously difficult for students because of the addition of three factors: challenges in understanding how objects behave when in the presence of a gravitational force field, conflicting sensory experiences, and discrepancies in the language used to describe those sensory experiences on vertically-moving situations.

Diagnosing Common Naïve Conceptions and Assessing Concepts in Newton's Laws

Because it was imperative for me to plan my instruction around what my students already knew and what naive conceptions they had, my students were accustomed to being given a variety of formative assessments throughout my teaching (in addition to the FCI that I administered early in the year). I would periodically assign diagnostic pre-test questions and then check their understanding later in the unit with parallel post-test questions, often drawn from various concept inventories, Diagnoser Tools type questions, or TIPERs. Student responses to questions such as the one shown from Diagnoser Tools (Figure 22.1) could give me some insight into what students believed about the relationship between force and motion.

Diagnoser Tools helped me to identify common areas with which my students struggled. It allowed me to map students' responses to numbered facets of knowledge, as shown in Table 22.1.

John's sister put the block on a very slippery table and gave it a shove. After she let go, the block slid across the table with an almost **constant speed**.

Which statement below best describes the **horizontal forces** on the block as it slides across the table to the right?

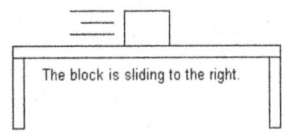
The block is sliding to the right.

[a] The force to the right is zero; the frictional force to the left is very small.

[b] The force to the right is greater than the frictional force to the left.

[c] A force of motion to the right is proportional to the speed.

[d] A constant force of motion keeps the block moving to the right.

Figure 22.1. Diagnoser Tools Pre-test Question.
(Source: Diagnoser Tools, 2016.)

Table 22.1. Answer Selections and Their Corresponding Facets for the Diagnoser Pre-test Question.

Selected Answer	Facet of Understanding (or Naive Conception) Associated with Each Selected Response
a	02: All forces are balanced when an object is moving at a constant speed in one direction. (Correct)
b	70: If an object is pushed or pulled and it remains at rest, there must be a force keeping it at rest that is larger than the pulling force. (Incorrect)
c	72: The excess force is proportional to the speed. (Incorrect)
d	71: The excess force is described as the "force of motion." (Incorrect)

(Source: Author, collected from Diagnoser Tools, 2016.)

In identifying students' prior knowledge and diagnosing naive conceptions, both my remedial and my advanced students displayed two prominent overall "commonsense" beliefs. First, students typically believed that objects only moved while under the influence of a force, and that that force gets transferred or imparted to help keep an object moving until it runs out. The idea, perhaps the most problematic driving naive conception in mechanics, is called the "impetus belief." The impetus belief was an early Galilean belief that an object must be supplied with a kind of internal, active force that propels an object forward (Hestenes et al., 1992). Even when students were asked to identify the object or source of the force field that they believed impelled an object, such as a tossed ball to continue to move, students would regularly refer to "invisible movers" or "ghost forces." While students were typically accepting of the concept of inertia after being exposed to it, it was not uncommon for students to still default to representing or describing inertia as a "kind of force." A second innate belief of many students was that an unbalanced or unopposed force would result in constant velocity motion (not a uniform acceleration, as is actually the case). This was likely because students have only ever lived in a world in which objects typically experience significant frictional or drag forces in opposition to their motion.

As students wrestled with these ideas, they engaged in metacognitive practices in whiteboard discussions, small group problem-solving, end of unit reflections, and assessments. Despite students' abilities to correctly solve mathematical problems or even to appropriately describe phenomena associated with force in the context of a focused activity, they would frequently and easily fall back into erroneous thinking when presented with tempting distractors that displayed pervasive naive conceptions. Clearly, there was something "visceral" and personal about Newton's 2nd law that my students did not fully comprehend. I needed to find a solution to help my students objectively challenge their own thinking in light of incontrovertible evidence.

Teaching Newton's Laws with Research-based Metacognitive Practices

Understanding the naive conceptions I was up against, I typically began my unit on Newton's Laws with a "bowling ball relay" (Photo 22.1) in which students attempted to quickly maneuver a bowling ball around a set of obstacles using a broom. Students kinesthetically demonstrated their naive conceptions and saw the consequences when they realized that a bowling ball continued to move freely with little resistance even after they had stopped pushing on it. For many students, this served as a discrepant event (O'Brien, 2010)—a jarring experience that made them realize that the world did not necessarily behave according to the mental

Photo 22.1. Student Maneuvers Bowling Ball Around Obstacles with a Broom.
(Photo credit: Rebecca Vieyra.)

model they used to predict its motion. This activity engaged students in a number of general metacognitive practices identified by Project CRISS (2016), including engaging background knowledge, using active learning strategies, and taking advantage of students' intuition on the topic. A follow-up discussion with students would reveal their surprise that no applied force was necessary to maintain the motion of an object.

Once students had verbalized their beliefs through dialogue following their first experience with the bowling ball relay, it was important to focus on building conceptual models through constructivist experiences while addressing naive conceptions through a variety of techniques. To address both the impetus belief as well as the belief that a net force results in constant velocity, not uniform acceleration, I provided students with an experience of pushing a hover ball down the a hallway. A hover ball is a commercially-available disk that uses a fan to blow air down out of the bottom of the disk so that it rests on a layer of air. The disk can easily slide across hard surfaces with very little friction when the fan is on. When the fan is off, it slides directly on the surface of the ground. Students were asked to push the hover ball down the hallway with the fan off and on, and to carefully notice both the type of motion that it displayed as well as the forces acting on it. To challenge the idea of impetus, students were asked to clearly label the origin of the force. Using whiteboarding techniques, students eventually came to consensus about the forces as displayed in Table 22.2.

Table 22.2. Force Diagrams Corresponding to Motion of Hover Ball.

	Hover Ball after push (on)	**Hover Ball after push (off)**
Force Diagram	Hover Ball with Normal Force (ground on ball) up, Weight (Earth on ball) down	Hover Ball with Normal Force (ground on ball) up, Friction (surface on ball) to the right, Weight (Earth on ball) down
Motion	Moves to the right at constant velocity.	Moves to the left, slowing down.

(Source: Author.)

Through class discussion, students usually resolved naive conceptions about impetus and the relationship between net force and motion with their classmates. Students themselves would verbalize that in neither case *after* the initial push was any kind of force present in the direction of motion. I then presented students with a variety of other experiences that involved forces and motion, and asked students to identify similarities between scenarios. Students typically deduced two principles: (1) objects with balanced forces did not change their motion (remained at constant velocity), and (2) objects with unbalanced forces did change their motion (accelerated). They learned when to adapt their thinking as they realized that Newton's Laws ultimately described two potential system types—balanced and unbalanced—and began to use that conceptual understanding to frame their thinking about varied scenarios. This strategy employed multiple metacognitive approaches identified by Marzano (2000), including identifying similarities and differences between the two scenarios, and using nonlinguistic representations to display and enhance understandings.

Although *discussions about the physics* concepts supported metacognitive practices, it was the *discussions about learning* the physics concepts and skills that were the metacognitive practice. Explicit discussions about *what* underlying principles they believed must be true, and *why* they believed they were true were very valuable. Rather than focusing only on getting the right answer or focusing on the content, discussions would often evolve into higher-level reflections on justification of knowledge ("How do you know what you are saying is accurate?") and an analysis of why students might be tempted to believe something "commonsense" rather than accurate ("What experiences have you had in your own life that support or conflict with what you observed in the lab and the conclusions you made?"). It was important to me to always conclude student-led discussions with some explicit reflection on how "commonsense beliefs" or "gut feeling" and "observations" were not always compatible with each other, and that it takes hard work to redefine mental models of understanding.

Learning and Knowledge Transfer

Despite such productive discussions centering around initial experiences with Newton's laws, the transfer of understanding from horizontal applied forces to "elevator-type problems" involving vertical forces was always difficult. Naive conceptions continued to arise in regard to beliefs about net force and motion: students tended to believe that an elevator moving at constant velocity upward *must* have a net upward force, and that an elevator moving at a constant velocity downward *must* have a net downward force. Although students typically understand early on that an object can move horizontally without any horizontal force acting upon it, they are often resolute that this is not possible for vertical situations. When asked to draw forces on a person in an elevator at various stages of the motion, I typically got two responses. One response would incorrectly show net forces during the constant velocity portion of an elevator, such as a longer upward force vector on a force diagram when moving upward at a constant velocity. The second type of response might actually show changes in the length of the gravitational force vector. One possible reason for this naive conception could stem from colloquial language use in regard to the terms "weight" and "weightlessness." Although weight is very simply defined as the pull of gravity on an object, we use the normal force of a scale acting upon us in the upward direction as a proxy for weight. Because the numbers on a scale do, indeed, change while standing on it during an elevator ride, students erroneously tend to believe that their weight does change, even if their mass does not. In the theoretical event that an elevator cable was to break and the elevator and the person inside were to enter into free-fall, we would typically describe the person as "weightless," despite the fact that the pull of gravity on the person had not changed at all. This difficulty is compounded by the fact that the human stomach is very sensitive to motion, and that students who have ever experienced a roller coaster or the like inherently *know* what it means to be weightless. In parallel, students *know* what it means to feel heavy at the bottom of a roller coaster loop, and therefore are wont to describe that their weight increased.

Students' experiences with vertical accelerations in roller coasters and elevators resulted in many students having very strong—and inaccurate—mental models associated with changing forces on their bodies. After nearly six years of struggling to find a solution that would give students an objective and personal experience that would likewise challenge their thinking, I happened upon an idea to develop a mobile accelerometer app to resolve some of those learning issues.

A Technological Solution: Mobile Accelerometers

In my first few years of teaching physics, I attempted to help my students understand vertical accelerations by taking them on mini field trips in the school elevator. However, the use of an analog bathroom scale only served to reinforce the naive conception that weight was indeed changing in the elevator, despite my explanations to the contrary. I later used a commercial electronic force plate upon which my students could stand and view the output force—listed as normal force, not as weight—on a computer screen. However, with a class of nearly 30 students, only a few students could have the personalized experience of seeing the normal force acting on their body change. When the elevator accelerated up, the students felt heavier, and when the elevator accelerated down, the students felt lighter. Clearly, to them, the downward force on their bodies was changing, and there was little I could do with the limited technology I had to help them change their minds about it.

I brought my concerns to my husband, a software engineer who had recently become interested in developing mobile apps. He recommended looking for an app that took accelerometer data from sensors built into all phones that would allow students to visualize the data on a graph. Doing so would allow them to have a personalized experience and to connect their physical sensations with the data on their own screens. The ability to make predictions about their experiences and then to use sensor-based data and graphs would allow them to check their own thinking against an objective source. This is possible because all phones have internal accelerometers in order to determine screen orientation based upon the angle of the device as held by the user, and for a variety of gaming applications. Inside of the accelerometer are three tiny prongs similar to flat springs that get bent under the force of gravity. As a result, when lying the mobile device flat, the x, y, and z coordinates each read 0, 0, and 9.8 m/s/s, respectively (which accounts for the acceleration of all objects on the surface of the Earth accelerating toward the center at that rate). Unfortunately, novice learners of physics generally do not have a global systems understanding of forces, and would expect a reading of 0, 0, 0 in their own frame of reference. It was clear that I needed to have an accelerometer app developed specifically for my use with high school students to reflect an understanding about forces, not accelerations.

My husband set to work developing an accelerometer app that would provide output in m/s/s, as well as a measure of the ratio between normal force and weight, a quantity typically termed "g-force." A g-force reading of "2" represents that the normal force on an object is equal to twice the magnitude of its weight. For a person in an elevator, this can be interpreted on a force diagram with an upward force arrow that is twice as long as the person's weight arrow. I engaged in designing the user interface so that it would be pedagogically appropriate for my students. The final app included not only a g-force meter, but a public suite of Android and iOS sensors tools for STEM education, which I named Physics Toolbox Sensor Suite (Vieyra Software, 2016) (Photo 22.2).

Using Physics Toolbox Sensor Suite

With Physics Toolbox Sensor Suite my students could now have their own personal or small group experiences with the elevator ride. Before riding in the elevator, students were first asked to generally investigate the relationship between the motion of their mobile device and the data that appeared on the screen. Students were asked to move around, spin, and shake their devices to observe the outcome. Students soon noticed that when the device was placed along the x, y, or z axis that one value always read "1," while the others read zero. This would lead to a discussion about the internal mechanisms of the smartphone and the meanings of g-force. Most students had some experiences with g-force in reference to race cars, roller coasters and fighter jets. Students were then asked to ride in an elevator upward, from one floor to the next. While placing their smartphone or tablet flat

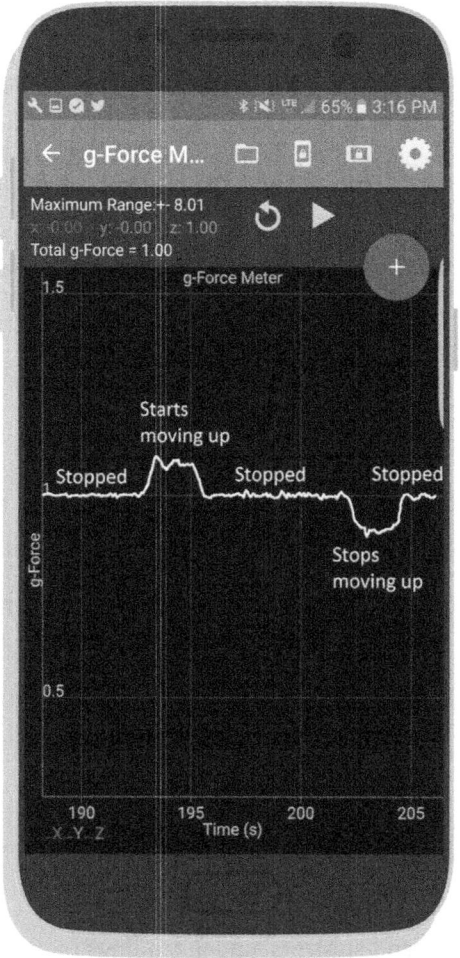

Figure 22.2. Physics Toolbox Sensor Suite Displaying Graph of G-forces in an Elevator.
(Photo by Rebecca Vieyra.)

on the floor of the elevator, they were to note their own bodily sensations and to connect them to the data they saw. Most students observed a graph as shown in Figure 22.2.

I helped my students to formalize their thoughts with an advanced organizer that broke up the elevator's motion into five parts. To draw the force diagrams, students used the bathroom scale in the classroom only to determine their weight, while being asked to justify what the reading on the scale actually meant. After riding on the elevator and recording the data with their smartphones, they used the values displayed on their graph of g-force to proportionally determine the strength of the normal force acting on their bodies (Table 22.3).

Using sensor data, students recognized that during the middle of the motion, the normal force *must* be indeed equal in length to the weight of the person, as the g-force read a value of "1." Additionally, students were very surprised to see a correlation between their own bodily sensations (i.e. normal, heavy, and light) were a direct result in changes in the normal forces acting upon their bodies, not changes in weight.

Table 22.3. Force Diagrams and Corresponding Motions and Sensations of an Elevator.

Elevator Motion Description for Ride UP	Rest	Speeds up	Moves up at constant velocity	Slows down	Rest
Force Diagram for Person in Elevator	Person with Normal Force (elevator floor on person) up and Weight (Earth on person) down, equal	Person with Normal Force up (longer) and Weight down	Person with Normal Force up and Weight down, equal	Person with Normal Force up (shorter) and Weight down	Person with Normal Force up and Weight down, equal
Sensation	Normal	Heavy	Normal	Light	Normal

(Source: Author.)

Transferring Understanding from the Classroom to Home and Field

With a mobile tool available to nearly all of my students with a smartphone or tablet, I was able to extend students' understandings in different environments. I encouraged my students to complete a variety of at-home projects, including a challenge to record a g-force of 0 in all dimensions at the same time. This can only be accomplished by allowing the device to be in free fall, so this was an assignment that required both family participation and significant safety instructions. A number of my students accomplished this task in the family living room or the student's bedroom over a sofa or large bed, while others were able to achieve this while leaping through the air in their dance studios or sports practices (Vieyra, Vieyra, Jeanjacquot, Marti, & Monteiro, 2015). Assignments such as these naturally led to a wide variety of discussions between students and their family members and friends, which reinforced their understandings about the relationship between force and motion, and even helped students to stand in the role of teacher while they helped others to overcome their naive conceptions or lack of understanding.

Because naive conceptions associated with force and motion proved so difficult to overcome, and because understanding the relationship between force and motion is so integral to understanding the physical sciences, I chose to include these concepts at an end-of-year assessment. As visiting an amusement park was something of an end-of-course ritual, I determined that this would be the perfect opportunity to return to these topics in a meaningful way. What resulted was a detailed lesson experience (Table 22.4) in which the mobile sensor was used to have students determine the types and quantities of forces acting on them at various points along the ride (Vieyra, Vieyra, Jeanjacquot, Marti, & Monteiro, 2015. In addition, students were asked to narrate how the forces on their bodies changed, and, most importantly, how Newton's laws accurately predicted and described these changes.

Table 22.4. Sampling of Amusement Park Physics Assignment.

Amusement Park Physics (sampling of assignment)

Motion, Forces, and Newton's Laws – Choose <u>any amusement park ride that causes you to feel changes in your apparent "weight."</u> Collect g-force data in three dimensions using Physics Toolbox Sensor Suite (or a comparable app with quantitative data), and save data as either as a screenshot or export the data as a .txt file and plot it using a graphical analysis tool. Your assignment must include the following pieces:
 	
    a. Image or sketch of ride with labels showing **three points** of study on the ride.
    b. Graph of 3-dimensional g-force vs. time. (Label the three points of study).
    c. Three *proportional, quantitative* force diagrams displaying forces at corresponding points on ride. Label actual values. Show calculations. *Note: This data can be very messy. It is up to you to decide what is important, valuable, and reasonable to display on your force diagrams.
    d. Description of how g-forces change – describe expected sensations and relate them to force diagrams (in part c).

(Source: Author.)

Metacognition in Summative Assessments

Near the end of the unit, but before the summative test, I again had my students take a comparable diagnostic question set similar to the one taken at the start of the unit. However, I required my students to explicitly reflect on their learning and their challenges. Students recorded their original answer (pre-answer) and selected the same or different post-answer before engaging in Peer Instruction (Mazur, 1996), a metacognitive process that encourages students to collaboratively come to consensus about an answer with justification before responding. Before the in-class session to check their answers, students reflected on the unit's learning targets, and I provided students with a list of common naive conceptions. I asked students to either identify the learning target that they believed the question was addressing, or to select a naive conception that they felt the question was meant to target. In addition, students were expected to engage their intuition by rating their own self-confidence on their post-answer (Table 22.5). During the review session in which each student used a "clicker" phone app to select an answer and transmit it to a website for whole-class display, students would mark their answers as correct or incorrect, and use the diagnostic tool as a guide for holistic unit reflection.

Table 22.5. Example Reflection Sheet with a Diagnostic Question, Pre- and Post-answer, Corresponding Learning Target, and Confidence Rating to be Completed by the Student Before Checking Answers.

Diagnostic Question	Pre-Answer	Post-Answer	Learning Target/ Naive conception	Confidence at Time of Post-Answer
John's sister put the block on a very slippery table and gave it a shove. After she let go, the block slid across the table with an almost constant speed. Which statement below best describes the horizontal forces on the block as it slides across the table to the right? [a] The force to the right is zero; the frictional force to the left is very small. [b] The force to the right is greater than the frictional force to the left. [c] A force of motion to the right is proportional to the speed. [d] A constant force of motion keeps the block moving to the right.				How I feel about this question: 😊 😐 ☹

(Source: Author.)

Following the clicker session, students identified the change in their pre- to post-test scores, and reflected on their overall growth by choosing to answer one of four questions (Table 22.6). It was during these reflections that I would frequently see the most insightful responses from students: delight from seeing

Table 22.6. End-of-Unit Reflection Questions.

For each unit, you may answer one or more of the following questions to demonstrate that you have reflected on the learning targets, your successes, and your challenges.
1. Based on your pre/post test, what learning target (or type of problem or concept) was the most difficult for you to master? Provide evidence from a class activity or assessment.
2. Give one example of a naive conception that you had about the unit either before or during the unit. Look at the list of common naive conceptions provided to you. Explain how you now know that this is not true.
3. Give one example of something that took you awhile to understand, but that you eventually understood. Explain what happened that caused it to "click" for you.
4. Give one example of something that you did not know or understand before we started the unit. Provide evidence from your pre-test, post-test, or assignment that demonstrates that you did not know or understand it before, but that you know it now.

(Source: Author.)

personal gains, frustration from continuing to be tempted by their own naive conceptions, and, sometimes, relief from identifying consistent (if even erroneous) patterns in their thinking. Although I often anticipated which students struggled and why, I occasionally would find a student who expressed that he or she had uncovered some element of their thinking that was a surprise to both of us.

Students frequently reflected that it was through the activities in which they confronted their naive conceptions through the analysis of data from mobile devices that helped them to really come to understand Newton's laws. Comparable to the study on conceptual change by Redish (1994), I saw for myself that students constructed, assimilated, or accommodated their mental models about force and motion when presented with new or

Conflicting data from their experiences in the elevator and on roller coasters. At the end of the course, most of my students had successfully overcome the temptation to resort back to naive conceptions that suggest that force is necessary for motion, that net force results in constant velocity, or that weight actually changes in an elevator. The formative and summative emphasis on metacognition served as a continual reminder to all of us in the classroom that learning is a process that can be arduous and recursive, and that these challenges are entirely normal but can be overcome.

HOW OTHERS CAN ADAPT THIS BEST PRACTICE

Metacognitive strategies are relevant to all forms of learning, not just STEM. However, discipline-specific educational research can inform specific practices that can help teachers to practically implement approaches and resources to tackle specific learning difficulties.

Identify Your Discipline-Specific Learning Challenges, Strategies, and Tools

Although metacognitive techniques suggested by Marzano (2000) and Project CRISS (2016) are effective for general learning, they do not always take into account the struggles that students might encounter with specific concepts that are hard to grasp because of complex or theoretical abstractions that are specific to a subject, such as the proportional reasoning needed in math to understand gear ratios in engineering, or the complex thermodynamic relationships in meteorology and environmental science. Likewise, learning in STEM goes far beyond concepts, and might include learned skills or even attitudes that cannot be easily quantified, compared, or measured. For example, a reasonable outcome for a technology course might be that girls and boys both demonstrate positive attitudes toward STEM careers. The metacognitive strategies required to help girls as well as boys think about their own competencies, motivation, and confidence in pursuing STEM careers are likely to look very different from the metacognitive strategies used to learn concepts.

Because metacognitive strategies typically emerge from educational research, this requires K–12 STEM educators to often move beyond their own collegial sphere into that of higher education. What follows are a number of suggestions for building up personal expertise in discipline-specific metacognition.

Join a professional learning network with a focus on metacognition

Look specifically for opportunities to deeply reflect on student learning as it is relevant to your own practice. Metacognition requires attention to oneself and to one's students as individuals, so it is important to seek out opportunities that allow the teacher to remain highly focused. Because metacognitive practices are so student-focused, these opportunities can often be found in more intimate professional environments, such as through mentoring experiences in which a colleague observes students' interactions in class and helps to review student work alongside the classroom teacher. For example, National Board Teacher Certification (http://www.nbpts.org/) includes one portfolio element that requires teachers to identify a single student and to follow his or her progress over the course of a unit. Opportunities such as these are a chance to evaluate students' mental models and plan metacognitive interventions. Small community-based programs such as the Math Teachers' Circles Network (http://www.mathteacherscircle.org/) bring together only a handful of teachers at a time throughout the year to intensely focus on metacognitive strategies for a single problem. When aiming to improve metacognitive practices in the classroom, quality is of far greater value than quantity.

Identify existing research-based teaching practices and tools relevant to your discipline

Visiting sites such as PhysPort (2016) and the Science Education Resource Center (2016), joining a professional society, and reading professional journals can provide a wealth of resources that connect educational research to the practicing K–12 teacher. PhysPort alone includes a number of stand-alone and interdisciplinary practices and tools.

- Minds-On Physics (MOP, 2016): *physics*
- Modeling Method of Instruction (Wells et al., 1995): *physics, chemistry, biology, astronomy, middle school integrated physical science and engineering*
- Peer Instruction (Mazur, 1996): *general STEM*

- PhET Interactive Simulations and teacher tips (PhET, 2016): *physics, chemistry, biology, Earth science, astronomy, mathematics*
- Physics and Everyday Thinking (Physics and Everyday Thinking, 2007): *physics*
- Physics Union Mathematics (Physics Union Mathematics, 2016): *physics, mathematics*
- Responsive Teaching in Science (Response Teaching in Science, 2013): *general STEM*

Make Student Thinking Visible and Explicit

Effective metacognitive practices are dependent upon both the teacher and the student recognizing what they know, what they do not know, and what their thought processes are. For this to happen, student thinking must be visible (able to be observed) and explicit (able to be described or quantified). The following are a number of ways that teachers can make student thinking visible and explicit.

Use assessments (concept test and diagnostic tools) to measure knowledge, skills, and attitudes about learning

Concept tests and diagnostic tools have been developed for many STEM fields. In addition to the discipline-specific content-based tools like the FCI in physics, there are a number of well-known tests that have relevance to all K–12 STEM fields, including the following:

- Assessment of Textbook Problem Solving Ability (Marx & Cummings, 2010)
- Attitudes and Approaches to Problem Solving Survey (Mason & Singh, 2010)
- Colorado Learning Attitudes about Science Survey (Adams et al., 2006)
- Concise Data Processing Assessment (Day & Bonn, 2011)
- Lawson Classroom Test of Scientific Reasoning (Lawson, 1978)

Engage students in Socratic dialogue and in representing their thinking in multiple ways

Asking students to verbalize and challenge each other's thinking is perhaps one of the most efficient ways for teachers to observe student learning. Whiteboarding is one effective approach that can be used to encourage this kind of dialogue. In addition, students can be asked to represent their thinking through diagrams, images, mathematical equations, graphs, skits, and through written narrative. All of these give teachers a lens into student thinking.

Ask students to reflect on their thinking processes

Explicit questions that are not about content, but about learning, help students to identify their own struggles and to self-regulate by making decisions about how they might do things differently to remediate. Continual self-awareness should be encouraged, such as using the "fist to five" method in which students share their perceived level of confidence on a topic with their teacher (with a fist representing zero confidence, and five fingers representing the highest level of confidence). Three-sided table tents can also be constructed on their desks for students to self-assess how they are doing on a project or problem, and can be directed toward teachers with messages such as "We're doing ok," "Please stop when you have a moment," or "Stop! Help now!" Other explicit techniques can include exit slips and reflection questions that pertain to their own struggles and growth in learning.

Using Sensors to Address Pervasive Challenges

A significant amount of effort from the educational technology world has gone into individualization and personalization of learning. Because of this, teachers have a wide variety of opportunities to find or modify tools to suit the specific needs of their students. In my own case, my students needed to make a personal connection between their technology, their data, and their bodily sensations while riding in an elevator or on a roller coaster. What resulted was the development of a mobile sensor app. While teachers are unlikely to have a software engineer at their own disposal, commercial sensor probeware abounds in many schools, and there are thousands of freely available sensor apps that can be used to enhance learning through metacognition and self-regulation. While possibilities with Physics Toolbox apps are shared below, any of the experiences can be undertaken using any equivalent kind of commercial or mobile sensor.

Because research shows the positive impact of using sensors and visualization tools (Thornton, 1989; Thornton et al., 1990; Vernier, 2007), teachers should strongly consider including technology and working with quantitative data in their instruction. Even students in conceptual courses can benefit from seeing changes in the relative value of quantities over time, either on a graph or in a data table. Furthermore, teachers might consider how the added benefit of using personal mobile devices can help students to see these changes in a variety of environments, not just the classroom or laboratory. As Physics Toolbox Suite has been extended to include sensors far beyond just the accelerometer, STEM teachers might find this or other sensor apps helpful.

For example, students often struggle with understanding the cause of the seasons, naively attributing the heat in the summer to Earth being closer to the sun, despite the fact it is winter in the opposite hemisphere. Students can use Physics Toolbox Light Meter (or the light meter setting of Physics Toolbox Sensor Suite), then take their mobile device outside and investigate what variables influence the reading on the screen, finding that both the overall brightness of the sky and the angle at which the light meter is placed have an impact. Analogously, students can use the light meter mode to see how light intensity changes as a function of the Earth's tilt. Using a light source to shine toward a standard classroom globe, students can place the smartphone's light sensor tangent to the surface of the earth in the northern and southern hemispheres to understand how each hemisphere receives a different amount of solar energy.

Students of mathematics can also use the light meter setting to see how the distance from a single light source, such as a bulb or the lamp from the smartphone flash, influences light intensity. Plotting the relationship between light intensity and distance will result in an inverse-square law. Using sensor tools to relate real data and observations to a novel function might help students to deepen their intuition about functions and their relevance to science and engineering. Teachers can make extended connections to STEM through experiments that measure the reflectivity of various types of surface vegetation versus blacktop or concrete, or even have students measure the luminescence of different concentrations of chemical solutions. Nearly all STEM disciplines can benefit from the inclusion of sensors to enhance student learning in some way.

CONCLUSION

Effective K-12 STEM teachers work to bridge the gap between educational research and teaching practice, teaching and learning, and from students' prior knowledge and naive conceptions to robust understandings. It is important for teachers to know that hidden beneath their everyday best practices are decades of research. At the heart of it all is metacognition. Frequently, learning challenges in STEM are associated with incorrect or incomplete prior knowledge, and teachers must be aware of and help students to reflect upon their own naive conceptions and mental models as they learn.

Teachers would do well to acquaint themselves, in particular, with discipline-specific educational research. Students frequently struggle to comprehend topics and abstractions unique to the discipline they are studying. Especially effective are metacognitive activities that encourage students to reflect on their own thinking and make comparisons to observations of the world. Sensor tools, especially on mobile devices that allow students to investigate the world in non-traditional learning spaces, are both prevalent and personal, and support students as they build and modify their understandings of the world.

WORKS CITED

Adams, W., Perkins, K., Podolefsky, N., Dubson, M., Finkelstein, N., & Wieman, C. (2006). New instrument for measuring student beliefs about physics and learning physics: The Colorado learning attitudes about science survey. *Physical Review Special Topics: Physics Education Research*, 2(1), 010101-1.

American Association of Physics Teachers. (2016). *Publications*. Retrieved September 28, 2016, from https://www.aapt.org/Publications/

American Modeling Teachers Association. (2016). *American Modeling Teachers Association*. Retrieved on September 28, 2016, from http://modelinginstruction.org/

Blackwell, L., Trzesniewski, K., & Dweck, C. (2007, January/February). Implicit theories of intelligence predict achievement across an adolescent transition: A longitudinal study and an intervention. *Child Development*, 78(1), 246-263.

Bransford, J., Brown, A., & Cocking, R. (1999). *How people learn: Brain, mind, experience, and school*. Washington, DC: National Academies Press.

Cushman-Patz, B. (2014). Using whiteboards to create a student-centered, collaborative classroom. In T. Spuck & L. Jenkins (Eds.), *Einstein fellows: Best practices in STEM education* (pp. 275-299). New York: Peter Lang Press.

Day, J., & Bonn, D. (2011). Development of the concise data processing assessment. *Physical Review Special Topics: Physics Education Research*, 7(1), 010114.

Diagnoser Tools. (2016). *Diagnoser Tools*. Retrieved September 28, 2016, from http://diagnoser.com/

Donovan, S., & Bransford, J. (2005). *How students learn: Science in the classroom*. Washington, DC: National Academies Press.

Harvard-Smithsonian Center for Astrophysics. (2016, April 21). *Resource: Minds of our own*. Retrieved from https://www.learner.org/resources/series26.html

Hestenes, D., Wells, M., & Swackhamer, G. (1992). Force concept inventory. *The Physics Teacher*, 30(3), 141-158.

Hieggelke, C., Kanim, S., Maloney, D., & O'Kuma, T. (2013). *TIPERs: Sensemaking tasks for introductory physics*. New York: Pearson.

Hieggelke, C., Maloney, D., O'Kuma, T., & Kanim, S. (2005). *E&M TIPERs: Electricity & magnetism tasks*. New York: Pearson.

Hieggelke, C., Maloney, D., & Kanim, S. (2011). *Newtonian tasks inspired by physics education research*. New York: Pearson.

Knight, R. (2002). *Five easy lessons: Strategies for successful physics teaching*. New York: Pearson.

Laws, P. W. (1988). Workshop physics: Replacing lectures with real experience. In *The conference on computers in physics instruction proceedings* (pp. 22-32). New York: Addison-Wesley.

Lawson, A. (1978). The development and validation of a classroom test of formal reasoning. *Journal of Research in Science Teaching*, 15(1), 11.

Lovett, M. (2008). Teaching Metacognition, 2008 Educause Learning Initiative annual meeting presentation. Retrieved from https://net.educause.edu/upload/presentations/eli081/fs03/metacognition-eli.pdf

Maciel, T. APS News. (2015). *Smartphones in the classroom help students see inside the black box*. Retrieved on September 28, 2016, from https://www.aps.org/publications/apsnews/201503/smartphones.cfm

Marx, J., & Cummings, K. (2010). *Development of a survey instrument to gauge students' problem-solving abilities*. Presentation at Physics Education Research Conference, Portland, OR.

Marzano, R. J. (2000). *Classroom instruction that works: Research-based strategies for increasing student achievement*. Alexandria, VA: Association for Supervision and Curriculum Development.

Marzano, R. J. (2009). *Designing & teaching learning goals & objectives*. Bloomington, IN: Marzano Research Laboratory.

Mason, A., & Singh, C. (2010). Surveying graduate students' attitudes and approaches to problem solving. *Physical Review Special Topics: Physics Education Research, 6*(2), 020124.

Mazur, E. (1996). *Peer instruction: A user's manual*. New York: Pearson.

McDermott, L. (1984). Research on conceptual understanding in mechanics. *Physics Today, 37*, 24–32.

McDermott, L. (1991). Millikan lecture 1990: What we teach and what is learned—closing the gap. *American Journal of Physics, 59*(4), 301–315.

Megowan-Romanowicz, C. (2016). Whiteboarding: A tool for moving classroom discourse from answer-making to sense-making. *The Physics Teacher, 54*(2), 83–86.

Minds on Physics. (2016). *Minds*On physics*. Retrieved September 28, 2016, from http://www.srri.umass.edu/mop

National Academy of Science. Board on Behavioral, Cognitive, and Sensory Sciences. (2016). *How people learn II: The science and practice of learning*. Retrieved September 28, 2016, from http://sites.nationalacademies.org/DBASSE/BBCSS/CurrentProjects/DBASSE_090054

National Research Council. (2012). *A framework for K–12 science education: Practices, crosscutting concepts, and core ideas*. Washington, DC: National Academies Press.

NGSS Lead States. (2013). *Next generation science standards: For states, by states*. Washington, DC: National Academies Press.

O'Brien, T. (2010). *Brain-powered science: Teaching and learning with discrepant events*. Arlington, VA: NSTA Press.

O'Kuma, T., Maloney, D., & Hieggelke, C. (2003). *Ranking task exercises in physics*. New York: Pearson.

PASCO. (2016). *PASCO*. Retrieved September 28, 2016, from https://www.pasco.com/

Physics and Everyday Thinking. (2007). Retrieved September 28, 2016, from http://cpucips.sdsu.edu/web/pet/

Physics Union Mathematics. (2016). *Physics union mathematics*. Retrieved from http://pum.rutgers.edu/

PhysPort. American Association of Physics Teachers. (2016, April 12). PhysPort: Supporting physics teaching with research-based resources. Retrieved from http://www.physport.org

Project CRISS. (2016). *Project CRISS*. Retrieved September 28, 2016, from http://www.projectcriss.com/

Redish, E. F. (1994). The implications of cognitive science for teaching physics. *American Journal of Physics, 62*(6), 796–803.

Responsive Teaching in Science. (2013). Retrieved September 28, 2016, from http://cipstrends.sdsu.edu/responsiveteaching/

SERC: Science Education Resource Center. (2016). *Teaching metacognition*. Retrieved September 28, 2016, from http://serc.carleton.edu/NAGTWorkshops/metacognition/teaching_metacognition.html

Singer, S. R., Nielsen, N., & Schweingruber, H. (2012). *Discipline-based education research: Understanding and improving learning in undergraduate science and engineering*. Washington, DC: National Academies Press.

Thornton, R. K. (1989). *Using the microcomputer-based laboratory to improve student conceptual understanding in physics*. Proceedings from Microcomputers in Physics Education symposium, Adana, Turkey.

Thornton, R. K., & Sokoloff, D. R. (1990). Learning motion concepts using real-time microcomputer-based laboratory tools. *American Journal of Physics, 58*, 858–866.

Vernier. (2007). *What the research says about the value of probeware for science instruction*. Retrieved from http://www.vernier.com/files/misc/vernier_whitepaper.pdf

Vernier. (2016). *Vernier*. Retrieved September 28, 2016, from http://www.vernier.com/

Vieyra, R., Vieyra, C., Jeanjacquot, P., Marti, A., & Monteiro, M. (2015, December). Turn your mobile device into a science laboratory space. *The Science Teacher, 82*(9), 32–40.

Vieyra Software. (2016). *Physics toolbox*. Retrieved September 28, 2016, from http://www.vieyrasoftware.net

Wells, M., Hestenes, D., & Swackhamer, G. (1995). A modeling method for high school physics instruction. *American Journal of Physics, 63*(7), 606–619.

Wenning, C. (2008). Dealing more effectively with alternative conceptions in science. *Journal of Physics Teacher Education Online, 5*(1).

Wenning, C., & Vieyra, R. (2015). Metacognition and self-regulation. In *Teaching high school physics*. Amazon Publishing.

CHAPTER TWENTY-THREE

Applications OF Satellite Imagery, Remote Sensing, AND Computer Visualizations

Observing the Earth and Visualizing the Future

JOHN D. MOORE

INTRODUCTION

Over the past 5 decades, a technological infrastructure has been developed in the geosciences. It is comprised of observation networks functioning in all four spheres (earth, oceans, atmosphere, and space). The ability to observe the planet in real time continues to grow every day. The applications of satellite imagery, remote sensing, and computer visualizations have contributed to better predictions and management of significant environmentally related events, and have acted as a catalyst in the development of advanced technologies. This has led to a better understanding of the Earth and its interactive systems.

The world continues to experience events that are often catastrophic in terms of loss of life and property—earthquakes, tsunamis, volcanic ash, wildfires, hurricanes, tornadoes, drought, flash floods, and oil spills.

> In the decades ahead, scientists predict that society will face issues such as the impact of climate change, land-use, and sustainability. The ability to observe and predict Earth's environment, including weather, space weather, and climate, and to improve the accuracy of those predictions in a complex society that is ever more dependent on environmental variability and change, has heightened the importance and value of environmental observations and information. These observations, and the predictions on which they are based, are now essential to many components of society—including national defense, industry,

policy-making bodies, and the people and institutions that manage natural resources—as well as to the comfort, health, and safety of the public. (National Research Council, 2003)

These Earth-related issues and events present serious challenges to our nation's economy and national security. It has become clear that "economic growth and human well-being wholly depend on the availability of adequate supplies of water for agriculture, energy production, transportation, ecosystem services, manufacturing, and waste management" (NSF Advisory Committee for Geosciences, 2009).

Why should the geosciences be part of the K–12 education paradigm? As we face future natural and human-generated hazards and disasters, the geosciences have a critical role to play in decision making, public awareness, and safety. The question then becomes: Why hasn't geoscience taken a more central role? There are many barriers to widespread availability of high-quality geoscience education.

> Because Earth Science is most often taught in middle school, it is not taught with much rigor and does not build on essential physics, chemistry, and biology concepts. Most states require only two or three science classes for high school graduation and Earth Science often does not qualify as a laboratory science, making this a less desirable elective. It is commonly offered as a remedial high school science class and there is no AP Earth System Science program to attract the highest achieving students. The end result is that fewer than 30% of students take any type of geoscience class in high school. (NSF Advisory Committee for Geosciences, 2010)

The development of national STEM (science, technology, engineering, math) education policy gives the geosciences another opportunity to penetrate the traditional science education paradigm of physics, chemistry, and biology.

The *Next Generation Science Standards* (NGSS), released in April 2013, include a focus on Earth and space sciences, and promote project- and problem-based learning (PBL) instructional strategies. Teachers using satellite and remote sensing technologies to incorporate imagery, data, and real-time observations in the classroom support these elements of the NGSS. The standards propound the practice of science, as opposed to rote memorization or traditional lecturing. The NGSS encourages states to adopt science standards that support research in the classroom.

One may question the capabilities of pre-college students to engage in authentic investigations. Indeed, some scientists have. Yet geoscience work conducted by my high school students have won two international "Educational Best Practices in Geographic Information Systems" awards (2003–2004) sponsored by the Intergraph Corporation. On two separate occasions, student research projects were selected to represent the United States in the Global Learning and Observations to Benefit the Environment (GLOBE) Program's Global Learning Exhibition (GLE). The first project was presented in Finland in 1998, and a second project was presented at GLEII in Croatia in 2003. This author and a student presented in Croatia as part of a GLOBE Croatia Land Cover Symposium (1999). Numerous students and I have collaborated as co-presenters at Institute of Electrical and Electronics Engineers (IEEE) Geoscience and Remote Sensing Society Conferences in Anchorage, Alaska, Toronto, Canada, and Toulouse, France. As a result of the presentation in Toronto, the University of Moscow restructured its entire geoscience department. A yearlong international study with GLOBE students from Denmark resulted in an invitation to present findings to the Denmark Ministers of Education, which had direct impact on Denmark's national curriculum.

Much of this work and the resulting student experiences have been documented as part of *Classroom Close-Up New Jersey*, an Emmy award-winning program sponsored by the New Jersey Education Association. Do not underestimate what students are capable of achieving when given the opportunity to explore issues that are relevant to them. Students view these opportunities as challenges and realize that these are the pathways to success and can provide direction for their future education and careers. When students are able to take responsibility for and participate in their educational and career path choices, they begin to visualize their future.

BACKGROUND

> *In order to teach any subject with the best success, the instructor should not only fully understand it, in all its principle and details, but should also clearly perceive what particular faculties of the mind are concerned in its acquisition and use ... never permit him to accept any statement as true which he does not understand. Let him learn not by authority but by demonstrations addressed to his own intelligence. Encourage him to ask questions and to interpose objections. Thus he will acquire that the most important of all mental habits, that of thinking for himself.*
>
> —FISH (1875, P. VI)

As teachers engage in authentic research and critical thinking strategies, a foundation for innovation is established. Innovation leads to best practices, and best practices often lead to further innovation—a desirable cycle. Teachers should possess strong content knowledge, understand the nature of their students, and expose students to learning communities where students ask questions, conduct investigations, and develop solutions:

> There is ample evidence that authentic research experiences are a highly effective strategy for engaging students and learners in the study of science and pursuit of degrees and careers in STEM fields. Geoscience-related research has always served as a compelling subject area that is intrinsically fascinating for most audiences. With the increasing societal relevance of environmental geoscience topics, the opportunities to use geoscience data and engagement

of students and citizens in geoscience research are increasing. (NSF Advisory Committee for Geosciences, 2010)

The application of satellite images, remote sensing, and computer visualizations has a documented history of providing opportunities for innovative instructional strategies that lead to higher levels of student achievement.

> There is a natural role for the geosciences to play in advancing broad STEM literacy and engaging students in STEM careers. Earth Science, defined broadly, is intrinsically fascinating for students and of clear relevance to their daily lives. Since it is often the first real science class to which students are exposed, it serves as an important "gatekeeper" for student attitudes toward science and science careers in general. (NSF Advisory Committee for Geosciences, 2010)

The remote sensing community has developed a robust infrastructure of Earth system data collectors ranging from ground and ocean sensors, Doppler radar, aircraft, ships, drones, and satellites that all monitor our planet on a continuous basis. The United States has a significant unclassified infrastructure investment through federal agencies such as the National Aeronautic and Space Administration (NASA), the National Oceanographic and Atmospheric Administration (NOAA), and the National Science Foundation (NSF). The Department of Defense and the intelligence communities have their own set of robust (though classified) resources. The National Center for Atmospheric Research (NCAR) and the University Center for Atmospheric Research (UCAR) are home to some of our nation's largest supercomputers. These computers are tasked with running atmospheric and other related Earth system models. Remote sensing data can be graphically represented through visualization models.

With the launch of Sputnik in 1957, satellites were introduced to the world. In 1960, the first commercial satellite was launched, which transmitted a weather image of Earth. The Television and Infrared Observation Satellite (TIROS) carried special television cameras that viewed Earth's cloud cover from a 450-mile orbit. By 1965, nine more TIROS satellites were launched. The era of satellite remote sensing of Earth was born. This technology revolutionized weather and storm prediction.

On December 7, 1972, perhaps the most famous photograph of the Earth, the *Blue Marble*—the first full disc photograph of Planet Earth—was taken by the crew of the Apollo 17 spacecraft at a distance of about 45,000 kilometers (28,000 mi). The first Geostationary Operational Environmental Satellite (GOES) was launched on October 16, 1975. The early GOES viewed the Earth only about 10% of the time. These satellites were in operation from 1975 to 1994. From April 13, 1994, to the present, the next generation of spacecraft (GOES I–M), have been in operation. GOES-8 was launched on April 13, 1994. The remaining spacecraft in this series (GOES 9–12) were successfully launched between 1994 and 2001. Using visual and infrared imagery, this generation of satellites views the Earth 100% of the time, taking continuous images and soundings. GOES satellites provide data for severe storm evaluation, information on cloud cover, winds, ocean currents, fog distribution, storm circulation, and snow melt.

In 1991, NASA launched the Earth Observing System (EOS; now called the Earth Science Enterprise), a comprehensive program to study the Earth as an environmental system. By using satellites and other tools to intensively study Earth, NASA's goal was to examine how humans affect natural processes and vice versa (NASA's Earth Observing System Project, 2013). These satellite remote sensors have resulted in improved weather forecasts, managing agriculture and forests, and other critical environmental information about the Earth system.

NASA and its international partners continue to operate and add to a 50-year-plus database through several Earth-observing satellites that closely follow one another along the same orbital "track." This coordinated group of satellites, constituting a significant subset of NASA's current major operating satellite missions, is called the Afternoon Constellation, or the A-Train for short. These satellites are in a polar orbit, crossing the equator northbound at approximately 1:30 p.m. local time, within seconds to minutes of each other. This allows near-simultaneous observations of a wide variety of parameters to aid the scientific community in advancing our knowledge of Earth-system science. The United States continues to design, build, and enhance world-class Earth observation systems to provide reliable geospatial intelligence, often in real time, to decision makers at all levels during environmental events or crisis.

In 2003, the U.S. Department of Labor identified Geospatial Technologies as one of the top three emerging growth industries. Geospatial Technologies consists of two major components: Geographic Information Systems (GIS) and Global Positioning Systems (GPS). GIS themes are developed based on data sets gathered through Remote Sensing technologies:

> GIS allows creators to view, understand, question, interpret, and visualize data in many ways that reveal relationships, patterns, and trends in the form of maps, globes, reports, and charts. A GIS helps you answer questions and solve problems by looking at the data in a way that is quickly understood and easily shared. (ESRI, 2013)

The geosciences scientific and educational policy communities (i.e., Office of Science and Technology Policy [OSTP], National Academy of Sciences, NSF, NASA, and NOAA) have developed strategic plans and technological frameworks that support and provide guidance for the development of geosciences education in pre-college education. In 2013, an announcement from OSTP, "Big Data for Earth System Science," supported funding for a research center that catalyzes innovative thinking in Earth system science by providing scientists a place and time for in-depth analysis, state-of-the-art computing capabilities, and collaborative tools invaluable for making sense of huge data sets. These Big Data projects will improve our understanding of issues such as species response to climate change, earthquake recurrence rates, and the next generation of ecological indicators. (Office of Science and Technology Policy Executive Office of the President, 2012)

NSF, NASA, and the NOAA have collaborated with scientific societies and organizations in the generation of peer-reviewed publications resulting in several literacy documents. What are commonly referred as "The Literacies" are short and to the point, though extremely robust in terms of content. They establish a baseline of underlying principles and concepts that the agencies feel all Americans should understand. The following documents have been developed and published: "Essential Principles and Fundamental Concepts for Atmospheric Literacy," "Ocean Literacy: The Essential Principles of Ocean Sciences," "Climate Literacy: The Essential Principles of Climate Science," "Earth Science Literacy Principles: The Big Ideas and Supporting Concepts of Earth Science," and "Energy Literacy: Essential Principles and Fundamental Concepts for Energy Education."

The literacy documents provide a framework in which to develop relevant content for the inclusion of the geosciences in a traditional academic track or a career technical education (CTE) program. The practical application of the documents' propounded skill sets can be applied to real-life problems and provide a robust and rigorous educational experience in terms of content. The need for public literacy in the geosciences becomes more evident each year as citizens are personally impacted by related events or learn about them through the media. The NSF acknowledges the need for public literacy as "Americans learn more about threats to the Earth, such as the peril of global climate change and the increasing frequency of natural and manmade hazards" (NSF Advisory Committee for Geosciences, 2009). The loss of human life, property, and resources is becoming more catastrophic in part due to changing population demographics (i.e., where people work and choose to live):

> The scientific questions being explored by geoscientists address issues that are thus among the most important facing the nation. Our future sustainability and prosperity will require a new interdisciplinary geoscience workforce that reflects the nation's diversity and has the capacity to develop innovative solutions to meet the challenges that lie ahead. But, a capable scientific community is not sufficient to solve these problems alone—all citizens need to understand the science of Earth processes well enough that they can make evidence-based decisions in their personal lives and communities, in response to continued evolution of coupled human and natural Earth Systems. (NSF Advisory Committee for Geosciences, 2010)

The American Meteorological Society agrees:

> Earth observations enable a wide range of routine activities and alert us to dangers from severe weather and other natural hazards. Furthermore, observations provide a long-term record that enables us to assess climate variability and change, and a rigorous basis for making predictions through model development, testing, and validation. These predictions enable informed risk-management decisions, identify new opportunities for advancement, and help us better understand the workings of our Earth system. (American Meteorological Society, 2012)

Developing career technical programs based on these frameworks provides students with the opportunity to engage in authentic learning experiences and research through exploring and building skill sets that can then be applied to careers that represent the twenty-first-century workforce. The alternative is to develop robust and rigorous academic geosciences course opportunities in traditional settings.

BEST PRACTICE

With the rapid advancement and availability of real-time satellite data to the K–12 classroom, teachers and students can engage in doing real science, in real time. The next generation of geoscientists will be able to remotely sense the entire planet from their workspace. These opportunities present a workforce paradigm of conducting continuous investigations and providing information and data in the same manner that scientists do in the workplace. In addition to the technological advances, there have been changes in instructional strategies that include scientific inquiry, authentic science, problem-based learning, and hands-on investigations. In science education, this is often referred to as *doing science like scientists do.*

The development of a Geoscience and Remote Sensing Laboratory (GRSL) in the classroom provides the opportunity to monitor, investigate, and conduct authentic research from a local to a global level and to address the environmental issues that potentially impact our natural world and society. A GRSL is designed to study our planet in its true state, fluid and dynamic and as a sum of integrated systems—the Earth System. Viewing the planet in this state helps students understand that studies and investigations of this nature may consist of thousands of ever-changing variables and often consist of very complex interrelationships. This becomes evident to students as they monitor and observe the natural phenomena around them.

In the flexible environment of a GRSL, teams of student investigators conduct qualitative and quantitative studies. Students are required to take the lead in presenting data, imagery, and updates as events that pertain to their research become real-time, current environmental events. Students essentially take on the role of a Principal Investigator (PI). The geoscience and remote sensing laboratory creates an educational version of what NASA refers to as an Earth Information System (EIS).

Students are regularly required to present the "status of Planet Earth" through what we call EarthCast. The EarthCast encourages students who are engaged in authentic research to present their findings to the class. Students are required to design a short presentation of 3–5 minutes, modeled after a broadcast media weather forecast. This focuses the students on presentation skills and public speaking, building confidence in the process. The EarthCast incorporates a student-teaching-student model that has proven successful. In addition, following a scientific-research community model, classrooms are "briefed" on a wide-ranging array of topics, thus gaining a more comprehensive view of how the planet functions, while seeing how these parts of the Earth system are interrelated. The integration of a GRSL into an existing Earth Science program or a new Earth Systems Science course allows students to acquire the necessary rigorous laboratory skills required by a college or university admissions office, while developing proficiency in technological skills using government and industry standards and analysis tools.

It should be noted that the author developed the concept of a GRSL as part of the instructional strategies used in the geospatial technologies program at a career and technical high school. Career majors are a course of study within the CTE environment. Typical of most CTE schools, students spend one-half of the school day, 5 days a week, for all 4 years in a career major course of study. Therefore, students have adequate time and opportunity to engage in long-term authentic studies and projects. In the design and development of a CTE program, one must understand the entry-level position requirements of the career, which include both academic and skill-based proficiencies. These can be obtained through research, meeting directly with potential employers, government and/or industry leaders, or what is recommended in most CTE communities: the development of an advisory board. However, it is also important to note that these strategies can be incorporated into traditional physics, chemistry, and biology courses and are not restricted to CTE programs.

The concept of developing a Geoscience Remote Sensing Laboratory resulted from examining what role environmental studies, geoscience and remote sensing, and geospatial technologies play in a CTE school environment. The class determined that what these studies contribute to the community is the ability to observe the Earth through the use of satellites, remote sensing tools, and computer visualizations. Furthermore, it was concluded that the studies to be conducted were not only of importance to the class, but also to society. It was also determined that the class would be engaged in a profession that is ever evolving, and often at a rapid pace. The GRSL was designed to examine and identify core skills that would transition along with rapid advances in technology, changing software, and public demand.

CTE programs often come with significant budget demands, because the CTE environment should mirror the workplace environment as closely as possible. In other words, it is more than an assignment that has a beginning and an end point. Keeping pace with current technologies can be financially demanding. In the twenty-first century, it is not unusual for technology to change within the students' 4-year high school experience. While other programs in our school were teaching specific skills and how to apply those to specific situations, it was understood that the geosciences' future career paths would require much more.

The GRSL was designed with the intent of replication; therefore, the author was cognitive of costs and required potential funding—or, in most cases, the lack thereof. The examination of the parallel evolution of science and technology applications for the K–12 school community played a role in solving this problem. The development and transfer of these technologies (i.e., computer systems and the Internet) greatly enhance the capabilities for students to access and utilize real-time data and satellite imagery. The Geoscience and Remote Sensing Laboratory was built with resources that are readily available to any instructor via the Internet and/or with very modest financial investments.

The GRSL functions as the workplace functions. A variety of components requiring different skill sets and levels of expertise exist, but all contribute to an overall purpose or product. Investigating our planet from a global perspective presents almost unlimited opportunities for students to engage in an area of research that is of personal interest. Topics include, but are not limited to, conservation of natural resources, weather, severe weather, space weather, climate, physical oceanography, wetlands, estuaries, marine mammals, forests, land cover, and wildlife migrations. The GRSL allows students to work independently or in collaborative groups. Dividing students' research into various disciplinary topics allows for the groups to gain a fundamental understanding of what is happening around the planet and how these topics relate to one another. Investigating these topics on a continuous basis allows students to see the relationship between what they are studying and what is happening in the real world.

For example, the beginning of the school year coincides with hurricane season, winter storms, tornadoes, and severe weather, including flash flooding and lightning. Students should have the opportunity of investigating these *as they happen*. Google Earth revolutionized the ability of students to use computer visualizations and other datasets. Using the datasets available through Google Earth and other developers, students can monitor earthquakes *in real time*, track weather systems, monitor coral reefs, access air-quality data, sea surface temperatures, and much more. The good news is that the availability of these data sets continues to grow every year.

NASA J-Track 3D and Eyes on Earth track all unclassified satellites in real time. Computer visualizations show the location and footprint of Earth-related satellites. Students have access to direct read-out data from the LandSat satellite through the United States Geological Survey (USGS) EarthNow Program. Through programs such as Journey North, students combine local field observations related to butterfly, hummingbird, bird migrations, and gray whales using GIS to create national data bases, thus tracking the "journey north." Students are able to combine satellite-derived sea surface temperatures (SSTs) with NOAA National Buoy Data Center real-time buoy data. The SatCam app developed by the University of Wisconsin allows students to make cloud observations from the ground while a satellite overpass is occurring. Using the very same cloud observations, students can participate in NASA's "S'COOL Project," providing NASA scientists with local ground observations to calibrate images received from orbiting satellite sensors. It is truly participatory science.

ADOPTION OF BEST PRACTICE

Engaging students in the observations of the Earth as a system allows students to view current events from a scientific point of view. There are few evenings in which the nightly news does not cover a story related to the geosciences. The GRSL capitalizes on the concept of the teachable moment.

In 2012, Superstorm Sandy hit the east coast of the United States, resulting in dramatic impacts to many states and essentially the entire northeast. NOAA's National Hurricane Center and National Weather Service, along with local broadcast media, provided critical information, data, and updates that supported communication between scientists tracking and analyzing the storm, emergency managers, and the public. Thanks to these protocols and emergency management programs, there was minimal loss of life. GRSL students were able to follow the event in real time, utilizing the same imagery and data available to scientists.

One of the instructional strategies used in the Geoscience and Remote Sensing Laboratory is student-directed learning. Have you ever had a student ask, "Why do we have to learn this?" This is a valid question; unfortunately, it goes unanswered too many times. Students are seeking relevance in their learning that goes beyond a test, graduation, or college acceptance. As an instructor or teacher, if you cannot answer that question, then perhaps your lesson or project should be reworked. Simply put, if it is not of relevance to the student, then why is it included in the curriculum? Having students take responsibility for their education has a powerful motivational impact.

The GRSL utilizes another unique instructional strategy, Space to Earth: Earth to Space (SEES). SEES was developed not only to represent physical data layering, but also as an educational instructional strategy to encourage and develop geospatial thinking in students. Based on data associated with latitude and longitude (GPS), students now have the ability to examine geo-referenced data sets derived on Earth, as well as from satellite instruments in orbit, thus enabling investigators to look from the top down (Space to Earth) or from the bottom up (Earth to Space).

The SEES model creates a vertical integration of data collected directly from the field, through remote sensing technologies on the ground and/or space platforms. Students have the opportunity to make observations and measurements on Earth and to enhance their data sets through the use of space-borne imagery. They can begin to observe Earth from both outside and *inside* the classroom. Keeping SEES in mind, students line up multiple data sets for the purpose of building a comprehensive model. In addition, using SEES (i.e., multiple geo-referenced data sets) promotes examining the Earth as a system. The advancement and availability of GIS technology and data provides an important component of this new instructional strategy in the K–12 community.

What about the ground component of student studies? The GLOBE Program was established in 1994. Successful programs like GLOBE demonstrate that "students learn science better when they are involved with the collection and manipulation of data" (NSF Advisory Committee for Geosciences, 2010). The GLOBE Program is supported by NASA, NOAA, the National Science Foundation, and the Department of State. GLOBE has grown into an international program being conducted in 112 countries with over 54,000 teachers trained in 24,000 schools and over 23 million student observations submitted. Working within the GLOBE community allows teachers and students to add an international component to their studies. The introduction of *scientific protocols* gives students insight into the nature of science. Following established scientific protocols, students set up study sites at their schools and collaborate on international studies. GLOBE's strategic priorities are to

> improve student understanding of environmental and Earth system science across the curriculum; contribute to scientific understanding of Earth as a system; build and sustain a global community of students, teachers, scientists, and citizens; and engage the next generation of scientists and global citizens in activities to benefit the environment. (The GLOBE Program, 2013)

Imagery, photographs, datasets, and computer visualizations of the Earth from space are all available to teachers and students on the Internet from a host of international space agencies and universities from around the world, much of it in real time or near-real time. Therefore, teachers and students have the ability to monitor the planet Earth from the perspective of space. This provides opportunities for authentic research, project-based learning, development and application of critical thinking skills, and it builds problem-solving skill sets.

CONCLUSION

In the late 1980s, a teacher from Pennsylvania and a group of students conducted a project to build a polar orbiting satellite receiver gaining access to direct readout data in the classroom. This had never been done, and it prompted the notion that satellite imagery and data can be used in the classroom. This, in turn, led to a Satellite and Education Conference in 1987. The Satellite Educators Association (SEA) was born. International networks of scientists, engineers, university faculty, pre-college faculty, and students collaborate with industry and government in the exchange of ideas, resources, and opportunities. The annual conference is held each year at California State University-LA with support from NASA (Jet Propulsion Laboratory), NOAA, and others.

As part of the Multinational Youth Studying Practical Applications of Climatic Events (M.Y.S.P.A.C.E.) team, pre-college teachers and students continue to gather to learn new skills, participate in a mini-research project, and develop goals for continued research and studies throughout the year. The SEA conference concludes each year with the students' presentation to the conference attendees, thus giving students firsthand experience in presenting findings to an audience of professionals. For the past several years, student delegations from China and England have participated, allowing the program to become a truly international study.

As part of these student-led geoscience research studies, high school teachers and students around the world are building and launching CubeSats. In 2013, the first CubeSat built entirely by high school students was launched into space marking yet another milestone in aerospace history. CubeSats are a type of nanosatellites for space research that usually have a volume of exactly one liter (10 cm cube) and a mass of no more than 1.33 kg. The Build, Launch, Utilize, and Educate sing CubeSats (BLUECUBE) Project, developed by partners that include Einstein Fellows including myself, is an education and workforce development effort utilizing real-world CubeSat missions to engage the pre-college community, as well as community college and university students creating a remote sensing pipeline with a foundation in engineering. BLUECUBE promotes increased pre-college student interest in STEM careers via direct student involvement in every phase of a CubeSat space mission: Mission Management and Operations, Spacecraft Construction and Testing, and Spaceflight Data Utilization. In addition, BLUECUBE seeks to create an organization that will hold/conduct/manage local, regional, national competitions around the concept of engaging in the production of a flight-ready spacecraft, with the ultimate goal of achieving the launch of said spacecraft, and the follow-up utilization of data produced by the payload that can be incorporated into a common body of study. BLUECUBE is a clearinghouse for funding opportunities, resources, and supplies, and would manage the creation and operation of nodes where collaborative pre- and post-flight activities can be developed and carried out. BLUECUBE creates pathways for teacher professional development with the goal of developing master STEM teacher leadership in the United States, which translates into authentic STEM learning experiences for students in the classroom. Students are directly involved in the acquisition of remotely sensed imagery and data.

After reading all this, only one question remains: "How to get started?" Resources are available through federal agencies, universities, and industry that provide sources of satellite imagery, real-time data, and computer visualizations. The author highly recommends the following resources for professional development in both the educational strategies introduced in this chapter and content knowledge:

- *The Earth System Science Education Alliance* (ESSEA) is a NASA, NSF, and NOAA-supported program implemented by the Institute for Global Environmental Strategies (IGES) to improve the quality of geoscience instruction for pre-service and in-service K–12 teachers. Begun in 2000, the program currently enjoys the participation of 40+ institutions, with over 3,000 teachers having completed an ESSEA course as of Fall 2009.
- *The American Meteorological Society* (AMS) Education Program promotes the teaching of atmospheric, oceanographic, and hydrologic sciences through pre-college teacher training and instructional resource material development. It also promotes instructional innovation at the introductory college course level, hence the K–13 designation for the program. All programs promote activity directed toward greater human resource diversity in the sciences AMS represents.
- *The Global Learning and Observations to Benefit the Environment* (GLOBE) program's vision is a worldwide community of students, teachers, scientists, and citizens working together to better understand, sustain, and improve Earth's environment at local, regional, and global scales. GLOBE's mission is to promote the teaching and learning of science, to enhance environmental literacy and stewardship, and to promote scientific discovery.

As presented in this chapter, the application of satellite imagery, remote sensing, and computer visualizations has a well-documented history and an established robust set of resources ready for immediate implementation. The future is a place where visionaries travel, and innovation is the United States' most valuable resource. Our nation requires that the next generation of Americans develop those skill sets, too. The Geoscience and Remote Sensing Laboratory presents opportunities for teachers and students to *visualize their future*. The advantage of using remotely sensed data and imagery is that much can be learned about geographic locations, many of which cannot be easily accessed. The entire surface of the Earth can be viewed, analyzed, and interpreted. Scientists, educators, and students have the capability of looking at the planet as a total global system—a sum of integrated systems, physically interacting in a constantly changing fluid and dynamic state.

The identification of geoscience data resources becomes the first challenge in student investigations or research questions. Therefore, research, problem-solving critical thinking, and innovative thinking skills are developed and utilized by students immediately. Working with students on developing innovative practices can often be frustrating to both the students and the teacher. However, the potential for success makes the work worthwhile. There are few topics that truly represent the integration of science, technology, engineering, and mathematics and directly impact life, property, economy, and national security. The applications of satellite imagery, remote sensing, and computer visualizations meet those requirements.

WORKS CITED

American Meteorological Society. (2012). *Earth observations, science, and services for the 21st Century: American Meteorological Society policy workshop report*. Washington, DC: American Meteorological Society.

ESRI. (2013). *What is GIS?* Retrieved October 20, 2013, from http://www.esri.com/what-is-gis/overview

Fish, D. (1875). *The complete arithmetic: Oral and written*. New York: Ivison, Blakeman, Taylor & Co.

The GLOBE Program. (2013, May 16). *Vision statement & mission*. Retrieved October 20, 2013, from http://www.globe.gov/about-globe/vision-statement-mission

NASA's Earth Observing System. (2013, March 19). *NASA's Earth Observing System project science office*. Retrieved October 20, 2013, from http://eospso.gsfc.nasa.gov/content/nasas-earth-observing-system-project-science-office

National Research Council. (2003). *Satellite observations of the Earth's environment: accelerating the transition of research to operations*. Washington, DC: National Academies Press.

NSF Advisory Committee for Geosciences. (2009). *GeoVision report*. Arlington, VA: National Science Foundation.

NSF Advisory Committee for Geosciences. (2010). *Strategic frameworks for education & diversity, facilities, international activities, and date & informatics in the geosciences*. Arlington, VA: National Science Foundation.

Office of Science and Technology Policy Executive Office of the President. (2012, March 29). *Obama administration unveils "Big Data" initiative: Announces $200 million in new R&D investments*. Retrieved October 20, 2013, from http://www.whitehouse.gov/sites/default/files/microsites/ostp/big_data_press_release.pdf

CHAPTER TWENTY-FOUR

Integrating Informal STEM Learning INTO Your Curriculum

REMY DOU AND TERRIE RUST

INTRODUCTION

Experts across the country tout science, technology, engineering, and math (STEM) education's preeminence when it comes to securing our nation's economic competitiveness and preparing our children for the future (National Research Council, 2010). Research has provided a plethora of in-school, effective strategies that facilitate STEM learning, but a critical component still deserves more attention. On average, Americans spend only 5% of their lives in school and only a small part of that on STEM learning. Yet learning never stops, and effective STEM education can occur during the remaining 95% of people's lives (Falk & Dierking, 2010). The majority of life occurs outside of school.

STEM learning can occur in a variety of out-of-school settings. It might occur while a family discusses food production pathways after being prompted by a child at the dinner table. A young girl might flip through TV channels and watch the latest episode of *SciGirls*, an award-winning, STEM-based children's program, or a young man might play an online game designed to engage players in the scientific process. A group of retirees might visit the exhibits at a nearby national park. A teacher might take his students on a field trip to interact with a math-based museum exhibit.

Learning experiences, like field trips and the other examples mentioned, are often referred to as informal learning. In the case of informal STEM learning (ISL), the potential for impact includes cognitive and affective repercussions.

These venues have unique ways of harnessing children's interest in science, technology, engineering, and mathematics. Informal learning environments can provide the joy of discovery—the making, the doing, the exploring. A COSMOS Corporation survey indicated that many scientists' most memorable ISL experiences were those that took place in aquariums, zoos, or museums (COSMOS Corporation, 1998).

The people, projects, and activities encountered in ISL environments are often tailored to complement the formal education sector—the classroom. A successful ISL outing can be engaging and a valuable learning experience, providing exactly the kind of exposure that can motivate students in the classroom to pursue STEM-related careers. Because of their rich variety, opportunities exist for nearly all educators to connect with informal educators or informal learning venues that enrich classroom lessons. The idea that field trips are the only way students can have these kinds of opportunities is antiquated. Field trips are just one of many ways in which formal and informal learning can overlap.

BACKGROUND

ISL often means different things to different people. A simple look at the variety of terms used to identify ISL makes that clear (e.g., out-of-school learning, informal education, free-choice learning, and public understanding of science). Its stakeholders include producers of STEM media (e.g., IMAX documentaries and children's television programming), education researchers, park rangers, museum specialists, afterschool program supporters, youth networks (e.g., Boy Scouts of America, Girl Scouts of the USA, and 4-H), citizen scientists, and many more. The diversity of these people and organizations makes agreeing on a universal definition for ISL quite difficult. The environments in which ISL takes place vary as well and can be as diverse as national parks, museums, family living rooms, zoos, virtual spaces, and so on. But, in general, and in a very broad way, ISL and ISL environments include places and activities that promote "learner choice, low consequence assessment, and structures that build on the learners' motivations, cultures, and competence ... informal environments provide a safe, nonthreatening, open-ended environment for engaging with science" (National Research Council, 2009, p. 47).

Because of its fragmented nature, ISL as a field did not coalesce until recently. Even in its current state, where collaborations and research have validated its impact, more can be done to further unite ISL's different stakeholders. These stakeholders often have different goals in mind. Some focus on producing children's media, others on reaching out to retirees, and still others focus on young families. Because of its breadth, ISL's history in the United States does not follow a clear trajectory, but its significance has strengthened over the years (see Appendix A, Table 24.1).

ISL venues are popular and have an immense impact on society at large. In 2011, an estimated 65.4 million visits were made to the Association of Science-Technology Centers' (ASTC) 365 science centers and museums in the United States (Association of Science-Technology Centers, 2011). Experiences in ISL venues have been shown to contribute to adult learning (Center for the Advancement of Informal Science Education, 2012). The National Science Teachers Association (NSTA) advocates the importance of ISL as part of its official position statements. The association advocates for the important role ISL institutions play in pre-K–12 education, teacher training, its impact on underrepresented minority populations, and creating a more STEM-literate citizenry. NSTA continues to tout further development of formal and informal collaborations (NSTA, 2012).

Informal learning experiences play an important role during American children's summer vacation time. The much-touted "summer learning loss," which often occurs between school-year terms, has been shown to recede when children are involved in out-of-school learning activities. Unfortunately, this has also been shown to disproportionately disadvantage students from families of low socioeconomic status, who cannot afford or do not have the support to involve their children in out-of-school learning experiences during the summer months (Alexander, Entwisle, & Olson, 2007). Integration of ISL into in-school time can help inspire and motivate all students, giving them access to the same resources as those visiting venues over the summer.

For educators, the benefits of ISL lie in the blurry line between formal and informal education. Yet pressure to focus on curriculum assessments places a burden on educators to convince administrators of the validity of informal outings (Kisiel, 2005; Kisiel, 2011). Fortunately, field trips and other informal collaborations align to academic standards more deeply than one might think. Museums and other providers of ISL activities understand these academic pressures and often build the standards into their curricula. In fact, ISL is often aligned to meet the scientific practice and concept knowledge requirements that are part of standards such as the *Next Generation Science Standards* (National Research Council, 2012) and the *Standards for Technological Literacy* (International Technology and Education Association, 2007). ISL research supports STEM-learner outcomes (see Appendix B).

Convincing school administrators of the importance of informal STEM learning experiences outside your school may throw up a small roadblock in the pursuit of student engagement. However, this can be further facilitated by locating ISL venues in your community or ISL resources that can be availed of within your classroom.

BEST PRACTICE

How does one begin to develop best practices in an area like informal STEM learning? For us, the authors of this chapter, connections between formal and informal learning began to form during elementary school. These connections enabled us to develop a love of learning, which continues to this day.

Terrie's introduction to informal education experiences came through her Brownie troop field trips to factories near where she grew up: the San Francisco Bay area. Discovering the marvel of how bread, candy, potato chips, cookies, and peanut butter were made provided lasting memories that impacted her far into adulthood. In addition to those experiences, throughout her K–12 education, her teachers scheduled field trips to the zoo, an observatory, museums, the theatre, an aquarium, and cultural festivals. During those pre-computer years, the only option for learning outside the classroom was to visit venues in person.

Remy's early experiences with informal STEM learning came primarily through watching televised programs on Discovery and Animal Planet, and Public Broadcasting Service (PBS) programs such as NOVA. (Many forms of television and radio media, like these, are funded in part through NSF grants.) These programs not only sparked Remy's interest in science, but also provided an impetus to begin reading STEM-themed fiction and non-fiction.

These formative experiences followed both of us into adulthood as we pursued careers in STEM education. The reminder of the influence these experiences had on our early learning led to our desire to include informal STEM learning in our curriculum.

Taking Students into the Field

In 2009, Remy received a grant from Toyota Tapestry to implement an aeroponics curriculum. His high school students began the process of developing an aeroponics lab where plants would grow without the use of soil (a fine, nutrient-rich mist is sprayed onto plant roots). This project-based botany course focused on a variety of botanical concepts, including the study of nature's aeroponically grown flora, epiphytes. Remy recognized that Florida's tropical climate perfectly nurtured epiphytic growth, so during the project's development he set out to find linkages to informal education venues that would enrich classroom lessons. Two in particular stood out because of their proximity and pertinence: Fairchild Tropical Botanic Garden and Everglades National Park.

As a science teacher residing in Florida, Remy tapped into the rich natural resources in his state for many opportunities to engage his students in ISL:

- Fairchild Tropical Botanic Garden provided a place for students to interact with experts, including some who worked on large-scale aeroponics systems.
- Everglades National Park presented his classes with nature's personal lab station. Students waded through wetland, walking into beautiful cypress domes under which epiphytes flourished. There, through guided questioning, Remy led students to consider what made these plants and environments ideal for soil-less growth.
- Because of his familiarity with the region, he connected with the owner and manager of the Everglades Outpost, a wildlife rescue center, where Remy's AP biology students learned firsthand about animal behavior.
- The students in his chemistry class visited Biscayne Bay, where they took salinity measurements from within the bay and on the edges of the bay facing the Florida Straits, providing hands-on data collection similar to what scientists would do.

Terrie's teaching focus naturally lent itself to opportunities for expanding student learning outside the classroom—first as a home economics and career education teacher to junior high and high school students in California and later, in Arizona, as a STEM teacher, providing technology education, engineering education, computer skills training, and career education to students in grades 5–8.

- Terrie took her high school career education students to a county career center to observe how career planning, training, and job placement occurred.
- Her high school home economics students spent a couple of hours at a local Kmart store purchasing all of the required equipment for their new home economics classroom kitchens. Many of the students commented on their newfound awareness of how much it costs to stock a kitchen with basic preparation tools. Since there was a budget to follow, her students also had to work out spending limits for the items they were assigned to collect—a valuable lesson, since they would be thinking about moving into their own apartments in a few years.
- One year, Terrie's eighth-grade classes (approximately 120 students) jointly toured the Palo Verde Nuclear Power Plant (outside Phoenix) and learned about alternative energy, radioactivity, and power consumption. The tour host shared the extreme safety measures that were taken to protect the public from leaks, airplanes crashing into cooling towers, and other hazards related to nuclear energy and nuclear power plants. Students came away with an appreciation for nuclear energy's benefits to their state and community.
- Terrie arranged for her female students to attend a variety of events designed to introduce girls to consider careers underserved by women and those requiring strong STEM skills: Sally Ride Science Festivals, DeVry

University's *her world* events, Challenger Space Center's Wings to Fly workshop, a tour of Honeywell Avionics, and a day-long workshop at Universal Technical Institute. The girls discovered new skills and career choices they were not aware existed.
- Terrie took a team of students creating a website called "The History of Technology in Peoria, Arizona" to the Peoria Historical Society Museum to meet with the curator and to collect research on their topic. The completed website was hosted by the City of Peoria for 7 years on their city website. In addition, the young webmasters received certificates of appreciation from one of the city's council members.
- Terrie advised a team of students who participated in the "Future City Competition," working with them after school during the months leading up to the city-wide competition. An engineer-mentor assigned to the team worked with the students on several occasions. At the competition, the team was awarded the Rich Goewy Community Awareness Award. This award recognized the student-designed city the judges felt they'd most like to live in.

Bringing the Experts to the Classroom

Although by definition ISL typically takes place "outside school walls," many ISL activities now play a major role on school campuses during school hours.

When field trips were not possible, either logistically or because of lack of funds, Terrie brought the ISL experiences to her students. A representative from Salt River Project (a Phoenix power company) shared with Terrie's students the technological history and design significance of the canal system created by Phoenix's earliest residents, the Hohokam Indian tribe. These canal designs are still in use today. Terrie introduced a lesson on solar energy with the help of a representative from Arizona Public Service (another Phoenix power company), who drove a solar car (hot rod version!) right up to the door of the classroom for students to sit in and examine while he explained how solar energy powered the vehicle, conserved fossil fuels, and reduced pollution.

When Remy's students could not visit an ISL environment, he brought the environment to the classroom. Experts from nature venues visited the class to share information with students. For his life science courses, a wildlife expert helped students explore animal adaptations. For chemistry students, a science demonstration guru returned year after year to dazzle students with liquid nitrogen experiments, getting them to think about the science behind his demonstrations.

Whatever the activity, commonalities ran through all of them. First, these activities enriched existing lessons and fit within the curricular guidelines of both Terrie's and Remy's courses. Second, these activities did not stand alone. Classroom lessons before each activity supported the event, and post-activity assignments built on what students learned. Third, they were feasible in terms of resources, some of which the schools provided, but some of which Terrie and Remy secured through grants, donations, and other forms of funding. Several funding suggestions are provided in Appendix D.

Even though many students are fortunate enough to experience ISL opportunities with their families, the fact remains that a large number would never have had those opportunities if it were not for the field trips, after-school programs, and specialized events offered through their schools. Additionally, ISL tools—including films, educational websites, PBS videos, and interactive programs—originally designed for use outside the classroom help engage students in a deeper learning of their subject matter within the classroom. The ease in accessing ISL resources such as these means that any teacher can expand student engagement and strengthen the curriculum, while students reap the benefits of classroom opportunities to interact with ISL resources they might not experience otherwise.

ADAPTING THE BEST PRACTICE

The opportunities that Remy and Terrie provided their students through informal learning were those that allowed students to build upon the foundations of classroom learning. There are many ways in which teachers can provide informal STEM learning to their students.

Field Trips

Museums

Most teachers are likely to be familiar with the primary types of museums located within their community. Some museums typically found in many mid-sized to large cities include science centers, art museums, historical museums, natural history museums, and children's museums. However, many others exist as well. Teachers would not want to overlook some of these specialized types: firefighting, computer, gem and mineral, railroad, insectariums, archaeology, aerospace, agriculture, living history, medical, industrial, arboretums, and nature centers. Teachers may discover that many of these STEM-themed museums exist in their communities. Check with your local tourism office for listings.

Museums *love* students! They are pretty fond of teachers, too, as evidenced by the number of museums that provide teachers with ready-made lesson plans, worksheets, and pre-trip and post-trip activities to make trip preparation easier. Many museums have formal programs for student groups and offer discounted rates or free entry. Watch for special exhibitions or films at your local museums, too, which may fit specifically with a unit within your curriculum.

Museums are popular. Yet when Terrie once surveyed her junior high students, she was saddened to hear many acknowledge that they had never visited a museum of any kind. Remy also recalled having rarely visited museums growing up. Clearly, providing this type of ISL to students is valuable in helping them connect real-world experiences with the classroom and expand their view of STEM.

Specific science venues

Observatories, zoos, wildlife parks, marine sanctuaries, nature centers, and aquariums are specific venues that focus on science topics. These venues allow students to see nature in its element. Unfortunately, they are not available in all cities; when available, though, they provide unique experiences for students to connect with science. Often overlooked are venues located at local colleges and universities. Visit the websites of colleges or universities near you to see if they offer specific science venues. Contact them directly to inquire about bringing students to tour their facility.

Farms and fisheries

Working farms, dairies, and fisheries make for wonderful educational field trips. Students of all ages enjoy learning how our food gets from the farm to our dining tables. Many of these locales allow students to participate in harvesting crops, milking cows, or catching fish; these are engaging hands-on experiences that bring science and biology to life. Some farms may be able to share how biotechnology plays a role in their food propagation and production, which is especially interesting for high school agriculture or environmental science students. Many farms, large and small, cater to school groups or public tours.

For teachers who live in major cities, these types of trips may seem out of reach. However, because of the increasing popularity of community gardens, students can still get their hands dirty and learn about the growing process. Gardens may even be located on top of a high-rise building rather than on the ground! Greenhouses may also exist in urban areas, providing another opportunity for this type of student engagement. Look for help in locating these opportunities by connecting with members of local horticultural societies.

Commercial or private fishing vessels located on seacoasts or waterways may be willing to host student tours. Fisheries, often called fish hatcheries, are areas whose aquatic populations are harvested for commercial purposes and are located both on-shore and off-shore. On-shore, contained fisheries (called fish farms), are primarily used for aquaculture, a technological process involving cultivating freshwater and saltwater populations under controlled conditions (*Aquaculture*, 2013).

Public tours are usually available. Some organizations will even let students fish and, for a small fee, will allow students to take their catches home with them.

National, state, and local parks

Almost 280 million people visit the nearly 400 national parks in the United States each year (*Welcome to visitor use statistics*, 2012). Two-and-a-half times as many visitors to national parks visit the more than 6,600 state parks each year (Walls, 2009). Add to those impressive numbers individuals who frequent their community's local parks, and you have a staggering number of outdoor enthusiasts. Visitors to parks often do more than enjoy the scenery, picnic areas, and hiking and biking trails: they observe and study the flora and fauna, learn about the park's geologic origins, and study the geography or archaeology of the area. Parks are popular sites for school group visits. National and state parks have educational programs provided by park rangers. Many local parks are within walking distance of schools and provide easy opportunities for students to get out of the classroom and take a closer look at the plants, insects, and "critters" that reside in them.

A field trip to a national or state park may be beyond the range of a day trip but can make for a fun multiple-day trip. Some schools offer "outdoor education" for their students (frequently reserved for fifth or sixth graders) and plan annual trips to national parks for these outings.

National research centers

The federal government funds a variety of research centers including the National Laboratories, Long Term Ecological Research centers, and local National Oceanic and Atmospheric Administration (NOAA) facilities. Many of these include educational outreach components, and some facilitate visits or student internships.

Factories

Factory tours provide unique learning opportunities. Students not only discover the processes necessary to move products from their raw state to a consumer-ready state, but they also get to see innovation in action. The machines themselves are a marvel to behold and serve to get students to think about technological and engineering design processes. Factories also provide students with a chance to learn about the different types of work responsibilities and jobs required for production and distribution of products. Another benefit for students is that they are often rewarded with free samples, depending upon the type of factory visited. (Terrie's childhood factory visits all resulted in the receipt of yummy samples!) Factories in students' communities provide one additional benefit: they allow students to see and learn about the types of manufacturing and technical jobs available to them right where they live.

Although some factories remain closed to outside groups for liability reasons, teachers should be aware that those that allow visits may require signed permission forms from students' parents or guardians.

Theatres

Theatre visits are another option that teachers can utilize for extending student learning. Terrie took her students to view IMAX films to further their exposure to technology and engineering. Einstein Fellow colleagues of Remy and Terrie have also used performing arts trips to introduce students to plays and dance programs with a STEM topic as the theme. Field trips to theatres can be an interdisciplinary activity, too, combining English language arts (ELA), history, or arts study with STEM course content when applicable.

After-School Programs

There are a number of opportunities for engaging students in after-school programs. Programs often take place at school sites, making them easy for students to attend. Some examples of after-school programs hosted at schools are robotics clubs, computer clubs, science clubs, engineering clubs, and environmental clubs. These usually have a teacher as an advisor, but some are sponsored by outside groups. After-school programs are often offered in conjunction with competitions like the Future City Competition that Terrie's students participated in. A similar and popular program is FIRST Robotics. After-school STEM-themed programs may also be offered at community recreation centers, Boys and Girls Clubs, through the Girl Scouts, in conjunction with university or college STEM programs, or at science centers or museums. Even if some of these programs do not involve the teacher, they are a great way to encourage students to participate informally in STEM activities.

ISL Resources Accessible for Classrooms

There are a variety of reasons teachers might not be able to take their students out of the school environment. Transportation issues, lack of funds to cover the costs, the unavailability of substitute teachers to cover a classroom, the case of an appropriate venue being too far from the school, administrative restrictions, or other excuses can put a damper on even the most "gung-ho" teachers' desire to provide some ISL experiences for their students. Fortunately, many engaging ISL resources can be adapted for use inside the classroom. As Terrie and Remy discovered, identifying the available speakers within one's community who can bring their organization's ISL activities to the school is one step in the right direction. Spending a brief time online to locate available teaching resources from some of the more well-known producers of STEM educational materials can also generate amazing results. A lot of material is available free or at a small cost. Intel, PBS, Discovery, NBC Learn, National Geographic Society, the Lemelson Center at the Smithsonian Museum of American History, the National Aeronautics and Space Administration (NASA), and the Library of Congress provide excellent resources. These resources range from websites, videos, online interactive digital learning experiences, television programs, simulations, science artifacts (such as moon rocks), digital artifacts, games, quizzes, walkthroughs, and much more.

Virtual Field Trips

One ISL resource that teachers should consider is virtual field trips. There are hundreds of virtual field trips students can take on the web. Some are simple; others are larger productions that include a variety of trips and also provide teaching resources. Virtual field trips afford opportunities for bringing the world into the classroom.

Details about these resources and others familiar to the authors can be found in Appendix C.

CONCLUSION

We, the authors of this chapter, represent just two of many educators who employ the principles in this chapter every school year. In fact, although the ISL activities described and listed here serve as replicable models, a plethora of other possibilities has been omitted. Attempting to cover everything would be overwhelming. Instead, the goal has been to communicate the research-based principles that support ISL implementation in the classroom and to present some inspirational examples.

The benefits of school-related informal STEM learning experiences extend beyond the classroom. These outings have the potential to inspire students to pursue deeper learning in the STEM fields. With the variety of outlets available, both in-person and virtual, there is something for every student, every teacher, and every school. Many individuals who pursue teaching as a career do so because of their passion to inspire and motivate students. That is exactly what ISL experiences can do. Linking these to classroom learning only strengthens the impact teachers already make.

APPENDIX A

Table 24.1. Major Events in Informal Science Learning.

	Major Events in Informal Science Learning
1957	National Science Foundation (NSF) conducts first studies of public knowledge of science.
1958	NSF creates program on "Public Understanding of Science."
1961	American Association for the Advancement of Science (AAAS) begins newsletter on "Understanding," linking science journalists, Hollywood film and television producers, mass communication researchers, adult educators, and museum staff.
1962	Founding of the Pacific Science Center in Seattle.
1969	Founding of the Exploratorium in San Francisco.
1983	NSF recreates Public Understanding of Science Program as Informal Science Education.
1988	Founding of the Visitor Studies Association.
1989	A grant awarded to the Association for Science-Technology Centers by the Institute for Museum and Library Services results in a series of articles called "What Research Says About Learning Science in Museums."
1991	An *International Journal of Science Education* special issue on informal science learning is published.
1992	The journal *Public Understanding of Science* is established.
1996	Funded by a consortium of federal agencies, the first major informal learning research grant is awarded to the Museum Learning Collaborative.
1997	A *Science Education* special issue on informal science learning is published.
1998	NSF-funded conference results in publication of *Free-Choice Science Education: How We Learn Science Outside of School* (Falk, 2001).
2001	Founding of the Center for Informal Learning and Schools.
2002	Free-choice/informal learning is added as a strand of graduate study in science and mathematics education in the College of Science at Oregon State University.
2004	Founding of the Learning in Informal and Formal Environments Center.
2005	Informal Science (http://www.informalscience.org) is launched to share evaluation and research on informal science learning environments.
2007	NSF funds the Center for Advancing Informal Science Education.
2008	NSF publishes *Framework for Evaluating Impacts of Informal Science Education Projects*.
2012	NSF recreates Informal Science Education as Advancing Informal STEM Learning.

(Source: Abbreviated and adapted from National Research Council, 2009, pp. 16–17.)

APPENDIX B

Examples of outcomes impacting the science learner include:

- Learners generate, understand, remember, and use concepts, explanations, arguments, models, and facts related to science.
- Learners manipulate, test, explore, predict, question, observe, and make sense of the natural and physical world.
- Learners experience excitement, interest, and motivation to learn about phenomena in the natural and physical world.
- Learners think about themselves as science learners and develop an identity as people who know about, use, and sometimes contribute to science (National Research Council, 2009).

Examples of outcomes impacting the technology education learner include:

- Learners understand how inventive and creative thinking helps people adapt to the natural world to meet human needs and wants.
- Learners identify the differences between the human-made and natural world, and the differences between science and technology.
- Learners gain a broader point of view of the importance of human creativity and innovation in the process of refining and improving existing technologies.
- Learners develop skills for solving complex technological problems experientially.
- Learners explore the interconnectedness of technology and other fields of study and how technological progress promotes the advancement of science and mathematics (ITEA, 2007).

APPENDIX C

A Sampling of ISL Resources

If you are new to ISL, this is an excellent podcast to get you started.

- Learning Science in Informal Environments http://media.nap.edu/podcasts/nax95informalsc.mp3

    This podcast looks at what, how, and where informal science learning takes place and the benefits it can provide to all people. The material presented is applicable to informal technology, engineering, and math learning as well. *A must-listen* for all teachers considering inclusion of ISL with their students. (11 minutes)

Competitions

- National Engineers Week Future City Competition (http://futurecity.org/)

  The Future City Competition is a national, project-based learning experience where students in sixth, seventh, and eighth grades imagine, design, and build cities of the future.

- FIRST LEGO League and FIRST Robotics (http://www.firstinspires.org/)

  FIRST competitions are focused on building an interest in science and technology through robotics. Three levels of competition are offered based on age groupings. The Junior FIRST LEGO League (JrFLL) is for students in grades K–3. The FIRST LEGO League is for students in grades 4–8. The FIRST Tech Challenge is for students in grades 7–12. The FIRST Robotics Competition (FRC) targets students in grades 9–12.

- National STEM Video Game Challenge (http://stemchallenge.org/)

  This is a multi-year competition for middle school and high school students. The aim is to increase student interest in STEM subjects and careers.

- National Academy of Engineering (NAE) at The National Academies-sponsored NAE Grand Challenges for Engineering (http://www.engineeringchallenges.org/) and NAE Grand Challenge K12 Partners Program (http://www.grandchallengek12.org/)

  With input from people around the world, an international group of leading technological thinkers was asked to identify the Grand Challenges for Engineering in the 21st Century. The NAE Committee on Engineering's Grand Challenges has identified 14 areas awaiting engineering solutions in the 21st century.

To locate competitions in a variety of STEM areas, type "STEM student competitions" into your browser.

Citizen science projects

- American Association of Variable Star Observers (AAVSO) (http://www.citizensky.org)

  Citizen Sky is the hub of bright variable star activities at the AAVSO. Currently these activities include binocular observing and digital single-lens reflex (DSLR) camera observing.

- Citizenscience.gov (https://www.citizenscience.gov/)

  New (2016) site crafted with the help of The Wilson Center designed to provide crowdsourcing and citizen science resources, including a catalog of projects, a toolkit, and a community to share best practices.

- Community Collaborative Rain, Hail, and Snow Network (CoCoRaHS) (http://ww.cocorahs.org)

  CoCoRaHS is a grassroots effort to measure precipitation in the backyards of citizens from the Atlantic to the Pacific. It's easy to join, takes only 5 minutes a day, and is a fun way to learn about this wonderful natural resource that falls from the sky. Citizen observations will provide an ever-clearer picture of the amount of precipitation that falls in our nation's backyards.

- Florida Museum of Natural History's Project Butterfly Wings (http://www.flmnh.ufl.edu/educators/resources/project-butterfly-wings/)

  Project Butterfly Wings participants monitor butterflies in gardens and natural areas and conduct regular surveys of the species visiting them. They enter this information into an online database linked to an interactive website, where other participants, the public, and University of Florida scientists can access the information. This project is aimed at students in grades 4–8.

- Scistarter (http://scistarter.com/index.html)

  For all of your citizen science needs, this website allows you to search for specific citizen science projects.

- Zooniverse—Real Science Online (http://www.zooniverse.org)

  Over one dozen citizen science projects await visitors of Zooniverse. Participants can help scientists explore the surface of the moon, determine where stars come from, and discover how galaxies form by examining quality space images. A variety of non-space-science activities exist, including whale communication, climate, and anthropological research.

Television programs

- Public Broadcasting Service (PBS) (http://www.pbs.org/video/)

  The Public Broadcasting Service is a non-profit public broadcasting television network in the United States. Many programs can be accessed online and are also available for purchase. Hundreds of curriculum-compatible titles are offered.

  A variety of the most popular children's programs focusing on STEM topics are broadcast on PBS. Some familiar names are *SciGirls*, *Design*

Squad, Cyberchase, Peep in the Big Wide World, and Fetch with Ruff Ruffman. Airing times vary from state to state, but many videos and clips can be found for free online.

- NOVA (http://www.pbs.org/wgbh/nova/)

  Award-winning STEM-themed topics are the focus of these programs. Many can be accessed online and are also available for purchase.

Simulations

- Froguts (http://www.froguts.com/)

  Virtual frog dissection

- EdHeads (http://edheads.org/?)

  Edheads is an online educational resource that provides free science and math games and activities that promote critical thinking. Choose from Simple Machines, Virtual Knee Surgery, or Stem Cell Heart Repair, among others. All activities meet state and national standards.

- NASA Educational Space Walk Simulations (http://www.nasa.gov/audience/foreducators/spacesuits/simulation/)

  This shuttle mission involves multiple spacewalks to repair the Hubble Space Telescope.

- Math Simulations (http://www.techtrekers.com/sim.htm)

  This site provides links to a variety of math simulations, including The Stock Market Game and Virtual Fractions.

- PhET (http://phet.colorado.edu)

  PhET hosts a large number of high-quality simulations that allow students to explore everything from changes in energy to probability. This award-winning project has received funding from the National Science Foundation and covers many STEM subjects.

Virtual field trips

- 4-H Virtual Farm (http://www.sites.ext.vt.edu/virtualfarm/main.html)

  Students can visit a horse farm, a beef or dairy farm, a poultry farm, a fish farm, and a wheat farm.

- Museo Galileo Institute and the Museum of the History of Science (http://www.museogalileo.it/en/explore/virtualmuseum.html)

Videos and animations reconstruct the historical contexts and thematic references of the objects on display, allowing for an exploration of the collections by subject—a treasure-trove for science lovers.

- The JASON Project (http://www.jason.org/)

  JASON and its partners create connections between students and scientists and researchers using multiple platforms and technologies in real time and near-real time, virtually and physically, to provide mentored, authentic science experiences. These experiences include award-winning, standards-based STEM curricula developed with NOAA, NASA, National Geographic Society, and others; after-school and out-of-school activities; camp experiences; and exploration programs for museums, aquariums, libraries, and community centers.

Locate additional virtual field trip offerings by visiting:
http://alicechristie.org/search/trips.html
http://www.theteachersguide.com/virtualtours.html#Museums
http://www.educationworld.com/a_tech/tech/tech071.shtml
http://www.pitt.edu/~poole/VirtualFieldTrips.html

Videos

- NASA eClips (http://www.nasa.gov/audience/foreducators/nasaeclips/index.html)

  Short, educational video segments to help students make real-world connections.

Podcasts (in audio or video format)

- Bytesize Science (http://www.youtube.com/user/BytesizeScience)

  An assortment of quick podcasts addressing a variety of science topics can be accessed here.

- The Wild Classroom (http://www.thewildclassroom.com/)

  A selection of podcasts on topics related to nature and biodiversity is provided.

- Engineering the Tools for Exploration (http://media.nap.edu/podcasts/nax68exploratio.mp3)

  In the popular mind, scientists and engineers have distinct job descriptions, but, in truth, the distinction is blurry. In the century ahead, engineers will continue to be partners with scientists in the great quest for understanding many unanswered questions of nature and the solar system.

- Grand Challenges for Engineering (http://media.nap.edu/podcasts/nax-53grandchall.mp3)

    See the details for this challenge under Competitions.

Radio programs

- Science and the Sea (http://www.scienceandthesea.org/index.php?option=com_content&task=category&sectionid=1&id=2&Itemid=10)

    This site provides a variety of 2-minute programs with a marine science focus.

Websites

- Exploratorium (http://www.exploratorium.edu/explore/websites)

    Interact with a variety of excellent online exhibits covering many interesting STEM fields.

- Design Squad Nation (http://pbskids.org/designsquad/)

    This website has it all: competitions, online simulations, project ideas, videos, and more. Design Squad began as a PBS television program that introduced young people to engineering design.

- The Lemelson Center for the Study of Invention and Innovation (http://invention.si.edu/home/)

    Located in the Smithsonian National Museum of American History, the Lemelson Center is dedicated to exploring invention in history and encouraging inventive creativity in young people. The site provides engaging activities, videos and podcasts, and teacher resources. The center also provides free invention videos to teachers to show in their classrooms, for just the cost of shipping.

- The Discovery Channel (http://www.discovery.com/)

    Links to Discovery Channel programs including *Mythbusters*, *Dirty Jobs*, *Frozen Planet*, *Planet Earth*, and *Stormchasers*.

- Discovery Kids (http://.discoverykids.com/)

    The Tell Me page offers a plethora of information for site visitors on a host of topics including science, space, health, and machines, to name a few.

- How Stuff Works (http://www.howstuffworks.com/)

    How Stuff Works explains thousands of topics. The site uses various media in its effort to explain complex concepts, terminology, and mechanisms including photographs, diagrams, videos and animations, and articles.

Many of the popular television programs for youth (mentioned in this appendix) have created websites providing additional exposure to the topics covered on the shows. In addition, many of the programs that air on PBS channels have accompanying websites that are worth checking out.

After-school programs

- Afterschool: Middle School and Science, Technology, Engineering and Math (STEM) (https://www.metlife.com/assets/cao/foundation/MSBrief_2010-2.pdf)

    This MetLife Foundation Afterschool Alert Issue Brief provides examples of some exemplary STEM afterschool programs.

Miscellaneous resources

- National Science Digital Library (https://nsdl.oercommons.org/)

    The National Science Digital Library provides high-quality online educational resources for teaching and learning, with current emphasis on STEM disciplines—both formal and informal, institutional and individual—in local, state, national, and international educational settings. NSDL does not hold content directly. Instead, it provides structured descriptive information (metadata) about web-based educational resources held on other sites by their providers.

- NBC Learn (http://www.nbclearn.com/portal/site/learn/about)

    NBC Learn is the educational arm of NBC News dedicated to providing resources for students, teachers, and lifelong learners. NBC Learn K–12 offers unique collections of video resources, primary sources, historic footage, images, mini-documentaries, and text resources.

- Library of Congress (https://www.loc.gov/)

    The Library of Congress's expansive collections provide teachers with primary source documents to support their curriculum. Digital libraries afford easy accessibility to myriad curriculum-ready resources.

APPENDIX D

Funding Sources

Wells Fargo Teacher's Partner Program provides funds up to $500. Applications are evaluated based on overall quality of project, educational benefits to students, integration of proposed project with school curriculum, and number of students who will benefit. Wells Fargo Teacher's Partner grant applications are reviewed frequently, so you'll get a response (accepted or declined) within a short time. These grants are available in many states with Wells Fargo branches. For details, Google "Wells Fargo Teacher's Partner Program" to see if these grants are available for your locale.

Kids in Need Teacher Grants provide K–12 educators with funding to provide innovative learning opportunities for their students. Individual Teacher Grant awards range from $100 to $500 and are used to finance creative classroom projects. The number of grants awarded varies from year to year, depending on the amount of funds being requested. Typically, 300–600 grants are awarded annually. Applications are accepted during the back-to-school season. For an application, visit http://www.kinf.org/grants/.

Career and Technical Education (CTE) Grants may be available if your course falls under the CTE umbrella in your district. Contact your school district's CTE Director to learn if there are any CTE grant funds that might be obtained.

Professional Organizations you might belong to may have scholarship or grant programs providing funds. Check with them directly. Terrie received funding from the national professional organization of her teaching specialty as well as from the state teacher organization she was a member of.

Parent/Teacher/Student Organizations (PTSOs) at schools often allow teachers to "apply" for funds for student trips or classroom learning opportunities or materials. Your school secretary can provide contact information for the treasurer of the organization who can then offer details about such funding.

WORKS CITED

Alexander, K. L., Entwisle, D. R., & Olson, L. S. (2007). Lasting consequences of the summer learning gap. *American Sociological Review, 72*(2), 167–180.

Aquaculture. (2013, April 17). Wikipedia. Retrieved October 15, 2013, from http://en.wikipedia.org/wiki/Aquaculture

Association of Science-Technology Centers. (2011). *2011 science center and museum statistics*. Retrieved May 7, 2013, from http://www.astc.org/about/pdf/Backgrounders/2011%20Science%20Center%20Statistics.pdf

Center for the Advancement of Informal Science Education. (2012, April 24). *ISE experiences help adults make informed decisions about new or changing science*. Retrieved May 7, 2013, from http://iseevidencewiki.org/index.php/ISE_experiences_help_adults_make_informed_decisions_about_new_or_changing_science

COSMOS Corporation. (1998). *A report on the evaluation of the National Science Foundation's Informal Science Education Program*. Arlington, VA: National Science Foundation. Retrieved from http://www.nsf.gov/pubs/1998/nsf9865/nsf9865.htm

Falk, J. H. (2001). *Free-choice science education: How we learn science outside of school*. New York: Teachers College Press.

Falk, J. H., & Dierking, L. (2010). The 95 percent solution. *American Scientist, 98*, 486–493.

ITEA. (2007). *Standards for technological literacy: Content for the study of technology* (3rd ed.). Reston, VA: International Technology Education Association.

Kisiel, J. (2005). Understanding elementary teacher motivations for science fieldtrips. *Science Education, 89*(6), 936–955.

Kisiel, J. (2011). *Why so hard? Gaining insights from schoolteachers and informal science education staff regarding teacher use of ISE resources*. Paper presented at the meeting of the National Association for Research in Science Teaching, Orlando, FL.

National Research Council. (2009). *Learning science in informal environments: People, places, and pursuits*. Washington, DC: National Academies Press.

National Research Council. (2010). *Rising above the gathering storm, revisited: Rapidly approaching category 5*. Washington, DC: National Academies Press.

National Research Council. (2012). *A framework for K–12 science education: Practices, crosscutting concepts, and core ideas*. Washington, DC: National Academies Press.

National Science Teachers Association. (2012). *NSTA position statement: Learning science in informal environments*. Retrieved May 7, 2013, from http://www.nsta.org/about/positions/informal.aspx

Walls, M. (2009). *Parks and recreation in the United States: State park system*. Washington, DC: Resources for the Future. Retrieved from http://www.rff.org/RFF/Documents/RFF-BCK-ORRG_State%20Parks.pdf

Welcome to visitor use statistics. (2012). Retrieved from https://irma.nps.gov/Stats/

Index

1998 FS144, 125
21st century skills, 305, 335
    Partnership for 21st Century Skills (P21), 20, 147, 163
    Standards, 305
4-H Virtual Farm, 558
5e curriculum design, 498
Accelerated Reader, 379
Active Critical Learning Principle, 286
active learning, 284, 452, 506, 513
Adamson, K., 372
adaptivity, 482–483, 487–488, 492–493
Adélie penguin team, 170
    interactive (penguin science) website, 170
Advanced Placement (AP) Chemistry, 153–155
after school, 39, 52, 55, 57, 61, 67, 79, 128, 131, 143, 159, 196, 265, 436, 459–460, 548–549, 552, 559, 561
    Afterschool: Middle School and Science, Technology, Engineering and Math (STEM), 561
    club, 159
    programs, 39, 549, 552, 561
Afternoon Constellation (A-Train), 533
Aguirre, M., 104

Ahnemann, O., 475
Ainsworth, S., 473
Airzee, S., 139
Albrizio, L., 71
Alderman, M. K., 88
Alexander, K. L., 562
Allen, A., 139
Allen, D. E., 441
alternate, 296–297, 364
    assessment, 157–158
alternative writing formats, 393
Altieri, J. L., 417
American Association for the Advancement of Science (AAAS), 4, 20, 41, 141, 242, 355, 371, 419, 554
American Association of Physics Teachers (AAPT), 159, 161, 508, 526–527
American Association of School Administrators (AASA), 228, 246
American Association of University Women (AAUW), 68
American Association of Variable Star Observers (AAVSO), 556
American Astronomical Society (AAS), 129, 131, 140
American Chemical Society (ACS), 159

Chemagination Challenge (Contest), 150, 153–154
American Geophysical Union (AGU), 41
American Meteorological Society (AMS), 535, 541–542
American Physical Society (APS), 159
Andersen, N., 350
Anderson, C., 371
Anderson, R. D., 139
Andes, S. M., 42
Andree, A., 179
Ansong, D., 104
Antarctica, 118, 168, 170–171
Apogee Components, 158, 161
appetizer project, 151
Applebaum, S., 352
application(s), 1–2, 7–10, 32–35, 39, 65, 98, 108, 116, 150, 154, 156, 168, 181, 185, 196, 204, 213–214, 230, 252–253, 261, 271, 303, 332, 389, 395, 449, 457, 464, 467, 469–470, 508–509, 516, 529, 532, 534, 537, 539, 541–542, 562
appropriateness, 386, 391–392
architecture, 148, 445, 451, 456, 458, 461, 467–468, 495
Argabright, S., 221
Argonne National Laboratory Educational Programs, 163
argumentation, 110–111, 328, 358, 402, 414, 489
Arizona Public Service, 548
Arizona State University, 55, 302, 507
Arnold, E. M., 299
Arsem, E., 140
art brüt, 451
Aspen Media Literacy Leadership Institute, 326
assessment, 24, 30, 77, 82, 84–85, 88, 93–95, 98, 104, 134, 145, 153, 157–158, 184, 206, 242, 256, 261, 268, 271, 286, 292, 303, 305, 315, 321, 337, 388–390, 392, 402, 446–447, 449–450, 452, 454–455, 466, 469–472, 481, 498–499, 506, 508, 510–511, 519, 523, 544–545
    Assessment & Teaching of 21st Century Skills (ACTS21s), 335

    formative, 98, 261, 271, 303, 305, 315, 498, 510
Association for Supervision and Curriculum Development (ASCD), 41, 396, 472
Association of Science-Technology Centers (ASTC), 545
Atkinson, R., 42
at-risk students, 75–78, 81, 86–87, 93, 103
attention focusing, 480–481, 493
attribution, 482, 488, 492–493
Aufderheide, P., 350
authentic, 87, 107–114, 116–119, 121–123, 125–130, 132–133, 135–138, 173, 250–251, 254–256, 259–260, 262, 269–270, 303, 327, 354, 356, 376, 432, 435, 449–450, 454, 466, 471, 492, 500, 530–531, 535–536, 539, 541, 559
    Authentic Science Rating Instrument (ASRI), 133, 137
    learning, 118, 135–136, 259, 535
    science, 108–110, 112–114, 116–117, 119, 121–123, 125–130, 132–133, 135–138, 535, 559
avatar, 289–290, 294, 296

backward design, 445–446, 449, 471–472
Bacon, E. E., 474
Bainbridge, W. S., 298
Baker, D., 371
Baker, S., 139
Balfanz, R., 88
Bandura, A., 473, 500
Barab, S. A., 351
Barba, P., 475
Barmer, A., 473
Barnosky, A. D., 473
Barros, R., 246
Barth, P., 246
Barton, A. C., 473
Barton, P., 139
Bascom, D., 141
Bashan, N., 141
Basu, S. J., 473
Battelle, 109, 139
Bean, T. W., 417
Beane, J. A., 441

Bearden, K. K., 200
Beckingham, B., 500
Beddington, J., 247
Beede, D. J., 71
Beier, M. E., 475
Bell, P., 473
Bell, S., 139
Bellogín, A., 161
Benchmarks for Science Literacy, 371, 400, 419
Bencze, J., 139
Bencze, L., 350
Benyus, J., 246
Berg, A. C., 442
Berger, I., 350
Berglund, K., 42
Berkowicz, J., 441
Berland, L., 373, 442
Bernstein, A., 473
Berrow, A., 139
Bhattacharyya, G., 298
Bianchi, S., 246
big ideas, 6, 449, 453, 534
Billig, S. H., 221–222
biodiversity, 213, 234, 239–241, 243, 245, 251, 258–259, 445, 452–456, 460–461, 465–466, 468, 470–471, 559
biodynamic farming/growing, 213
bioethics, 287
biology, 1, 11–14, 101–103, 110, 131, 172–173, 181–183, 185, 188–190, 194, 196–197, 225, 231–233, 236, 258, 270, 287–288, 294, 335, 337, 378, 383–384, 431–433, 454, 507, 522–523, 530, 536, 547, 550
biomimicry, 183, 230
Bjork, R., 474
Blackawton, P., 139
Blair, C., 139
Blank, M. J., 442
Blumenfeld, P., 473
Blust, R., 221
board meetings, 66, 316–317
    meeting in the round, 317
Board on Science Education (BSE), 42–43, 140, 195, 200
Bögeholz, S., 351
Bohannon, J., 298

Bohr, Niels, 16–17
Bonney, R., 141, 277
Borko, H., 474
botany, 289, 421–422, 429–430, 440, 546
Botov, I., 501
Bouillion, L. M., 442
Boyle's Law, 336, 340
Brain Awareness Week, 159
Bransford, J. D., 20
Breiner, J. M., 442
Brice, A., 71
Bricker, L. A., 473
Brickhouse, N., 299
Bridgeland, J. M., 442
broad fields, 6
broadening participation, 447, 451
Bronx Green Machine, 231
Brophy, J. E., 88
Brown, A. L., 20, 322
Brown, B. A., 417, 419
Brown, J. S., 473
Brown, S. D., 299
Brumer, M., 141
Brundtland, G. H., 222
Bruner, J. S., 20
Buck Institute for Education (BIE), 148, 150, 160–161, 442
Bullock Mann, F., 473
Burns, R., 42
business, 23–24, 28, 38, 40, 45–46, 55, 57, 62–63, 66, 69–70, 114, 118, 147–148, 204, 211–213, 215–218, 259, 298, 333, 434, 438, 440, 452
Bustamante-Donas, J., 141
Butler, D. L., 500
Buxton, C., 372
Bybee, R. W., 221, 277, 417
Bytesize Science, 559

Cabral, A., 141
Calabrese Barton, A., 443
Calderón, M., 371
Calkins, S. D., 105
Callahan, B. E., 352
Canaday, D., 373, 419
Cancer Research Institute (CRI), 156

Cantador, I., 161
Cardenas, V., 104
Career and Technical Education (CTE), 49, 69, 72, 346, 562
Carey, G., 299
Carey, K., 104
Carlos y el Zorrillo, 406–407
Carlsen, W., 277, 371
Carlson, J., 500
Carnevale, A. P., 442
Carrillo, R., 372
Carter, L., 350
Casey, R., 139
Castano, C., 350
Ceballos, G., 473
Center for Media Literacy (CML), 326
Center for Ocean Sciences and Education Excellence SouthEast (COSEE SE), 35, 41
Center for Public Education (CPE), 105, 246
Center for the Advancement of Informal Science Education (CAISE), 545
Centers for Disease Control (CDC), 217, 221
Chabot Space and Science Center, 67
Challenger Space Center, 57, 68, 548
Change the Equation (CEq), 28, 41–42, 221, 428
Charles' Law, 336, 340–342
Charney, J., 139
Chemers, M. M., 473
chemiluminescence, 155
chemistry, 1–2, 11–17, 101–103, 144, 150–155, 159, 181, 193, 197, 209, 302, 317, 335, 337, 339, 344–345, 378, 383, 430, 432–433, 436, 507, 509, 522–523, 530, 536, 547–548
ChemMatters, 154
Cheung, R. W., 162
Chinn, C., 139
Chivian, E., 473
Chowa, G. A. N., 104
Chrystalbridge, M., 221
citizen science, 251–252, 254, 272, 298, 493–494, 557
  projects, 39, 133, 251, 291, 488, 556–557
Clark, R. E., 162
Classroom Close-up New Jersey, 531
Cleary, T. J., 500
Clements, R., 246
cloning, 188, 287
Cocking, R. R., 20, 322
Coger, R., 71
Coletta, S., 139
collaboration, 1–2, 19, 32–33, 38–39, 65, 68, 111–113, 123, 127, 134, 143–144, 147, 151–152, 166, 170, 174–175, 187, 209, 227, 249–256, 258–265, 267–271, 283–286, 292, 295, 297, 306, 308, 310, 312, 321, 329, 335, 357, 416, 424, 428, 440, 460, 462, 544–545
Collazo, T., 372
Colley, K., 42
Collins, A., 473
Committee on Science, Engineering, and Public Policy (COSEPP), 71, 161
communication, 1, 9, 32–33, 93, 132, 143–144, 147, 237, 249–250, 252–255, 259, 262, 264–267, 270–271, 296, 326, 328, 333, 335, 346–347, 356, 369, 376–377, 380, 384, 387, 390, 394, 400, 415–416, 538, 554, 557
Community Collaborative Rain, Hail, and Snow Network, 557
Community Supported Agriculture (CSA), 212
competitions, 39, 118, 132, 145–146, 150, 155, 157, 159, 218, 291, 298, 346, 432, 434, 540, 552, 556, 560
complex texts, 379, 390
comprehensible input, 404
computational thinking, 30–31, 111, 306, 415, 480, 495–497
computer visualizations, 529, 532, 536–537, 539, 541–542
Comulada, W., 299
concept tests, 523
Cone, N., 372
Considine, D., 350
constellation cube, 123
content specific, 262, 315, 377, 389
contextualize, 2, 356, 400–401, 416, 454
Cook, L. A., 277–278
cooperative learning, 98, 146–147, 317, 327, 469, 506
Copenhagen, 15–17
core content, 3, 6, 453
Cornell notes, 337, 339
COSMOS Corporation, 544

Crawford, M., 373
creativity, 1, 32, 47, 107–108, 143–144, 147, 155, 215, 225, 227, 250, 252, 259, 270, 289, 291, 298, 325, 335, 345, 348, 358, 406, 421, 423–424, 448, 455, 555, 560
credible sources, 393
Crippen, K., 501
critical thinking, 1, 31–32, 94, 97, 147, 150, 166, 219, 250, 252–253, 259, 270, 335, 381, 399, 455, 503, 531, 542, 558
    skills, 287, 296, 327–328, 334, 395, 539
Cronin, J., 140
Cronshey, R. G., 221
crosscutting concepts, 5–6, 9, 18, 111, 336, 414, 445–446, 453, 458, 464
cross-curricular, 204, 207, 209, 271, 382, 436
cross-disciplinary, 3, 6, 18, 270
Crouch, C. H., 322
Crute, I., 247
Crutzen, P., 473
Cruz, N. I., 443
CubeSat, 540
    BLUECUBE (Build, Launch, Utilize, Educate CubeSats) Project, 540–541
    Mission Management and Operations, 540
    Spacecraft Construction and Testing, 540
    Spaceflight Data Utilization, 540
Culligan, T., 200
cultural traditions, 447
curriculum, 3, 6–9, 15, 18, 20, 32, 39, 41, 45, 61, 65, 75, 98, 117, 126, 128, 147, 149–150, 156, 158–159, 167, 169–170, 185–186, 190, 194, 197–198, 203–204, 207, 209–211, 217–221, 250, 252, 262–263, 267, 284, 287, 326–328, 349–350, 377–378, 396, 412, 415, 426, 428–429, 433, 436, 440–441, 450–453, 456, 463–465, 472, 483, 487, 498–500, 506, 509, 531, 538–539, 543, 545–546, 549–550, 557, 561–562
    fused curriculum, 3
    spiraling, 9
Custer, R., 73

Dana Foundation, 159
Daniels, H., 105
Darling-Hammond, L., 88, 179

data sets, 170, 174, 291, 298, 497, 533–534, 537, 539
Davis, K., 179
Dawson, V. M., 350
de Brey, C., 473
de Laet, M., 73
Deantoni, T., 179
Deci, E. L., 500
Deeper Learning Initiative, 147
Department of Energy (DOE), 117
Design Squad Nation, 560
design thinking, 147, 160
    Design Thinking for Educators, 160
Deustua, S., 140
Devaney, L., 71
Dewey, J., 20, 442
Diagnoser Tools, 507, 510–511
Dierking, L., 563
Diez-Palomar, J., 475
differentiate, 47, 103, 358, 361, 363, 391, 457, 503
digital citizenship, 254, 264
digital-age literacy, 147
DiIulio Jr, J. J., 442
discipline referrals, 96, 102–103, 125
discourse, 207, 328, 353–371, 402–404, 411, 416–417, 507
Discovery Kids, 560
discovery-based learning, 147
District of Columbia Public Schools (DCPS), 336–337
diverse student groups, 336
diversity, 47, 69, 243, 245–246, 296, 434, 450, 453, 456, 465, 534, 541, 544
Dixon, P., 179
D-Lab MIT Youth Outreach, 160
Donnelly, L., 351
Donnelly, M., 88
Dorph, R., 42
Drake, S., 42
Dresner, M. W., 179
Drew, C., 140
dropout rates, 76, 91
Dubner, J., 201
Dubuffet, Jean, 451
Duch, B. J., 441
Duchovnay, B., 200

Duguid, P., 473
Duncan, S., 299
Duncan, T., 201
Dunlop Velez, E., 473
Dunlosky, J., 473
Dunn, K., 475
Dupont Science Essay Challenge, 150

Earth Information System (EIS), 536
Earth Observing System (EOS), 533
earth science, 121, 123, 225, 230, 523, 530, 532–534, 536
    Earth Science Enterprise, 533
Earth System Science Education Alliance (ESSEA), 541
Earthwatch Educator Fellowships, 176
Edelson, D., 140
Edelstein, W., 442
EdHeads, 558
EdLeader, 21, 147
Education for Life and Work, 423
Education for Sustainable Development (ESD), 207
educational, 1, 6, 8, 10, 24, 28, 30, 46, 48, 65, 69–70, 91–93, 95–97, 144, 146–147, 159–160, 166, 168, 170, 172–173, 184, 186, 203–204, 206, 212, 216–217, 219, 226, 229, 231, 238, 241, 250–251, 255, 260, 271, 327–329, 347, 355, 378, 400, 404, 407, 416, 426–427, 432, 439, 445, 447, 478, 504–508, 521–522, 524–525, 530–531, 533–534, 536, 538, 541, 549–551, 553, 558–559, 561–562
educational technologies, 327–328
Eger, J., 442
Ehrlich, P. R., 473
Einstein, Albert, 88, 96, 104, 143, 158, 407
electronic mass media, 329
Elementary and Secondary Education Act (ESEA), 26, 428, 446
elementary STEM, 26, 28, 39
Ellerson, N., 246
emulation, 483, 485–486, 495–496
engineer, 12, 28, 34–35, 38–39, 46, 60–61, 86, 112–113, 116, 128, 132–133, 136, 157–159, 172, 174, 196, 219, 250, 259, 283, 289, 294, 325, 332, 355, 363–364, 366, 370, 393, 415, 430, 436, 438, 453, 500, 515, 524, 531, 540, 548, 556, 559
engineering, 1–6, 11, 18–19, 23–25, 29–31, 33–35, 37–41, 45–46, 48–49, 58, 60–61, 67–68, 71, 93, 109, 111, 113, 116, 118, 123, 132, 136, 143–147, 149, 156, 158, 167, 183–184, 186, 189–190, 197, 203–204, 207, 209, 218–219, 223, 229–230, 249–250, 259, 271, 287, 306, 316, 325, 332, 336, 338, 340, 347, 350, 353–357, 359, 365–366, 375–378, 382, 391, 400, 407, 410–411, 414–415, 417, 422–423, 425, 428, 432, 434–436, 441, 445–446, 451, 453, 458, 464, 466, 469, 477, 503, 505, 507, 509, 521–522, 525, 530, 540, 542–544, 547, 551–552, 555–556, 559–561
    challenges, 145, 158, 422
    design process, 37, 184, 197, 411, 415, 551
    Engineering the Tools for Exploration, 559
    mentality, 332
English Language Learners (ELL), 78, 91, 336–337, 378, 394
English Learners (EL), 353–354, 399, 401, 406, 410, 416, 447
Entertainment Software Association (ESA), 282, 299
Entwisle, D. R., 562
environmental education, 205–207, 226, 241, 426
Environmental Protection Agency (EPA), 220
Epstein, D., 42
Erduran, S., 351
Erlenmeyer flask, 341
essential questions, 6, 9, 449, 453–454, 458, 464–465, 470
Evagorou, M., 350
Evangelista, Gay, 49
Evans, G., 247
Everglades National Park, 546–547
    Everglades Outpost, 547
Every Student Succeeds Act (ESSA), 428
evidence-based, 4, 328, 331, 505, 507, 534
Exploratorium, 554, 560
Exploravision, 150, 153–155
expository text, 377–378, 416
Eyes on Earth, 537
Eyler, J. S., 221

INDEX | 571

Fairweather, J., 221
Falk, J. H., 563
Felton, M., 442
FemGineers, 61
Ferreira, M., 105
field experiences, 165, 228
field trips, 52, 55, 57, 62, 99, 228, 431, 433, 440, 449, 451, 462, 469, 515, 543–546, 548–550, 552
  virtual, 253, 272, 553
Fiesta Bowl Aerospace Challenge, 68
Filippenko, A., 141
Finn, J. D., 88
FIRST Robotics, 291, 552, 556
  Lego League (FLL), 159, 556
  Lego League Jr (FLLjr), 159, 556
  Robotics Competition (FRC), 159, 556
  Tech Challenge (FTC), 556
Fish is Fish, 404
Fish, D., 542
Fisher, D., 416, 418
Fisher, J. Richard, 117
Fleischman, H., 140
Florida Museum of Natural History's Project Butterfly Wings, 557
fluency, 267, 356, 381, 394
Foehr, U., 247
formulating, 108, 381
Foster, A., 299
foundations, 147, 224, 241, 335, 440, 549
Fraquelli, L., 139
Fraser, J., 246
Frayer Model, 386
Frayn, M., 21
Frazier, R., 443
Froguts, 558
Fulbright Memorial Fund, 80
Full Speed Ahead! How Fast Things Go, 406, 408
Fulu, I. A., 162
function, 6, 33, 82–85, 87–88, 138, 190, 229–230, 355, 359, 451, 455–458, 460, 464–465, 467, 471, 525, 536–537
  function notation, 82–85, 87
functional illiteracy, 329
funding, 1, 3, 35, 38, 55, 57, 62–67, 91, 93, 118, 136, 204, 211, 213–214, 219–220, 228, 237, 268, 440, 533, 537, 540, 549, 558, 562
Furtak, E., 373

Galaxy Zoo, 133–135
gallery walks, 316–317, 467
gaming, 48, 281–285, 287, 289–298, 328, 516
Garcia, A., 473
Garcia, B. F., 473
Gardner, A., 500
Garet, M. A., 179
Gay-Lussac's Law, 336, 342–343
Gee, J. P., 299, 418
GEMS Club, 62, 72
gender, 45, 47–50, 59–60, 63, 65–66, 68, 70–71, 76, 92, 158, 210, 264–265, 288, 404, 413
  equity, 45, 50, 59–60, 65–66, 68, 70–71
  stereotypes, 48
general education development (GED), 79
Geographic Information System (GIS), 230, 530, 533
Geology of Mankind, 453
geoscience, 186, 529–542
  Geoscience and Remote Sensing Laboratory (GRSL), 535, 537–538, 541
Geostationary Operational Environmental Satellite (GOES), 532
Gibbons, P., 371, 418
Giles, D. E., Jr., 221
Gillman, L., 299
Girls Exploring Technology (GET), 46, 50–65, 68
Girls in Engineering and Technology, 67
Girls, Math & Science Partnership, 67
GK-12, 187
Glied, S., 201
global collaboration, 249–254, 256, 259, 262, 267–268, 271
global education, 251–253, 255
Global Learning and Observations to Benefit the Environment (GLOBE), 530–531, 539, 541
  Croatia Land Cover Symposium, 531
Global Learning Exhibition (GLE), 530
Global Positioning Systems (GPS), 230, 533, 539

global STEM education continuum, 252–253, 270
globalization, 69, 251
Godfray, H., 247
Goldhaber, G., 140
Goldman, R., 162
Gomez, L. M., 442
Gomez-Zwiep, S., 373
Good, T. L., 88
Gopnik, A., 42
Gormly, K., 72
Grace, M., 351
grading rubrics, 293, 393
graduation rates, 76
Graham, S., 500
graphic organizer, 368, 390, 408, 413, 471
Gray, C. J., 221
greenhouse effect, 121–122
Greenwood, D. A., 473
Gresch, H., 351
Groh, S. E., 441
Grow, P. L., 442
growth model, 93, 96, 103
Guevara, E., 105
guided inquiry, 190, 405, 414, 469, 487
guided reading, 391
Gupta, V., 372
Guthrie, J. T., 397, 418, 501
Gutiérrez, M. V., 443, 475

habits of mind, 284
Hackett, G., 299
Haddad, L., 247
Haiku poetry, 339
Hall, M., 372
Hallinen, J., 21, 43
Hancock, M. P., 200
hands-on activities, 68, 178, 327, 340, 356, 431, 448
    Hands-On Universe (HOU), 117
Hardy, I., 371
Harkness, S. S., 442
Harris, H. R., 500
Hartley, K., 501
Hartley, L., 141
Hartry, A., 42

Hashweh, M. Z., 179
Hasselhorn, M., 351
Hayes-Wood, Katreena, 49
Hazari, Z., 443
Heim, M., 140
Heimlich, J., 246
Heisenberg, Werner, 16–17
Helm, B., 247
Helms, J. D., 221
Henriksen, D., 442
Henry, B., 88
her world, 54, 57, 548
Hewson, P., 179
higher-order thinking, 111, 283
Hill, C. C., 72
Hmelo-Silver, C., 139
Hoff, D. L., 72
Hofferth, S., 247
Hofshire, L., 222
Hogan, K., 371
Holiday Chemistry, 159
Holland, W., 299
Holschuh, J., 351
home project, 8
homeopathic, 114, 213
Hoody, L., 247
Hopkins, B., 371
Hopkins, C. A., 221
Hopkins, K., 72
Hopstock, P. J., 140
How Stuff Works, 561
Howes, E., 352
Hu, I., 473
Huggins, A., 373
Hull, J., 105
Hunter, A., 179
Hussar, W., 473
Hyde, A., 105
Hynd, C., 351

IDEO Design, 160
Illinois Mathematics and Science Academy, 159
imagination, 87, 143–144, 155, 168, 211, 223, 294, 296, 421, 484
immersion programs, 166

iMovie, 126
Industrial Revolution, 1, 46, 332
Informal STEM Learning (ISL), 298, 543, 545–546, 549, 553–554
innovation, 1, 24–26, 28, 30, 32, 40, 46–47, 143–146
  Innovation Lab, 147
innovators, 23, 40, 143, 146–147, 151, 325
inquiry-based, 32, 93, 97, 148, 153, 166, 168–170, 173–174, 177, 190, 194, 210, 251, 269–271, 283, 353–355, 357, 370, 399, 401, 407, 411–412, 415–416, 479
inspiration, 143–144, 151, 422, 458, 461, 553
inspire, 3, 8, 24–25, 34, 68, 87, 143–145, 158, 230, 249, 285, 294, 395, 426, 432, 459–460, 507, 545, 553
Institute for Global Environmental Strategies (IGES), 541
Institute of Electrical and Electronics Engineers (IEEE) Geoscience and Remote Sensing Society, 531
integrated curriculum, 20
interactive website, 170, 507, 557
interdisciplinary, 1–6, 9, 12–13, 18, 20, 24, 28, 30–31, 34, 40, 154–155, 176, 178, 208, 327, 358, 377, 400, 522, 534, 552
Intergraph Corporation, 530
International Science & Engineering Fair (ISEF), 132
International Technology and Engineering Educators Association (ITEEA; formerly ITEA), 41, 377
intrinsically motivated, 146, 382

Jackman, W. S., 221
Jacobs, B., 299
Jenkins, H., 299
Johnson, C. C., 442
Johnson, D. W., 162
Johnson, R. T., 162
Jorgenson, O., 442
Joyce, C., 200

K-12, 3–6, 9, 18, 25–26, 28, 33, 40, 60, 65, 86, 91, 107, 111, 118–119, 143–144, 146–147, 155, 159–160, 165, 176–177, 183, 186–187, 195, 197, 203–204, 206–207, 219, 226, 229, 241–242, 306, 325, 355, 400, 425, 445–446, 477–478, 506, 508–509, 522–523, 525, 530, 535, 537, 539, 541, 545–546, 561–562
Lab Partners, 147
Lab Wiki, 160
Laboratory, 147
Keane, S. P., 105
Keeley, P., 42
Kellert, S., 247
Kemeny, J. G., 418
Kena, G., 473
Kerby, S., 104
Kids in Need Teacher Grants, 65, 562
Kirschner, P., 162
Kisiel, J., 563
Kitsantas, A., 501
Kitt Peak, 118, 129–130
Klawe, Maria, 48
Kleine, L., 105
Kliebard, H. M., 21
Kloetzer, B., 371
Klosterman, M., 351
Koebler, J., 140
Koehler, C. M., 442
Kohler, R., 21, 43
Kolbert, E., 473
Kopkowski, C., 179
Kovalik, S., 72
Krajcik, J. S., 141, 418, 473
Krasny, M. E., 443
Kress, G., 474
K-W-L charts, 387, 413

La Fe Policy Research & Education Center, 105
laboratory-based research opportunities, 165
Ladson-Billings, G., 89
Lam, S.-f., 162
Land, S. M., 444
Landes, N., 500
Larmer, J., 162, 442
Latzke, M., 373, 419
Laurie, R., 350
Laursen, S., 179
Lauscher, H. J. N., 500

Lave, J., 418, 474
Law, J. S., 162
Lawrence Livermore National Laboratory (LLNL), 177
Lawrence, D., 247
Lawrence, R., 162
learning methodologies, 148, 252, 259
Lee, H., 140
Lee, O., 371–372, 418, 474
Lee, S. W.-Y., 201
Lemke, J. L., 418
Lemov, D., 372
Lent, R. C., 418
Lent, R. W., 299
Leopold, A., 247
LeRoy, K., 372
Lewis, E., 371
Lewis, S., 372
Lexile Level, 379
Lhamon, C., 372
Library of Congress, 392, 396, 426, 553, 561
Lieberman, G., 247
lifelong learners, 86–87, 561
line of best fit, 82–84
Link, E., 141
Lips, H., 72
literacy, 3–5, 10, 26, 28, 34, 41, 147, 176, 186, 209, 285, 326–327, 329–335, 337–338, 346–347, 356, 360, 370, 375–378, 380–384, 387–390, 393–396, 399–405, 411–412, 414, 416, 425, 446, 532, 534, 541, 545
   content literacy, 383
   disciplinary literacy, 377, 399
   media literacy, 326–327, 329–330, 334, 337–338, 346–347
Llewellyn, D., 372
Loike, J. D., 201
Lord, L., 299
Los Alamos Energy and Environmental Science Academy, 174
Lotus Effect, 190–193
Loucks-Horsley, S., 179
Louv, R., 247
Love, N., 179
Lowery, P., 299
Lucore, Rebecca, 47
Luykx, A., 372, 418, 474

machine tools, 157–158
Maddux, J., 299
Maerten-Rivera, J., 372–373
Mahoney, G., 72
Mainardi, D. S., 200
Malhotra, B., 139
Maltese, A. V., 42
Manwaring, R., 104
Manzo, A. V., 418
Manzo, U. C., 418
Marine Advanced Technology Education (MATE), 35, 41
Markowitz, D., 140
Marsh, E. J., 473
Martinez-Pons, M., 501
Marx, R. W., 141, 372
Marzano, R. J., 89, 527
mass media, 327, 329, 346
Massachusetts Institute of Technology (MIT), 35, 332
   MIT Media Lab, 147, 160
massively multiplayer online games (MMOGs), 296
mastery, 77, 79, 97, 169, 270–271, 283, 286, 290–292, 296–297, 308, 448, 488, 495, 500
Materials Research Science and Engineering Centers (MRSEC), 186
math, 1–4, 14, 17–19, 28–31, 33, 36, 39, 41, 45–52, 60, 62, 67–68, 70–71, 75–76, 78–82, 85–87, 93, 109–111, 113, 116, 118, 123, 127, 134, 136, 138, 143–144, 148, 156, 159, 167, 176, 183–184, 196–197, 203–204, 207, 209–210, 218–219, 223, 228, 231, 241–242, 249, 259, 271, 283, 293, 298, 302, 305–309, 316, 320–321, 325, 329, 333, 340–341, 353, 375–378, 380–383, 386–387, 389–390, 393, 395, 400, 403, 415, 422, 424–425, 428, 432–435, 441, 445, 477–478, 500, 503, 505, 511, 521–523, 525, 530, 542–544, 554–555, 558, 561
   simulations, 558
Math Science Partnership (MSP), 118
mathematical or computational model, 340
mathematician, 28, 86, 113, 389, 393, 500
mathematics, see: math
Mathews, M., 140
Mayer, R., 474

Mazur, E., 322, 527
McBride, Y., 105
McCaffrey, T., 42
McCarthy, B., 442
McCarthy, D., 442
McClarty, K. L., 105
McCrae, B., 350
McDaniel, M., 474
McFarland, J., 473
McGonigal, J., 299
McKeon, D., 105
McKeown, R., 221
McNees, P., 72
McTighe, G., 474
McTighe, J., 21, 323, 473, 475
meaning making, 402, 405, 414, 450, 454
Medina, J., 322
Melaville, A., 442
Melton, M., 442
mental disciplinarians, 7
mental models, 358, 504, 507, 509, 514–515, 521–522, 525
mentor, 39, 48, 50, 52, 57–58, 61, 63, 70, 117–118, 130, 132, 144, 152–153, 158, 176, 188, 190, 350, 434, 548, 559
   mentoring, 45, 50, 57, 103, 130, 167, 449, 522
   mentorship, 86
Mercandante, B., 299
Mergendoller, J. R., 162
metacognition, 405, 503–505, 507, 519, 521–522, 524–525
   metacognitive prompts, 470, 482–483, 486–487, 494–495
   metacognitive monitoring, 480–481, 487
Metz, K., 140
Meyer, D., 442
Meyer, S. J., 222
Michaels, S., 42, 140, 322, 351, 372
Microsoft in Education's Innovative Schools Program, 147
Millar, R., 501
Miller, H. C., 141
Miller, J., 201
Miller, R. T., 42
Mitchell, S., 72
mobile devices, 508, 521, 524–525

modalities, 356, 415, 448, 451–452, 469–471
modeling, 82, 84, 110, 203–204, 218–219, 230, 302, 381–382, 386–387, 394, 410, 451, 466, 471, 483–486, 495–496, 506–507, 522
   Modeling Physics Workshop, 302
   with mathematics, 82
Modi, K., 72
Moeller, K., 371
Moeller, M., 141
Moje, E. B., 418
Monzó, L., 372
Moore, J. C., 141
Morelli, B., 140
Moreno, R., 474
Moret, B., 72
Morgan County Farmers Market, 214
Morison, K. B., 442
Mortimore-Smith, S., 299
Mountain View Solar, 214
Moursund, D., 163
Mueller, M., 351
Muir, J., 247
multi-disciplinary, 3, 12
Multinational Youth Studying Practical Applications of Climatic Events (M.Y.S.P.A.C.E.), 540
Multiple Routes Principle, 286
Mundry, S., 179
Murie, C., 372
Murphy, D., 105
Murphy, S., 105
Murphy, T., 42
Museo Galileo Institute and the Museum of the History of Science, 558
Musu-Gillette, L., 473
Myers, A., 441
Myth Busters, 126

nanotechnology, 86, 183, 189–190, 192, 194–197, 199
Nastasi, B. K., 371
Nathan, M. J., 473
National Academies Press, 21, 42, 71, 88–89, 140, 163, 179, 200, 247, 278, 299, 322–323, 351, 372–373, 397, 419, 443, 474, 526–527, 542, 563

National Academy of Engineering (NAE), 422, 556
  NAE Grand Challenges for Engineering, 556
National Academy of Sciences, 71, 527, 533
National Aeronautics and Space Administration (NASA), 99, 156, 165, 176, 254, 291, 553
  eClips, 559
  Educational Space Walk Simulations, 558
  J-Track 3D, 537
  NASA/IPAC Teacher Archive Research Program (NITARP), 118
  S'COOL Project, 538
National Alliance for Partnerships in Equity (NAPE), 59
National Assessment of Educational Progress (NAEP), 184, 400
National Center for Atmospheric Research (NCAR), 532
National Center for Education Statistics, 42–43, 140, 398, 418, 443, 447, 473
National Center for Women in Information Technology (NCWIT), 59
National Chemistry Week, 150, 152–153
National Commission on Excellence in Education, 163
National Council of Teachers in Mathematics (NCTM), 41
National Council of Teachers of English (NCTE), 285, 419
National Education for Sustainability K12 Student Learning Standards, 207
National Endowment for the Humanities (NEH) Summer Grants Workshops, 178
National Engineers Week, 158
  Future City Competition, 61–62, 65, 548, 552, 556
National Environmental Education Act (NEEA), 206
National Environmental Education Foundation (NEEF), 206
National Girls Collaborative Project (NGCP), 60, 65, 68
National Institute of Health (NIH), 118, 156
National Marine Educators Association, 35, 41

National Mole Day Foundation, 153
National Oceanic and Atmospheric Administration (NOAA), 165, 176, 196, 291, 551
  National Buoy Data Center, 538
  National Hurricane Center, 538
  National Weather Service, 538
  Teacher at Sea, 37, 41, 176
National Radio Astronomy Observatory (NRAO), 114, 117, 129, 132
National Research Centers, 551
National Research Council (NRC), 4, 21, 31, 42, 89, 140, 163, 165, 179, 195, 200, 247, 278, 299, 323, 372–373, 419, 423, 443, 474, 527, 542, 554, 563
National Science Board (NSB), 24, 325, 347, 351
National Science Digital Library (NSDL), 41, 561
National Science Foundation (NSF), 3, 24, 26, 41, 75, 118, 165, 168, 174, 176, 183, 185–186, 196, 297, 349, 532, 539, 554, 558
  NSF Advisory Committee for Geosciences, 530, 542
National Science Teachers Association (NSTA), 41, 150, 200, 278, 376, 408, 412, 417, 545, 563
  NSTA/Toshiba Exploravision Contest, 154
National STEM Video Game Challenge, 291, 298, 556
National Wildlife Federation – Cool School Challenge, 220
Natural Resource Conservation Service (NRCS), 220
Nature Conservancy, 226, 247
NBC Learn, 553, 561
NC (North Carolina) STEM Learning Network, 21
Neigeborn, L., 139
Nemeroff, M., 139
Nemet, F., 352
New Jersey Education Association, 531
New Tech Network, 21
Newmann, F., 140

Next Generation Science Standards (NGSS), 5, 18–19, 26, 35, 41, 111, 184, 195, 229, 242, 249, 286, 290, 305, 336, 350, 355, 400, 445, 509, 530, 545
Nicolescu, B., 43
Nist, S., 351
Nixon, R. S., 474
No Child Left Behind (NCLB) Act of 2001, 242, 428, 443, 446, 474
nomenclature, 294, 296
Norris, S. P., 351
North American Association of Environmental Education (NAAEE), 206, 222
NOVA, 546, 558
    NOVA ScienceNow, 130–131

O'Connor, C., 372
O'Malley, K. J., 105
O'Neill, T., 443
Oasis Elementary, 50, 53, 55, 58
Obama, Barack, 1
Obama, Michelle, 230
ocean literacy, 34–35, 41, 534
Oemig, P. A., 419
Office of Science and Technology Policy (OSTP), Executive Office of the President, 533
Ohio Department of Education, 30, 43
Oil City, 119, 125, 128–132
Olin College, 147
Olson, L. S., 562
open-ended questions, 47, 94, 98, 455
OpenRocket, 158
Orphanos, S., 179
Orr, D., 247
Osborne, J., 351, 501
Osborne, M., 179
Osei-Akoto, I., 104
Otero, V., 373
outside of the textbook, 87, 389

Pack, H., 140
Palmer, T. M., 473
Pandey, V. C., 222
Papson, S., 162
parental involvement, 93, 95–97, 99

Partners in Learning Network, 163
partnerships, 38–40, 59, 63–64, 66, 128, 136, 147, 183, 187, 204, 250, 260, 421, 423–424, 426–429, 434, 436, 439–441
Pashler, H., 474
Pathways to Science, 179
Paul Mock's Greenhouses, 213
Pavelis, G. A., 221
Pawellek, C., 141
Payne, R., 89
pedagogical strategies and techniques, 376
peer instruction (PI), 304, 519, 522
Pelczar, M., 140
Penfield, R., 372
Pennypacker, C., 140
Penuel, W. F., 179
Peoria Historical Society Museum, 548
Peoria Unified School District (PUSD), 49
Perales, L., 72
Perlmutter, S., 140
perseverance, 146, 151, 155, 160
Peters, E. E., 501
Peters-Burton, E. E., 501
Pew Research Center, 140
Phelps, M., 72
PhET, 523, 558
Phillips, K. B., 72
Phillips, L. M., 351
Physics Central, 163
Physics Quest, 159
Physics Toolbox, 516–517, 524–525
Pickert, K., 72
Piery, L., 474
Pinal, A., 373, 419
Pinnell, M., 221
Piro, J., 443
Pitt, J., 222
Pittsburgh Regional Science & Engineering Fair, 123
place-based learning, 426–427, 447
Plastic Paradise: The Great Pacific Garbage Patch, 455–456
Platten, P., 500
Plucker, J. A., 442
podcasts, 555, 559–560
PolarTREC, 117–118, 177

Pop, M. M., 179
Poris, M., 299
poverty, 76, 81, 91–92, 115, 119, 428, 436
Preparing the Next Generation of STEM Innovators: Identifying and Developing our Nation's Human Capital, 325
President's Council of Advisors on Science and Technology (PCAST), 3, 25, 143
press release, 128
Pressley, M., 371, 419
Pretty, J., 247
Primack, R. B., 474
primary sources, 392, 396, 561
Pringle, R. M., 473
problem solving, 1, 31–32, 86, 94, 107, 122, 143–145, 209, 219, 227, 250–252, 255–256, 258–259, 269–271, 333, 335, 381, 401, 414, 511, 523, 539, 542
problem-based learning, 18, 94, 104, 148, 252, 426–427, 469, 530, 535
professional development (PD), 2, 11–12, 18, 27–28, 35, 40, 59, 86–87, 93, 128, 130, 165–168, 173, 177–178, 181, 184, 186–187, 195, 204, 263, 269, 302, 392, 432, 472, 506, 541
professional growth, 160, 166
Programme for the International Student Assessment (PISA), 325
Project CRISS, 506, 513, 521
project-based, 32, 37, 39, 209, 212, 252, 271, 331, 546
    curriculum, 170, 207, 209
        learning (PBL), 2, 31–32, 93, 97–98, 122, 145, 148–150, 159, 256, 259, 349, 426–427, 471, 539, 556
            Avalon School, 159
            DaVinci School, 160
            Edutopia PBL, 160
            Greenwich Public School, 159
            High Tech High School (HTH), 159
            Illinois Mathematics and Science Academy (IMSA), 159
            Pacific Education Institute, 159
Psychosocial Moratorium Principle, 286
Public Broadcasting Service (PBS), 546, 557
public service announcement (PSA), 346

Purzer, S., 371
Putman, R. T., 474

Quest to Learn (Q2L), 284
questioning skills, 312
quests, 116, 284, 286, 290, 293–294, 469

Racing Extinction, 453, 461–462
radio programs, 560
Rahm, J., 141
Rajesh, H., 372
rate of change, 82–85
Rathbun, A., 473
Rawson, K. A., 473
reading, 16–17, 23, 61, 76, 133, 147, 170, 175, 227–228, 234, 236, 297, 318, 325, 329–330, 332–333, 340, 377–383, 385–391, 394–396, 400–401, 403–407, 409–414, 416–417, 451, 471, 478, 482, 487, 505, 516–517, 522, 525, 541, 546
    active readers, 386–387
    close reading, 339, 390
    Content Area Reading Inventory (CARI), 388
    leveled reading, 396
    pre-reading questions, 381
    post-reading strategies, 392–393
    Raygor Readability Test, 386
    unrehearsed oral reading, 394
readability, 378, 386, 388, 391
    readability tests, 386
Ready, Set, Science!, 304
realia, 409
real-world, 3, 24, 28–35, 37–40, 82, 97–98, 108, 110, 112, 116, 136, 138, 148, 156, 168, 173–175, 207–209, 216, 218–219, 251, 270, 328, 359, 415, 426, 432, 447, 509, 540, 550, 559
Rebull, Luisa, 130
Reiser, B., 373
reluctant learners, 87
remote sensing, 529–533, 535–542
remotely operated vehicle (ROV), 29, 35
Renaissance Charter High School for Innovation, 231
Research Experiences for Teachers (RET), 34, 86, 117, 166, 179, 181–182

research paper, 124–126, 155, 384, 392–393
research project, 25, 86, 98, 111, 117–118, 121–127, 136, 150, 176, 188–190, 192, 195, 213, 455, 460, 489, 530, 540
Resnick, L., 372
Reveles, J. M., 419
rewards, 286, 288–289, 291, 294
Rice, D. C., 419
Richardson, N., 179
Richmond, M., 141
Rideout, V., 247
Rittmayer, A. D., 475
Ritz, Stephen, 231
Rizzi, R., 221
Roberts, D., 247
Roberts, D. A., 443
Roberts, L. F., 141
Robinson, G., 141
Robinson, S., 247
Rock, D. A., 88
RockSim, 158
Rogers, R., 299
Rohrer, D., 474
role model, 36, 47–48, 60–61, 63, 70–71, 81, 103, 116, 118, 136, 144, 449
Root-Bernstein, M. M., 443
Root-Bernstein, R. S., 443
Rose Petal Effect, 183, 192, 197–198
Rotheram-Borus, M., 299
Route 21, 43
Rube Goldberg, 157
    Machine Challenge, 157
Rueda, R., 372
Rufo-Tepper, R., 299
Russell, S. H., 200
Rust, T., 72
Ryoo, K., 417

S²TEM Centers SC (South Carolina), 43
Sadler, P. M., 443
Sadler, T. D., 351–352, 443
safe environment, 78, 87, 356
Salen, K., 299
Sally Ride Science Festivals, 68, 547
Salmond, K., 72
Salomon, G., 475
Salt River Project, 548

Sanchez, M., 371
Sanders, M., 21, 43
Santau, A., 372–373
Sarkar, S., 443
Satellite Educators Association (SEA), 540
satellite imagery, 291, 529, 537, 540–542
Savery, J. R., 443
scaffolding, 6–7, 9, 13, 98, 357–358, 361, 370, 380, 455, 460, 471
Scarloss, B., 201
Schiavelli, M., 141
Schimpl-Neimanns, B., 105
Schmitt, M., 105
Schneider, R., 141
Schoenberg, J., 72
School of Rock (SOR), 177
Schraw, G., 501
Schultz, K., 299
Schwartz, A. T., 21
Schweingruber, H. A., 42, 140, 322, 351
science, 1–6, 9–11, 14–19, 23–40, 45–52, 57–58, 60, 62, 67–68, 70–71, 93–95, 99–104, 107–123, 125–133, 135–138, 144–145, 149–150, 152–154, 156, 158–160, 165–179, 181–186, 188–189, 193–198, 203–204, 207–212, 218–219, 223, 225, 228–231, 239–240, 242, 249–250, 256, 259–260, 269, 271, 283, 286, 290–291, 294, 297, 306, 316, 325–337, 339, 342, 344–347, 349–350, 353–359, 375–376, 378, 382, 384–385, 389, 395, 399–406, 408–409, 411–416, 421–422, 424–426, 428–430, 432, 434–436, 438, 440–441, 445–448, 451, 460, 477–479, 482–483, 486–488, 492–495, 500, 503, 505, 507–509, 519, 521–523, 525, 530–539, 541–546, 548–550, 552–561
    competitions, 132, 145, 150, 157, 434
    fair, 119, 123, 128, 154, 156–157, 173, 175
    inquiry, 117
    space science, 6, 14, 116, 119, 121, 123, 249, 453, 530
    symposiums, 156
Science and the Sea, 560
Science Education Partnership Award (SEPA), 118
Science Teacher and Researcher (STAR), 176

Science, Mathematics, Engineering and
    Technology (SMET), 2
Science, Technology, Engineering and
    Mathematics (STEM), see: STEM
scientific and engineering practices, 336,
    355–356, 359, 445–446, 453, 464
scientific literacy, 10, 186, 329–332, 335, 402
scientific practices, 130, 184, 195, 283, 291,
    354, 355, 402, 405, 412
scientist, 10, 12, 16–17, 27–28, 32, 34, 37–39,
    48, 60, 86, 104, 108–113, 116–118, 120,
    124–126, 128, 130–136, 143, 149, 151, 154,
    156, 168–171, 173–177, 181–185, 188,
    191–192, 195–198, 229, 236, 250–251,
    257, 283, 295, 325, 332–334, 348, 350, 355,
    363–365, 367, 369–370, 393, 412, 421, 431–
    432, 434, 445, 453, 458, 460, 467, 483–485,
    487–493, 495, 500, 504, 529–530, 534–535,
    538–542, 544, 547, 557, 559–560
Scistarter, 291, 557
Scott, B., 351
Sea Perch, 35, 37, 41
Sea Studios Foundation, 475
Secada, W., 372
self-regulated learning, 94, 477, 479, 481–483,
    487, 500
sensors, 504, 508, 515–516, 524–525, 532–533,
    538
    accelerometer, 515–516, 524
service learning, 203–205, 209–210, 214–220,
    426–428
Seymour, E., 179
Shaffer, D. W., 299
Shanahan, C., 419
Shanahan, T., 419
Shape of Life, 460
Shapira, D., 141
Shapiro, A., 299
Shapley, K., 201
Shea, L., 373
Sheehy, K., 73
Shelley, B. E., 140
Shelton, T. L., 105
Shemwell, J. T., 373
Sherer, M., 299
Sherin, B. L., 475
Shields, P., 42

Shouse, A. W., 42, 140, 322, 351
Silver, E., 246
Silverstein, S. C., 201
Simic-Muller, K., 475
Simmons, M., 352
Simon, S., 351
simulations, 339, 448, 523, 553, 558, 560
    simulation software, 158
Sinatra, G. M., 501
Sinclair, H. R., Jr., 221
Six Flags America Roller Coaster Design
    Contest, 159
sixth extinction, 445, 453, 460–461
Slavin, R., 371
Slezak, M., 141
smartphone, 346, 357, 504, 508–509, 516–518,
    525
Smith, J. F., 443
Smith, L. K., 474
Smith, N., 442
Smithsonian, 251, 392, 396
    Air and Space Museum, 156
    American Art Museum, 144
    Lemelson Center for the Study of
        Invention and Innovation, National
        Museum of American History, 553, 560
    Newsdesk, 163
Sobel, D., 475
social occupations, 8
social studies, 6, 27, 33, 172, 207, 270, 415,
    421–426, 428–429, 436–438, 440–441
Society of Women Engineers (SWE), 60
socioeconomic, 76–77, 91–92, 227, 229, 545
socioscientific issues (SSI), 327
Socratic dialogue, 505, 507, 523
Sodian, B., 371
Sofer, W., 139
Soil Conservation Service, 205
Soloway, E., 141
Songer, N. B., 140
Sonnert, G., 443
Space to Earth: Earth to Space (SEES), 538
SpaceCAD, 158, 163
Sperling, E., 350
Spillane, N. K., 21
Spitzer Space Telescope, 117, 130–131
Spuck, T., 129, 131, 171, 526

Sputnik, 2, 40, 241, 532
Squire, K., 299
stakeholders, 23, 28, 38–40, 127, 214–215, 219–220, 256, 261–262, 266–268, 425, 440, 544
standards, 4, 18–19, 25–26, 30, 81, 98, 114, 122–123, 126–127, 167, 204, 206–208, 238, 250–252, 255, 259–260, 262–263, 270–271, 305, 307, 355, 431, 440, 446, 449, 464, 468, 479, 536, 545, 558
  Common Core State Standards (CCSS), 305, 377, 400
  Framework for K–12 Science Education, 4, 18, 111, 195, 229, 306, 355, 400, 425, 445–446, 509
  National Science Education Standards, 242, 330
  Next Generation Science Standards (NGSS), 5, 18–19, 26, 35, 41, 111, 184, 195, 229, 242, 249, 286, 290, 305, 336, 350, 355, 400, 445, 509, 530, 545
  Standards for Mathematical Practice, 81–82, 85, 87, 305
  Standards for Technological Literacy (STL), 545
standards-based, 28–31, 34, 40, 77, 225, 284, 559
Stanton, T. K., 443
Star Trek, 115
State Farm Youth Advisory Board, 214, 220
Stein, R., 246
Steiner, Rudolph, 213
Steinkuehler, C., 299
STEM (science, technology, engineering, mathematics), see: individual subject listings
  community, 126, 286
  education, 4, 23–26, 28–30, 39–41, 45, 59, 67, 144, 209, 219, 241, 249–250, 252–253, 270, 296, 380, 425, 441, 445, 472, 478, 516, 543, 546, 553
  fields, 4–5, 23, 26, 28, 40, 45, 48–49, 61, 67–68, 70–71, 109, 168, 170–171, 174, 326, 332, 348–349, 415, 425, 435, 505, 508, 523, 531, 553, 560
  literacy, 3, 5, 26, 28, 209, 330, 332–333, 346–347, 375–377, 383, 395–396, 446, 532
  proficiency, 333
STEM-based identities, 289
STEM-capable, 3
STEM-proficient, 3, 40
stemstories, 60
Stenson, C. M., 221
Stevens, F., 475
Steward, S. B., 443
Stiles, K., 179
Stimson, Rufus W., 8
Stoddart, T., 373, 419
Strage, A. A., 222
Straits, W., 373
Streitmatter, J., 73
student attitudes, 94, 242, 532
student-centered classroom, 316
Suarez, E., 373
Suitts, S., 475
Sungur, S., 105
Supernova Cosmology Project, 117
Supovitz, J. A., 179
sustainability, 203–205, 207–211, 213, 215–219, 226, 252, 327, 466–468, 529, 534
  education, 205, 207, 209
Swango, C. J., 443
Sweedyk, E., 73
Sweller, J., 162

Taasoobshirazi, G., 501
Tai, R. H., 42
Tan, E., 443
Task Force on American Innovation, 43
Taylor, J., 500
teacher preparation, 28, 87, 93, 117, 382, 395, 508
Teachers Experiencing Antarctic and Arctic (TEA) program, 168
teaching and learning, 9, 17, 30, 127, 146–147, 203, 252, 256, 263, 269–271, 404, 423, 432, 438, 446, 470, 525, 541, 561
teaching practices, 166, 423, 522
Team America Rocketry Challenge, 158, 163
teamwork, 98, 151, 155, 194, 292, 395
Techbridge, 67
technical text, 386
technology, 1–6, 18, 23–26, 29, 33, 35, 39–41, 45–53, 55–62, 67–71, 82, 85, 93, 97,

109–110, 113, 116, 136, 138, 143, 146, 149, 154–156, 167, 174–175, 177, 183–186, 195–197, 203–204, 209, 218–219, 223, 230, 249, 253–254, 259, 262, 264, 266–268, 271, 283, 298, 316, 325, 328–332, 334–335, 337, 346, 350, 353, 363, 367, 375–378, 380, 391–392, 400, 415, 422, 425, 428, 432–436, 441, 445–446, 466, 477–478, 503–505, 507–509, 515, 522, 524, 530, 532–534, 536–537, 539, 542–545, 547–548, 550, 552, 554–556, 561
Technology, Life, Careers, 50–51, 62
Tekkaya, C., 105
Television and Infrared Observation Satellite (TIROS), 532
television programs, 553, 557, 561
Texley, J., 43
The Belgrade Charter, 206, 221
The Discovery Channel, 560
The JASON Project, 559
The Wild Classroom, 559
Thibodeau, P., 141
Things that Float and Things that Don't, 406, 409, 411, 415
Thomas, J. W., 141, 163, 279
Thomas, M. M., 418
Thomas, S., 247
Tidball, K. G., 443
tiered sentence, 360–362
Tiffany-Morales, J., 42
Togioka, S., 443
Topor, D. R., 105
Torres, R., 299
Toulmin, C., 247, 398
Toyota International Teacher Program, 178
trade books, 391, 399, 401, 404–406, 408–409, 411–417
Trail, J., 201
Traill, S., 444
Trangle, S., 247
transfer, 5–9, 305, 315, 377, 384, 430, 449–450, 454–466, 470–472, 497, 510, 514, 518, 537
Traphagen, K., 444
Trudel, A. R., 105
Trumbull, D. J., 141
Tsai, C.-C., 351–352
T-STEM Network, 21

Tsupros, N., 21, 43
T-Tauri stars, 130
Turner, E. E., 475
Tyler, R. W., 21
Tyson, Neil deGrasse, 130–131

U.S. Bureau of Labor Statistics, 27, 43
U.S. Census Bureau, 92, 105
U.S. Department of Education, 28, 43, 353, 428
U.S. Department of Labor, 222, 533
U.S. Geological Survey (USGS), 170, 538
  EarthNow Program, 538
U.S. Partnership for Education for Sustainable Development, 222
University Center for Atmospheric Research (UCAR), 532
Udvar-Hazy Museum, 156
underrepresented, 3–4, 46, 58–59, 144, 445–446
  minorities, 104, 545
Understanding by Design, 6, 303, 305, 449, 452, 465, 467, 470
underwater robotics, 37
United Nations Educational, Scientific, and Cultural Organization (UNESCO), 206
University of Wisconsin-Madison, 182, 189, 192, 194

Valanides, N., 141
Van Scotter, P., 500
Vanosdall, R., 442
Venville, G., 350
Verhoeff, T., 163
verification inquiry, 414
vermiculture, 215, 221
vertical alignment, 9
videos, 69–70, 97, 169, 350, 383, 391, 396, 432–433, 472, 549, 553, 558–561
Virginia Department of Education, 373
virtual handshake, 267
visualize, 48, 132, 290, 293, 381, 390, 504, 508, 516, 531, 533, 541
vocational education, 8
Vomvoridi-Ivanovic, E., 475
Von Aufschnaiter, C., 351

Vorgelegt, V., 105
Vygotsky, L. S., 419

Wagner, T., 146, 163
Walls, M., 563
Walter Reed National Military Medical Center, 156
Wassersug, R., 141
Weber, K., 73
Wegner, E., 474
Wei, R. C., 179
Wellage, G., 140
Wells Fargo Teacher's Partner Program, 64–65, 562
Wells, N., 247
Wenglinsky, H., 201
Westbrook, A., 500
whiteboarding, 301–308, 312, 315–318, 320–322, 507, 513, 523
whiteboards, 301–304, 307, 309–311, 313–315, 317–322, 329, 457
Wigfield, A., 418, 501
Wiggins, A. W., 475
Wiggins, G. P., 21, 323
Wiggins, J., 474
Wilke, R. A., 179
Wilkinson-Flicker, S., 473
Willingham, D. T., 473
Wimmer, J. J., 474
Wise, A. E., 444
WKQW-96.3 Radio, 140
Wolozin, L., 299
Women in Education Proactive Network (WEPAN) Knowledge Center, 60
Woodrow Wilson Senior High, 336–337
Woolfolk, A., 180
word maps, 385–386
word walls, 385
workforce, 1, 3–5, 23, 25–26, 38, 40, 45–48, 70, 109, 144, 204, 219, 221, 227, 241, 270, 326, 425, 534–535, 540
World of Warcraft (WOW), 281, 296
World Water Monitoring Day, 150, 152
Wormeli, R., 89
Write-to-Learn (WTL), 388–389
Wu, Y.-T., 351–352
Wynn, C. M., 475

Yamagata, J., 443
Yamagata, L., 443
Yellowstone National Park, 228, 231–233, 235, 237
Yim, P.-s., 162
Yocco, V., 246
Yoon, K. D., 180
Yoon, K. S., 201

Zeidler, D., 351–352
Zeldin, A. F., 449, 475
Zemelman, S., 105
Zhang, J., 473
Zimmerman, B. J., 501
Zimmerman, E., 299
Zimmerman, H. T., 444
Zion, M., 141
Zohar, A., 352
Zometool, 455, 475
Zoology Brüt, 445, 450–454, 456, 458–462, 465–466, 468–470, 472
Zooniverse, 133, 291, 299, 557
Zwiep, S., 373
Zwiers, J., 373

Critical Pedagogical Perspectives

M. Cathrene Connery and
Greg S. Goodman, *General Editors*

Educational Psychology: Critical Pedagogical Perspectives is a series of relevant and dynamic works by scholars and practitioners of critical pedagogy, critical constructivism, and educational psychology. Reflecting a multitude of social, political, and intellectual developments prompted by the mentor Paulo Freire, books in the series enliven the educator's process with theory and practice that promote personal agency, social justice, and academic achievement. Often countering the dominant discourse with provocative and yet practical alternatives, Educational Psychology speaks to educators on the forefront of social change and those who champion social justice.

For further information about the series and submitting manuscripts, please contact:

M. Cathrene Connery | Greg S. Goodman
cconnery@salisbury.edu | gsgoodman17@gmail.com

To order other books in this series, please contact our Customer Service Department at:

(800) 770-LANG (within the U.S.)
(212) 647-7706 (outside the U.S.)
(212) 647-7707 FAX

Or browse online by series at:

www.peterlang.com

Made in the USA
Monee, IL
28 April 2026